Lecture Notes in Physics

Edited by J. Ehlers, München, K. Hepp, Zürich
R. Kippenhahn, München, H. A. Weidenmüller, Heidelberg
and J. Zittartz, Köln
Managing Editor: W. Beiglböck, Heidelberg

116

Mathematical Problems in Theoretical Physics

Proceedings of the International Conference on Mathematical Physics
Held in Lausanne, Switzerland
August 20 – 25, 1979

Edited by K. Osterwalder

Springer-Verlag
Berlin Heidelberg New York 1980

Editor

Konrad Osterwalder
Mathematik
ETH-Zentrum
CH-8092 Zürich

ISBN 3-540-09964-6 Springer-Verlag Berlin Heidelberg New York
ISBN 0-387-09964-6 Springer-Verlag New York Heidelberg Berlin

Library of Congress Cataloging in Publication Data. International Conference on Mathematical Physics, Lausanne, 1979. Mathematical problems in theoretical physics. (Lecture notes in physics; 116) Bibliography: p. Includes index. 1. Mathematical physics--Congresses. I. Osterwalder, K., 1942- II. Title. III. Series. QC19.2.I538 1979 530.1'5 80-12140

Printing and binding: Beltz Offsetdruck, Hemsbach/Bergstr.
2153/3140-543210

Foreword

This volume contains the proceedings of the International Conference on Mathematical Physics held at the Swiss Federal Institute of Technology in Lausanne (EPFL), August 20 - 26, 1979. This conference continued the tradition of the meetings held in Moscow (1972), in Warsaw (1974), in Kyoto (1975), and in Rome (1977). The next conference is scheduled for the summer of 1981 in Berlin.

The symbol $M \cap \Phi$ which has been chosen as a trade mark for all these conferences denotes the intersection of Mathematics and Physics. It indicates the hope and the intention that these conferences should not only be an occasion to communicate results, but also an opportunity to intensify and to demonstrate the close contact between various fields of mathematics and of physics. To this purpose, the main lectures of this conference were all organized as review talks, covering a broad area of research, and, with as few exceptions as possible, parallel sessions were avoided.

The major topics of this conference were Schrödinger Operators, Statistical Mechanics, Quantum Field Theory, Gauge Theory, Dynamical Systems, Supersymmetry and C*-Algebras. It is clear that this is not all there is in Mathematical Physics, and there are certainly other areas of physics which have lively contact with mathematics which could not be included here because of time limitations.

There were two major review talks on each of the topics Schrödinger Operators, Statistical Mechanics, Quantum Field Theory and Gauge Theory. Furthermore there was a long session (half a day or more) of shorter communications on each one of the seven topics. These sessions were planned and set up by session organizers who were completely free to invite their speakers, to accept or turn down contributed papers, and to organize the part of these proceedings which belongs to their respective session.

This volume contains with one exception all the main lectures and most of the smaller contributions. Unfortunately, no manuscript of Prof. I.M. Singer's talk was available. The interested reader is advised to look instead at Singer's contribution to the Proceedings of the 1979 Cargèse summer school on Recent Developments in Gauge Theories.

The conference was sponsored by the International Mathematical Union, the U.S. National Science Foundation (travel support for American participants through a grant to Rutgers University), Le Fonds National Suisse de la Recherche Scientifique, Le Troisième Cycle de la Physique en Suisse Romande, La Fondation Herbette de la faculté des sciences de l'Université de Lausanne, l'Université de Genève, Eidg. Technische Hochschule Zürich, l'Ecole Polytechnique Fédérale de Lausanne.

It is a pleasure to thank these organizations for their generous financial and moral support of the conference.

The members of the International Advisory Board were as follows:

H. Araki, V.I. Arnold, M.F. Atiyah, F.J. Dyson, G. Gallavotti, R. Haag, A. Jaffe, A. Janner, E. Lieb, J. Lebowitz, A. Martin, Y. Neeman, H.M. Nussenzveig, L. O'Raifeartaigh, K. Osterwalder, C. Piron, R. Raczka, D. Ruelle, B. Simon, Y. Sinai, R. Stora, W. Thirring, G. Velo, A. Verbeure, W. Zimmermann.

The members of the Conference Committee were:

W. Thirring, A. Martin, K. Osterwalder, Ph. Choquard.

In the name of all the participants I would like to thank everybody who supported the efforts to make this conference a success: the speakers and authors of the contributions to these proceedings, the session organizers, the members of the advisory board and of the conference committee, the organizers and all the secretarial and administrative staff members of the host institution, and in particular the man without whose enthusiasm, tireless devotion and great organizational skill this conference never would have taken place: Prof. Ph. Choquard.

December 1979

Konrad Osterwalder
Editor

Table of Contents

Session Organized by E. BREZIN

QUANTUM FIELD THEORY

Main Lectures

Session Organized by J. FRÖHLICH

RECENT DEVELOPMENTS IN QUANTUM SCATTERING THEORY (*)

J.M. COMBES (**)

INTRODUCTION

Let me first recall what is considered in elementary text books as the basic ansatz of quantum scattering theory. For a particle in an external field vanishing at infinity the ingoing and outgoing solutions of the Schrödinger equation satisfy a "radiation" asymptotic condition :

(1) $$\phi_{\pm}(x,k) \underset{|x|\to\infty}{\approx} e^{ikx} + |x|^{\frac{1-n}{2}} e^{\pm i|k||x|} f(\Omega)$$

where n is the dimension of configuration space and Ω the angular scattering parameter. It was already known in 1928 [1] that for Coulomb forces the phase of the scattered wave had to be modified as $|k||x| \longrightarrow |k||x| + |k|^{-1} \ln(4|k||x|)$ but that nevertheless it was still legitimate to interpret $f(\Omega)$ as the scattering amplitude. By stationary phase arguments the time evolution of wave-packets should then also obeys an asymptotic condition

(2) $$\begin{cases} \psi(t) \underset{t\to\infty}{\approx} \psi_{in}(t) + \psi_{out}(t) \\ \lim\limits_{t\to\pm\infty} \psi_{\substack{in\\out}}(t) = 0 \end{cases}$$

(**) Département de Mathématiques, Université de Toulon and Centre de Physique Théorique II, CNRS, Marseille

Postal Address : Centre de Physique Théorique
CNRS - Luminy - Case 907
F-13288 MARSEILLE CEDEX 2 (FRANCE)

Although these statements are deliberately vague from a mathematical point of view, they are essential for the physical interpretation of scattering theory. It may seem surprising that the mathematical literature is so poor concerning the analysis of (1) and (2) apart from some notable exceptions like the papers of Green and Lanford [42] and Ikebe [43]. During the last decade very much more attention has been paid to the "operatorial" formulation of scattering theory in terms of wave-operators, the connection with the original scattering wave approach being more and more hidden by the abstract formalism. This divorce of the theory with physical intuition is certainly responsible for the lack of definite progress in some important domains like multichannel systems or analytic scattering theory with long-range forces. So I will not hesitate to describe as main results of the last two years the analysis by Agmon [2] of (1) in Besov spaces and the geometrical phase space analysis by Enss [3] of (2). The techniques developped in these works seem to provide new powerful tools of investigation ; they reconcile a strong mathematical technology, which is more or less implicitely pseudo-differential operator theory, with our classical intuition of scattering phenomena. It is not so surprising that their basic strategy which is essentially simultaneous localization in x and k space can be so fruitful since one expects that for large times scattering wave-packets should propagate along classical trajectories. Some abstract preliminary formulations of related properties have been known for a long time like e.g. decay properties and Ruelle's characterizations of continuum states [4], or the so-called limiting absorption principle. However it is only rather recently that powerful enough methods have been developped and used for more systematic investigations [5] [6] [7] [8] [9], culminating with the works of Agmon and Enss. We will come back to these works and to some reasonable speculations about the new possibilities they open after a more general review of the present status of scattering theory and the evolution of ideas in the last years.

II. TIME-DEPENDENT SCATTERING THEORY

Modern formulation of the abstract scattering theory attempts to cope with larger applications than just ordinary quantum potential scattering. For this reason the usual Möller definition of wave-operators needs to be slightly extended. We denote by $\mathcal{H}$, H (resp. $\mathcal{H}_0$, H_0) the Hilbert space and Hamiltonian describing states and dynamics of the perturbed (resp. free) system. Let J be a bounded linear mapping from $\mathcal{H}_0$ to $\mathcal{H}$ playing the role of an identification operator : a state $\phi \in \mathcal{H}_0$ of the free system is identified with the state $J\phi \in \mathcal{H}$ of the perturbed system. The choice of such an identification operator is a matter of physical consideration and is not necessarily unique ; J need not be bijective as we will see in some applications. The wave-operators $W_\pm(H, H_0; J)$ are then defined by the following strong limits when they exist [10], [11]:

$$W_\pm(H, H_0; J)\phi = \operatorname*{s\cdot lim}_{t \to \pm\infty} e^{iHt} J e^{-iH_0 t} \phi$$

In most applications the domains of $W_\pm(H, H_0; J)$ are contained in $P_{0,ac}\mathcal{H}_0$, the subspace of absolute continuity of H_0 consisting of those $\phi \in \mathcal{H}_0$ such that $\langle P_0(\lambda)\phi, \phi\rangle$ is an absolutely continuous function (here $P_0(\cdot)$ denotes the spectral family of H_0). One says that the wave-operators <u>exist</u> if

$$W_\pm(H, H_0; J) = \operatorname*{s\cdot lim}_{t \to \pm\infty} e^{iHt} J e^{-iH_0 t} P_{0,ac} \tag{3}$$

exist or in other words if $P_{0,ac}\mathcal{H}_0 = \mathcal{D}(W_\pm(H, H_0; J))$

Local wave-operators $W_\pm(H, H_0; \Delta; J)$ are obtained by restricting the domain of $W_\pm$ to a spectral subspace $P_0(\Delta)\mathcal{H}_0$ for H_0 corresponding to a closed set $\Delta \subset \mathbb{R}$. The properties of generalized wave-operators (intertwining, chain rule, etc...) are similar to those of the usual ones ; however isometry only holds under the additional requirement that the identification operator is asymptotically isometric in the following sense :

$$\lim_{t \to \pm\infty} \| J e^{-iH_0 t} P_{0,ac} \phi \| = \| \phi \| \tag{4}$$

One says that $W_{\pm}(H, H_0; J)$ are complete if $W_{\pm}(H,H_0;J)P_{0,ac}\mathcal{H}_0 = P_{ac}\mathcal{H}$ with obvious notations . This is known to imply unitarity of the S-matrix operator

$$S(H,H_0;J) = W_+^*(H,H_0;J)\,W_-(H,H_0;J)$$

provided wave-operators are isometric.

The following criteria relating existence and completeness is due to T. Kato

Proposition 1

Assume J is surjective. If both $W_{\pm}(H,H_0;J)$ and $W_{\pm}(H_0,H;J^*)$ exist then $W_{\pm}(H,H_0;J)$ is complete.

If furthermore J (resp. J^*) is asymptotically H_0 (resp. H) isometric then $S(H,H_0;J)$ is unitary.

The first example of a scattering system in the above sense is of course potential scattering for a single-particle. Here $\mathcal{H} = \mathcal{H}_0 = L^2(\mathbb{R}^3)$ and $H_0 = -\Delta$, $H = -\Delta + V$. If V is "$-\Delta$ relatively bounded" in the sense of operators or forms then H can be defined as a self-adjoint operator (provided V is also real) in an almost unique way from a physical point of view. Then it is natural to take $J = 1$. However for potentials having nasty singularities in a bounded domain some other choices can be more convenient for technical purposes (see Example 2.4 below). Let us mention some other relevant examples :

Example 2.1 : Obstacle Scattering

Let Ω be a bounded regular obstacle in $\mathbb{R}^n$, $\mathcal{H} = L^2(\mathbb{R}^n \setminus \Omega)$, $\mathcal{H}_0 = L^2(\mathbb{R}^n)$ and J the orthogonal projection operator on $\mathcal{H}$. If one takes $H_0 = -\Delta$ and $H = -\Delta +$ boundary conditions on $\partial\Omega$ then $W_{\pm}(H,H_0;J)$ exist and are isometric [12].

Example 2.2 : Scattering by stationary external metrics and Yang-Mills potentials ([13] [14]) :

Consider an external static metric on $\mathbb{R}^n$, $\underline{g}(x) = \{g_{k\ell}(x)\}_{1\le k,\ell\le n}$ an external Yang-Mills potential $A(x) = \{A_k(x)\}_{1\le k\le n}$ where each $A_k(x)$ is an endomorphism of an internal symmetry space $\mathcal{V}$ of a quantum spinorial particle and a potential V . The Hamiltonian of the particle in these external fields of forces is given by :

$$H = -\frac{1}{\sqrt{g}}(\partial_k + iA_k)\sqrt{g}\, g_{k\ell}(\partial_\ell + iA_\ell) + V$$

with $g = \det \underline{g}$.

If one assumes that the metric is asymptotically flat and locally regular, e.g. :

(5i) $$g_{k\ell} - \delta_{k\ell} \in \mathcal{A}(\mathbb{R}^n)$$

and that the external potentials vanish at infinity, e.g.

(5ii) $$A_k \in \mathcal{A}(\mathbb{R}^n) \otimes \mathcal{E}nd\, \mathcal{V}$$

(5iii) $$V \in \mathcal{A}(\mathbb{R}^n) \otimes \mathcal{E}nd\, \mathcal{V}$$

then it is natural to compare the large time behaviour of this system with the free one in $\mathcal{H}_0 = L^2(\mathbb{R}^n)$ under the time evolution with generator $H_0 = -\Delta$. The following identification operator can then be used :

$$(J\phi)(x) = g^{-1/4}(x)\,\phi(x)$$

Under assumption (5i) J is asymptotically isometric.

Example 2.3 : Multichannel Scattering

Consider for simplicity a 3 particle non relativistic system interacting through two-body potentials V_{ij} such that pairs (ij) can form only one bound-state with binding energy $-E_{ij}$. The Hamiltonian acting on $\mathcal{H} = L^2(\mathbb{R}^6)$ is, denoting by T the kinetic energy operator in the center of mass frame :

$$H = T + V$$

$$V = \sum_{1 \le i < j \le 3} V_{ij}$$

Let

$$\mathcal{H}_0 = L^2(\mathbb{R}^6) \oplus \Big(\bigoplus_{1 \le i < j \le 3} L^2(\mathbb{R}^3_{ij}) \Big)$$

and

$$H_0 = \begin{pmatrix} T \\ -\Delta_{12}/2\mu_{12} + E_{12} \\ -\Delta_{13}/2\mu_{13} + E_{13} \\ -\Delta_{23}/2\mu_{23} + E_{23} \end{pmatrix}$$

describe the "free" evolution in the direct sum $\mathcal{H}_0$ of channel Hilbert spaces. Here μ_{ij} denotes the reduced mass of particle k and center of mass of i and j, $(i,j,k) = (1,2,3)$ and Δ_{ij} the Laplacian with respect to the relative coordinates of k and the center of mass of i and j. The following identification operator is frequently used (10) (15) :

$$J \begin{pmatrix} \phi \\ \phi_{12} \\ \phi_{13} \\ \phi_{23} \end{pmatrix} = \phi + \phi_{12} \otimes \xi_{12} + \phi_{13} \otimes \xi_{13} + \phi_{23} \otimes \xi_{23}$$

where ξ_{ij} is the normalized bound-state wave function of the pair (i,j). This identification operator is asymptotically isometric ; this follows immediately from the well-known asymptotic orthogonality of channels. The "channel" wave-operators of Ekstein and Jauch are obtained from $W_\pm(H, H_0; J)$ by

$$W_\pm^\alpha = W_\pm(H, H_0; J) P_\alpha J^*$$

where P_α, $\alpha = 1,2,3,4$, is the orthogonal projection operator on the channel subspace α of $\mathcal{H}_0$ and the adjoint J^* of J is given by

$$J^* \phi = \begin{pmatrix} \phi \\ \langle \phi, \xi_{12} \rangle \otimes \xi_{12} \\ \langle \phi, \xi_{13} \rangle \otimes \xi_{13} \\ \langle \phi, \xi_{23} \rangle \otimes \xi_{23} \end{pmatrix}$$

It is easy to see that unitarity of the multichannel S-matrix $S_{\alpha\beta} = (W_+^\alpha)^* W_-^\beta$ is equivalent to unitarity of $S(H, H_0; J)$. However the criteria given above for completeness does not apply to this situation. For if we consider for example a state which is asymptotically in a two-particle channel, e.g. :

$$e^{-iHt}\phi \xrightarrow[t\to\pm\infty]{} e^{-i(-\Delta_{12}/2\mu_{12}+E_{12})t}\phi_{12}$$

then $e^{iH_0t}J^* e^{-iHt}\phi$ has obviously no strong limit ; in other words $W_\pm(H,H_0;J^*)$ does not exist for multichannel systems. So new criteria and identification operators are needed ; P. Deift and B. Simon show in [51] the convenience of a "geometrical" choice.

We now come to criteria for existence of wave-operators generalizing a well-known result stated originally by J.M. Cook and S.T. Kuroda. The form given below originates from works of M. Schecter [16], B. Simon [17] and T. Kato [18]. The original result of Cook called "Proof of a Lemma due to Eckstein" applied to one-body scattering with a potential $V\in L^2(\mathbb{R}^3)$.

Theorem 1

Assume $\langle J\phi, H\psi\rangle - \langle JH_0\phi,\psi\rangle = \langle A\phi, B\psi\rangle\ \forall\phi\in\mathcal{D}(H_0),\ \psi\in\mathcal{D}(H)$ where

i) A is a linear operator from $\mathcal{H}_0$ to a Banach space $\mathcal{K}$, $\mathcal{D}(A)\supset\mathcal{D}(H_0)$

ii) B is a linear operator from $\mathcal{H}$ to $\mathcal{K}^*$ and B is H-bounded.

If $\phi\in\mathcal{D}(H_0)$ satisfies

$$\int_T^\infty \|Ae^{-iH_0t}\phi\|\,dt<\infty \quad \text{for some } T<\infty$$

then

$$\phi\in\mathcal{D}(W_+(H,H_0;J))$$

Example 2.4 (19)

Let us consider scattering of a quantum particle by a potential V which is short range in the sense

$$|V(x)|\leq C|x|^{-1-\varepsilon},\ |x|<R,\ \varepsilon>0$$

but can be nasty locally in the sense that one only requires existence of a self-adjoint operator H satisfying

$$H\phi = -\Delta\phi + V\phi \quad \forall\phi\in C_0^\infty(\mathbb{R}^3\setminus\{|x|<R\})$$

By the above theorem $W_\pm(H, H_0; J)$ exists where J is $1-\chi_R$ with χ_R a smooth characteristic function of the sphere of radius R. Now since $\|\chi_R e^{-iH_0 t}\phi\| \underset{t\to\pm\infty}{\longrightarrow} 0$ it follows that

$$W_\pm(H, H_0; J) = W_\pm(H, H_0; 1)$$

Example 2.5 : The Hack-Kuroda Theorem

Let H be an N - particle Hamiltonian with two-particle interactions $V_{ij}(x) = O(|x|^{-1-\varepsilon})$, $|x| \to \infty$, $\varepsilon > 0$, and $V_{ij} \in L^2_{loc}(\mathbb{R}^3)$ Then channel wave-operators exist. Many other applications with proofs can be found in Volume III of the Reed-Simon series [20]. For potential scattering they allow to deal with more or less important local singularities and short range forces. For long-range potentials it is well-known that the definition of wave-operators has to be modified in order to adapt divergent phases [1]. However Cook-Kuroda method can be adapted as was shown by many authors and more recently in the culminating work of L. Hörmander [21] whose result is described below.

Hörmander defines admissible perturbations of the Laplacian Δ as follows

(6) $$V(x, D) = \sum_{|\alpha| \leq 1} (V_\alpha^L + V_\alpha^S) D^\alpha = V^L + V^S$$

where (denoting by α an n-index and D^α the corresponding differential operator) :

1) $(1+|x|)^{-m} V_\alpha^S \in L^2(\mathbb{R}^n)$ for some m and

2) $|D^\beta V_\alpha^L(x)| \leq C(1+|x|)^{-m(\beta)}$, $|\beta| \leq 3$

where $m(0), --, m(3)$ are positive and $m(1) + m(3) < 4$.

Under these conditions Hörmander shows

Theorem 2

Let V be an admissible perturbation of Δ and W be a solution of the Hamilton-Jacobi equation ;

$$V^L\left(\frac{\partial W}{\partial \xi}, \xi\right) + \xi^2 - \frac{\partial W(\xi, t)}{\partial t} = 0$$

Then the modified wave-operators

$$W_{\pm}(-\Delta+V,-\Delta)=s\cdot\lim_{t\to\pm\infty} e^{iHt}e^{-iW(t,D)}$$

exist and are isometric operators intertwining $-\Delta$ and $-\Delta+V$.

Hörmander's proof uses stationary phase arguments and applies to a more general class of systems whose generator of evolution is given by a pseudo-differential operator.

Despite the wide applicability of Cook-Kuroda method for existence proofs it is not very convenient for a time-dependent proof of asymptotic completeness via the Proposition 1. The main reason is that it is usually not possible to estimate decay properties of $e^{-iHt}\phi$ for a dense set of ϕ's in a sufficiently explicit way. Enss's method remedies to this using simply the fact that according to Ruelle [4] the particle ultimately leaves the region where the interaction is important. Before Enss's discovery the only systematic investigation of completeness by time-dependent methods was the so-called Kato-Birman Theory. It originated in 1957 with a beautiful paper of Kato [22] which can be considered as the first building stone of abstract scattering theory ; a more detailed history can be found in [20]. We only mention here its more important statements and recent applications. The following proposition is the optimal version given by D.B. Pearson [23] of the main completeness result :

Proposition 2

Assume $HJ \supset JH_0+V$ where $V\in B_1(\mathcal{H}_0,\mathcal{H})$ (trace class operators from $\mathcal{H}_0$ to $\mathcal{H}$). Then $W_{\pm}(H,H_0;J)$ and $W_{\pm}(H_0,H;J^*)$ exist and are adjoint to each other. In particular if J is onto then $W_{\pm}(H,H_0;J)$ is complete.

In this form the result is not very useful since in most applications (e.g. potential scattering) the perturbation V does not even have point spectrum. This statement however is of wide applicability once it is supplemented by the "Invariance principle" for wave-operators. The form given below is sufficient for most applications of interest :

Proposition 3

Let f be a real piecewise monotone function such that f' is locally positive and of bounded variation. Then under Pearson's conditions one has

$$W_{\pm}(f(H),f(H_0);J)=W_{\pm}(H,H_0;J)$$

Although the mathematical meaning of the Invariance principle is quite obvious from the spectral theorem since a self-adjoint operator and its functions have the same generalized eigenvectors, its physical interpretation only rests on a paper of J.D. Dollard [24] relating it to invariance of the S-matrix under change of origin and scale on time and energy. Let me recall one of its deeper consequence, namely the equality of the S-matrices for quantum and acoustical scattering [25] by the same medium.

Among applications, let us mention :

Example 2.0 : Potential Scattering, S.T. Kuroda [29]

If $V(x) \in L_1 \cap L_2(\mathbb{R}^3)$ then for c large enough $(H+c)^{-1} - (H_0+c)^{-1}$ is trace-class. This result is rather weak since it requires a stronger decay of V than necessary.

Example 2.1 : Completeness of wave-operators for obstacle Scattering P. Deift-B. Simon 26 , Jensen-Kato 27

For this problem one takes $f(\lambda) = e^{t\lambda}$ and show that $Je^{-tH} - e^{-tH_0}J$ is trace-class. A related problem concerns decoupling of finite singularities for potential scattering [26] [28] ; these singularities can be decoupled by an artificial Dirichlet boundary condition on a bounded surface which encloses them. Such a perturbation is equivalent to an obstacle and a trace condition is satisfied for $H = -\Delta + V$ and $\hat{H} = -\Delta + V$ plus Dirichlet boundary condition provided the singularities of V are not too bad in the sense that $-\Delta \leq C_1(-\Delta + V + C_2)$ for some constants C_1 and C_2 . Then only the behaviour of V outside the surface is relevant for the question.

Example 2.2 : -Scattering by external static metric and Yang-Mills potentials. R. Schräder 14

If n is the dimension of space R. Schräder shows that under conditions (5i) (5ii) (5iii), one has $(H+1)^{-\ell}J - J(H_0+1)^{-\ell} \in \mathcal{B}_1(\mathcal{H}_0, \mathcal{H})$ for $2\ell > n$

- Spectral shift function, Krein's Theorem

One important virtue of the Kato-Birman theory is that it leads to very useful properties of the total phase shift as expressed by a theorem of Krein. The total phase-shift $\delta(\lambda)$ is defined as follows : since the S-operator $S(H, H_0; J)$ commutes with H_0 one has a direct sum decomposition

$$S(H, H_0; J) = \int^{\oplus} S(\lambda) d\lambda$$

Then formally :

(7)
$$e^{2i\delta(\lambda)} = \mathrm{Det}\, S(\lambda)$$

Existence of $\delta(\cdot)$ can be shown under rather general conditions [12]. Although (7) defines $\delta(\cdot)$ modulo 2π it is possible to show that under these existence conditions $\delta(\cdot)$ can be chosen as a continuous function. The following theorem of Krein relates $\delta(\cdot)$ to the so-called "spectral shift function" and leads to many interesting properties :

Theorem (M.G. Krein 30)

Let K_0 and K be self-adjoint operators on a Hilbert space $\mathcal{H}$ such that $K = K_0 + V$ with $V \in B_1(\mathcal{H})$. Then there exists a function $\xi \in L^1(-\infty,+\infty)$ such that

1) $\|\xi\|_{L_1} \leq \|V\|_{B_1(\mathcal{H})}$

2) $\int_{-\infty}^{+\infty} \xi(\lambda)\, df^{-1}(\lambda) = \mathrm{tr}\left(f^{-1}(K) - f^{-1}(K_0)\right)$ for all $f \in C^2(\mathbb{R})$ with $f' > 0$, $f(+\infty) = +\infty$

3) $\int_{-\infty}^{+\infty} |\xi(\lambda)|\, df^{-1}(\lambda) \leq \|f^{-1}(K) - f^{-1}(K_0)\|_{B_1(\mathcal{H})}$

4) $\xi(\lambda) = \pi^{-1} \lim_{\varepsilon \downarrow 0} \arg \det\left(1 + V(K - \lambda + i\varepsilon)^{-1}\right)$

5) $e^{-2i\pi\xi(\lambda)} = \det S(\lambda)$

Here again the full power of the theorem rests on an invariance principle stating that if f is a twice differentiable function on the spectrum of K and K_0 such that $f' > 0$ and $f(+\infty) = +\infty$ then (with obvious notations) :

(9)
$$\xi(f(K), f(K_0); f(\lambda)) = \xi(K, K_0; \lambda)$$

This implies existence of the spectral shift for pairs of operators to which Kato-Birman theory can be applied. For such pairs it is usually possible using property 4) to show continuity of $\xi(\cdot)$ and then according to 5) the spectral shift is equal to $-\pi^{-1}\delta(\cdot)$ plus a constant multiple of π. Then using 1) 2) 3) one can obtain very useful estimates on the total phase shift and its variations and on the corresponding physical quantities, e.g. cross sections. Let us mention some applications :

1) Asymptotic expansions

For the choice $f(\lambda) = e^{t\lambda}$ one obtains from formula 2) in Krein's theorem :

$$-\beta\int_{-\infty}^{+\infty} \xi(\lambda)e^{-t\lambda}\,d\lambda = \operatorname{tr}(e^{-tH} - e^{-tH_0}) \tag{10}$$

By a theorem of Laplace transforms of asymptotic expansions (see e.g. [31]) asymptotic expansions of $\xi(\lambda)$ at large "energies" λ are related to asymptotic expansions of $\operatorname{tr}(e^{-tH} - e^{-tH_0})$ at small "temperature" t. For Laplace-Beltrami operators such low temperature expansions can be derived in particular form the works of H. McKean and I.M. Singer [32] or P.B. Gilkey [33]. The following asymptotic expansions have been obtained

Example 2.1 : Obstacle Scattering (V.S. Buslaev 34 , A. Majda-J. Ralston 35)

For $n \geq 3$ dimensions :

$$\frac{1}{2\pi}\delta(\lambda) = \frac{(4\pi)^{-n/2}}{\Gamma(\frac{n}{2}+1)}\lambda^{\frac{n}{2}}V - \frac{(4\pi)^{-\frac{(n-1)}{2}}}{\Gamma(\frac{n-1}{2}+1)}\lambda^{\frac{n-1}{2}}\frac{S}{4}$$
$$-\frac{(4\pi)^{-n/2}}{\Gamma(\frac{n}{2})}\lambda^{\frac{n}{2}-1}\int_{\partial\Omega}\frac{H}{6}\,dS + O(\lambda^{\frac{n-3}{2}})$$

Here V is the volume of the obstacle Ω, S the area of the boundary $\partial\Omega$ and H the mean curvature function of $\partial\Omega$. For $n = 2$ one has

$$\frac{1}{2\pi}\delta(\lambda) = \frac{\lambda}{4\pi}V - \frac{\lambda^{1/2}}{4\pi}S + O(1)$$

Example 2.2 : Scattering by external metrics and Yang-Mills potentials (R. Schräder 14)

For $n \geq 3$ dimensions :

$$(2\pi)^{\frac{n}{2}-1}\,\delta(\lambda) = \sum_{m=0}^{2} \frac{\lambda^{\frac{n}{2}-m}}{\Gamma(1+\frac{n}{2}-m)}\, B_n + O(\lambda^{\frac{n}{2}-2})$$

where

$$B_0 = \int (\sqrt{g}-1)\, d^n x \ \dim \mathcal{V}$$

$$B_1 = \frac{1}{6}\, \mathrm{tr}_{\mathcal{V}} \int (-R_{ijij} - 6V)\sqrt{g}\, d^n x$$

$$B_2 = \frac{1}{360}\, \mathrm{tr}_{\mathcal{V}} \int \Big(-12 R_{ijij,kk} + 5 R_{ijij} R_{klkl} - 2 R_{ijik} R_{ljkk} + 2 R_{ijkl} R_{ijkl} + 60 V R_{ijij} + 180 V^2 - 60 V_{;kk} - 30 F_{kl} F_{kl} \Big) \sqrt{g}\, d^n x$$

Here R_{ijkl} is the Rieman tensor, F_{kl} the field strength tensor and standard notation for covariant derivatives has been used. Notice that for potential scattering $(g=1)$ the expansion starts at $m=1$ with the usual Born term. Notice also that the form of B_0 given here generalizes the obstacle scattering first term. Finally it is interesting to remark (as proved in [13]) that the expansion is only in terms of gauge invariant quantities since the S-matrix is invariant itself ; for example the term B_1 involves the total curvature and the potential. It would be possible to push the expansion farther using Gilkey's result [33] ; the next term B_3 would contain for example 43 structurally different terms !

These asymptotic expansions are of interest in many respects. First of all from the point of view of inverse scattering theory they show how one can recover some invariants of the "scatterer" (obstacle, metric or external fields) from high-energy experiments. Second as notices by A. Majda and J. Ralston [35] there is a close analogy between the asymptotic expansion for obstacle scattering and those for the number of eigenvalues of the corresponding internal problem ; these two authors conjecture existence of a Levinson theorem for decaying modes (or resonances) and raise the academic but fascinating problem of a one to one correspondence between these decaying modes and internal eigenvalues.

2) Monotonicity and sign of the phase shifts

From Krein's theorem and

$$\mathrm{tr}(e^{-\beta H} - e^{-\beta H_0}) = -\xi \int e^{-\beta \lambda}\, \xi(\lambda)\, d\lambda$$

it follows that the total phase shift has the sign of the left hand side trace. In particular the total phase shift is negative for positive potentials and obstacle scattering. I refer to T. Kato [36] for more refined investigations on the sign and variation of the phase shift as a function of the perturbation. Let me simply mention that the determination of the sign of the above trace is a subject of great mathematical and physical interest which has deserved a lot of attention these last years after the appearance of Kato's distributional inequality, the concept of semi-group domination and its relation to Feynman path integrals. A beautiful application of these ideas has been made by R. Schräder in the case of scattering by Yang-Mills potentials. From Kato's inequality for magnetic fields and Yang-Mills potentials (B. Simon [37] , H. Hess, R. Schräder and D. Uhlenbock [38]) he argues that Yang-Mills potentials act repulsively on scalar particles, whereas from the paramagnetic conjecture of H. Hogreve, R. Schräder, R. Seiler [39]) ;

$$\mathrm{tr}\,(e^{-\beta H_A^{\mathrm{spin}\,1/2}} - e^{-\beta H_0}) \geq 0$$

it should follow that they act attractively on spin 1/2 particles. This conjecture however has been proved only up to now at very low and very high energies [39].

Enss' method ([3] , [41])

Consider potential scattering with short range forces in the following sense

$$\| |x|\, V(-\Delta+1)^{-1} \chi(|x| \geq R) \| \xrightarrow[R\to\infty]{} 0$$

As (2) suggests one expects that for large positive times, $e^{-itH}P_{ac}\phi$ contains only outgoing waves so that

(11) $$(W_-(H,H_0)-1)e^{-iHt}P_{ac}\phi \underset{t\to+\infty}{\longrightarrow} 0$$

Then asymptotic completeness holds by the following argument : assume that $\phi \in R(W_-)^\perp \cap P_{ac}\mathcal{H}$. Then $e^{-iHt}\phi$ also does

But then :

$$\|W_-(H,H_0)e^{-iHt}\phi - e^{-iHt}\|^2 = \|W_-(H,H_0)e^{-iHt}\phi\|^2 + \|\phi\|^2$$

tends to zero so that $\phi = 0$.

The original Enss' proof [3] of (11) uses a decomposition by hand of phase space leading to (2) ; essential in his proof is Ruelle's characterization of continuum states [4] . I present here a more algebraic proof also due to Enss, using an idea of E. Mourre [9] according to which the set of outgoing states is roughly the subspace $P(D\geq 0)\mathcal{H}$ where $D = XP+PX$ is the generator of the dilation group. The basic ingredients are

i) $e^{iHt}\frac{D}{t}e^{-iHt}P_{ac} \to 4H$ (converges in strong resolvent sense)

ii) $(W_-(H,H_0)-I)\varphi(H)P(D\geq 0)$ is compact [9] when φ is a positive function with compact support on an interval $[a,b]$, $a>0$. So assume $\phi \in P_{ac}\mathcal{H}$ and $\varphi(H) = H$ for some φ of the above type ; this is true for a dense set in $\mathcal{H}$. Then

$$(W_-(H,H_0)-I)e^{-iHt}\phi = (W_-(H,H_0)-I)\varphi(H)\{P(D\geq at) + P(D\leq at)\}e^{-iHt}\phi$$

By the compactness stated in ii) and since $P(D\geq at)$ tends strongly to zero the first term has a zero limit. The second one tends to $(W_-(H,H_0)-I)\varphi(H)P(H\leq a)e^{-iHt}\phi$ which is zero ; this completes the proof of (11).

Various applications of Enss' method to more general systems can be found in B. Simon's paper [41] .

III. STATIONARY SCATTERING THEORY

The basic problem of the stationary approach is the so-called "limiting absorption principle" that is for a suitable Banach space $\mathfrak{X}$ satisfying the continuous dense inclusions

$$\mathfrak{X} \hookrightarrow \mathcal{H} \hookrightarrow \mathfrak{X}^*$$

the existence in some operator topology of the limits in $\mathcal{B}(\mathfrak{X},\mathfrak{X}^*)$:

$$R(\lambda \pm i0) = \lim_{\varepsilon \downarrow 0^+} (H-\lambda \mp i\varepsilon)^{-1} \tag{12}$$

If λ belongs to the continuous spectrum of H it is well-known that the right-hand side of (12) has no limit in $\mathcal{B}(\mathcal{H})$ although it can be defined as an unbounded operator. The relevance of this problem for Scattering Theory comes in particular from the fact that scattering cross-sections involve such boundary values. For example for one particle potential scattering one has at real energies :

$$\sigma(|k|) = \lim_{\varepsilon \downarrow 0^+} \operatorname{Im} \langle Ve^{ik.x}, (H-k^2-i\varepsilon)^{-1} Ve^{ik.x} \rangle$$

If $Ve^{ik.x}$ belongs to a Banach space $\mathfrak{X}$ such that (12) holds then the cross-section is finite. More precise estimates on the way the limit is reached in (12) can provide bounds, asymptotic behaviour and dispersion relations for amplitudes.

The limiting absorption principle also plays an essential role in the analysis of regularity and asymptotic properties for solutions of the (two-Hilbert space) Lippman-Schwinger equation :

$$v_\pm(E) = Ju(E) - R(E \pm i0)(HJ - JH_0)u(E) \tag{13}$$

where $u(E)$ is a given solution of $(H_0 - E)u(E) = 0$. The second term on the right-hand side of (13) plays the role of the scattered wave ; it is an element of $\mathfrak{X}^*$ satisfying in addition ingoing or outgoing boundary conditions, which in an abstract way correspond to belonging to the ranges of $R(E \pm i0)$. In the special case of potential scattering such boundary conditions should be equivalent to Sommerfeld's radiation condition and then lead to (1). This connection is explicited in a beautiful work of S. Agmon and L. Hörmander [41] . Among earlier analysis let me mention the pioneering works of J. Green and O. Lanford [42] and T. Ikebe [43] .

Finally the limiting absorption principle enters the mathematical theory of "weak" wave-operators obtained by an Abel-limit procedure

$$\widetilde{W}_{\pm}(H,H_0;J) = s\cdot\lim_{\varepsilon\downarrow 0} \pm\varepsilon\int_0^{\pm\infty} e^{-\varepsilon|t|}\, e^{iHt} J e^{-iH_0 t}\, dt$$

$$= s\cdot\lim_{\varepsilon\downarrow 0} \varepsilon\int (H-\lambda\pm i\varepsilon)^{-1} J (H_0-\lambda\mp i\varepsilon)^{-1}\, d\lambda$$

Notice that in general one simply has

(14) $$W_{\pm}(H,H_0;J) \subset \widetilde{W}_{\pm}(H,H_0;J)$$

Weak local wave-operators $\widetilde{W}_{\pm}(H,H_0;\Delta;J)$ can also be defined by restricting the λ-integral in (14) to the interval Δ. One has the following abstract existence and completeness theorem for $\widetilde{W}_{\pm}(H,H_0,\Delta,J)$ which originates in more or less different forms from works of T. Kato [10], M. Schecter [44] and I. Sigal [45]. Its proof depends on the general theory developed by T. Kato and S.T. Kuroda [46].

<u>Theorem</u>

Let X (resp. X_0) $\subset \mathcal{H}$ (resp. $\mathcal{H}_0$) be an H (resp. H_0) dense Banach space in the sense that there exists no proper closed subspace of $\mathcal{H}$ (resp. $\mathcal{H}_0$) invariant under H (resp. H_0) and containing X (resp. X_0). Assume that

1) The limiting absorption principle holds for H_0 on some $\mathcal{X}_0 \hookrightarrow \mathcal{H}_0$ containing X_0 and for all $\lambda\in\Delta$.

2) Let $\Delta^+ = \{z=\lambda+i\mu \;, 0\le\mu\le a \;, \lambda\in\Delta\}$. Then there exists a function $F(z), z\in\Delta^+$, continuous in the strong operator topology of $B(X,X_0)$.

3) $$(H-z)^{-1}\phi = J(H_0-z)^{-1}F(z)\phi \qquad \forall\phi\in X$$

Then there exists $Z_+(\Delta)\in B(\mathcal{H}_0,\mathcal{H})$ intertwining H and H_0 such that

strong Abel. lim. $e^{itH} J e^{-itH_0} Z_+(\Delta) = P(\Delta)$

This theory has been applied [46] to completeness of wave-operators for the one-body problem with potentials $V(x) = O(|x|^{-1-\varepsilon})$ as $|x|\to\infty$.

Here it is essential that the decay of the potential is isotropic. In a recent paper it has been shown by D.R. Yafaev [47] that break down of completeness is possible in 2 dimensions for potentials satisfying :

$$|V(x)| \leq C(1+|x_1|)^{-\alpha_1}(1+|x_2|)^{-\alpha_2}$$

with $\alpha_1+\alpha_2>1$. D.R. Yafaev argument is based on a nice Born-Oppenheimer like analysis and is intuitively rather convincing of the generality of this phenomenon.

Among other applications of abstract Stationary Scattering Theory let me mention the attempts to treat the N-body problem in this framework by K. Yajima [48] for $N=3$ and by I. Sigal [45] . However for the general method to work it is necessary that in addition to some reasonable assumptions on bound-states and quasi-bound-states the potentials $V_{ij}(x)$ decay like $|x|^{-2-\varepsilon}, \varepsilon>0$ in order to avoid threshold problems ; such a restriction seems to be inherent to these approaches and to indicate their limitations. I will not review here the general theory of smooth perturbations which furnishes a perturbative way to verify the hypothesis of abstract stationary theory ; I refer to [20] , [49] for its review and applications. Rather I would like to insist on the constructive approach of S. Agmon [2] [5] and S. Agmon and L. Hörmander [41] in Sobolev and Besov spaces. Let me first recall the earlier result [5] that the limiting absorption principle holds for one-body short-range potential scattering in the auxiliary spaces $\mathcal{H}=L^2_s \ \forall s>\frac{1}{2}$ with

$$L^2_s=\left\{\phi\ ,\ (1+|x|^2)^{s/2}\phi\in L^2(\mathbb{R}^n)\right\}$$

The restriction $s>\frac{1}{2}$ is strongly related to trace properties on (n-1) dimensional momentum hypersurfaces (e.g. the energy shell $k^2=E\neq 0$) of Fourier transforms of elements in L^2_s . For this reason one cannot reach the limit $s=\frac{1}{2}$. So it is natural to ask whether there is an optimal auxiliary space such that

$$L^2_{\frac{1}{2}}\supset\mathcal{H}\supset L^2_s \qquad \forall s>\frac{1}{2}$$

for which the limiting absorption principle would hold. Another question is how to adapt long range forces to the analysis in [5] . The first question is answered in [41] when $V=0$ but can be immediately extended to short range forces by the method of [5] . The results are as follows : let

$$\mathcal{B}^*=\left\{u\ ,\ \sup_{R>1}\frac{1}{R}\int_{|x|<R}|u(x)|^2dx<\infty\right\}$$

the dual of the Besov space . Then

1) The limiting absorption principle holds in $\mathcal{B}$; the limit in (12) is a weak limit.

2) The outgoing (resp. ingoing) solutions of the Schrödinger equation

(15) $$(-\Delta+V-E)\,u=\phi \quad , \quad \phi\in\mathcal{B}$$

that is elements in $\mathcal{B}^*$ of the form

(16) $$u=R(E+i0)\,\phi$$

(resp. $u=R(E-i0)\,\phi$) satisfy generalized radiation conditions in the following sense : Let $\Phi(x,\xi)$ be a symbol in $(\mathbb{R}^n\setminus\{0\})\times\mathbb{R}^n$ homogeneous of degree zero in x , continuously differentiable n times in x and once in ξ , satisfying

$$|D_x^\alpha D_\xi^\beta\,\varphi(x,\xi)|\leq C \quad , \ |\alpha|\leq n,\ |\beta|\leq 1,\ |x|=1$$

Consider the associated pseudo-differential operator

$$\varphi(x,D)f(x)=(2\pi)^{-n}\int\varphi(x,\xi)\hat{f}(\xi)\,e^{ix\cdot\xi}\,d\xi$$

Then if $\varphi(2\xi,\xi)=0$ and $\varphi(-2\xi,\xi)\neq 0$ for $\xi^2=E\neq 0$, a solution $u\in\mathcal{B}^*$ of $(-\Delta+V-E)\,u=\phi\in\mathcal{B}$ is outgoing if and only if

(17) $$\frac{1}{R}\int_{|x|<R}|\varphi(x,D)u|^2\,dx \underset{|x|\to\infty}{\longrightarrow} 0$$

The Sommerfeld radiation condition corresponds to the particular choice

$$\varphi(x,D)=\frac{\partial}{\partial|x|}-i\sqrt{E}$$

associated to the symbol

$$\varphi(x,\xi)=\frac{x}{|x|}\,\xi-\sqrt{E}$$

3) Outgoing and ingoing elements of $\mathcal{B}^*$ are (for a fixed energy) asymptotically orthogonal in the sense

$$\lim_{R\to\infty}\frac{1}{R}\int_{|x|<R}\bar{u}(x)\,v(x)\,dx=0$$

4) Finally outgoing solutions have a radial asymptotic expansion at infinity in the sense that if u satisfies (15) and (16), then there exists a square integrable

function f on the energy sphere such that

$$\lim_{R\to\infty} \frac{1}{R}\int |u(x) - |x|^{\frac{1-n}{2}} e^{i\sqrt{E}|x|} f(\tfrac{x}{|x|})|^2 dx = 0$$

Notice that for potentials $V \in \mathcal{B}$ (which require decay faster than $|x|^{-2}$) this last result implies the validity of (1) in a precise sense. For potentials decaying faster than Coulomb solutions of the Lippman-Schwinger equation (13) still obey a decomposition like (1) if instead of pure plane waves one uses some suitable superpositions belonging to $\mathcal{B}^*$.

The main content of the recent (unpublished) work of S. Agmon [2] is to extend these properties to long-range forces. The basic idea is that since a priori L^2_Δ estimates for $(-\Delta + V - z)$ for short range forces as derived in [5] are uniform in z in any compact set outside the zero energy one should be able to adapt a sufficiently small long-range perturbation. This program however requires for obvious reasons pseudo-differential operator calculus, whereas elliptic methods and Fourier analysis were enough for the short range case.

My conviction that these progress open a new era in Scattering Theory does not come only from the fact that they provide with the help of some classical and powerful P.D.E. techniques some optimal results for one-body systems ; as an example the limiting absorption principle in Besov space gives almost immediately the last result of K. Chadan and A. Martin [50] on finiteness of the cross-section for potentials such that $|x|^2 V(x)$ only have logarithmic decay. The main virtue of this approach certainly lies in the fact that in contrast to the traditional integral equation approach to scattering, these new techniques are linear in both the unperturbed H_0 and the potential V in a way which is perfectly well adapted to an intuitive analysis of the phase space properties of scattering systems. This is certainly a great virtue with respect to more complicated situations where the complexity of integral equation methods has considerably slowed down any progress. This is the case of course for the N-body problem and it is very fortunate that new geometric ideas as those which have been advocated recently by P. Deift and B. Simon [51] seem to fit admirably with this type of analysis.

To conclude this section let me mention another application of the theory, namely low energy expansions of the S-matrix and time decay properties. The results of A. Jensen and T. Kato [6] (see also J. Rauch [7] for related results) state that as mappings between the weighted Sobolev spaces $\mathcal{H}^1_{-\Delta}(\mathbb{R}^3)$ and $\mathcal{H}^{-1}_{\Delta}(\mathbb{R}^3)$ for suitable $\Delta > \frac{1}{2}$ depending on the decay of the potentialand the order of the expansion, one has

$$(H - z)^{-1} = -z^{-1} B_{-2} - i z^{-1/2} B_{-1} + B_0 + i z^{1/2} B_1 + \cdots$$

$$e^{itH} - P_0 - \sum_{j=1}^{N} e^{-it\lambda_j} P_j = -(\pi i t)^{-1/2} B_{-1} + (4\pi i)^{-1/2} t^{-3/2} B_1 + \cdots$$

In these formulas $B_{-2} = P_0$ the projection operator on the zero eigenspace of H, P_j is the projection operator for H associated to the eigenvalue λ_j and B_{-1} depends on P_0 and the zero energy resonance wave-function that is a solution of $H\phi = 0$ belonging to $\mathcal{H}^1_{-1}(\mathbb{R}^3)$ but not to $L^2(\mathbb{R}^3)$. So B_{-2} and B_{-1} do not appear for generic couplings. As a consequence the S-matrix $S(\lambda)$ has a low-energy expansion in $\mathcal{B}(L^2(S_2))$, where S_2 is the unit sphere in $\mathbb{R}^3$:

$$S(\lambda) = \Sigma_0 + i\lambda^{1/2}\Sigma_1 - \lambda\Sigma_2 + \cdots$$

Here Σ_0 is equal to one when there is no zero energy resonance and Σ_1 is related as usual to the scattering length.

This concludes this review of the present status of Scattering Theory. It contains however important omissions like the beautiful work of K. Yajima on the classical limit of the S-matrix or H. Kitada and Y. Saito works on long-range potentials ; fortunatley some of these gaps will be filled in the afternoon session.

REFERENCES

[1] W. GORDON, über den stoss zweier punktladungen nach der Wellenmechanik Z. Phys. 48, 180 (1928).

[2] S. AGMON, Lecture Notes, Salt Lake City (1978).

[3] V. ENSS, Asymptotic completeness for quantum mechanical potential scattering, Commun.Math.Phys. 61, 285 (1979).

[4] D. RUELLE, A remark on bound states in potential scattering theory, Nuovo Cimento A 61, 655 (1969).

[5] S. AGMON, Spectral properties of Schrödinger operators and Scattering Theory, Ann. Scuola Norm. Sup. Pisa, II, 2, 151 (1975).

[6] A. JENSEN and T. KATO, Spectral properties of Schrödinger operators and time decay, Duke Math. J. 46 (1979).

[7] J. RAUCH, Local decay of scattering solutions for Schrödinger equation, Commun.Math.Phys. 61, 149 (1978).

[8] L. HÖRMANDER, The existence of wave-operators in scattering theory, Math. Z. 146, 169 (1976).

[9] E. MOURRE, Link between the geometrical and the spectral transformation approaches in scattering theory, Commun.Math.Phys. 68, 91 (1979).

[10] T. KATO, Two-space scattering theory with applications to many-body problems, J. Fac. Sci. Univ. Tokyo, 24, 503 (1977).

[11] M. SCHECTER, Completeness of wave-operators in two Hilbert spaces, Ann.Inst. H. Poincaré, XXX, 109 (1979).

[12] T. KATO, Some recent results in scattering theory, Univ. of California Preprint, Berkeley (1979).

[13] P. COTTA-RAMUSINO, W. KRÜGER and R. SCHRADER, Quantum scattering by external metrics and Yang-Mills potentials, to appear in Ann. Inst. H. Poincaré.

[14] R. SCHRÄDER, High Energy behaviour for non-relativistic scattering by stationary external metrics and Yang-Mills potentials, CERN preprint TH 2698 (1979).

[15] C. CHANDLER, A.G. GIBSON, N-body quantum scattering theory in two Hilbert spaces, Journ.Math.Phys. 19, 1610 (1978).

[16] M. SCHECTER, A new criterion for scattering theory, Duke Math. J., 44, 863 (1977).

[17] B. SIMON, Scattering theory and quadratic forms : on a theorem of Schecter, Commun.Math.Phys. 53, 151 (1977).

[18] T. KATO, On the Cook-Kuroda criterion in scattering theory, Commun.Math.Phys. 67, 85 (1979).

[19] J. KUPSH, W. SANDHAS, Möller operators for scattering on singular potentials, Commun.Math.Phys. 2, 147 (1966).

[20] M. REED, B. SIMON, Scattering Theory (Methods of Modern Mathematical Physics, vol. III), Academic Press (1979).

[21] L. HÖRMANDER, The existence of wave-operators in scattering theory, Math. Z. 146, 69 (1976).

[22] T. KATO, On finite dimensional perturbations of self-adjoint operators, J.Math.Soc. Japan 9, 239 (1957).

[23] D.B. PEARSON, A generalisation of Birman's trace theorem, J. Funct.Anal. 28, 182 (1978).

[24] J.D. DOLLARD, Interpretation of Kato's invariance principle in scattering theory, J. Math. Phys. 17, 46 (1976).

[25] P. LAX and R. PHILLIPS, Scattering Theory , Academic Press, New York, 1967.

[26] P. DEIFT and B. SIMON, On the decoupling of finite singularities from the question of asymptotic completeness, J. Funct. Anal. 23, 218 (1976).

[27] A. JENSEN and T. KATO, Asymptotic behaviour of the scattering phase for exterior domains, Commun. in P.D.E 3, 12, 1105 (1978).

[28] M. COMBESCURE and J. GINIBRE, Scattering and local absorption for the Schrödinger operator, J. Funct. Anal. 29, 54 (1978).

[29] S.T. KURODA, Perturbations of continuous spectra by unbounded operators, I,II. J. Math. Soc. Japan, 11, 247 (1959).

[30] M. G. KREIN, On the trace formula in the theory of perturbation, Mat. Sbornik 33 (75), 597 (1953).

[31] G. DETSH, Introduction ot the theory and applications of the Laplace transform, Springer, Berlin (1974).

[32] H. McKEAN and I.M. SINGER, Curvature and the eigenvalues of the Laplacian, J. Diff. Geometry, 1, 43 (1967).

[33] P.B. GILKEY, Recursion relations and the asymptotic behaviour of the eigenvalues of the Laplacian, Princeton University Preprint (1977).

[34] V.S. BUSLAEV, Dokl.Akad. Nauk SSSR, 197, 999 (1971) ; Soviet Math. Dokl., 12, 999 (1971).

[35] A. MAJDA and J. RALSTON, An analogue of Weyl's theorem for unbounded domains, I, II. Duke Math. Journal 45, 183 and 45, 513 (1978).

An Epilogue, III, University of California Preprint (1979).

[36] T. KATO, Monotonicity theorems in scattering theory, Hadronic Journal 1, 134 (1978).

[37] B. SIMON, Kato's inequality and the domination of semi-groups, J. Funct. Anal., 32, 97 (1979).

[38] H. HESS, R. SCHRÄDER and D. A. UHLENBROCK, Duke Math. Jour., 44, 893 (1977).

[39] H. HOGREVE, R. SCHRÄDER and R. SEILER, Nuclear Phys. B142, 525 (1978).

[40] B. SIMON, Phase space analysis of simple scattering systems. Extensions of some work of Enss. To appear in Duke Math. Jour.

[41] S. AGMON and L. HÖRMANDER, Asymptotic properties of solutions of differential equations with simple characteristics, Journal d'Analyse, 1976.

[42] T. GREEN and O. LANFORD III, Rigorous derivation of the phase shift formula, J. Math. Phys. 1, 139 (1960).

[43] T. IKEBE, Eigenfunction expansions associated with the Schrödinger operators and their application to scattering theory, Arch. Rat. Mech. Anal., 5, 1 (1960).

[44] M. SCHECTER, Completeness of wave-operators in two Hilbert spaces, Ann.Inst. H. Poincaré, XXX 2, 109 (1979).

[45] I. SIGAL, Scattering theory for many-particle systems I, II, Preprint ETH Zürich

[46] T. KATO and S.T. KURODA, The abstract theory of scattering, Rocky Mtn.J.Math. 1, 127 (1971).

[47] D.R. YAFAEV, On the break-down of completeness in potential scattering, Commun.Math.Phys. 65, 167 (1979).

[48] K. YAJIMA, An abstract stationary approach to 3-body scattering, J.Fac.Sc. Univ. of Tokyo, Sec.1A, 25, 109 (1978).

[49] J.M. COMBES, Scattering theory in quantum mechanics and asymptotic completeness, Proceedings of the AMP Conference, Rome 1977, Springer Lecture Notes.

[50] K. CHADAN and A. MARTIN, Boundedness of total cross-section in potential scattering, Poster session of this Conference.

[51] P. DEIFT and B. SIMON, A time-dependent approach to the completeness of multiparticle quantum systems, Com.Pure Appl.Math. 30, 573 (1977).

SCHROEDINGER OPERATORS WITH ELECTRIC OR MAGNETIC FIELDS

W. Hunziker

Institut für Theoretische Physik
ETH Hönggerberg, 8093 Zürich, Switzerland

The Zeeman- and Stark effect are first examples of quantum mechanical perturbation theory. Nevertheless it has taken half a century to develop an adequate mathematical description. Here we summarize the results of a systematic effort in recent years, notably by Avron, Herbst and Simon [2,35].

I. MAGNETIC FIELDS

1. Kato's inequality

A key to Hamiltonians with general external magnetic fields B(x)=rot a(x) is Kato's distributional inequality [13,57 II]:

$$-\Delta |u| \leq \mathrm{Re}\,(\mathrm{sgn}\, u)\,(i\nabla + a(x))\,u \quad , \tag{1}$$

which holds in any dimension if $u \in L^1_{\ell oc}$ and $(i\nabla + a)u \in L^1_{\ell oc}$. Here a(x) is a real C^1-vector field and

$$(\mathrm{sgn}\, u)(x) = \begin{cases} \overline{u(x)}\,|u(x)|^{-1} & \text{if } u(x) \neq 0 \\ 0 & \text{if } u(x) = 0 . \end{cases}$$

With this inequality the following theorem has an easy proof [13, 57 II]:

<u>Theorem 1</u> Let a(x) be a real C^1-vector field on R^n and $0 \leq V_1 \in L^2_{\ell oc}(R^n)$. Then the operator

$$(i\nabla + a)^2 + V_1 = (p - a)^2 + V_1 \tag{2}$$

is essentially selfadjoint on $C_0^\infty(R^n)$.

More recently, Kato's inequality was recognized as the infinitesimal form of an inequality between semigroups [11,18,19]:

Theorem 2 Let A and B be selfadjoint operators on $L^2(M,d\mu)$ and bounded from below. Then the following 3 statements are equivalent:

Semigroup domination

$$|e^{-tB}u| \leq e^{-tA}|u| \tag{3}$$

for all u and $0 \leq t < \infty$.

Resolvent domination

$$|(B - z)^{-1}u| \leq (A - z)^{-1}|u| \tag{4}$$

for all u and $-\infty < z < \sigma(A)$.

Kato's inequalities

$$(f,A|u|) \leq \mathrm{Re}\,(f, (\mathrm{sgn}\,u)\,Bu) \tag{5a}$$

$$(f,A|u|) \leq \mathrm{Re}\,(f, (\mathrm{sgn}\,u)\,Au) \tag{5b}$$

for all $0 \leq f \in Q(A)$ = form domain of A and all $u \in D(B)$ resp. $u \in D(A)$. Here $(f,A|u|)$ is understood in form sense.

Remarks

(i) L^2 is used to have a natural absolute value map $u \to |u|$. This is generalized in [11] and applied to particles in external Yang-Mills fields [11,16] .

(ii) (3) implies that exp (-tA) is positivity preserving:

$$e^{-tA}u \geq 0 \text{ for } u \geq 0 \quad \longleftrightarrow \quad |e^{-tA}u| \leq e^{-tA}|u| \text{ for all } u .$$

This is equivalent to (5b) [18].

(iii) As an application of (1) and theorems 1 and 2 we note that

$$|e^{-t((p-a)^2+V_1)}u| \leq e^{-tp^2}|u| . \tag{6}$$

Resolvent domination is convenient to estimate perturbations. Let V be a multiplication operator of relative bound α with respect to A . Then (4) implies

$$\| V (B - z)^{-1} \| \leq \| V(A - z)^{-1} \| ,$$

and letting $z \to -\infty$ we conclude that V has relative bound $< \alpha$ with respect to B. Combining theorem 1 with (6) we thus obtain a large class of Hamiltonians with magnetic fields [13, 57 II]:

Theorem 3 Let $a(x)$ and $V_1(x)$ satisfy the hypothesis of theorem 1. If V_2 is a real multiplication operator of relative bound < 1 with respect to Δ, then

$$H (a) = (p - a)^2 + V_1 + V_2 = H_o(a) + V \tag{7}$$

is essentially selfadjoint on $C_o^\infty(R^n)$, bounded below and

$$|e^{-tH(a)} u| \leq e^{-tH(o)} |u| \tag{8}$$

for all u and $0 \leq t < \infty$.

Remarks

(i) In the following we always assume $H(a)$ to be of this type. Other results on selfadjointness are found in [4, 14, 15, 20, 57 II,59], and references to earlier work in the notes to section X.4 of [57 II]. In particular, there is a form-analogue of theorem 3 allowing $a \in L^2_{\ell oc}$, $0 \leq V_1 \in L^1_{\ell oc}$ and V_2 of form-bound < 1 relative to Δ [4,20].

(ii) It was remarked by Nelson (quoted in [19]) that (8) follows by inspection from the functional integral representation of $\exp(-tH(a))$ [10, 20, 61]. Closely related is a simple proof of (8) using the Trotter product formula and the fact that in one dimension $a(x)$ can be gauged away [20].

2. Diamagnetism

For identical particles, (8) also holds in the subspace of totally symmetric wave functions. It expresses the diamagnetic effect of an arbitrary magnetic field on a system of spinless particles [17, 18]:

Theorem 4 For $H(a)$ given by (7),

$$\inf \sigma(H(a)) \geq \inf \sigma (H(0)) \tag{9}$$

$$\mathrm{Tr}\, e^{-\beta H(a)} \leq \mathrm{Tr}\, e^{-\beta H(0)} . \tag{10}$$

<u>Proof</u> (9) is evident from resolvent domination and can also be proved directly by Kato's inequality [17]. An elementary argument for ground states is given in [61]. (10) follows from

<u>Lemma 1</u> Let A, B be bounded, positive operators on $L^2(M, d\mu)$ satisfying $|Au| \leq B\,|u|$ for all u. If B is trace class then A is trace class and $\operatorname{Tr} A \leq \operatorname{Tr} B$.

We remark that $\exp(-\beta H(0))$ is trace class for suitable V_1. Alternatively, one should formulate theorem 3 for particles in a box.

By the Golden-Thompson-Symanzik inequality the (field-independent) classical free energy is a lower bound for the quantum free energy in arbitrary fields. Other classical bounds and classical limits are discussed in [10].

3. Spectral- and Scattering Theory

Here (8) is exploited using the following compactness criterion [52, 56]:

<u>Lemma 2</u> Let A, B be bounded operators on $L^2(M, d\mu)$ satisfying $|Au| \leq B\,|u|$ for all u. If B is compact, then A is compact.

As a consequence, $F\,(i + H(a))^{-1}$ is compact for any multiplier $F \in C_0$. For $a = 0$ this is well known and easy to prove. For $a \neq 0$ it then follows from resolvent domination and lemma 2. This is the essential ingredient of

<u>Lemma 3</u> (i) [53] All points $E \in \sigma_{ess}(H(a))$ arise from "states at ∞" in the sense that

$$\|(H(a) - E)\, u_n\| \to 0 \tag{11}$$

for some sequence $u_n \in C_0^\infty$ with $\|u_n\| = 1$ and $u_n(x) = 0$ for $|x| < n$.

(ii) [58, 57 III] If u is orthogonal to all eigenvectors of $H(a)$, then there exist two sequences $t_n \to \pm\infty$ such that

$$\|F\, e^{-it_n H(a)}\, u\| \to 0 \tag{12}$$

for any $F \in C_0$.

Scattering Theory (12) is the starting point in Enss' proof of asymptotic completeness [54], which is extended in [60] to homogeneous magnetic fields. Similar results can be obtained by Agmon's methods [4]. For general a(x), resolvent domination allows to apply the Kato-Birman theory to the pair H(a), $H_0(a)$ under the same conditions as in the case a = 0 [4].

Essential spectrum In general there is a diamagnetic effect analogous to (9) on the bottom $\Sigma(a)$ of the essential spectrum of H(a) :

$$\Sigma(a) \geq \Sigma(0) \tag{13}$$

Proof: Setting $E = \Sigma(a)$ in (11) and using Kato's inequality we obtain

$$(|u_n|, H(0)|u_n|) \leq (u_n, H(a)\, u_n) .$$

Since $|u_n| \to 0$ weakly, $\Sigma(0) \leq \liminf (|u_n|, H(0)|u_n|) \leq \lim (u_n, H(a)u_n)$ $= \Sigma(a)$.

According to (11) only the behaviour of V(x) and a(x) at ∞ is relevant for the essential spectrum. Since a(x) is gauge-dependent it is desirable to have some criteria in terms of the magnetic field itself, which can be introduced as the commutator

$$B_{k\ell} = a_{k,\ell} - a_{\ell,k} = i\,[\Pi_k, \Pi_\ell] ,$$

where $\Pi = p - a$. By the Schwarz inequality

$$\langle B_{k\ell} \rangle \leq \langle \Pi_k^2 + \Pi_\ell^2 \rangle \leq \langle H_0(a) \rangle . \tag{14}$$

As an application, assume that a single field component $B_{k\ell} \to B > 0$ as $|x| \to \infty$. Then by (11) and (14)

$$B \leq \inf \sigma_{ess} (H_o (a)) ,$$

an inequality which is saturated for homogeneous field. In particular $H_0(a)$ has purely discrete spectrum if some $B_{k\ell} \to \infty$ as $|x| \to \infty$. More sophisticated "magnetic bottles" are discussed in [4]. It is clear from (11) that these results are not affected by the presence of a potential V(x) vanishing at ∞ . For such a case the effect of a magnetic field on the spectrum is shown in Fig. 1:

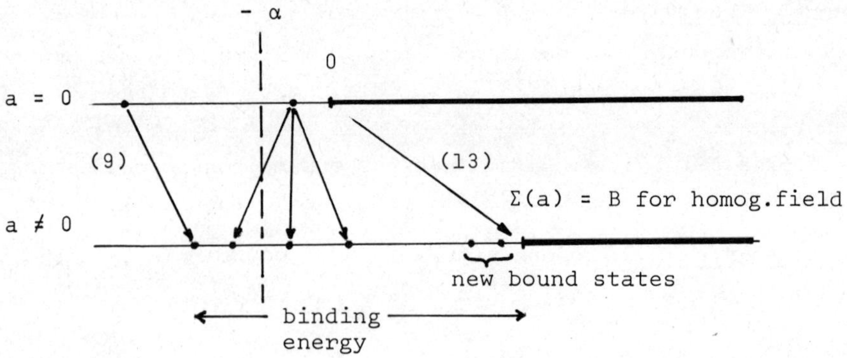

In the following we summarize results of various generality concerning negative energy bound states (number, stability, perturbation expansion) and enhanced binding effects in magnetic fields (binding energy, new bound states, strong binding in high fields).

4. Negative Energy Bound States

Here we assume $V(x) \to 0$ at ∞. For any $\alpha \geq 0$ let $N_\alpha(a)$ be the total multiplicity of all eigenvalues $< -\alpha$ of $H(a)$. $N_\alpha(a) > N_\alpha(0)$ is possible since some eigenvalues may decrease (as in first order Zeeman effect, see Fig. 1). However, by resolvent domination and Lemma 1 the Birman-Schwinger bound

$$N_\alpha(a) \leq (4\Pi)^{-2} \int d^3x d^3y \, |V(x)| \, \frac{e^{-2\sqrt{\alpha}|x-y|}}{|x-y|} \, |V(y)| \tag{15}$$

is independent of a(x) [4,10]. In [4] the same is proved for the Cwickel-Lieb-Rosenbljum bound

$$N_0(a) \leq C_n \int d^n x \, |V(x)|^{n/2} \tag{16}$$

which holds in $n \geq 3$ dimensions. The reason is that (apart from the value of c_n) (16) is a classical phase volume and thus independent of a(x). An immediate consequence of these estimates was noted in [4,10]:

Theorem 5 The Lieb-Thirring estimate of the stability of matter [55] extends to matter in arbitrary magnetic fields, and by a remark of Fröhlich (given in [4]) also to matter coupled to a quantized radiation field.

The stability of any eigenvalue $E(0) < 0$ of $H(0)$ for $a(x) = \frac{1}{2} B \wedge x$ (homogeneous field) results from the continuity properties of the re -

solvent as $B \to 0$ [4]:

$$\underbrace{(z - H(a))^{-1} = (z - H_0(a))^{-1}}_{\text{strongly (not norm) continuous as } B \to 0} + \underbrace{(z - H(a))^{-1} V(z - H_0(a))^{-1}}_{\text{norm continuous as } B \to 0}$$

for z in the resolvent set of $H(0)$. By integration over a small circle around $E(0)$ it follows that the corresponding spectral projection is norm continuous as $B \to 0$. Therefore $E(0)$ is the limit as $B \to 0$ of a group of eigenvalues $\{E_k(a)\}$ of $H(a)$ with the same total multiplicity.

The Rayleigh-Schrödinger perturbation series has been studied in particular for

$$H(B) = \frac{1}{2}(p - \frac{1}{2} B \wedge x)^2 - \frac{1}{r} .$$

A proof of Borel summability (including atoms) has been announced [2,7]. For the ground state energy perturbation series

$$E_0(B) \sim \sum_{n=0}^{\infty} a_{2n} B^{2n}$$

there is a convincing (but not fully rigorous) argument [1] that

$$a_{2n} = (-1)^{n+1} \left(\frac{4}{\pi}\right)^{5/2} \pi^{-2n} (2n + \frac{1}{2})! \, (1 + O(\frac{1}{n}))$$

which is very well supported by the numerical values of a_{2n} up to $n = 100$ [1].

5. Enhanced Binding

The following inequality of Lieb holds for arbitrary V and $a(x) = \frac{1}{2} B \wedge x$ [4]:

$$\inf \sigma (H(a)) - B \leq \inf \sigma (H(0)) . \tag{17}$$

In the case $V(x) \to 0$ at ∞ (Fig. 1) this says that the ground state binding energy is enhanced in homogeneous magnetic field. Another form of (17) is the "paramagnetic inequality"

$$\inf \sigma (\tilde{H}(a)) \leq \inf \sigma (\tilde{H}(0)) \tag{18}$$

where $\tilde{H}(a)$ is the spin 1/2 Hamiltonian $H(a) + \vec{B}\,\vec{\sigma}$. (18) has been extended in 2dimensions to polynomial B's [8]. However, there is also a counterexample [9] in two dimensions which shows that (18) cannot hold for general a's and V's [12].

Following [4] we conclude our review of magnetic fields with a discussion of the 1-particle Zeeman Hamiltonian

$$
\begin{aligned}
H(B,\lambda) = &- \frac{d^2}{dz^2} + \lambda V\,(x^2 + y^2\ ;\ z) \\
&- \Delta_{xy} + \frac{B^2}{4}(x^2 + y^2) - BL_z\ ,
\end{aligned}
\tag{19}
$$

assuming $B \geqslant 0$. In each angular momentum subspace $L_z = m\ \varepsilon$ $\{0, \pm1, \pm2 \ldots\}$ the bottom line of (19) (as an operator on $L^2(R^2)$) has purely discrete spectrum with a lowest eigenvalue

$$\Sigma_m = B\,(\,|m| - m + 1)$$

and a corresponding harmonic oscillator eigenfunction $f_m(x,y)$. Hence if $V(x) \to 0$ at ∞ , $H(B,\lambda)$ has essential spectrum $[\Sigma_m, \infty)$ in the sector $L_z = m$. For trial states of the form $u = f_m(x,y)\ g(z)$,

$$\langle H(B,\lambda)\rangle_u = \Sigma_m + \left\langle -\frac{d^2}{dz^2} + \lambda\, V_m \right\rangle_g \quad \text{with} \tag{20}$$

$$V_m(z) = \int dxdy\ |f_m(x,y)|^2\ V(x^2+y^2,\ z)\ ,$$

which exhibits the <u>one-dimensional aspect</u> of the problem for $B \neq 0$ and leads to the following effect: if $V \leqslant 0$ and not identically zero, the same is true for all V_m . Then for any m and $\lambda > 0$, $\langle H(B,\lambda)\rangle_u < \Sigma_m$ if $g(z) \sim \exp\,(-\,\alpha|z|)$ with α sufficiently small. Therefore if $B > 0$ and $\lambda > 0$, $H(B,\lambda)$ has infinitely many discrete eigenvalues below $\Sigma_0 = B$ corresponding to $m = 0,1,2\ldots$, and infinitely many embedded eigenvalues with $m < 0$ - even if $H(0,\lambda)$ has no eigenvalues at all (V short-range, λ small). Another peculiar feature of (19) is that the ground state may have $m \neq 0$ and may be degenerate. That this cannot happen for $V = -1/r$ is proved in [7]. By the same mechanism, a homogeneous magnetic field can produce extra bound states of negative ions [3,7].

From (20) it also follows that the ground state binding energy $B - E_0(B)$ for $V = -1/r$ diverges as $B \to \infty$, since the Coulomb potential is catastrophic in one dimension. In this case, scaling $\vec{x} \to B^{1/2}\vec{x}$ gives the unitary equivalence

$$H(B,1) \sim B\,H(1, B^{-1/2}) ,$$

i.e. a weak coupling problem for $B \to \infty$. This is used to derive the asymptotic law [2,6]:

$$B - E_0(B) =$$

$$[\tfrac{1}{2}\ln B - \ln\ln B - 0{,}05796 + O(\frac{\ln\ln B}{\ln B})]^2$$

in extreme magnetic fields ($B \gg 1 \sim 10^9$ Gauss).

We end with the remark that the 2-particle Zeeman problem does <u>not</u> reduce to the 1-particle problem (19). Translation covariance in homogeneous magnetic field and the analogue to the separation of the center of mass are discussed in [5]. This analysis forms the basis for the theory of the Zeeman effect in atoms [7].

II. ELECTRIC FIELDS

6. Stark Hamiltonians

Qualitatively, the Stark effect in hydrogen should be regarded as the "free fall" in homogeneous electric field, perturbed by the Coulomb potential. This is the point of view in

<u>Theorem 6</u> [57 IV] For real f

$$H_0(f) = -\Delta - fx_1$$

is essentially selfadjoint on $C_0^\infty(R^3)$ and has spectrum $(-\infty, +\infty)$ if $f \neq 0$. If $V \in L^2_{\ell oc}(R^3)$ and $V(x) \to 0$ for $|x| \to \infty$ then V is compact relative to $H_0(f)$. Hence, if V is also real,

$$H(f) = -\Delta - fx_1 + V$$

is essentially selfadjoint on $C_0^\infty(R^3)$ and has spectrum $(-\infty, +\infty)$ for $f \neq 0$.

If $f \neq 0$, scattering theory for the pair $H(f)$, $H_0(f)$ reflects the fact that the electric field accelerates the particle like $x_1 \sim t^2$ as $t \to \pm\infty$. Therefore it suffices for $V(x)$ to fall off faster than $x_1^{1/2}$

as $x_1 \to +\infty$. In particular, the Coulomb potential has short range in this sense. From the many results [21, 33, 47, 49, 50, 60] we quote an example of [60]:

Theorem 7 Let $f \neq 0$. If $V(1+x_1^2)^{\varepsilon+1/4}$ is compact relative to $H_0(f)$ for some $\varepsilon > 0$, then the wave operators

$$\Omega_{\pm} = \text{s-}\lim_{t \to \pm\infty} e^{itH(f)} e^{-itH_0(f)}$$

exist and are complete in the strong sense: $(\text{Ran}\ \Omega\pm)^{\perp}$ = subspace spanned by the eigenvectors of $H(f)$. The eigenvalues of $H(f)$ have finite multiplicities and no point of accumulation.

Absence of eigenvalues for $f \neq 0$ was proved for hydrogen by Titchmarsh [46] and more generally in [21]. Sufficient conditions are, essentially:

(i) If $V(x)$ is smooth in some right half space $a < x_1 < \infty$ where the total force satisfies

$$-\frac{\partial V}{\partial x_1} + f > \varepsilon > 0 \quad ,$$

then any solution of $H\psi = E\psi$ vanishes identically in $a<x_1<\infty$.

(ii) If, in addition, $V(x)$ satisfies the hypothesis of a unique continuation theorem (e.g. $V(x)$ locally bounded on $R^3 \setminus S$, where S is a closed set of measure zero such that $R^3 \setminus S$ is connected), then $\psi(x) = 0$ for all x .

As a consequence of this result and theorem 7, $H(f)$ is unitarily equivalent to $H_0(f)$ for a large class of potentials including the Coulomb potential [33] .

7. Stark Resonances

We have just seen that all bound states of $H(0)$ generally dissolve into the continuum when the electric field is turned on. Their reappearance as $f \to 0$ can be described not only in terms of spectral concentration [25, 40, 41, 57 IV] but more concisely by resonances in the framework of dilation analyticity. For the separable case of hydrogen this is discussed in [27, 28, 30]. Here we summarize the more general result of [34] which can be extended to multiparticle systems [36].

By a real dilation $x \to e^{\Theta}x$, $H(f)$ is unitarily equivalent to

$$H(\theta,f) = U(\theta)H(f)\,U^{-1}(\theta)$$

$$= e^{-2\theta}p^2 - e^{\theta}fx_1 + V(e^{\theta}x_1) \qquad (21)$$

$$= H_0(\theta,f) + V(\theta)\ , \text{ where}$$

$$U(\theta) \quad : u(x) \to e^{3\theta/2}\, u(e^{\theta}x)\ . \qquad (22)$$

We assume dilation analyticity of V in the following sense:

For some $n < 1$, $V(\theta)(1 + p^{2n})^{-1}$ is compact and analytic in θ in a strip $|\mathrm{Im}\ \theta| < \gamma$. (23)

The Stark effect is now described in terms of spectra of the family $H(\theta,f)$ with $f \in R$ and $|\mathrm{Im}\ \theta| < \gamma$, starting from the unperturbed spectrum of $p^2 + V$ ($\mathrm{Im}\ \theta = 0$, $f = 0$):

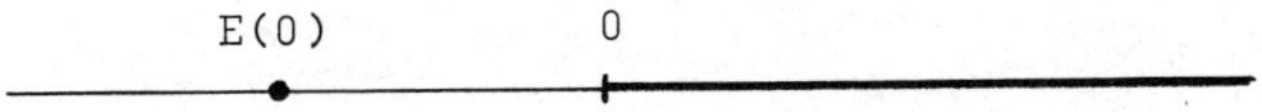

(Fig.2)

For simplicity $E(0)$ is assumed to be a nondegenerate eigenvalue. The next step is to take $\mathrm{Im}\ \theta \neq 0$ (we choose $\mathrm{Im}\ \theta > 0$), still keeping $f = 0$:

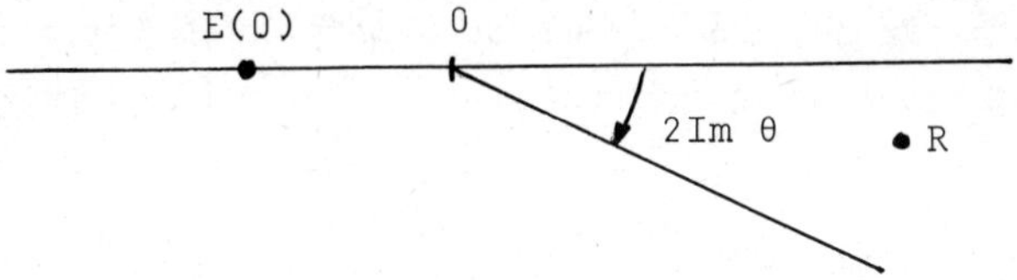

(Fig.3)

This is the result of [51]: The continuum swings be an angle $2\ \mathrm{Im}\ \theta$. The rest of the spectrum is discrete and independent of θ , since the eigenvalues are analytic in θ and independent of $\mathrm{Re}\ \theta$ by unitary equivalence. In particular, $E(0)$ remains an eigenvalue as long as $|\mathrm{Im}\ \theta| < \pi/2$. New ("resonance") eigenvalues R may appear in the sector swept by the continuum but nowhere else.

Now let $0 < \mathrm{Im}\ \theta < \pi/3$ and $f \neq 0$. Then the spectrum becomes purely discrete and independent of θ :

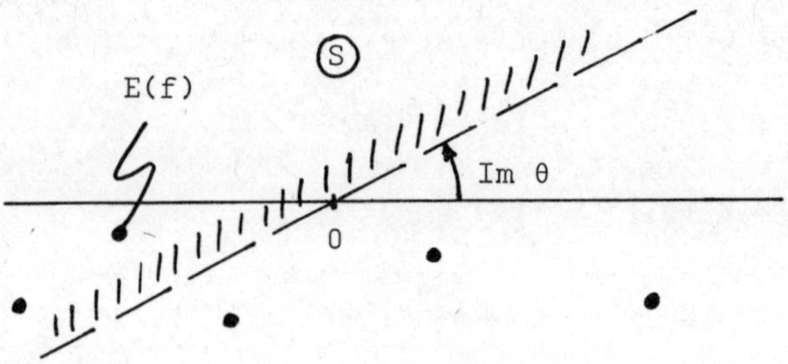

(Fig. 4)

While the value of Im θ is no longer visible in the spectrum, it still plays an important role for stability: in the shaded halfplane S the eigenvalues are stable as functions of f . In particular there is a "Stark resonance" $E(f) \to E(0)$ as $f \to 0$, which has an asymptotic expansion to all orders in f :

$$E(f) \sim \sum_{n=0}^{\infty} a_n f^n \quad . \tag{24}$$

Since E(f) is independent of θ , this is precisely the Rayleigh-Schrödinger series calculated for the perturbation $-fx_1$ of $p^2 + V$.

Finally, if we set $\text{Im}\,\theta = 0$, the spectrum jumps back to the continuum $(-\infty,+\infty)$ of the Stark Hamiltonian H(0,f). However, there is a direct relation between the spectrum of H(θ,f) shown in Fig. 4 and H(0,f) :

$$(u,(z - H(0,f))^{-1}\upsilon) = (U(\bar{\theta})u \ , \ (z-H(\theta,f))^{-1} \ U(\theta)\upsilon) \ . \tag{25}$$

Here u and υ are from a dense set of dilation-analytic vectors, and (25) is first established for z in the set $S \cap \{\text{Im}\, z > 0\}$ and sufficiently far from its boundary, where both resolvents exist. By analytic continuation in z it then follows that:

(i) For $f \in R$ and $\text{Im}\,\theta > 0$ the spectrum of H(θ,f) is in the lower half plane and in fact strictly below the real axis if the Stark Hamiltonian H(0,f) has no eigenvalues (Fig. 4).

(ii) The matrix elements $(u,(z-H(0,f))^{-1}\upsilon)$ have a meromorphic continuation from $\text{Im}\, z > 0$ into $\text{Im}\, z < 0$ with poles at the eigenvalues shown in Fig. 4.

The clue to these results is the explicit "free fall-propagator" written as

$$P^t = \exp(-ite^{-\theta}H_0(\theta,f))$$

$$= e^{-it^3\alpha f^2/12}\, e^{itfx_1/2}\, e^{-it\alpha p^2}\, e^{itfx_1/2}$$

where $\alpha = e^{-3\theta}$. For $\mathrm{Im}\ \alpha \leq 0$ (i.e. $0 \leq \mathrm{Im}\ \theta \leq \pi/3$) and real f, P^t is a strongly continuous contraction semigroup mapping $\mathcal{S}$ into $\mathcal{S}$ (This defines the generator $\sim H_0(\theta,f)$ as the closure of its restriction to $\mathcal{S}$). Via Laplace transform we obtain the bound

$$\| (z - H_0(\theta,f))^{-1} \| \leq d(z)^{-1}$$

for $z \in S$ (Fig. 4), where $d(z)$ is the distance of z from the boundary of S. For $f \neq 0$ and $\mathrm{Im}\ \alpha < 0$ there is a stronger estimate

$$\|P^t\| \leq e^{t^3 f^2 \mathrm{Im}\alpha/12},$$

which falls off faster than any exponential as $t \to \infty$. As a result, $H_0(\theta,f)$ has empty spectrum for $f \neq 0$ and $0 < \mathrm{Im}\ \theta < \pi/3$.

To study the effect of the perturbation $V(\theta)$ we consider

$$V(\theta)\, P^t = e^{-it^3\alpha f^2/12}\, e^{itfx_1/2}\, \overbrace{V(\theta)(1 + p^{2n})^{-1}}^{\text{compact}}$$

$$\cdot \underbrace{(1 + p^{2n})\, e^{-it\alpha p^2}}_{\|\ \| \leq \text{const}\,(1 + t^{-n})\ \text{for}\ \mathrm{Im}\,\alpha < 0}\, e^{itfx_1/2}$$

The compact factor turns the strong continuity in f, θ and t into norm continuity, so that the Laplace transform $K(z,\theta,f) = V(\theta)(z - H_0(\theta,f))^{-1}$ is

(i) compact

(ii) analytic in (z,θ)

(iii) norm convergent to zero as $d(z) \to \infty$

(iv) norm continuous as $f \to 0$

for $z \in S$, $f \in R$ and $0 < \mathrm{Im}\ \theta < \min(\phi, \pi/3)$. We add that (i) and (ii) extend to all z by the resolvent equation for $(z - H_0(\theta,f))^{-1}$. The result of Fig. 4 now follows from analytic Fredholm theory [57.I] and the representation

$$(z - H)^{-1} = (z - H_0)^{-1} + (z - H_0)^{-1}K(1 - K)^{-1} ,$$

in which (just as in section 4) the last term is norm continuous as $f \to 0$ for $z \notin \sigma(H(\theta,0))$.

8. Perturbation Theory

Borel summability of (24) is proved for hydrogen in [27, 28], and for the general case in [36]:

Theorem 8 Let V satisfy condition (23), $0 < \alpha < \min(\gamma,\pi/2)$, and let $E(f)$ be the Stark resonance corresponding to a nondegenerate eigenvalue $E(0) < 0$ of $p^2 + V$. Then, for some $R_\alpha > 0$, $E(f)$ is analytic in the region $-\alpha < \arg f < \pi + \alpha$, $f < R_\alpha$ where it satisfies a strong asymptotic estimate

$$\left| E(f) - \sum_{n=0}^{N-1} a_n f^n \right| \leq C^N N! \, |f|^N \tag{26}$$

for all N.

In [36] this is actually proved for multiparticle systems and for certain degenerate cases (including hydrogen). As a consequence of theorem 8, the Borel transform $B(z) = \Sigma a_n z^n/n!$ is analytic in the union of the circle $|z| < C^{-1}$ with the sector $\pi/2 - \alpha < \arg z < \pi/2 + \alpha$. For sufficiently small $f > 0$ the Stark resonance $E(f)$ can then be obtained by inverse Borel transform:

$$E(f) = \omega \int_0^\infty dt \, B(\omega t f) \, e^{-\omega t} ,$$

where $\omega \neq 0$ is choosen in $\pi/2 - \alpha < \arg \omega < \pi/2$. This answers the puzzling question how the real perturbation series (24) can determine the complex resonance $E(f)$.

If the Stark Hamiltonian has no eigenvalues, then the resonance $E(f)$ has a width

$$\Gamma(f) \equiv -2 \operatorname{Im} E(f) > 0$$

for real $f \neq 0$, which by (26) vanishes faster than any power of f as $f \to 0$. In this case the perturbation series (24) is divergent for any $f \neq 0$. In fact $\Gamma(f)$ determines the leading term of a_n as $n \to \infty$ [23, 35, 36] :

<u>Theorem 9</u> If $V(x) = V(-x)$ in addition to the hypothesis of theorem 8, then $a_n = 0$ for n odd and

$$a_{2n} = -\frac{1}{\pi}\int_0^R \Gamma(x)\, x^{-(2n+1)}\, dx + O(R^{-2n}) \tag{27}$$

for some $R > 0$ and $n \to \infty$.

<u>Proof</u> Since $E(f)$ is nondegenerate, $V(x) = V(-x)$ implies $\overline{E(f)} = E(-\bar{f})$. Let K be the semicircle in $\mathrm{Im}\, z \geq 0$ with basis $-R \leq z \leq R < R_\alpha$. Then (27) follows by expanding both sides of the identity

$$E(f) = \frac{1}{\pi i}\int_K dz\, z\, E(z)(z^2-f^2)^{-1}$$

in powers of f .

---

9. Hydrogen

The general theory of Stark resonances is illustrated by detailed qualitative and quantitative results in the case $V = -1/r$, in particular for the ground state resonance $E_0(f)$. Of the many numerical calculations [22, 24, 26, 32, 39, 43, 44, 45] of Stark energy shifts and lifetimes only [22, 39] refer directly to resonances in the dilation-analytic sense. Rayleigh-Schrödinger coefficients a_n have been computed to high order ($n \sim 100$) by algebraic schemes [42, 44]. The qualitative results concern the asymptotic behaviour of $E_0(f)$ as $f \to 0$ [22, 23, 26, 31, 43, 44, 48] or $f \to \infty$ [22, 23] and of a_n as $n \to \infty$ [22, 23, 44]. We quote from [22], where $E_0(f)$ is calculated to high accuracy in the range $10^{-2} < f < 10^{20}$. The results are given in atomic units, i.e. for the ground state resonance of

$$H(f) = -\frac{1}{2}\Delta - fx_1 - r^{-1} .$$

<u>$f \to 0$:</u>

$$\Gamma_0(f) = \frac{4}{f}\, e^{-\frac{2}{3f}}\, [1 + Af + Bf^2 + O(f^3)] \tag{28}$$

$$A = -8.916 \;,\; B = 25.57 .$$

This is accurate to 1 part in 10^4 up to $f = 10^{-2}$. The leading term is known as "Oppenheimers formula" after an incorrect result in [38] and can be obtained as a tunneling probability in WKB-approximation [37,48]. This procedure to calculate $\operatorname{Im} E_0(f)$ is justified and extended in [31].

<u>$n \to \infty$:</u> As a consequence of (27) and (28)

$$a_{2n} = -\frac{6}{\pi}\left(\frac{3}{2}\right)^{2n}(2n)!\left[1+\frac{A}{3n}+\frac{2B}{9n(2n-1)}+O(n^{-3})\right]$$

which is accurate to 1 part in 10^4 for $n > 12$.

<u>$f \to \infty$:</u>

$$\arg E_0(f) = -\frac{\pi}{3}\left[1+\frac{1}{2\ell n f}+O\left(\frac{\ell n\,\ell n f}{(\ell n f)^2}\right)\right]$$

$$|E_0(f)| = \frac{1}{2}\left(\frac{f}{2}\,\ell n f\right)^{3/2}\left[1+\frac{8}{3}\,\frac{\ell n\,\ell n f}{\ell n f}+O\left(\frac{1}{\ell n f}\right)\right], \tag{29}$$

accurate to $\sim 1\%$ at $f = 10^{20}$. The asymptotic regimes (28) and (29) are joined by the numerical values for $E_0(f)$ in the large intermediate range $10^{-2} < f < 10^{20}$ [22].

References for Magnetic Fields

[1] Avron J.E.; Adams B.G.; Čížek J.; Clay M.; Glasser M.L.; Otto P.; Paldus J.; Vrscay E.: "The Bender-Wu formula, SO(4,2) dynamical group and the Zeeman effect in hydrogen"; subm. to Phys.Rev.Lett.

[2] Avron J.E.; Herbst I.W.; Simon B.: Phys. Rev. Lett. 62A, 214-216 (1977)

[3] Avron J.E.; Herbst I.W.; Simon B.: Phys. Rev. Lett. 39, 1068-1070 (1977)

[4] Avron J.E.; Herbst I.W.; Simon B.: Duke Math. J. 45, 847-883 (1978)

[5] Avron J.W.; Herbst I.W.; Simon B.: Ann. Phys. 114, 431-451 (1978)

[6] Avron J.E.; Herbst I.W.; Simon B.: "The strongly bound states of hydrogen in intense magnetic fields"; Phys.Rev. A, to appear

[7] Avron J.E.; Herbst I.W.; Simon B.: "Schrödinger Operators with Magnetic Fields. III. Atoms in magnetic fields"; to be submitted to Commun.math.Phys.

[8] Avron J.E.; Seiler R.: Phys. Rev. Lett. 42, 931-933 (1979)

[9] Avron J.E.; Simon B.: "A counterexample to the paramagnetic conjecture"; submitted to Phys. Lett.

[10] Combes J.M.; Schrader R.; Seiler R.: Ann. Phys. 111, 1-18 (1978)

[11] Hess H.; Schrader R.; Uhlenbrock D.A.: Duke Math. J. 44, 893-904 (1977)

[12] Hogreve H.; Schrader R.; Seiler R.: Nucl. Phys. B 142, 525-534 (1978)

[13] Kato T.: Israel J. Math. 13, 135-148 (1972)

[14] Kato T.: Integral Eqn. and Operator Th. 1, 103-113 (1978)

[15] Schechter, M.: J. Functional Analysis 20, 93-104 (1975)

[16] Schrader R.; Seiler R.: Commun. math. Phys. 61, 169-175 (1978)

[17] Simon B.: Phys. Rev. Lett. 36, 1083-1084 (1976)

[18] Simon B.: Indiana Univ. Math. J. 26, 1067-1073 (1977)

[19] Simon B.: J. Functional Analysis 32, 97-101 (1979)

[20] Simon B.:
J. Operator Theory 1, 37-47 (1979)

References for Electric Fields

[21] Avron J.E.; Herbst I.W.:
Commun. math. Phys. 52, 239-254 (1977)

[22] Benassi L.; Grecchi V.:
"Resonances in Stark Effect and Strongly Asymptotic Approximants";
to appear in J. Phys. B

[23] Benassi L.; Grecchi V.; Harrell E.; Simon B.:
Phys. Rev. Lett. 42, 704-707 (1979)

[24] Brändas E.; Froelich P.:
Phys. Rev. A 16, 2207-2210 (1977)

[25] Conley C.C.; Rejto P.A.:
"Spectral concentration II: General theory" in:
Perturbation Theory and its Applications in Quantum Mechanics (C.H. Wilcox ed.)
Wiley, New York 1966

[26] Damburg R.J.; Kolosov V.V.:
J. Phys. B 9, 3149-3157 (1976), and
J. Phys. B 11, 1921 (1978)

[27] Graffi S.; Grecchi V.:
Lett. Math. Phys. 2, 335-341 (1978)

[28] Graffi S.; Grecchi V.:
Commun.math.Phys. 62, 83-96 (1978)

[29] Graffi S.; Grecchi V.:
J. Phys. B 12, L 265-267 (1979)

[30] Graffi S.; Grecchi V.; Simon B.:
"Complete Separability of the Stark Problem in Hydrogen";
to appear in J. Phys. B

[31] Harrell E.; Simon B.:
"The Mathematical Theory of Resonances Whose Widths are Exponentially Small";
manuscript 1979

[32] Hehenberger M.; Mc Intosh H.V.; Brändas E.:
Phys. Rev. A 10, 1494-1506 (1974)

[33] Herbst I.W.:
Math. Zeitschrift 155, 55-71 (1977)

[34] Herbst I.W.:
Commun.math. Phys. 64, 279-298 (1979)

[35] Herbst I.W.; Simon B.:
Phys. Rev. Lett. 41, 67-69 (1978)

[36] Herbst I.W.; Simon B.:
"Dilation Analyticity in Constant Electric Field II: N-Body Problem, Borel Summability";
manuscript 1979, to be submitted to Commun.math.Phys.

[37] Lanczos C.:
Z. Physik 68, 204-232 (1931)

[38] Oppenheimer J.R.:
Phys. Rev. 31, 66-81 (1928)

[39] Reinhardt W.P.:
Int. J. Quant. Chem. Symp. 10, 359-367 (1976)

[40] Rejto P.A.:
Helv. Phys. Acta 43, 652-667 (1970)

[41] Riddel R.C.:
Pacific J. Math. 23, 377-401 (1967)

[42] Silverstone H.J.:
Phys. Rev. A 18, 1853-1864 (1978)

[43] Silverstone H.J.:
"Asymptotic relations between the energy shift and ionization rate in the Stark effect in hydrogen";
preprint 1979

[44] Silverstone H.J.; Adams B.G.; Čížek J.; Otto P.:
"Asymptotic formula for the perturbed energy coefficients and calculations of the ionization rate by means of high order perturbation theory for hydrogen in the Stark effect";
preprint 1979

[45] Silverstone H.J.; Koch P.M.:
"Calculation of Stark effect energy shifts by Padé approximants to Rayleigh-Schrödinger perturbation theory";
preprint 1979

[46] Titchmarsh E.C.:
"Eigenfunction Expansions Associated with Second Order Differential Equations", Part II.
Oxford University Press 1958

[47] Veselič K.; Weidmann J.:
Math. Zeitschrift 156, 93 (1977)

[48] Yamabe T.; Tachibana A.; Silverstone H.J.:
Phys. Rev. A 16, 877-890 (1977)

[49] Yajima K.:
"Spectral and scattering theory for Schrödinger operators with Stark-effect";
to appear in J. Fac. Sci. Univ. Tokyo

[50] Yajima K.:
"Spectral and scattering theory for Schrödinger operators with Stark-effect, II";
preprint 1978

Other References

[51] Aguilar J.; Combes J.M.:
Commun.math.Phys. 22, 269-279 (1971)

[52] Dodds P.G., Fremlin D.H.:
"Compact operators in Banach lattices";
submitted to J. Functional Analysis

[53] Enss V.:
Commun. math. Phys. 52, 233-238 (1977)

[54] Enss V.:
Commun. math. Phys. 61, 285-291 (1978)

[55] Lieb E.H.:
Rev. Mod. Phys. 48, 553-569 (1976)

[56] Pitt L.:
"A compactness condition for linear operators in function spaces";
submitted to J. Operator Theory

[57] Reed M., Simon B.:
"Methods of Modern Mathematical Physics". Vol. I-IV;
Academic Press, New York

[58] Ruelle D.:
Nuovo Cimento 61 A, 655-662 (1969)

[59] Schechter M.:
"Spectra of Partial Differential Operators";
North Holland, Amsterdam 1971

[60] Simon B.:
Duke Math. J. 46, 119-168 (1979)

[61] Simon B.:
"Functional Integration and Quantum Physics";
Academic Press, New York 1979

[62] Thirring W.:
"Lehrbuch der Mathematischen Physik. 3. Quantenmechanik von Atomen und Molekülen";
Springer Verlag, Wien, New York, 1979

A NEW METHOD FOR ASYMPTOTIC COMPLETENESS

Volker Enss
Fakultät für Physik der Universität
D-4800 Bielefeld 1, F.R. Germany

We present a geometric proof of asymptotic completeness for scattering systems which follows with mathematical rigor the physical intuition of the space-time behavior of a state. As an example we will treat here quantum mechanical short-range potential scattering [3]. The Hilbert space is $\mathcal{H} = L^2(\mathbb{R}^\nu)$. The decomposition $\mathcal{H} = \mathcal{H}^p(H) \oplus \mathcal{H}^{cont}(H)$ corresponds to the point - resp., continuous spectral subspaces of the self-adjoint Hamiltonian $H = H_o + V$, $H_o = -(1/2m)\Delta$, V a multiplication - or pseudodifferential (velocity dependent force) operator.

Asymptotic completeness (strong form) means that any state orthogonal to the bound states $(\mathcal{H}^p(H))$ is an asymptotically free scattering state, i.e., for any $\Psi \in \mathcal{H}^{cont}(H)$

$$\lim_{\tau\to\infty} \sup_{t\geq 0} \| [\exp(-iHt) - \exp(-iH_o t)] \exp(-iH\tau)\Psi\| = 0 \ ;$$

or equivalently,

$$\Psi \in \operatorname{Ran} \Omega^{out}, \ \Omega^{out} = \text{s-}\lim_{t\to\infty} \exp(iHt)\exp(-iH_o t) \ ; \tag{1}$$

or $$\lim_{\tau\to\infty} \| (\Omega^{out} - \mathbb{1}) \exp(-iH\tau)\Psi \| = 0 \ ; \tag{2}$$

and similarly for the past. This implies absence of a singular continuous spectrum. Actually it is sufficient to show (2) for a dense set of vectors $\Psi \in \mathcal{H}^{cont}(H)$ and for a suitably chosen sequence of times τ_n; that is what we will do below.

We impose on the potential a local condition to avoid local absorption [8] and decay assumptions. Let $g, \bar{g}$ be functions in $C_o^\infty(\mathbb{R})$, typically one may think of them as smoothed characteristic functions, g(H) etc. as smoothed projections. By $F(|\vec{x}| \leq R)$ etc. we denote the multiplication with the characteristic function of the indicated region.

We restrict the singularities of the potential by assuming *local compactness*

$$F(|\vec{x}| \leq R)g(H) \text{ is compact } \forall\, g,\ R < \infty \ . \tag{3}$$

This assumption is very weak, it holds if $H_o + V$ is defined as operator-sum, form-sum, Friedrichs extension, or ... . See [6] for a physical discussion, [1] for sufficient conditions on V.

Next we require that far away from the scatterer the potential becomes weak:

$$\| [(H+i)^{-1} - (H_o+i)^{-1}]F(|\vec{x}| \geq R)\| =$$
$$= \| (H+i)^{-1} V(H_o+i)^{-1}F(|\vec{x}| \geq R)\| \to 0 \quad \text{as} \quad R \to \infty \quad . \tag{4}$$

(3) and (4) together are equivalent to:

$$g(H) - g(H_o) \text{ is compact for all } g \quad . \tag{5}$$

To single out *short-range interactions* we have to specify the decay rate, e.g., in (4), but the following weaker requirement is sufficient: For all $g, \bar{g}$

$$\|\bar{g}(H)\ V\ g(H_o)\ F(|\vec{x}| \geq R)\| =: h(R) \in L^1([0,\infty)) \quad . \tag{6}$$

$V = (1 + |\vec{x}|)^{-1-\varepsilon}$ is a short-range potential in any dimension ν.

We consider the dense set of vectors $\Psi \in H^{cont}(H)$ whose compact energy support does *not* include zero. Thus, for each Ψ there is a $g \in C_o^\infty(\mathbb{R}\backslash\{0\})$ such that

$$g(H)\Psi = \Psi \quad . \tag{7}$$

An argument related to the ergodic theorem shows (corollary on p.343 in [10]) that for any $\Psi \in H^{cont}(H)$ there is a sequence of times $\tau_n \to \infty$ such that

$$\text{w} - \lim_{n\to\infty} \exp(-iH\tau_n)\Psi = 0 \quad . \tag{8}$$

To prove (2) we will show for any $\varepsilon > 0$:

$$\| (\Omega^{out} - \mathbb{1}) \exp(-iH\tau_n)\Psi\| < \varepsilon \quad \text{as } n \to \infty \,. \tag{9}$$

As a first step we have to wait until the state will leave an arbitrily big region: for any $R < \infty$ (to be chosen later depending on ε)

$$\|F(|\vec{x}| \leq 2R)\ \exp(-iH\tau_n)\Psi\| =$$
$$\|F(|\vec{x}| \leq 2R)\ g(H)\ \exp(-iH\tau_n)\Psi\ \| \to 0 \quad \text{as} \quad n \to \infty \tag{10}$$

by local compactness (3) and (8) [11,1]. Furthermore, by (5) and (8)

$$\|g(H_o)\ \exp(-iH\tau_n)\Psi - \exp(-iH\tau_n)\Psi\| =$$
$$\| [g(H_o) - g(H)]\ \exp(-iH\tau_n)\Psi\ \| \to 0 \quad \text{as} \quad n \to \infty \quad . \tag{11}$$

Thus the component of the state with small kinetic energy which is annihilated by $g(H_o)$ i.e., the component with velocities in a neighborhood of zero, disappears.

Choose $\bar{g}(\omega) \in C_0^\infty(\mathbb{R})$ such that $\bar{g}(\omega)\, g(\omega) = g(\omega)$; then by the intertwining property

$$\begin{aligned}&(\Omega^{out} - \mathbb{1})\, g(H_o)\, \exp(-iH\tau_n)\Psi \\ &= \bar{g}(H)(\Omega^{out} - \mathbb{1}) g(H_o)\, \exp(-iH\tau_n)\Psi \\ &\quad + [\bar{g}(H) - \bar{g}(H_o)]\, g(H_o)\, \exp(-iH\tau_n)\Psi \;.\end{aligned} \tag{12}$$

The second term on the RHS vanishes as $n \to \infty$ by (5) and (8). We use a simple extension of Cook's method to estimate the first term. For any pair $g,\bar{g}$:

$$\begin{aligned}&\|\bar{g}(H)\ (\Omega^{out} - \mathbb{1})\ g(H_o)\Phi\| \\ &\leq \int_0^\infty dt\ \|\bar{g}(H)\ V\ g(H_o)\ \exp(-iH_o t)\Phi\| \;.\end{aligned} \tag{13}$$

We know already from (10) that at time τ_n, $t = 0$, the state $g(H_o)\ \exp(-iH\tau_n)\Psi$ is far away where the potential is weak. But the integral (13) will be small only if the state moves farther away for $t > 0$ under the *free* time evolution. This is true for the outgoing part Ψ_n^{out} of the state $g(H_o)\ \exp(-iH\tau_n)\Psi = \Psi_n^{out} + \Psi_n^{in}$.

A classical free particle moves away from the origin in the future if $\vec{p} \cdot \vec{x} \geq 0$. Guided by the classical intuition we will restrict $\vec{x}$ depending on $\vec{p}$, we use a phase space decomposition of the quantum state. One can split a state Φ into pieces $\Phi_{j,\vec{c}}$ having strict $\vec{p}$-space localization in small neighborhoods of $\{\vec{p}_j\}$ and being concentrated in $\vec{x}$-space near $\vec{c} \in \mathbb{Z}^\nu$. Then $\Phi^{out} = \Sigma\ \Phi_{j,\vec{c}}$ where $\vec{p}_j \cdot \vec{c} \geq 0$. Alternatively one can use a smoothed projection onto the positive spectral halfspace w.r.t. the self-adjoint operator $\vec{x}\cdot\vec{p} + \vec{p}\cdot\vec{x}$.

Typically in any such decomposition the outgoing (incoming) part Ψ_n^{out} (Ψ_n^{in}) has the following properties: (a) the compact kinetic energy support is contained in $E > (m/2)v_{min}^2 > 0$; (b) up to rapidly decreasing tails it is localized in $|\vec{x}| > 2R$ and $\vec{p}\cdot\vec{x} + \vec{x}\cdot\vec{p} \geq 0$ (≤ 0); (c) $\|\Psi_n^{out/in}\| \leq 2\,\|\Psi\|$.

The following estimate on the "propagation into the classically forbidden region" can be shown using direct calculation [3], stationary phase method [10,12], spectral theory [4], commutators [7], Mellin transforms [9]:

$$\begin{aligned}&\|F(|\vec{x}| \leq R + v_{min}t)\ \exp(-iH_o t)\Psi_n^{out}\| \\ &\leq A_N(1 + t + R)^{-N} \quad \forall\, N,\ t \geq 0 \;.\end{aligned} \tag{14}$$

The cutoff at R instead of 2R takes care of the quantum tails.

Now we are ready to finish the estimate (13) for $\Psi_n^{out} = g(H_o)\Psi_n^{out}$:

$$\int_o^\infty dt\,\|\bar{g}(H)\ V\ g(H_o)\ \exp(-iH_o t)\Psi_n^{out}\|$$
$$\leq 2\,\|\Psi\|\int_o^\infty dt\,\|\bar{g}(H)\ V\ g(H_o)\ F(|\vec{x}|\geq R + v_{min}t)\|$$
$$+\,\|\bar{g}(H)\ V g(H_o)\|\int_o^\infty dt\,\|F(|\vec{x}|\leq R + v_{min}t)\ \exp(-iH_o t)\Psi_n^{out}\|.$$

Using (14) and the decay assumption (6) this expression is smaller than any $\varepsilon > 0$ for R big enough. Thus asymptotically Ψ_n^{out} has a free future time evolution and it lies in the range of Ω^{out}.

It remains to show that the incoming component vanishes as $n \to \infty$. An analogous argument shows that any incoming part would have evolved freely in the past (relative to τ_n) and would have had to come in from even farther away:

$$[g(H_o)\ \exp(-iH\tau_n)\Psi]^{in} \approx [g(H_o)\ \exp(-iH_o\tau_n)\Psi]^{in}$$
$$\approx [g(H_o)\ \exp(-iH_o\tau_n)\ F(|\vec{x}| \geq R + v_{min}\tau_n)\Psi]^{in} \to 0$$

because $\|F(|\vec{x}| \geq R + v_{min}\tau_n)\Psi\| \to 0$. This finishes the outline of the proof of asymptotic completeness.

A different version [7] uses that (3) and (6) implies compactness of $\bar{g}(H)\ V\ g(H_o)$ and one shows that $\bar{g}(H)(\Omega^{out} - \mathbb{1})\ g(H_o)$ is compact on the positive spectral subspace of $\vec{p}\cdot\vec{x} + \vec{x}\cdot\vec{p}$.

The method can be extended to treat long-range potentials [4], the Dirac-equation and very general free Hamiltonians [12], classical wave equations [12,2], two cluster scattering for multiparticle systems as a special case of multichannel scattering [5], and many other applications.

References

1. W. O. Amrein, V. Georgescu: Helv. Phys. Acta 46, 635-658 (1973).
2. P. Cotta-Ramusino, W. Krüger, R. Schrader: Scattering by external metrics and Yang-Mills potentials, preprint F.U. Berlin, 1979.
3. V. Enss: Commun. Math. Phys. 61, 285-291 (1978).
4. ———: Ann. Phys. (N.Y.) 119, 117-132 (1979).
5. ———: Commun. Math. Phys. 65, 151-165 (1979).
6. R. Haag, J. A. Swieca: Commun. Math. Phys. 1, 308-320 (1965).
7. E. Mourre: Commun. Math. Phys. 68, 91-94 (1979).
8. D. B. Pearson: Commun. Math. Phys. 40, 125-146 (1975).
9. P. A. Perry: Mellin transforms and scattering theory, preprint Princeton Univ., in preparation.
10. M. Reed, B. Simon: Methods of modern mathematical physics, Vol. III. Scattering theory, New York, London: Academic Press 1979.
11. D. Ruelle: Nuovo Cimento 61A, 655-662 (1969).
12. B. Simon: Duke Math. J. 46, 119-168 (1979).

PATHOLOGICAL SPECTRAL PROPERTIES

D. B. Pearson
Department of Applied Mathematics,
University of Hull, Hull, ENGLAND.

A particle moving in a straight line encounters a series of obstacles. At the i'th obstacle (i = 1, 2, 3, ... labelled from left to right) there is probability q_i of reflection and p_i of transmission; $q_i + p_i = 1$. For example, coming from the (i + 1)'th obstacle to the i'th there will be probability q_i of reflection back to the (i + 1)'th and p_i of transmission to the (i - 1)'th. Define the coefficient of reflection σ_i by $\sigma_i = q_i/p_i$.

For a pair of obstacles, labelled say 1 and 2, the joint transmission probability is

$$p = p_1p_2 + p_1(q_2q_1)\,p_2 + p_1\,(q_2q_1)^2\,p_2 + \dots = p_1p_2/(1 - q_1q_2).$$

Hence the joint coefficient of reflection is

$$\sigma = \frac{q}{p} = \frac{1-p}{p} = \sigma_1 + \sigma_2.$$

For a sequence of n obstacles, the reflection coefficient is $\sigma = \sum_1^n \sigma_i$ - the σ's are additive. If n = ∞ there will be recurrence (i.e. a particle projected to the right will return to its starting point with probability one) if and only if $\sum_1^\infty \sigma_i = \infty$.

Consider now a quantum mechanical particle moving in a one-dimensional potential V(r) $(-\infty<r<\infty)$ made up of an infinite series of "bumps". For a single bump, define the transfer matrix $\begin{pmatrix} M_{11} & k^{-1}M_{12} \\ kM_{21} & M_{22} \end{pmatrix}$ transforming $\begin{pmatrix} \psi \\ \psi' \end{pmatrix}$ to the left of the bump into $\begin{pmatrix} \psi \\ \psi' \end{pmatrix}$ to the right, for solutions of the Schrödinger equation $-\psi'' + V\psi = k^2\psi$ $(k^2>0)$. The transmission and reflection probabilities for a single bump, as a function of energy k^2, may be expressed in terms of the transfer matrix to give a reflection coefficient

$$\sigma = \frac{q}{p} = \frac{1}{4}(M_{21}^2 + M_{11}^2 + M_{22}^2 + M_{12}^2 - 2).$$

For a pair of bumps, labelled (1) and (2) and separated by distance N, the joint transfer matrix becomes

$$\begin{pmatrix} M_{11}^{(2)} & k^{-1}M_{12}^{(2)} \\ kM_{21}^{(2)} & M_{22}^{(2)} \end{pmatrix} \begin{pmatrix} \cos Nk & k^{-1}\sin Nk \\ -k\sin Nk & \cos Nk \end{pmatrix} \begin{pmatrix} M_{11}^{(1)} & k^{-1}M_{12}^{(1)} \\ kM_{21}^{(1)} & M_{22}^{(1)} \end{pmatrix}.$$

This leads to a combined reflection coefficient of the form

$$\sigma(k) = a(k) + b(k)\cos 2Nk + c(k)\sin 2Nk.$$

For large N, $\sigma(k)$ is a rapidly oscillating function of k, with mean value (averaging over a range of k large compared with $^1/N$) given by

$$\overline{\sigma}(k) = a(k) = \sigma_1 + \sigma_2 + 2\sigma_1\sigma_2.$$

Writing $\overline{\sigma} = \frac{1}{2}(e^{\overline{\lambda}} - 1)$, $\sigma_i = \frac{1}{2}(e^{\overline{\lambda}_i} - 1)$ we have $\overline{\lambda} = \overline{\lambda}_1 + \overline{\lambda}_2$. I.e. $\overline{\lambda}$ is additive. Since $\overline{\lambda} = \log(1 + 2\overline{\sigma})$, it follows that $(1 + 2\overline{\sigma})$ is multiplicative.

This suggests that, for a sequence of n bumps and in the limit as $n \to \infty$, we look at

$$\lim_{n\to\infty} \frac{\sigma^{(n)}(k)}{\prod_{i=1}^{n}(1 + 2\sigma_i(k))},$$

$\sigma^{(n)}$ being the combined reflection coefficient for the first n bumps. This limit may be shown to exist in the sense of distributions and converge to a measure $d\mu$ provided the distance N_i between consecutive bumps goes to infinity sufficiently rapidly. In other words, for large n and on averaging over energies, $\sigma^{(n)}(k)$ behaves like $\prod_{i=1}^{n}(1 + 2\sigma_i(k))$.

If $g_i \to 0$ as $i \to \infty$ (g_i = height of the i'th bump) the infinite product $\prod_{i=1}^{\infty}(1 + 2\sigma_i(k))$ converges if and only if $\sum_{i=1}^{\infty}\sigma_i(k)$ converges, and the condition for this is $\sum_{i=1}^{\infty} g_i^2 < \infty$. If $\sum_{i=1}^{\infty} g_i^2 < \infty$ (and the N_i increase sufficiently rapidly) then the measure $d\mu$ is absolutely continuous, but if $\sum_{i=1}^{\infty} g_i^2 = \infty$ then

(1) $d\mu$ is singular continuous with respect to Lebesque measure.

(2) $\lim_{n\to\infty}\sigma^{(n)}(k) = \infty$ for almost k (w.r.t. Lebesque measure). For large n, the transmission probability will be very small, except for a small range of energies - i.e. we have an almost totally reflecting barrier, which however is almost "transparent" at the right energies.

(3) $H = -\frac{d^2}{dr^2} + V$ in $L^2(0, \infty)$ has purely singular continuous spectrum. (The same applies in $L^2(-\infty, \infty)$ if the bumps are also made to extend in the negative r direction. Regarding V(r) as a spherically symmetric 3-dimensional potential, $H_\ell = -\frac{d^2}{dr^2} + V(r) + \frac{\ell(\ell + 1)}{r^2}$ in $L^2(0, \infty)$ has singular continuous spectrum in each partial wave subspace ℓ.

Proofs of the above results can either be found in [1] or follow the ideas and methods of [1]. A further reference on singular continuous spectrum is [2], which gives an abstract characterisation of the singular continuous subspace in potential scattering.

1 . Pearson, D., Comm. Math. Phys. 16, 30-36 (1978).

2 . Sinha, K., Ann. Inst. Henri Poincaré 26, 263 (1977).

PROOF THAT THE H^- ION HAS ONLY ONE BOUND STATE: A REVIEW, A NEW RESULT, AND SOME RELATED UNSOLVED PROBLEMS

Robert Nyden Hill
Department of Physics, University of Delaware
Newark, Delaware, 19711/USA

The H^- ion, made up of a proton and two electrons, has long been known to have at least one bound state below the lowest breakup threshold. Variational calculations of sufficient accuracy to demonstrate this appeared shortly after the discovery of the Schrodinger equation. A rigorous demonstration that additional bound states do not exist below the lowest breakup threshold first appeared in 1977.[1,2] The present paper will review that proof, prove the new result that the three electron system H^{--} has no bound states below the continuum in the spin 3/2 sector, and discuss related unsolved problems.

Review

The internal Hamiltonian H_{int} for two electron atomic systems such as H^-, He, or Li^+ can be written in the form

$$H_{int} = H_0 - 2\gamma\nabla_1 \cdot \nabla_2 + Z^{-1}V \qquad (1)$$

where

$$H_0 = - \nabla_1^2 - 2r_1^{-1} - \nabla_2^2 - 2r_2^{-1}, \qquad (2)$$

and

$$V = 2|\vec{r}_1 - \vec{r}_2|^{-1} \qquad (3)$$

Here $\vec{r}_1$, $\vec{r}_2$ give the electronic positions relative to the nucleus, $\gamma = m/(m + M) = \mu/M$, m and M are respectively the electronic and nuclear masses, μ is the reduced mass, and Z is the nuclear charge in units of the electronic charge. The length unit is the reduced mass Bohr radius $\hbar^2/(\mu Ze^2)$ and the energy unit is the hydrogenic binding energy $Z^2\mu e^4/(2\hbar^2)$. The notation agrees with reference 2 except that the length and energy units differ by factors of Z and Z^2 respectively.

The time-independent Schrodinger equation can be solved exactly for H_0. The coulomb repulsion V between the electrons, and the Hughes-Eckart term $-2\gamma\nabla_1 \cdot \nabla_2$ must, however, be handled via approximate methods which produce rigorous bounds. The basic tool used is a well known comparison theorem:

Theorem 1: Let $H^{(1)}$ and $H^{(2)}$ be two essentially self-adjoint (Hermitian) Hamiltonians whose discrete eigenvalues below the bottom of the essential spectrum can be characterized by the familiar variational principal $E = \min \langle\psi|H|\psi\rangle/\langle\psi|\psi\rangle$, with the minimization for excited states carried out subject to the constraint that $|\psi\rangle$ be orthogonal to preceding eigenvectors. Denote the ordered eigenvalues of $H^{(i)}$ by $E_1^{(i)} \leqslant E_2^{(i)} \leqslant \dots \leqslant E_n^{(i)} \leqslant \dots \leqslant E_{ess}^{(i)}$, where $E_{ess}^{(i)}$ is the energy at which the essential spectrum (if any) begins. Assume $\langle\psi|H^{(1)}|\psi\rangle$ is defined for all vectors $|\psi\rangle$ for which $\langle\psi|H^{(2)}|\psi\rangle$ is defined. Then if $\langle\psi|H^{(1)}|\psi\rangle \leqslant \langle\psi|H^{(2)}|\psi\rangle$ holds for all admissable state vectors $|\psi\rangle$, $E_n^{(1)} \leqslant E_n^{(2)}$ holds for all n, and $E_{ess}^{(1)} \leqslant E_{ess}^{(2)}$. Upper bounds on the number of bound states below the continuum (or lower bounds to eigenvalues) are obtained from Theorem 1 by letting $H^{(2)}$ be the original Hamiltonian with $H^{(1)}$ a more tractable approximate Hamiltonian.

The lowest threshold for breakup, into a hydrogen atom in its ground state plus a free electron at rest at infinity, lies at -1 for both H_{int} and H_0. It follows from Hunziker's theorem that the continuous spectrum begins at -1 for both H_{int} and H_0. Because H_0 has an infinite number of bound states below -1, the approximate Hamiltonian $H^{(1)}$ must contain enough of the original repulsive V to push all but one of this infinity of bound states up to -1; in particular, the approximation to V must couple to all of the levels of H_0 below the continuum. Stated another way, the approximation to V must contain enough of the original V to preserve shielding: If one electron is in a hydrogenic ground state with the second electron far out, the far out electron must see, after the replacement of V by its approximation, a potential which cannot support an infinite number of bound states.

The approximation to V is chosen to have the form $V^{1/2}PV^{1/2}$ where P is a projection operator.[3] The fact that a projection operator such as P cannot increase the length of a vector such as $V^{1/2}|\psi\rangle$ implies that $\langle\psi|V^{1/2}PV^{1/2}|\psi\rangle \leqslant \langle\psi|V|\psi\rangle$. Thus $H^{(1)}$, obtained from $H^{(2)} = H_{int}$ by replacing V by $V^{1/2}PV^{1/2}$, satisfies the hypothesis of Theorem 1. P is constructed as follows. Let Q_1 be a projector whose range is the space spanned by all states of the form $\phi_1(r_1)\chi(r_2)$ where ϕ_1 is the hydrogenic ground state and χ is arbitrary. Let G_1 be the generalized inverse of $Q_1V^{-1}Q_1$. Let $P_1 = V^{-1/2}G_1V^{-1/2}$. Then P_1 is a projection operator. Let P_2 be the projection operator obtained by interchanging the coordinates of particles 1 and 2 in P_1. P is chosen to be the projection onto the span of the ranges of P_1 and P_2. $V^{1/2}PV^{1/2}$ then couples symmetric states of the form

$$\psi_S(\vec{r}_1, \vec{r}_2) = f(\vec{r}_1)\phi_1(r_2) + \phi_1(r_1)f(\vec{r}_2), \tag{4}$$

and couples antisymmetric states of the form

$$\psi_A(\vec{r}_1, \vec{r}_2) = g(\vec{r}_1)\phi_1(r_2) - \phi_1(r_1)g(\vec{r}_2). \tag{5}$$

The space spanned by states of the forms ψ_S and ψ_A is denoted by $\mathcal{S}_{\parallel}$; the orthogonal complement of $\mathcal{S}_{\parallel}$ is called $\mathcal{S}_{\perp}$. States in $\mathcal{S}_{\perp}$ are not coupled by $V^{1/2}PV^{1/2}$.

Consider now the case of infinite nuclear mass, for which there is no Hughes-Eckart term ($\gamma = 0$). The spaces $\mathcal{S}_{\parallel}$ and $\mathcal{S}_{\perp}$ are then reducing spaces for

$$H^{(1)} = H_0 + V^{1/2}PV^{1/2} \tag{6}$$

The spectrum of $H^{(1)}$ on $\mathcal{S}_{\perp}$ starts at -1/2, which lies above the bottom of the continuum for the full problem at -1. The Schrodinger equation for $H^{(1)}$ on $\mathcal{S}_{\parallel}$ reduces to the one particle Schrodinger equations

$$(I + |\phi_1\rangle\langle\phi_1|)^{1/2}H_S(I + |\phi_1\rangle\langle\phi_1|)^{1/2}|f\rangle = (E + 1)(I + |\phi_1\rangle\langle\phi_1|)f\rangle \tag{7}$$

and

$$H_A|g\rangle = (E + 1)(I - |\phi_1\rangle\langle\phi_1|)|g\rangle \qquad (8)$$

for $|f\rangle$ and $|g\rangle$, where H_S and H_A are given by

$$H_S = (I + |\phi_1\rangle\langle\phi_1|)^{1/2}\{-\nabla^2 - 2r^{-1} + \tfrac{1}{2}|\phi_1\rangle\langle\phi_1|$$
$$+ Z^{-1}[U^{1/2}(I - M)^{-1}U^{1/2} - \tfrac{16}{35}|\phi_1\rangle\langle\phi_1|]\}(I + |\phi_1\rangle\langle\phi_1|)^{1/2} \qquad (9)$$

and by

$$H_A = (I - |\phi_1\rangle\langle\phi_1|)[-\nabla^2 - 2r^{-1} + Z^{-1}U^{1/2}(I + M)^{-1}U^{1/2}](I - |\phi_1\rangle\langle\phi_1|) \qquad (10)$$

Here U is the ordinary potential

$$U(r) = [\int \tfrac{1}{2}|\vec{r} - \vec{r}'||\phi_1(r')|^2 d^3\vec{r}']^{-1}, \qquad (11)$$

I is the identity, and M is a certain nonnegative Hilbert-Schmidt integral operator which is defined and analyzed in reference 2. For large r, $U(r) = 2r^{-1} + O(r^{-3})$ and M is $O(e^{-r})$. Thus (7) and (8), which can be regarded as one-particle Schrodinger equations for the outer electron, have incorporated shielding of the nuclear charge by the inner electron to the extent that the outer electron sees a potential at large distances which cannot support an infinite number of bound states. This potential at large distances behaves like r^{-3} instead of the physically correct r^{-4} because polarization is not properly treated when the inner electron is forced to remain in the hydrogenic ground state. Nevertheless, the approximation is good enough to count the number of bound states correctly. Standard methods for one particle operators show that, for $Z = 1$, H_S has one bound state and H_A has none. It follows that the H^- ion, in infinite nuclear mass approximation with coulomb interactions only, has at most one bound state below the continuum for the full problem in the singlet sector (spin 0, symmetric spatial wave function), and no bound states below the continuum for the full problem in the triplet sector (spin 1, antisymmetric spatial wave function).

When the nuclear mass is finite, the problem is complicated by the fact that the Hughes-Eckart term $-2\gamma\nabla_1\cdot\nabla_2$ couples the spaces $\mathcal{S}_{\parallel}$ and $\mathcal{S}_{\perp}$. This difficulty can be overcome by replacing the part of $H_0 - 2\gamma\nabla_1\cdot\nabla_2$ which couples $\mathcal{S}_{\perp}$ to itself by a lower bound B; a reduction to one particle equations like (7) and (8) is then possible. In this way it can be shown that there is at most one bound state for $\gamma < \gamma_S \approx 0.1736263618$, which includes the physical H^- ion.

A New Result

Now that rigorous results have been obtained for the number of bound states of the two electron atomic system H^-, it is natural to ask about the number of bound states of the three electron atomic systems H^{--} and He^-. The relevant Hamiltonian, in infinite nuclear mass approximation with coulomb interactions only, is

$$H_3 = H_{03} + Z^{-1}(V_{12} \otimes I_3 + V_{23} \otimes I_1 + V_{31} \otimes I_2) \qquad (12)$$

where

$$H_{03} = -\nabla_1^2 - 2r_1^{-1} - \nabla_2^2 - 2r_2^{-1} - \nabla_3^2 - 2r_3^{-1}. \qquad (13)$$

Here $V_{ij} \otimes I_k$ is the tensor product of the coulomb repulsion $V_{ij} = 2|\vec{r}_i - \vec{r}_j|^{-1}$ in the space of electrons i and j with the identity I_k in the space of electron k. The lowest threshold for breakup, into a two electron system in its lowest state plus a free electron at rest at infinity, is at the bottom of the spectrum of the two electron system. This threshold is known exactly only for the quartet sector of H^{--}. In the quartet sector (spin 3/2, antisymmetric spatial wave function), triplet H^-, which has no bound states below the bottom of its continuum at -1, is left behind when one electron moves off to infinity. Thus the continuum starts at -1 for the quartet sector of H^{--}.

It will now be shown that H^{--} has no bound states below -1 in the quartet sector. An approximate Hamiltonian $H_3^{(1)}$, which has no fewer bound states than H_3, can be obtained by replacing $V_{ij} \otimes I_k$ in H_3 by

$$V'_{ij} = (V_{ij}^{1/2} P V_{ij}^{1/2}) \otimes [I_k - (|\phi_1\rangle\langle\phi_1|)_k] + V_{ij} \otimes (|\phi_1\rangle\langle\phi_1|)_k \tag{14}$$

where $(|\phi_1\rangle\langle\phi_1|)_k$ is the projector $|\phi_1\rangle\langle\phi_1|$ in the space of electron k. V'_{ij} couples only states which have at least one electron in the hydrogenic ground state ϕ_1. In the quartet sector, these states have the form

$$\psi_A(\vec{r}_1, \vec{r}_2, \vec{r}_3) = h(\vec{r}_1, \vec{r}_2)\phi_1(r_3) + h(\vec{r}_2, \vec{r}_3)\phi_1(r_1) + h(\vec{r}_3, \vec{r}_1)\phi_1(r_2) \tag{15}$$

where $h(\vec{r}_1, \vec{r}_2) = - h(\vec{r}_2, \vec{r}_1)$ and $\int \phi_1(r_1)h(\vec{r}_1, \vec{r}_2)d^3\vec{r}_1 = 0$. Denote the space spanned by states of the form (16) by $\mathcal{S}_{\parallel}^{3A}$, and let $\mathcal{S}_{\perp}^{3A}$ be the orthogonal complement of $\mathcal{S}_{\parallel}^{3A}$ in the quartet sector. $\mathcal{S}_{\parallel}^{3A}$ and $\mathcal{S}_{\perp}^{3A}$ are then reducing spaces for

$$H_3^{(1)} = H_{03} + Z^{-1}(V'_{12} + V'_{23} + V'_{31}). \tag{16}$$

The bottom of the spectrum of H_{03} on $\mathcal{S}_{\perp}^{3A}$ starts at -3/4, which is above the bottom of the continuum in the quartet sector. The Schrodinger equation for $H_3^{(1)}$ on $\mathcal{S}_{\parallel}^{3A}$ reduces to the two particle Schrodinger equation

$$[P_{\phi\perp 1} \otimes H_{A2} + H_{A1} \otimes P_{\phi\perp 2} + (P_{\phi\perp 1} \otimes P_{\phi\perp 2})V_{12}(P_{\phi\perp 1} \otimes P_{\phi\perp 2})]|h\rangle = (E + 1)(P_{\phi\perp 1} \otimes P_{\phi\perp 2})|h\rangle \tag{17}$$

where H_{Ai} is the one particle triplet sector Hamiltonian H_A of equation (10) in the space of electron i and $P_{\phi\perp j} = I_j - (|\phi_1\rangle\langle\phi_1|)_j$. Because V_{12} is repulsive and H_A has no bound states below 0, the Schrodinger equation (18) has no bound states below -1. It follows that H^{--}, in infinite nuclear mass approximation with coulomb interactions only, has no bound states below the continuum in the quartet sector.

Related Unsolved Problems

Rigorous upper bounds for the number of bound states of H^{--} in the doublet sector (spin 1/2, spatial wave function belongs to the mixed representation of the permutation group on three objects), and for the number of bound states of He^- in the doublet and quartet sectors, have not yet been obtained. The major difficulty arises from the fact that the ground state energy $E_1^{(2)}$ and the ground state wave function Φ_1 are not known exactly for the two electron system left behind when one electron moves off to infinity. If rigorous upper bounds to the number of bound states of these systems are to be obtained via the methods outlined above, it appears necessary to introduce Φ_1 into the approximate Hamiltonian $H^{(1)}$ in such a way that the continuum for $H^{(1)}$ starts at $E_1^{(2)}$ just as for the original Hamiltonian H_3. This can in fact be done, and a one particle Schrodinger equation of the form

$$H_{1p}|\psi\rangle = (E - E_1^{(2)})|\psi\rangle \qquad (18)$$

obtained, where the one particle Hamiltonian H_{1p} has the bottom of its essential spectrum at 0, has no fewer bound states than the original Hamiltonian H_3, and depends on Φ_1. The analysis of H_{1p} is now under way. It will be interesting to see how accurate an approximation to Φ_1 is needed to obtain a rigorous determination of the number of bound states of H_{1p}.

A number of unsolved problems remain for the two electron atomic system. A variational calculation by G. W. F. Drake[4] has shown that the H^- ion has a second bound state, with an energy no higher than -0.25070 in our units, in the unnatural parity sector, where the continuum begins at -0.25. The methods used to show that H^- has only one bound state below -1 should also be effective in settling the question of the number of bound states in this sector.

Information on the number of bound states of the two electron atomic system as a function of Z and γ is also incomplete. The single bound state below the continuum possessed by H_S disappears for sufficiently small Z: The Schwarz inequality implies that $\langle\psi|(|\phi_1\rangle\langle\phi_1|)|\psi\rangle = |(\langle\phi_1|U^{-1/2})(U^{1/2}|\psi\rangle)|^2 \leqslant \langle\phi_1|U^{-1}|\phi_1\rangle\langle\psi|U|\psi\rangle = \frac{35}{32}\langle\psi|U|\psi\rangle$. This plus the fact that all eigenvalues of the integral operator M are nonnegative and less than 1 implies that, for $Z < 32/35$, the Hamiltonian with the potential $- 2r^{-1} + (\frac{1}{2}Z^{-1} + \frac{35}{64})U(r)$ has at least as many bound states as H_S. It is easy to see that this potential has no bound states below the continuum for sufficiently small Z, which implies that H_S has no bound states below the continuum for sufficiently small Z, and in turn that the original two electron atomic system has no bound state below -1 for $\gamma = 0$ and Z sufficiently small. The value of Z at which the bound state disappears is not known.

The values $Z = 1$, $\gamma = 1/2$ arise for two electrons plus a positron. The number of bound states of this system is also not known with certainty, but the question could be settled via the methods used for H^- if a less drastic approximation is used to handle the Hughes-Eckart term. An improved method of handling the Hughes-Eckart term should also permit treating the proton-electron-positron system; the existing proof[5] that this system has no bound state is incomplete because part of the argument rests on a numerical calculation for which error bounds were not computed.

References

1. R. N. Hill, Phys. Rev. Lett. 38, 643-646 (1977).
2. R. N. Hill, J. Math. Phys. 18, 2316 (1977).
3. The method used here generalizes a method introduced in N. W. Bazley, Proc. Natl. Acad. Sci. U. S. A. 45, 850-853 (1959) and Phys. Rev. 120, 144-149 (1960).
4. G. W. F. Drake, Phys. Rev. Lett. 24, 126-127 (1970).
5. I. Aronson, C. J. Kleinman, and L. Spruch, Phys. Rev. A 4, 841-846 (1971).

Number of eigenvalues of many-body Hamiltonians and Efimov's effect

I. M. Sigal

Department of Mathematics
Princeton University

1. In this talk I present a new method of obtaining estimates of the number of bound states of many-body systems with potentials close to critical. The estimates discussed below expose a striking phenomenon: existence of infinite number of bound states in certain short-range, many-body systems. Note that for all attractive potentials and many mixed ones the coupling constants can always be adjusted in such a way that the phenomenon occurs. The effect was experimentally observed in three nucleon scattering and in solid states. In the last case the role of particles was played by impurities. In the first case the forces between particles are close to or exactly critical, while in the second case they can easily be adjusted.

All our considerations are illustrated in the case of three particles. The Plank constant $\hbar$ is set to 1.

2. In order to formulate rigorously our results we need

Definition. A short-range, pair potential V_ℓ will be called critical iff $h_\ell \equiv (2\mu_\ell)^{-1}\Delta + V_\ell \geq 0$, where μ_ℓ is the reduced mass for the pair ℓ, and the equation $h_\ell\psi = 0$ has a solution of the form $\psi = (-\Delta)^{-1}|V_\ell|^{1/2}\phi$, where $\phi \in L^2(\mathbb{R}^3)$.

Remark. Since $h_\ell \geq 0$, such a solution is nondegenerate and is not an eigenvector of h_ℓ [5], we call it a quasibound state of h_ℓ.

Theorem 1. Let H be the Schrödinger operator of a three-body system with pair potentials $V_\ell \in L^p \cap L^q(\mathbb{R}^3)$, $p > 3/2 > q$. Let the masses of the particles obey the condition [7], which for the sake of space we write here only in the case $m_i = m_j : m_i/m_k > .9$. Then if the potentials V_{ik} and V_{jk} approach critical points, the number of eigenvalues of H increases to infinity and becomes infinite as V_{ik} and V_{jk} reach critical points.

Theorem 2. Let $m_k = m_j$, $m_i m_k^{-1} > .9$ and the potentials V_{ik} and V_{jk} be spherically symmetric and satisfy $\int_0^\infty |V_\ell(n)|n\,dn < \infty$. Then the number of bound states (isolated eigenvalues counting multiplicities) of H has the following asymptotic behavior as the potentials V_{ik} and V_{jk} approach critical points:

$$N = -\frac{0{,}0107}{\pi}(m_i/m_k - 0{,}9)^{3/2}\ell n\,\rho + \text{uniformly bounded term}, \qquad (*)$$

where $\rho = \max_{s=i,j} |\gamma_{sk}(0)|$, $\gamma_\alpha(k) = k \cdot \cot(\delta_\alpha(k)+k)$, $\delta_\alpha(k)$ is the s-wave phase shift for the pair α . Note that $\gamma_\alpha(0) < 0$ and small if and only if V_α has a shallow bound state; in this case $\gamma_\alpha(0) \simeq -\sqrt{\varepsilon_\alpha}$, where $-\varepsilon_\alpha$ is the bound state energy.

Corollary. If all three potentials approach critical points then Theorems 1 and 2 are true for all masses m_1 , m_2 and m_3 .

Indeed, the restriction on the masses holds always for some $(ijk) = (123)$.

Remark. An expression analogous to (*) can be obtained for a nonspherically-symmetric case as well. Here $\gamma_\alpha(k)$ should be replaced by $\det_p(1+V_\alpha(h_0+k^2)^{-1})$.

3. The physical idea behind our approach is that a system of three particles behaves in many ways as a system of two of the particles connected by an effective, attractive interaction, produced by the exchange of the third particle. We show that if this third particle has quasibound states with each of the other two particles, then the effective interaction is long-range, namely $\propto -|R|^{-2}$, at infinity. The effective interaction described above is somewhat analogous to the interaction via an exchange by virtual particles in quantum field theory with the square root of the energy of the third particle above (times the reduced mass) playing a role of the mass of the virtual particles.

4. Below $N(A,\lambda)$ denotes the number of isolated eigenvalues (counting the multiplicities) of an operator A , which are less than λ and Δ_x stands for the Laplacian in the variable x . A three-body Hamiltonian in the center-of-mass frame and in the Jacobi coordinates, say, for pair (12), $r = m_1x_1+m_2x_2/m_1+m_2 - x_3$, $R = x_1 - x_2$, has the form $H = -(2m)^{-1}\Delta_r - (2\mu)^{-1}\Delta_R + V$, where $m^{-1} = (m_1+m_2)^{-1}+m_3^{-1}$, $\mu^{-1} = m_1^{-1} + m_2^{-1}$.

5. The proof of Theorem 1 is based on the following three propositions, which are given without proofs. Propositions 1 and 2 do not assume the potentials to be critical.

Proposition 1. Let $u:\mathbb{R}^3 \longrightarrow L^2(\mathbb{R}^3)$ have an obvious smoothness, required to make the expressions below meaningful, and $\|u(R)\|_r = 1 \ \forall R \in \mathbb{R}^3$. Then $N(H,\lambda) \geq N(H_u,\lambda)$, where

$$H_u = (2\mu)^{-1}\Delta_R + \phi(R,u) \ , \ \phi(R,u) \equiv \langle (Hu)(R),u(R)\rangle_r$$
$$= \langle H_{BO}(R)u(R),u(R)\rangle_r - (2\mu)^{-1}\langle \Delta_R u(R),u(R)\rangle_r \ .$$

Here $H_{BO}(R) = -(2m)^{-1}\Delta_r + V(R)$ is a Born-Openheimer Hamiltonian on $L^2(\mathbb{R}^3,dr)$.

The next proposition describes u for which $\phi(R,u)$ has a long-range negative tail. Note that we have shown [6] that the ground state $u_{BO}(R)$ of $H_{BO}(R)$, the Born-Openheimer Anzatz which comes from disregarding the term $-(2\mu)^{-1}\langle\Delta_R u(R),u(R)\rangle_r$ in $\phi(R,u)$, leads to $\phi(R,u_{BO})$ with a positive tail. Henceforth, we assume, for simplicity, particles 1 and 2 to be identical. We set $U = V_{13} = V_{23}$ and $h = -(2n)^{-1}\Delta + U$, $n^{-1} = m_1^{-1} + m_3^{-1}$.

<u>Proposition 2</u>. Let $u(R) = T_{(1/2)R}v^{(R)} + T_{-(1/2)R}v^{(R)}$, where v is the solution of the equation

$$(h+E)v + v(R)U = 0 , \tag{1}$$

normalized in such a way that $\|u(R)\|_r = 1$, and $(T_R\phi)(r) = \phi(r+R)$. Then

$$\phi(R,u) = -E(R) + c_\alpha|R|^{-2} + \theta|R|^{-3} , \tag{2}$$

where $E(R)$ is the unique solution to the consistency equation for (1):

$$1 + [(h+E)^{-1}V](R) = 0 \tag{3}$$

and $c_\alpha = [2(1+e^{-\alpha})]^{-2}(1+(1-2\alpha+2\alpha^2/3)e^{-\alpha})m_3(m_3+m_1)^{-1}$, $\alpha=\sqrt{2nE}|R|$. Here θ is uniformly bounded in R.

<u>Proposition 3</u>. If the potential U is critical, then (3) has the unique solution of the form $E(R) = \alpha^2|R|^{-2} + \kappa|R|^{-3}$, where $\sqrt{2n}\,\alpha = .3216...$ is the unique solution to $x = e^x$ and κ is uniformly bounded in R.

6. The next lemma shows that $u(R)$ is, in fact, an approximate ground state of the Hamiltonian $H(R) = -(2n)^{-1}\Delta_r + V(R)$, where $(2n)^{-1}$ comes from $(2m)^{-1} + .5(2\mu)^{-1}$, which replaces $H_{BO}(R)$ in our case.

<u>Lemma 1</u>. Let $E_0(R)$ be the lowest eigenvalue of $H(R)$. Then

$$0 \le E(R) - E_0(R) \le \text{const } |R|^{-3} . \tag{4}$$

<u>Remark</u>. Estimates (2) and (4) can be proved without solving (1) and (3). It can be shown that only the asymptotics of v as $|R| \longrightarrow \infty$ is important.

7. The proof of Theorem 2 is obtained from Propositions 1, 2 and

<u>Proposition 3'</u>. If the potential U is spherically-symmetric (not necessarily critical) then the unique solution of (3) has the form $E(R) = \alpha(\gamma|R|)^2|R|^{-2} + \eta|R|^{-3}$, where $\sqrt{2n}\alpha(s)$ satisfies $x = e^{-x} - s$, $\gamma = \lim_{k\to 0} k\cot(\delta(k)+k)$, $\delta(k)$ is the s-wave phase shift for h and η is uniformly bounded in R.

Recall that: $\gamma > 0 \Longleftrightarrow h \geq 0$ and has no quasibound states; $\gamma = 0 \Longleftrightarrow h$ has quasibound states; $\gamma > 0 \Longleftrightarrow h \ngeq 0$ and has no quasibound states. Here quasibound states = solutions of $h\psi = 0$ such that $\psi \notin L^2(\mathbb{R}^3)$ and $|V|^{1/2}\psi \in L^2(\mathbb{R}^3)$.

<u>Remark</u>. Properties of $(h+E)^{-1}$, needed to prove Propositions 3 and 3', are described in [11] ([9,10] contain related formulae) for noncentral, critical potentials and in [12] for general spherically symmetric potentials.

8. Derivation of (1). We begin with the exact eigenvalue equation for $H(R)$:

$$(H(R)+E(R))u(R) = 0 \quad (5) .$$

Write ($f_R = T_R f$): $$u(R) = v^{(R)}_{(1/2)R} + s^{(R)}_{-(1/2)R} \quad (f_R = T_R f) .$$

(5) becomes: $(h+E)v + U_R v + (h_R+E)s_R + Us_R = 0$, where $h_R = -(2\mu)^{-1}\Delta + V_R$.

Approximation: $$U_R v \simeq v(-R)U_R \quad \text{and} \quad Us_R \simeq s(R)U .$$

Approximate equation: $$(h+E)v + s(R)U + (h_R+E)s_R + v(-R)U_R = 0 .$$

Decouple this equation and write the consistency condition. Dropping from $[(h+E)^{-1}V](\pm R)$ the terms of the higher order in $|R|^{-1}$ leads to $v(r) = v(-r)=s(r)$. In this approximation the decoupled equations are identical to (1).

<u>Literature Comments</u>. It was discovered by V. Efimov [2] that the number of bound states of a system of three identical particles with finite-range, attractive interactions such that there are no bound states in the two-body subsystems, grows as the potentials approach critical points. The arguments presented by Efimov are based on replacing the potential in a three-body problem in question by a boundary condition at the origin. R. Amado and J. Noble [1] have given an alternative mathematical proof, made rigorous by D. R. Yafaev [9] (Yafaev, however, does not treat the question of the growth of the number of bound states). Their proof is based on the study of the homogeneous Faddeev equation and is rather mathematically involved.

In [6] we give another, variational approach to the problem in question. Theorem 2 is contained in this paper. Similar ideas were advanced in A. Fonseca et al [3]. Different methods of estimation of the ground state energy of $H(R)$ (see Proposition 3 and Lemma 1) were proposed by M. Klaus and B. Simon [4] and B. Simon [8].

The author is indebted to W. Hunziker and B. Simon for useful discussions and to K. Hepp and W. Hunziker for their hospitality at the Department of Theoretical Physics, ETH, Zurich, where the first draft of this talk was written. The work was partly supported by USNSF Grant MCS-78-01885.

Appendix: Proof of Proposition 1. Let u satisfy the conditions of Proposition 1 and f_n be eigenvectors of H_u. Then the functions $\psi_n = u(r,R)f_n(R)$ are orthogonal and $\langle H\psi_n,\psi_n\rangle = \langle H_u f_n,f_n\rangle$. This via the variational principle implies that $N(H,\lambda) \geq N(H_u,\lambda)$.

Supplement: N-body case. There are many cases in which we expect the existence of infinite number of eigenvalues in short-range, N-body systems. In the simplest situation we have a proof of this fact:

Theorem. Consider an N-body, short-range system of which the bottom, Σ, of the continuous spectrum is defined by a break up into three clusters c_1, c_2, c_3. Assume two of the three clusters, say, c_1 and c_2, contain each just one particle. Further, assume that the subsystems $c_1 \cup c_3$ and $c_2 \cup c_3$ have quasibound states at their two-cluster threshold Σ. Then the whole system has infinite number of bound states.

References

1. R. Amado and J. Noble. Phys. Lett. 35B (1971), 25; Phys. Rev. D5 (1972), 1992-2002.

2. V. Efimov. Phys. Lett. 33B (1970), 563; Sov. J. Nucl. Phys. 12 (1971), 589.

3. A. C. Fonseca, E. F. Redish, and P. E. Stanley. Efimov Effect in an Analytically Solvable Model. Univ. of Maryland (1979).

4. M. Klaus and B. Simon. Ann. Inst. H. Poincaré, to appear.

5. M. Klaus and B. Simon. In preparation.

6. Yu. N. Ovchinnikov and I. M. Sigal. Ann. Phys., in press.

7. I. M. Sigal. Unpublished.

8. B. Simon. J. Funct. Annal., to appear.

9. D. R. Yafaev. Math. Sb. 94 (1974), 567-593.

10. A. Jensen and T. Kato, Preprint, 1979, Berkeley.

11. M. Loss and I. M. Sigal. In preparation.

12. V. De Alfaro and T. Regge. Potential Scattering, North-Holland (1965).

ON THE EXPONENTIAL FALL OFF OF WAVEFUNCTIONS AND ELECTRON DENSITIES

M. Hoffmann-Ostenhof+, Institut für Theoretische Physik der Universität Wien

T. Hoffmann-Ostenhof, Institut für Theoret. Chemie und Strahlenchemie d. Univ. Wien

R. Ahlrichs, Institut für Physikalische Chemie der Universität Karlsruhe

J. Morgan, Department of Physics, Princeton University, New Jersey, USA

1. Introduction

We give a short survey on our results obtained on the exponential decay of sub-continuum atomic wavefunctions and electron densities [1] - [4]. Let ψ denote an L^2-solution of the Schrödinger equation of an n-electron atom

$$(H-E)\ \psi(x_1,\ldots,x_n) = 0\ , \quad x_i \in \mathbb{R}^3 \quad (1 \le i \le n) \tag{1}$$

where

$$H = -\sum_{i=1}^{n} \frac{\Delta_i}{2} - \sum_{i=1}^{n} \frac{Z}{r_i} + \sum_{\substack{i,j=1\\ i<j}}^{n} \frac{1}{r_{ij}}\ ; \quad r_i = |x_i|\ , \quad r_{ij} = |x_i - x_j| \tag{2}$$

denotes the Hamiltonian in the infinite nuclear mass approximation, Z is the nuclear charge and E the eigenvalue corresponding to ψ. ψ is assumed to be real and normalized to 1.

The k-electron density ρ_k $(1 \le k \le n)$ is defined by

$$\rho_k(x_1,\ldots,x_k) = \int_{\mathbb{R}^{3(n-k)}} |\psi(x_1,\ldots,x_n)|^2 dx_{k+1}\ldots dx_n\ . \tag{3}$$

Decay properties of such wavefunctions and electron densities are interesting from a mathematical and physical point of view: There is a close relationship between spectral properties of the concerned Hamiltonian and asymptotic properties of sub-continuum wavefunctions resp. densities. Furthermore it will be seen that concepts like nuclear screening and the one-electron picture - on which part of our understanding of atomic structure is based - enter in a natural way. Unfortunately, but not unexpected, the situation with respect to correlation effects is less transparent.

Our main tool in the derivations of upper and lower bounds to wavefunctions and densities are comparison theorems for differential inequalities.

It should be noted that there is a different powerful approach for rather general situations by Deift et al. [5]. However for atoms in the infinite nuclear mass

+) Supported by Fonds zur Förderung der wissenschaftlichen Forschung in Österreich, Project No. 3655 approximation the methods outlined here lead to somewhat more detailed results.

2. Comparison Theorems

The derivation of the exponentially decreasing bounds to the k-electron densities is based on the following two comparison theorems:

Let Ω be an open subset of $\mathbb{R}^n$ and let f, g satisfy

(i) $f,g \in C^0(\bar{\Omega})$; $f,g \geq 0$ in Ω (and $f,g \underset{|x|\to\infty}{\to} 0$ for unbounded Ω),

(ii) $g \leq f \quad \forall\, x \in \partial\Omega$.

Theorem 2.1. (e.g. [5]): Let f, g obey (i), (ii) and let further

(iii) $(-\Delta + W_1)g + F_1 \leq 0$
$(-\Delta + W_2)f + F_2 \geq 0$ in the weak sense in Ω

(iv) $W_1 \geq W_2$, $F_2 \leq F_1$ and (v) $W_2 \geq 0$.

Then $f \geq g$ in all of Ω.

Theorem 2.2. [6]: Suppose $f > 0$ a.e. in Ω, let f, g obey (i), (ii) and let further

(iii) $(-\Delta + W_1)g \leq 0$
$(-\Delta + W_2)f \geq 0$ in the weak sense in Ω

(iv) $W_1 > W_2$ a.e. in Ω and (v) $\Delta f,\ \Delta g \in L^1(\Omega)$.

Then $f \geq g$ in all of Ω.

3. Schrödinger Inequalities and Exponential Decay of k-electron Densities ($1 \leq k \leq n$)

It will be shown that $\sqrt{\rho_k}$ satisfies a differential inequality which has essentially the structure of the Schrödinger equation of a k-electron system, where the absolute value of the energy is replaced by the sum of the first k successive ionization potentials ε_i ($1 \leq i \leq k$): Let

$$H^{(n-i)} = \sum_{j=i+1}^{n} \left(-\frac{\Delta_j}{2} - \frac{Z}{r_j}\right) + \sum_{\substack{j,\ell=i+1 \\ j<\ell}}^{n} \frac{1}{r_{j\ell}} \qquad (1 \leq i \leq n) \tag{4}$$

denote the Hamiltonian of the i-fold ionized n-particle system and let $E_0^{(n-i)}$ denote the ground state energy of $H^{(n-i)}$ in the appropriate symmetry subspace (induced by the symmetry behavior of ψ), then

$$\varepsilon_1 = E_0^{(n-1)} - E\ , \qquad \varepsilon_i = E_0^{(n-i)} - E_0^{(n-i+1)} \qquad (2 \leq i \leq n) \tag{5}$$

and hence $\sum_{i=1}^{n} \varepsilon_i = |E|$.

Theorem 3.1. (Schrödinger inequalities)

$$(-\frac{\Delta_1}{2} + \varepsilon_1 - \frac{Z}{r_1}) \sqrt{\rho_1(x_1)} \leq 0$$

$$\sum_{i=1}^{k} (-\frac{\Delta_i}{2} + \varepsilon_i - \frac{Z}{r_i} + \sum_{j>i}^{k} \frac{1}{r_{ij}}) \sqrt{\rho_k(x_1,\ldots,x_k)} \leq 0 \; ; \quad (2 \leq k \leq n-1) \tag{6}$$

$$(H-E)|\psi| \leq 0 \; .$$

The inequalities are to be understood in the weak sense.

Proof: We sketch the proof for ρ_1. For the case $k \geq 2$ the procedure is analogously and particularly for $|\psi|$ (6) is an immediate consequence of Kato's distributional inequality $-|\psi|\Delta|\psi| \leq -\psi\Delta\psi$. Due to (1) and (4) we have

$$\int \psi(-\frac{\Delta_1}{2} - \frac{Z}{r_1} + H^{(n-1)} + \sum_{j=2}^{n} \frac{1}{r_{1j}} - E)\psi \, dx_2 \ldots dx_n = 0 \; . \tag{7}$$

The variational principle and (5) yield

$$\int \psi(H^{(n-1)} - E)\psi \, dx_2 \ldots dx_n \geq \varepsilon_1 \, \rho_1(x_1) \; . \tag{8}$$

Furthermore via Cauchy-Schwarz's inequality

$$-\sqrt{\rho_1}\,\Delta_1 \sqrt{\rho_1} \leq -\int \psi \, \Delta_1 \, \psi \, dx_2 \ldots dx_n \; . \tag{9}$$

Neglecting the positive electronic repulsion term $(n-1) \int \frac{\psi^2}{r_{12}} dx_2 \ldots dx_n$ in (7) we obtain inequality (6) with the aid of (8) and (9).□

For the following we suppose $\varepsilon_1 \leq \varepsilon_2 \leq \ldots \leq \varepsilon_n$. This is just an experimental fact, a rigorous proof is so far missing.

Theorem 3.2.

$$\sqrt{\rho_1(x)} \leq C(1+r)^{\frac{Z}{\sqrt{2\varepsilon_1}} - 1} e^{-\sqrt{2\varepsilon_1}\, r} \tag{10}$$

$$\sqrt{\rho_k(x_1,\ldots,x_k)} \leq C \, S \prod_{i=1}^{k} (1+r_i)^{\frac{Z}{\sqrt{2\varepsilon_i}} - 1} e^{-\sqrt{2\varepsilon_i}\, r_i} \tag{11}$$

where C is a constant and S acts as a symmetrizer.

Proof: For ρ_1 the proof is straightforward: Choose $\Omega = \{x \in \mathbb{R}^3 \,|\, r \geq c\}$, c sufficiently large in Theorem 2.1. Since $\sqrt{\rho_1}$ obeys (6) we obtain the upper bound (10) by a suitable solution f of $[-\Delta + 2(\varepsilon - Z/r)]f \geq 0$. Inequality (11) is proven via a recursive procedure. We indicate the main steps for $\sqrt{\rho_2}$: Let $\Omega = \{(x_1,x_2) \,|\, r_1, r_2 \geq c\}$, c sufficiently large in Theorem 2.1. Due to (6) $\sum_{i=1}^{2} [-\Delta_i + 2(\varepsilon_i - Z/r_i)]\sqrt{\rho_2} \leq 0$. The suitable function f obeying this inequality in the other direction yields the desired bound (11) for $\sqrt{\rho_2}$ in Ω. To prove that f fulfills the boundary condition (ii) and to verify (11) in the region $\{(x_1,x_2) \mid r_1 \geq c,\ r_2 \leq c\}$

$$\rho_2(x_1,x_2) \leq k_\delta \max_{|x_1-x_1'|<\delta} \rho_1(x_1') \qquad (\delta > 0) \tag{12}$$

is used. (12) follows from a so called subsolution estimate by Trudinger [7]. For ρ_1 the already obtained upper bound is used.□

For 1-electron systems the r.h.s. of (10) shows the right asymptotic behavior. For the n-electron case one should expect an effective nuclear charge in the pre-exponential factors of (10) and (11). This is in accordance with the physical picture that k electrons at large distance from the nucleus "see" a nucleus screened by the remaining n-k electrons.

4. Screening Effect in the Decay of the k-electron Densities

4.1. Upper Bounds

To obtain results on screening effects a refinement of the Schrödinger inequalities is necessary, where the so far neglected electronic repulsion terms become important.

Theorem 4.1.1.

$$\sqrt{\rho_1(x)} \leq C\,(1+r)^{\frac{Z-(n-1)}{\sqrt{2\varepsilon_1}} - 1}\, e^{-\sqrt{2\varepsilon_1}\, r} \tag{13}$$

$$\sqrt{\rho_k(x_1,\ldots,x_k)} \leq C\, S \prod_{i=1}^{k} (1+r_i)^{\frac{Z-(n-k)}{\sqrt{2\varepsilon_i}} - 1}\, e^{-\sqrt{2\varepsilon_i}\, r_i}, \quad (2 \leq k \leq n) . \tag{14}$$

Proof: The proof is similar to that of Thm 3.1. To prove (13) we split the Hamiltonian by $H = -\Delta_1/2 - Z/r_1 + \bar{H}^{(n-1)}(x_1)$. The variational principle leads to

$$\int \psi\, \bar{H}^{(n-1)}(x_1)\psi\, dx_2 \ldots dx_n \geq \bar{E}_o(x_1)\, \rho_1(x_1) \tag{15}$$

with $E_o(x_1)$ denoting the bottom of the spectrum of $\bar{H}^{(n-1)}(x_1)$ in the appropriate symmetry subspace. By a perturbation argument [8] $\bar{E}_o(x_1)$ has the asymptotic expansion

$$\bar{E}_o(x_1) = E_o^{(n-1)} + \frac{n-1}{r_1} + O\left(\frac{1}{r_1^2}\right) . \tag{16}$$

(15) and (16) lead to

$$\left(-\frac{\Delta_1}{2} + \varepsilon_1 - \frac{Z-(n-1)}{r_1} - \frac{d}{r_1^2}\right) \sqrt{\rho_1(x_1)} \le 0 ; \qquad d > 0 ,$$

for sufficiently large r_1. Application of Theorem 2.1 completes the proof of (13). For $\sqrt{\rho_k}$ the proof is again recursive.□

A sharper result can be shown for $\sqrt{\rho_2}$ by taking into account the so far neglected term $\frac{1}{r_{12}} \sqrt{\rho_2}$:

Theorem 4.1.2.

$$\sqrt{\rho_2(x_1,x_2)} \le C\, S[(1+r_1)^{\alpha}(1+r_2)^{\beta}\, e^{-\sqrt{2\varepsilon_1}\, r_1}\, e^{-\sqrt{2\varepsilon_2}\, r_2}]$$

where

$$\alpha = \frac{Z-(n-2)+1/2}{\sqrt{2\varepsilon_2}} , \qquad \beta = \frac{Z-(n-1)}{\sqrt{2\varepsilon_1}} .$$

4.2. Lower Bounds

In the following we consider the mathematical ground state $\psi_o(x_1,x_2)$ of a 2-electron atom. It is well known that $\psi_o(x_1,x_2) > 0$ a.e.. To derive a positive lower bound to ψ_o, $\psi_o(x_1,x_2) > 0$ for $r_1, r_2 < \infty$ will be necessary. However this is an immediate consequence of

Theorem 4.2.1. [9] Let $\psi(x)$ $(x \in \mathbb{R}^{3n})$ be a solution of equation (1) and let $\psi(x_o) = 0$. Then ψ has both signs in any arbitrary small neighborhood of x_o.

Let $\rho_1(x)$ denote the 1-electron density corresponding to ψ_o. The asymptotic behavior of ρ_1 is described by

Theorem 4.2.2. (See also [10])

$$\sqrt{\rho_1(x)} \lesseqgtr C_{\pm}\, (1+r)^{\frac{Z-1}{\sqrt{2\varepsilon_1}} - 1}\, e^{-\sqrt{2\varepsilon_1}\, r} .$$

Proof: We define $u(x_1) = \int \phi(x_2)\psi_o(x_1,x_2)dx_2$, where ϕ denotes the ground state of the ionized system, i.e. $(-\Delta_2/2 - Z/r_2)\phi(x_2) = -\frac{Z^2}{2}\phi(x_2)$. By Theorem 4.2.1 and Cauchy-Schwarz's inequality $0 < u(x_1) \le \sqrt{\rho_1(x_1)}$. Using (1) we get

$$\left(-\frac{\Delta_1}{2} + \varepsilon_1 - \frac{Z}{r_1}\right) u(x_1) + \int \frac{\phi\psi_o}{r_{12}}\, dx_2 = 0 . \tag{17}$$

It can be shown that for sufficiently large r_1

$$\int \frac{\phi\psi}{r_{12}}\, dx_2 \le \frac{1}{r_1} u(x_1) + \frac{c}{r_1^{3/2}}\, r_1^{\frac{Z-1}{\sqrt{2\varepsilon_1}} - 1}\, e^{-\sqrt{2\varepsilon_1}\, r_1} \tag{18}$$

where we used the upper bound (13) for $\sqrt{\rho_1}$. Combining (17) and (18) a differential inequality for u results. Application of Theorem 2.1 in its inhomogeneous version leads to the desired lower bound for ρ_1.□

Unfortunately the lower bound to $\psi_o(x_1,x_2)$ is less satisfactory than the result for ρ_1.

Theorem 4.2.3.

$$\psi_o(x_1,x_2) \geq C_\delta \, e^{-(\sqrt{2|E|} + \delta)R} \; ; \qquad (\delta > 0)$$

with $R = \sqrt{r_1^2 + r_2^2}$ and δ arbitrarily small.

Proof: Let $\Omega = \{(x_1,x_2) \,|\, R > d > 0\}$, d sufficiently large. By Theorem 4.2.1 $\psi|_{\partial\Omega} > 0$ and Theorem 2.2 yields the above lower bound.

References

[1] M. Hoffmann-Ostenhof, T. Hoffmann-Ostenhof, Phys. Rev. A16, 1782 (1977)

[2] T. Hoffmann-Ostenhof, M. Hoffmann-Ostenhof, R. Ahlrichs, Phys. Rev. A18, 328 (1978)

[3] T. Hoffmann-Ostenhof, J. Phys. A12, 1181 (1979)

[4] R. Ahlrichs, M. Hoffmann-Ostenhof, T. Hoffmann-Ostenhof, J. Morgan, in preparation

[5] P. Deift, W. Hunziker, B. Simon, E. Vock, Commun. Math. Phys. 64, 1 (1978)

[6] T. Hoffmann-Ostenhof, J. Phys. A, in press

[7] N.S. Trudinger, Ann. Scuola Norm. Sup. Pisa (3) 27, 265 (1973)

[8] J. Morgan, B. Simon, submitted to Int. J. Quantum Chem.

[9] M. Hoffmann-Ostenhof, T. Hoffmann-Ostenhof, B. Simon, submitted to J. Phys. B

[10] W. Thirring, Lehrbuch der Mathematischen Physik 3, Springer (1979)

EXACT RESULTS FOR CONFINING POTENTIALS

H. Grosse
Institute for Theoretical Physics
University of Vienna, Austria

Abstract

We summarize recent results obtained together with A. Martin on properties of confining potentials as they are relevant in particle physics.

1. Physical Motivation

After the discovery of two of the so-called "New Particles" it was suggested that these states represent the ground state and first excited state of a quark-antiquark pair interacting via a potential going to infinity at infinity. The usual procedure is to start with a certain potential and to adjust a few parameters; then one makes predictions concerning the spectrum, leptonic decay widths and electric and magnetic transition probabilities [1]. Starting from Coulomb plus linear potential the existence of P states below the 2S state and of a D state above the 2S state was conjectured and found experimentally.

We asked how stable are certain features which one observes, like the level order or relations between the decay widths, against changes of the potential. So we determined classes of potentials showing these features and on the other hand classes giving opposite results [2,3].

Furthermore, since very recently a second heavy quark system has been detected experimentally, one is interested in comparing relevant quantities of two systems assuming that the same potential acts between the constituents.

2. Energy Levels

We consider the Schrödinger equation for a radial symmetric potential V(r) and ask for a class of potentials which give the "canonical" level order:

$$E_{1S} \leq E_{1P} \overset{A)}{\leq} E_{2S} \overset{B)}{\leq} E_{1D} \overset{A)}{\leq} E_{2P} \quad .$$

Note that A) become equalities for the Coulomb case, while B) becomes equal for the harmonic oscillator; furthermore, since the angular momentum operator is positive one has in general $E_{nS} < E_{nP} < E_{nD} < \dots$.

THEOREM 1: Let $W(r) = \frac{d}{dr}(r^2 V(r))$ then

A) $\quad (\frac{d}{dr})^2 W(r) \geq 0 , \quad - \infty < W(0) \leq 0 \Rightarrow E_{2S} \geq E_{1P} , E_{2P} \geq E_{1D} \dots$

The main idea of the proof is to start from the degenerate case and to follow the energy difference as a function of a parameter. Using differential equation methods, the Virial theorem, the Feynman Hellmann theorem and a kind of moment inequality (which follows from the node structure of the wave functions involved) one derives monotonicity of that energy difference. To have B) fulfilled one needs concavity in r^2:

THEOREM 2:

B) $\quad (\frac{d}{dr^2})^2 W(r) \leq 0 , \quad - \infty < \frac{W(r)}{r}\Big|_{r=\infty} \Rightarrow E_{2S} \geq E_{1D}$.

To illustrate conditions A) and B) let us give examples

$$V(r) = \int_{-1}^{2} d\alpha \, r^{\alpha} \, \varepsilon(\alpha) \, \rho(\alpha) \, ; \qquad \rho \geq 0 , \qquad \varepsilon = \alpha / |\alpha| .$$

We have worked out a larger class giving the canonical order together with counter-examples [3]. From theorems 1 and 2 we learn that the dominant contribution should not rise stronger than r^2 while potentials like $-r^{-3/2}$ should be avoided.

Assuming that the same potential describes two systems differing only in the mass one might compensate the decrease of levels due to an increase in mass by increasing the angular momentum. Using the Min-Max principle together with the elementary inequality

$$\int_0^\infty dr \, u'^2 \geq \int \frac{dr \, u^2}{4r^2} \qquad \text{for} \quad u(0) = 0$$

one arrives at the bounds (for $M > m$):

$$E_n(M, \ell = \tfrac{1}{2}(\sqrt{M/m} - 1)) \leq E_n(m, \ell = 0) , \quad \forall n .$$

Using the node structure and continuity an improvement has been obtained:

THEOREM 3: $V \varepsilon A) \Rightarrow E_1(m, \sqrt{M/m} - 1) \leq E_1(m, 0)$.

To relate the left hand side to physical observables one uses concavity properties:

From the Min-Max principle it follows that the ground state energy is concave in $\ell(\ell+1)$. We have obtained concavity in ℓ themselves for special cases: for $V = r^2+\lambda\delta V$, $\delta V \in B$) in first order in λ; it holds asymptotically for large ℓ, $V \in B$); and it holds for r^α, $0 < \alpha \leq 2$. More generally, we established monotonicity of the difference of the first two energy levels:

THEOREM 4: $\left(\frac{d}{dr^2}\right)^2\left(r^2\frac{dV}{dr}\right) \leq 0 \Rightarrow \frac{d}{d\ell}(E_2(\ell) - E_1(\ell)) < 0$.

Applying the above considerations to the J/ψ and Υ system gives a bound on the quark mass difference $M_b - m_c \geq 3.29$ GeV.

3. The Wave Functions at the Origin $\psi_n(0)$ $(\ell = 0)$

Leptonic decay probabilities are proportional to $|\psi_n(0)|^2$; experimentally $\Gamma_{\psi\to e^+e^-} \simeq 4.8$ keV $> \Gamma_{\psi'\to e^+e^-} \simeq 2.1$ keV. A result of A. Martin [4] states:

THEOREM 5: $\left(\frac{d}{dr}\right)^2 V(r) \gtreqless 0 \Rightarrow |\psi_2(0)|^2 \gtreqless |\psi_1(0)|^2$.

To get this result one uses a continuity argument, the node structure and the relation

$$u'^2(0) = \int_0^\infty dr\, u^2 \frac{dV}{dr} , \qquad u(r) = r\,\psi(r)$$

obtained from the Schrödinger equation. So from the data an overall concave potential is preferred against an overall convex one.

The generalization to higher states is more difficult to prove: it holds within the WKB approximation, it holds in first order perturbation theory around the linear potential and it is true asymptotically

$$\left(\frac{d}{dr}\right)^2 V(r) \gtrless 0 , \qquad \frac{dV}{dr} > 0 \qquad \psi_n(0) \xrightarrow[n\to\infty]{} \begin{cases} \infty \\ 0 \end{cases} .$$

The mass dependence of $\psi_n(0)$ is of obvious interest:

THEOREM 6: $\left(\frac{d}{dr}\right)^2 V(r) \leq 0 \Rightarrow \lim_{r\to 0} \left|\frac{u_1}{u_2}\right|^2 > \frac{m_1}{m_2} \quad \forall\, \ell$.

$u_{1,2}$ denote the wave functions to mass $m_{1,2}$. Application of such a bound excludes charge ±2/3 for the bottom quark [5]. From a generalization only partial answers are known.

One might ask for the mass dependence of the probability for a particle to be present within a sphere of radius R. For the ground state one finds [5]:

THEOREM 7: $\frac{dV}{dr} > 0 \Rightarrow \frac{\partial}{\partial m}\int_0^R dr\, |u_1|^2 > 0$.

4. Inverse Problem for Confining Potentials [6]

For a fixed angular momentum (especially $\ell = 0$) we have proved:

THEOREM 8: Given a sequence E_n and $u_n'(0)$ belonging to a confining potential which fulfills

$$V \in L^1_{loc}\,, \qquad \int_R^\infty dr\, \frac{V'^2}{V^{5/2}} < \infty\,, \qquad \int_R^\infty dr\, \frac{V''}{V^{3/2}} < \infty \quad \text{for some } R\,, \Longrightarrow$$

$V(r)$ is uniquely determined.

We have related that problem to a simpler one, which is the inverse problem for the same equation with different boundary conditions:

$$- w_n'' + (V(r) - \varepsilon_n)\, w_n = 0\,, \qquad w_n'(0) = 0\,.$$

The usual Gelfand-Levitan procedure is applicable to that problem. The relationship between both problems has been found by studying

$$R(E) = \left(\frac{d}{dr} u_E(r)/u_E(r)\right)\Big|_{r=0}$$

for a solution $u_E(r)$ defined by a WKB decay at infinity proving that $R(E)$ is Herglotz and admits a representation like

$$R(E) = \lim_{N\to\infty} \left\{ \sum_{n=1}^{N} \frac{|u_n'(0)|^2}{E_n - E} - \frac{2}{\pi} \sqrt{\frac{E_N + E_{N+1}}{2}} \right\}.$$

The data determine $R(E)$ but then

$$- R^{-1}(E) = \sum_{n=1}^{\infty} \frac{|w_n(0)|^2}{\varepsilon_n - E}$$

gives the information necessary for the second problem.

We have also considered the inverse problem where there is given $E_1(\ell)$ the ground state energy as a function of ℓ.

5. Regge Trajectories for Confining Potentials [3]

One tries to integrate the Schrödinger equation for complex angular momenta with $\mathrm{Re}\,\ell > -1/2$. The problem is that $E_n(\ell)$ might be non-analytic if the derivative

$$\frac{dE_n}{d\ell} = \int_0^\infty dr\, \frac{u_n^2}{r^2} \Big/ \left(\int dr\, u_n^2\right)$$

blows up.

One knows from the relation

$$\mathrm{Im}\,\lambda \int_0^\infty dr \frac{|u_n|^2}{r^2} = \mathrm{Im}\,E_n(\ell) \int_0^\infty dr\, |u_n|^2 , \qquad \lambda = \ell(\ell+1)$$

that $E_n(\ell)$ cannot have poles or essential singularities.

We have restricted ourselves to pure power potentials and have adapted a method developed by Loeffel and Martin for the anharmonic oscillator to exclude branchpoint singularities in the complex λ plane:

THEOREM 9: Let $V = r^\alpha$, then $E_n(\ell)$ can be continued to complex ℓ and is analytic in $\mathrm{Re}\,\ell > -1/2$. Furthermore we have a representation like

$$E_n(\lambda) = A + B\lambda + \frac{\lambda}{\pi} \int_{-\infty}^{-1/4} \frac{d\lambda'\, \mathrm{Im}\,E_n(\lambda')}{\lambda'(\lambda'-\lambda)} , \qquad \mathrm{Im}\,E_n > 0 .$$

References

[1] J.D. Jackson, C. Quigg, J.L. Rosner, Proc. 19th Int. Conf. on High Energy Physics, Tokyo, 1978 (Physical Soc. Japan, Tokyo, 1979) p. 391

[2] A. Martin, New Particles or "Why I believe in Quarks", Lectures given at the 15th Int. School on Subnuclear Physics, Ettore Majorana Centre for Scientific Culture, Erice 1977; CERN TH 2370

[3] H. Grosse, A. Martin, Exact Results on Potential Models for Quarkonium Systems, CERN TH 2674, to appear in Phys. Rep. (1979)

[4] A. Martin, Phys. Letters 67B (1977) 330

[5] C. Quigg, J.L. Rosner, Quantum Mechanics with Applications to Quarkonium, Fermilab Pub-79/22-THY (1979)

[6] H. Grosse, A. Martin, Nucl. Phys. B148 (1979) 413

THE QUASI-CLASSICAL LIMIT OF QUANTUM SCATTERING THEORY

Kenji Yajima

Department of Mathematics
University of Tokyo
Hongo, Tokyo, 113 Japan

1. Introduction, Theorem.

The aim of this note is to discuss the quasi-classical limit of quantum scattering operator and its relation to classical mechanical scattering theory. Let $H^h = -(\hbar^2/2m)\Delta + V(x)$ be a Hamiltonian for a quantum mechanical particle of mass m and $H(x,\xi) = \xi^2/2m + V(x)$ be the corresponding Hamiltonian for a classical particle. $\hbar = h/2\pi$, $h > 0$ is the Planck's constant. We assume that the potential $V(x)$ satisfies the following assumption.

ASSUMPTION(A). (1) $V(x)$ is a real valued infinitely differentiable function on R^n.
(2) For any multi-index α, there exist constants $C_\alpha > 0$ and $m(\alpha) > |\alpha| + 1$ such that

$$(1.1)\quad |(\partial/\partial x)^\alpha V(x)| \leqq C_\alpha (1 + |x|)^{-m(\alpha)}.$$

Under this assumption the following are well-known in quantum scattering theory (Agmon [1] and Kuroda [6]).

(1) H^h is a selfadjoint operator in the Hilbert space $L^2(R^n)$ and hence generates a one parameter unitary group $\exp(-itH^h/\hbar)$.

(2) Let $H_0^h = -(\hbar^2/2m)\Delta$ be the free Hamiltonian. Then the wave operators

$$W_\pm^h = \underset{t\to\pm\infty}{s\text{-}\lim} \exp(itH^h/\hbar) \exp(-itH_0^h/\hbar)$$

exist and are complete; hence the scattering operator

$$S^h = (W_+^h)^* W_-^h$$

is a unitary operator on $L^2(R^n)$.
The corresponding result for classical mechanics is proved by Hunziker [5] and Simon [7].
(1) For any $(a, \eta) \in \Gamma = R^n \times (R^n \setminus \{0\})$, there exists a unique solution $(x_\pm(t,a,\eta), p_\pm(t,a,\eta))$ of the Hamilton equation

$$\dot{x} = \partial H/\partial p\ ,\quad \dot{p} = -\partial H/\partial x$$

such that as $t \to \pm\infty$,

(1.2) $\quad |x_\pm(t,a,\eta)-t\eta-a| \to 0, \quad |p_\pm(t,a,\eta)-\eta| \to 0$

The wave operator $W_\pm^{cl}$ in classical mechanics is defined on Γ by the equation

$$W_\pm^{cl}(a,\eta) = (x_\pm(0,a,\eta), p_\pm(0,a,\eta)).$$

$W_\pm^{cl}$ is an infinitely differentiable canonical mapping on Γ.

(2) There exists a closed null set $e \subset \Gamma$ such that

$$W_-^{cl}(\Gamma \setminus e) \subset W_+^{cl}(\Gamma).$$

The scattering operator S^{cl} in the classical mechanics is defined as

$$S_-^{cl} = (W_+^{cl})^{-1} W_-^{cl}$$

on the initial set $\Gamma \setminus e$. S^{cl} is an infinitely differentiable canonical mapping on $\Gamma \setminus e$. We write as

$$S^{cl}(a,\eta) = (a_+(a,\eta), \eta_+(a,\eta)).$$

We study the asymptotic behavior of S^h as $h \to 0$ in momentum space representation on the coherent state $f_a^h = \exp(-ia\cdot\xi/\hbar)f(\xi)$ in terms of classical mechanics S^{cl}.

$$\hat{S}^h = \mathcal{F}^h S^h (\mathcal{F}^h)^*,$$

$\mathcal{F}^h$ is the Fourier transform:

$$(\mathcal{F}^h u)(\xi) = (2\pi\hbar)^{-n/2} \int \exp(-ix\cdot\xi/\hbar)u(x)\, dx.$$

To state our theorem, we introduce two set $e(a)$ and $e(a)^{ex}$ for $a \in R^n$.

$$e(a) = \{\eta \in R^n \setminus \{0\};\ (a,\eta) \in e\} \cup \{0\}$$

Clearly $e(a)$ is a closed set of R^n and is a null set for almost all $a \in R^n$.

$$e(a)^{ex} = e(a) \cup \{\eta \notin e(a) : \det(\partial\eta_+/\partial\eta)\ (a,\eta) = 0\}.$$

$e(a)^{ex}$ is a closed set of R^n. Since $|\eta_+| = |\eta|$, $e(a) = e(a)^{ex}$ if $n = 1$.

<u>THEOREM</u>. Let Assumption (A) be satisfied. Let $a \in R^n$ and let $e(a)^{ex}$ be defined as above. Suppose that $f \in L^2(R^n)$ has support in the exterior of $e(a)^{ex}$. Then for any $\eta \in R^n$ there exist only finite number of $\eta_j \in$ supp. f such that $\eta_+(a,\eta_j) = \eta$ and the following relation holds:

$$\lim_{h\downarrow o} \| (\hat{S}^h f^h)(\eta) - \sum_j \{\exp i(S(a,\eta_j)/\hbar - a_+(a,\eta_j)\cdot\eta/\hbar - \pi \mathrm{Ind}\gamma(a,\eta_j)/2)\} |\det(\partial\eta_+/\partial\eta)(a,\eta_j)|^{\frac{1}{2}} f(\eta_j) \| = 0.$$

Here the summation is taken over all η_j's such that $\eta = \eta_+(a,\eta_j)$; $\mathrm{Ind}\gamma(a,\eta_j)$ is the Keller-Maslov's index of the path $(x_-(t,a,\eta_j), p_-(t,a,\eta_j))$, $-\infty < t < \infty$, with respect to the obvious manifold being considered; $S(a,\eta_j)$ is the difference of the action integrals:

$$S(a,\eta_j) = \lim_{\substack{s\to-\infty \\ t\to\infty}} \{s\eta_j^2/2m - t\eta^2/2m + \int_s^t L(\dot{x}_-(\sigma,a,\eta_j), x_-(\sigma,a,\eta_j))d\sigma\} ,$$

where $L(\dot{x},x) = m\dot{x}(t)^2/2 - V(x(t))$ is the Lagrangian of the system.

For a smooth function $\Psi(\xi)$, we write as $f_\Psi^h(\xi) = \exp(-i\Psi(\xi)/\hbar)f(\xi)$ (we call this type of wave packet quasi-classical wave function). Then for smooth functions $P(\xi)$ and $Q(x)$ with compact supports we see

$$\lim_{h\downarrow o} (f_\Psi^h(\xi), P(\xi)f_\Psi^h(\xi)) = (f(\xi), P(\xi)f(\xi));$$

$$\lim_{h\downarrow o} (f_\Psi^h(\xi), Q(x)f^h(\xi)) = (f(\xi), Q(\partial\Psi/\partial\xi)f(\xi)).$$

By virtue of these relations, we may think that the quasi-classical wave function f_Ψ^h represents, in the limit $h = 0$, an ensemble of classical particles on the Lagrangian manifold $(\partial\Psi/\partial\xi, \xi)$ in the phase space $R^n \times R^n$ with momentum distribution density $|f(\xi)|^2 d\xi$ (see also Dirac [3]). Thus by taking f_a^h in the theorem, we choose a quasi-classical wave function such that, inthe limit $h - 0$, it represents classical particles concentrated at the configuration $x = a$ with momentum distribution density $|f(\xi)|^2 d\xi$. Then the incoming wave function at $t = -\infty$ represents particles like $x(t) \sim t\eta + a$, $p(t) \sim \eta$, where $\underline{a}$ is fixed and the distribution density of η is given by $|f(\eta)|^2 d\eta$. The theorem says that then the outgoing wave function can be approximated (as $h \downarrow 0$) by an incoherent superposition of the quasi-classical wave functions each of which represents at $t = \infty$ the particles $x(t) \sim t\eta + a_+(a,\eta_j)$, $p(t) \sim \eta$ with the density $|f(\eta_j)|^2 d\eta_j$ (with additional factor $\exp(-i\pi \mathrm{Ind}\gamma(a,\eta_j)/2)$). Moreover at $h = 0$ this incoherent superposition turns to coherent one, since

$$-(\partial/\partial\eta)\,(S(a,\eta_j)-a_+(a,\eta_j)\cdot\eta) = a_+(a,\eta_j)$$

and $a_+(a,\eta_j) \neq a_+(a,\eta_k)$ if $\eta_+(a,\eta_j) = \eta_+(a,\eta_k)$, $j \neq k$ by the canonical property of the classical scattering operator S^{cl}.

2. Outline of the proof.

Starting point of the proof of the theorem is the following

LEMMA 2.1. Let $a \in R^n$ and $f \in L^2(R^n)$. Then for any $\delta > 0$,

$$\lim_{h\downarrow o}\ \sup_{|t|\geq\delta}\ \|\exp(-itH_0^h/h)\,(\mathcal{J}^{h*}f_a^h)\,(x) - (m/i|t|)^{n/2}$$

$$\exp(im(x-a)^2/2t\hbar)f(m(x-a)/t)\| = 0.$$

The important point of Lemma 2.1 is that the approximation of $\exp(-itH_0^h/h)\,(\mathcal{J}^{h*}f_a^h)$ given by the second term is the uniform approximation with respect to the time $|t|>\delta$. This kind of approximation is impossible for the wave function $\exp(-itH_0^h/\hbar)(\exp(ix\cdot\xi/\hbar)$ $f(x))$.

If we assume that the support of the function $f(\xi)$ is outside of the origin, then for large $|t|$, the support of $f(m(x-a)/t)$ is far outside the effective range of scattering, where classical particles move almost freely. There the classical WKBJ-method does work well and we can prove the following lemma.

LEMMA 2.2. Let $f \in L^2(R^n)$ have support $K \subset R^n\,\{0\}$. Then there exists a constant $R > 0$ such that the following statements hold.

(1) For any $t < -R$, the mapping $K \ni \eta \to x_-(t,a,\eta)$ is a diffeomorphism.

(2) If $\eta = \eta(-R,a,x)$ is determined by the equation $x = x_-(-r,a,\eta)$, then

$$\lim_{h\downarrow o}\ \|\exp(iRH^h/\hbar)W_-^h\mathcal{J}^{h*}f_a^h(x) - \exp(-in\pi/4 + iS_-^{-R}(x)/\hbar)$$

$$|\det(\partial x_-(-R,a,\eta)/\partial\eta)|^{-\frac{1}{2}}f(\eta)\| = 0,$$

where for $\eta = \eta(t,a,x)$,

$$S_-^t(x) = \lim_{s\to-\infty}\left(\int_s^t L(\dot{x}_-(\sigma,a,\eta),x_-(\sigma,a,\eta))d\sigma + s\eta^2/2m\right).$$

From time $-R$ the wave packet gets into the effective scattering region. But if supp f is taken sufficiently small, there exists time $T > 0$ such that for $t > T$ it gets through the region and the mapping supp $f \ni \eta \to X_-(t,a,\eta)$ is again a diffeomorphism. In the scattering region we use Fujiwara's quasi-classical fundamental solution [4] and the stationary phase method [2] ; for $t > T$ we use classical WKBJ-method again. We get

LEMMA 2.3. Let $f(\xi)$ have sufficiently small support in the outside of $e(a)^{ex}$. Then there exists $T > 0$ such that for $t > T$, the mapping supp f $\eta \to x_-(t,a,\eta)$ is a diffeomorphism. Let $\eta \in$ supp f be determined by $x = x_-(t,a,\eta)$. Then

$$\lim_{h \downarrow 0} \sup_{t > T} \| \exp(-itH^h/\hbar) W_-^h \mathfrak{F}^{h*} f_a^h(x) - \exp(-in\pi/4 - i\operatorname{Ind}\gamma_t/2$$

$$+ iS^t(x)/\hbar) \det(\partial x_-(t,a,\eta)/\partial \eta)|^{-\frac{1}{2}} f(\eta)\| = 0,$$

where $\operatorname{Ind}\gamma_t$ is Keller-Maslov index of $\{(x_0(\sigma,a,\eta), p_-(\sigma,a,\eta)) - \infty < \sigma < t\}$.

By definition $\hat{S}^h = \lim_{t \to \infty} \exp(it\xi^2/2m\hbar) \mathfrak{F}^h \exp(-itH^h/\hbar) W^h \mathfrak{F}^{h*}$. Then by Lemma 2.3 and stationary phase method, we get the statement of the theorem.

REMARK 2.4. Similar results hold for long range scattering for Schrödinger equation and the Dirac spinor in the static electro-magnetic field [9] , [10]

[References]

1. Agmon, S.; Ann. Scu. Nor. Pisa, Ser IV, 2.2, 151-218 (1975).
2. Asada, K. and D. Fujiwara; Japan. J. Math. 4, 299-361 (1978).
3. Dirac, P. A.; Principles in Quantum Mechanics, 5 ed., Oxford.
4. Fujiwara, D.; to appear in J.d'Analyse Math.
5. Hunziker, W.; Commun. Math. Physics 8, 283-299 (1968).
6. Kuroda, S. T.; J. Math. Soc. Japan 25, 75-104 (1973).
7. Simon, Bl; Commun. Math. Physics 23, 37-48 (1973).
8. Yajima, K., to appear in Commun. Math. Physics.
9. Yajima, K.; preprint, Univ. of Tokyo.
10. Yajima, K.; in preparation.

SEMICLASSICAL QUANTUM MECHANICS FOR COHERENT STATES

George A. Hagedorn*
The Rockefeller University
New York, New York U.S.A.

Abstract: For certain Gaussian states we present a simple approximate evolution which is asymptotic to the Schrödinger evolution as $\hbar \to 0$. In 3 or more dimensions our error estimates are uniform in time if the potential is suitably chosen. Consequently, our methods apply to scattering theory. The approximate evolution is obtained by using the Trotter Product Formula and the second order Taylor expansion of the potential about the center of the wave packet.

In this brief note we wish to exhibit some simple ideas which can be used to study the $\hbar \to 0$ limit of quantum mechanics. Detailed proofs of theorems which use these ideas may be found in [1].

Our analysis deals with motion of certain normalized Gaussian wave functions $\psi_\alpha(A,B,\hbar,a,\eta,x)$ which are defined as follows:

Definition: Let A and B be complex $n \times n$ matrices with the following properties:

A and B are invertible; (1)

BA^{-1} is symmetric ([real symmetric] + i[real symmetric]); (2)

$\operatorname{Re} BA^{-1} = \frac{1}{2}[(BA^{-1}) + (BA^{-1})^*]$ is strictly positive definite; (3)

$(\operatorname{Re} BA^{-1})^{-1} = AA^*$. (4)

Then for all $a \in \mathbb{R}^n$, $\eta \in \mathbb{R}^n$, $\alpha \in \mathbb{R}$, and $\hbar > 0$ we define

$$\psi_\alpha(A,B,\hbar,a,\eta,x) = (2\pi)^{-n/4}\, \hbar^{-n\alpha/2} (\det A)^{-\frac{1}{2}} \exp\{-(4\hbar^{2\alpha})^{-1} \langle (x-a), BA^{-1}(x-a)\rangle + i\langle \eta,(x-a)\rangle /\hbar\}.$$

The choice of the square root in this definition depends on the context and must always be specified.

*Partially supported by NSF Grant PHY78-08066.

One should think of $\psi_\alpha(A,B,\hbar,a,\eta,x)$ as having its position localized near a and its momentum near η. The position and momentum widths are heuristically given by the matrices $\hbar^\alpha\,(AA^*)^{\frac{1}{2}}$ and $\hbar^{1-\alpha}(BB^*)^{\frac{1}{2}}$, respectively.

We let $H(\hbar) = H_0(\hbar) + V = -(\hbar^2/2m)\Delta + V$ be our quantum Hamiltonian. The gradient of V at x is denoted by $V^{(1)}(x)$, and the Hessian $\frac{\partial^2 V}{\partial x_i\,\partial x_j}$ is denoted by $V^{(2)}(x)$. With this notation we can state the finite time result of [1]:

<u>Theorem</u>: Suppose $1/3 < \alpha < 2/3$, $V \in C^2(\mathbb{R}^n)$, $|V(x)| \leq C_1 e^{Mx^2}$, and $V \geq -C_2$. Given any $R > 0$, assume there exists $\beta > 0$, such that $\|V^{(2)}(x) - V^{(2)}(y)\| \leq \beta|x-y|$ whenever $|x| < R$ and $|y| < R$. Let $a_0 \in \mathbb{R}^n$, $\eta_0 \in \mathbb{R}^n$, and let A_0 and B_0 satisfy (1)-(4). Then for each $T > 0$ and each positive $\lambda < \mathrm{Min}\,\{3\alpha - 1,\, 2-3\alpha\}$, there exist C and $\delta > 0$ such that $\hbar < \delta$ implies

$$\|e^{-itH(\hbar)/\hbar}\,\psi_\alpha(A_0,B_0,\hbar,a_0,\eta_0,\cdot)$$

$$-e^{iS(t)/\hbar}\,\psi_\alpha(A(t),B(t),\hbar,a(t),\eta(t),\cdot)\| < C\hbar^\lambda, \tag{5}$$

whenever $|t| < T$ (det $A(t)$ is never zero, and the branch of the square root $(\det A(t))^{-\frac{1}{2}}$ is determined by continuity in t). Here $[A(t),B(t),a(t),\eta(t),S(t)]$ is the unique solution to the system of coupled ordinary differential equations:

$$\frac{d\eta}{dt}(t) = -V^{(1)}(a(t)), \tag{6}$$

$$\frac{da}{dt}(t) = \eta(t)/m, \tag{7}$$

$$\frac{dA}{dt}(t) = i\hbar^{1-2\alpha}\,B(t)/2m \tag{8}$$

$$\frac{dB}{dt}(t) = 2i\hbar^{2\alpha-1}\,V^{(2)}(a(t))A(t), \tag{9}$$

$$\frac{dS}{dt}(t) = (\eta(t))^2/2m - V(a(t)) \tag{10}$$

with $A(0) = A_0, B(0) = B_0, a(0) = a_0, \eta(0) = \eta_0$, and $S(0) = 0$.

Furthermore, the differentials $\frac{\partial a(t)}{\partial a(0)}$, $\frac{\partial a(t)}{\partial \eta(0)}$, $\frac{\partial \eta(t)}{\partial a(0)}$, and $\frac{\partial \eta(t)}{\partial \eta(0)}$ exist,

and

$$A(t) = \frac{\partial a(t)}{\partial a(0)} A(0) + \tfrac{1}{2} i \hbar^{1-2\alpha} \frac{\partial a(t)}{\partial \eta(0)} B(0) \tag{11}$$

$$B(t) = \frac{\partial \eta(t)}{\partial \eta(0)} B(0) - 2 i \hbar^{2\alpha - 1} \frac{\partial \eta(t)}{\partial a(0)} A(0) . \tag{12}$$

Remarks: 1. Note that there are no problems with caustics in the above theorem. The Maslov index [4] is contained in the choice of the square root $(\det A(t))^{-\frac{1}{2}}$. The index appears explicitly in Yajima's approach [5] to the classical limit.

2. Our results are closely related to those of Heller [2] and Hepp [3]. However, for suitable potentials in 3 or more dimensions, the above theorem can be modified to handle scattering theory (see [1]).

The first ingredient in the proof of our theorem is the Trotter Product Formula. We approximate $e^{-itH(\hbar)/\hbar}\, \psi_\alpha(A(0),B(0),\hbar,a(0),\eta(0),\cdot\,)$ by $\left[e^{-itH_0(\hbar)/N\hbar}\, e^{-itV/N\hbar}\right]^N \psi_\alpha(A(0),B(0),\hbar,a(0),\eta(0),\cdot\,)$ for some large N .

Next, we replace $e^{-itV/N\hbar}\, \psi_\alpha(A(0),B(0),\hbar,a(0),\eta(0),\cdot\,)$ by $e^{-itW_{a(0)}/N\hbar}\, \psi_\alpha(A(0),B(0),\hbar,a(0),\eta(0),\cdot\,)$, where $W_a(x) = V(a) + \langle V^{(1)}(a),(x-a)\rangle + \frac{1}{2} \langle (x-a),\, V^{(2)}(a)(x-a)\rangle$.

We can now exactly compute

$$e^{-itH_0(\hbar)/N\hbar}\, e^{-itW_{a(0)}/N\hbar}\, \psi_\alpha(A(0),B(0),\hbar,a(0),\eta(0),\cdot\,)$$

$$= e^{iS_N(1)/\hbar}\, \psi_\alpha(A_N(1),B_N(1),\hbar,a_N(1),\eta_N(1),\cdot\,) .$$

Iterating the approximation procedure, we replace the next factor $e^{-itV/N\hbar}$ by $e^{-itW_{a(1)}/N\hbar}$ to obtain $e^{iS_N(2)/\hbar}\, \psi_\alpha(A_N(2),B_N(2),\hbar,a_N(2),\eta_N(2),\cdot\,)$, etc.

In general for $1 \leq n \leq N$, we have

$$\eta_N(n) = \eta(0) - \sum_{j=1}^{n} V^{(1)}(a_N(j-1))t/N .$$

$$a_N(n) = a(0) + \sum_{j=1}^{n} \eta_N(j)\ t/Nm ,$$

$$S_N(n) = \sum_{j=1}^{n} \left[(\eta_N(j))^2/2m - V(a(j-1))\right] t/N ,$$

$$A_N(n) = A(0) + \sum_{j=1}^{n} i\hbar^{1-2\alpha} B(j)\ t/2Nm ,$$

$$B_N(n) = B(0) + \sum_{j=1}^{n} 2i\hbar^{2\alpha-1} V^{(2)}(a_N(j-1))A_N(j-1)\ t/N .$$

As $N \to \infty$, $A_N(n), B_N(n), a_N(n), \eta_N(n)$, and $S_N(n)$ approach $A(nt/N)$, $B(nt/N)$, $a(nt/N), \eta(nt/N)$, and $S(nt/N)$, where $A(\cdot), B(\cdot), a(\cdot), \eta(\cdot)$, and $S(\cdot)$ satisfy eqns. (6)-(10). Eqns. (11) and (12) follow from eqns. (6)-(10) by standard ordinary differential equations methods.

Since we have fairly explicit information on the behavior of $A(t)$ and $A_N(n)$, we can prove that our wave packets remain sufficiently narrow in x-space for $|t| < T$. Thus we can estimate the errors incurred by the replacement of V by its second order Taylor approximations W_a. This error estimate implies the theorem (see [1]).

To extend the theorem to infinite times we require that V and some of its derivatives fall off sufficiently rapidly. This fall off allows us to control the Taylor expansion errors even though the x-space width diverges as $|t| \to \infty$. See [1] for the details.

References

1. Hagedorn, G.A.: Semiclassical Quantum Mechanics I: The $\hbar \to 0$ Limit for Coherent States. Commun. Math. Phys. (to appear).

2. Heller, E. J.: Classical S-matrix Limit of Wave Packet Dynamics. J. Chem. Phys. 65 , 4979-4989 (1976).

3. Hepp, K.: The Classical Limit for Quantum Mechanical Correlation Functions. Commun. Math. Phys. 35, 265-277 (1974).

4. Maslov, V.P.: Théorie des Perturbations et Méthodes Asymptotiques. Paris : Dunod 1972.

5. Yajima, K.: The Quasi-Classical Limit of Quantum Scattering Theory. Commun. Math. Phys. (to appear).

RANDOM SCHRÖDINGER OPERATORS

SOME RIGOROUS RESULTS

Hervé Kunz
Laboratoire de Physique Théorique
E.P.F.L.
14,Avenue de l'Eglise Anglaise
LAUSANNE, Suisse

Bernard Souillard
Centre de Physique Théorique
Ecole Polytechnique
91128 PALAISEAU, France

I. Introduction

We report here results on a class of finite difference Schrödinger operators with stochastic potentials. Our hamiltonian is then

$$H(V) = -\Delta + V$$

which is an operator on $\ell^2(\mathbb{Z}^\nu)$; Δ is the discretized Laplacian defined by

$$(\Delta\psi)(x) = \sum_{y:|y-x|=1} \psi(y) - 2\nu\psi(x)$$

for $\psi \in \ell^2(\mathbb{Z}^\nu)$ and the potential V act as a multiplication operator

$$(V\psi)(x) = V(x)\,\psi(x)$$

The potential V will be random. In order to deal with the essential features of the problem, we will stick here to the simplest case. The $V(x)$, $x \in \mathbb{Z}^\nu$, will be independent random variables with an absolutely continuous distribution $\rho(V)\,dV$, whose density ρ will be bounded and with compact support.

However, our results adapt to more general situations and the more complete statements and proofs will be found in the paper by the authors.

This model is the model used by Anderson and other authors since, to discuss transport and localization properties of various disordered systems. Simple physical realizations of such a model are :

- harmonic oscillators with nearest neighbour coupling and random masses.
- electrons scattered by impurities creating an effective random potential $V(x)$ at the point x , in the tight binding approximation.

There are four main problems of interest in the study of such systems :

- Determine the spectrum $\Sigma(H)$ with probability one. Physically, this means to find the support of the density of states, in particular to find if there are gaps or not.
- Find the nature of the spectrum, with probability one. The case where it is pure point corresponds to what is called in the physical litterature the presence of localized states, whereas the case where it is continuous, corresponds to the presence of delocalized states.
- Determine the decay properties of the pseudo-eigenfunctions. One would like to know for example if we have exponential decay when the spectrum is pure point.
- Finally, one is interested by the transport properties of such a system. This depends more precisely on the physical interpretation of the model. In the case of oscillators, the quantity of interest is the heat conduction, whereas it is the electrical conductivity which matters in the case of electrons.

Up to now, the only results which can be considered as rigorous from the mathematical point of view, which have been obtained, are concerned with the one-dimensional situation. Apart from some work on the presence of gaps, the following results have been obtained :

1. Beginning with Borland [1] in 1963, a series of contributions by various authors (Ishii, Lebowitz-Casher, Pastur) has led to a proof of the absence of absolutely continuous spectrum.
2. In 1977, Goldstein-Molchanov-Pastur [2] have proven that the spectrum is pure point.
3. In 1978, Molchanov [3] proved that the wave functions decay exponentially.

The methods used to prove such results are however all very specific to one-dimensional situations, and do not give a clue to an understanding of the multidimensional case.

Our method of approach is not one-dimensional from the beginning and brings the problem into a kind of problem of continuous spin system in statistical mechanics and suggest a precise connection between the phase transition in statistical mechanics and the transition from point to continuous spectrum.

Presently, however, it is only in one dimension that we have answered all the main questions mentioned before : we get a new proof that the spectrum is pure point, that the wave functions, decay exponentially and furthermore the static conductivity is proved to vanish, which is a new result.

II. General results

The spectrum $\Sigma(H)$ of this class of hamiltonians is simply the algebraic sum of the spectrum of the Laplacian and the spectrum of the potential

Theorem 1 With probability one

$$\Sigma(H) = [0, 4\nu] + \operatorname{supp}\{\rho\}$$

A simple corollary of this theorem is that the hamiltonian has no isolated eigenvalues, with probability one.

Our next goal is to obtain useful criteria for localization. They are provided by the following results.

Let e_α be the eigenvalues of the operator H restricted to the box Λ, with zero boundary conditions and $\frac{\psi_\alpha(x)}{\|\psi_\alpha\|}$ denote the associated normalized eigenvector. Let $D(r;A)$ be defined by

$$D(r;A) = \sup_{\Lambda,x} < \sum_{e_\alpha \in A} \frac{|\psi_\alpha(x)\,\psi_\alpha(x+r)|}{\|\psi_\alpha\|^2} >_\Lambda$$

where A is an interval of energy.

Then we have

Theorem 2 If $\sum_{r \in \mathbb{Z}^\nu} D(r;A) < \infty$, then with probability one, the spectrum of H is pure point in the interval A.

It is interesting to note that when $H = -\Delta$, we have on the opposite : $\inf_r D(r;A) > 0 \quad \forall A \subset \Sigma(H)$. It seems therefore that the presence of an absolutely continuous part in the spectrum of H is associated with the existence of long range order in the correlation function $D(r;A)$

A criterion for the presence of continuous spectrum is provided by the following :

Theorem 3 If the so-called participation ratio

$$R_\Lambda(A) = < \sum_{e_\alpha \in A} \frac{|\psi_\alpha(0)|^4}{\|\psi_\alpha\|^4} >_\Lambda$$

is such that

$$\lim_{\Lambda \uparrow \mathbb{Z}^\nu} R_\Lambda (A) = 0$$

then with probability one, H has continuous spectrum in A.

The proof of these two theorems is based on Ruelle's theorems [4], which make precise the intuitive idea that a wave packet will diffuse if and only if the spectrum of the hamiltonian is continuous.

Concerning the transport properties, we have obtained the following criterion for the static conductivity.

Let $\sigma_\Lambda(\omega)$ be the frequency (ω) dependent conductivity, given by the Green-Kubo formula, then if the static conductivity of the infinite system is defined by

$$\sigma = \lim_{\omega \to 0^+} \lim_{\Lambda \uparrow \mathbb{Z}^\nu} \frac{1}{\omega} \int_0^\omega \sigma_\Lambda(\omega')\, d\omega'$$

we have

Theorem 4

If $\sum_{r \in \mathbb{Z}^\nu} |r|^2 D(r;R) < \infty$, then $\sigma = 0$

This result shows that exponential localization, for example, implies zero conductivity.

We have seen in this way that it is possible to say something about the nature of the spectrum or the transport properties if we can get enough information about some wave function correlation functions, like $< \sum_{e_\alpha \in A} \frac{\psi_\alpha(x)\, \psi_\alpha(y)}{\|\psi_\alpha\|^2} >_\Lambda$ of the hamiltonian restricted to a finite box Λ . We are faced now with the problem of finding explicit expressions for such quantities. It is possible to do so, quite generally, and in fact to bring the result in a form which bears some resemblance to expectation values of correlation functions in a lattice system of unbounded spins. The result is given by the following:

Theorem 5

Let $<f>_\Lambda = < \sum_{e_\alpha} f(e_\alpha, \frac{\psi_\alpha}{\|\psi_\alpha\|}) >_\Lambda$

Then $<f>_\Lambda$ is given by the expression :

$$<f>_\Lambda = \int de \int \prod_{x \in \Lambda} \rho\left[e + \frac{(\Delta\psi)(x)}{\psi(x)}\right] \exp - \sum_{x \in \Lambda} \psi_x^2 \; J(\psi)\, f(e,\psi) \prod_{x \in \Lambda} d\psi_x$$

where

$$J(\Psi) = 2 \prod_{x\in\Lambda} \Psi_x^{-3} \left(\sum_{x\in\Lambda} \Psi_x^2\right)^2 \Big/ \sum_{\tau_\Lambda} \prod_{(x,y)\in\tau_\Lambda} \Psi_x \Psi_y \Big/$$

In this last expresssion, the sum runs over all the spanning trees τ_Λ of Λ .

The idea behind the proof of this result is that the probability distribution of V induces a probability distribution on Ψ to be an eigenvector, of $H_\Lambda(V)$, of corresponding eigenvalue e .

III. The one-dimensional case

By means of the criteria discussed and the formalism settled, it is possible to give a rather detailed analysis of the model in one dimension. The results obtained can be summarized in the following theorem.

Theorem 6

a) $D(r;A) \leq c(A) \exp - \lambda(A)|r|$

for any compact subset A of R, with $\lambda(A) > 0$, and $\lambda(A)$, $c(A)$ independent of Λ.

b) for almost all V , $H(V)$ has pure point spectrum

c) for almost all V , the eigenvectors are exponentially decaying

d) for almost all V , the static conductivity σ is zero.

The results b), c), d) follow essentially from a). The exponential bound on $D(r;A)$ results from a detailed analysis of the correlation function $< \sum_{e_\alpha\in A} \frac{\Psi_\alpha(x)\Psi_\alpha(x+r)}{\|\Psi_\alpha\|^2} >_\Lambda$ by means of some kind of transfer operator, familiar in statistical mechanics of one-dimensional systems.

Moreover, it is possible, using this technique, to give explicit expressions for the average spectral function, its average absolute value and the participation ratio.

References

1 R.E. Borland, Proc.Roy.Soc. A 274, 529 (1963)

2 I. Ya Goldstein, S.A. Molchanov, L.A. Pastur, Functs.Anal.y. Prilozhen 11, 1 (1977)

3 S.A. Molchanov, Math. USSR Izvestija 42 (1978) Transl.12,69 (1978)

4 D. Ruelle, Nuovo Cimento A 61 655 (1969).

ON THE SCATTERING PROBLEM WITH INDEFINITE METRIC

Xia Daoxing

Fudan University, Shanghai, China

Several physicists try to get over the difficulty of divergence in the theory of quantum fields by using indefinite metric. The scattering theory concerning indefinite metric may be formulated in the following.

Let $\mathcal{H}$ be a Hilbert space with inner product (x,y), $x,y \in \mathcal{H}$, $\mathcal{H} = \mathcal{H}_+ \oplus \mathcal{H}_-$, $P_\pm : \mathcal{H} \to \mathcal{H}_\pm$ be the projection and $J = P_+ - P_-$. We introduce an indefinite metric to the space $\mathcal{H}$ by giving the definition $[x,y]=(Jx,y)$. The Hamiltonian H is an operator, self-adjoint with respect to the indefinite metric $[x,y]$, and $H = (H_- + \eta\Gamma^*)P_- + (\Gamma + H_+)P_+$, where $\Gamma : \mathcal{H}_+ \to \mathcal{H}_-$ is a bounded linear operator, $H_\pm : \mathcal{H}_\pm \to \mathcal{H}_\pm$ is the self-adjoint operator and $\eta = -1$ in this case. The space of all physical states is denoted by $\mathcal{H}'$. A vector $x \in \mathcal{H}$ is in $\mathcal{H}'$, iff there is an analytic vector-valued function $x(\cdot)$ which is analytic both in $\theta\lambda>0$ and in $\theta\lambda<0$, satisfies the equation $(H-\lambda)x(\lambda) = x$ and has the property that for every $a \in \mathcal{H}$, there is a function $f_a(\cdot) \in L^1$ such that

$$(x(\lambda),a) = \frac{1}{2\pi i} \int \frac{f_a(\omega)d\omega}{\omega-\lambda} .$$

Let $H' = H\big|_{\mathcal{H}'}$. We shall consider $\mathcal{H}_+$ as the space of all asymptotic states. Let H_0 be the free Hamiltonian in $\mathcal{H}_+$. The wave operator $W_\pm = W_\pm(H',H_0) : \mathcal{H}_+ \to \mathcal{H}'$ is defined by

$$\lim_{t\to\pm\infty} \| e^{-itH'} W_\pm \psi - e^{-itH_0} \psi \| = 0 .$$

Under certain conditions, we proved the existence and the unitarity of the scattering operator $S = W_-^{-1} W_+$; in addition to these, we gave an expression of scattering operator. We notice that if the wave operators $W_\pm(H_+,H_0)$ of the pair of Hamiltonian in the Hilbert space $\mathcal{H}_+$ exist and are unitary, then $W_\pm(H',H_0) = W_\pm(H',H_+)\, W_\pm(H_+,H_0)$ and $S = W_-(H_+,H_0)^{-1} S_1 W_+(H_+,H_0)$, where $S_1 = W_-(H',H_+)^{-1} W_+(H',H_+)$. But $W_\pm(H_+,H_0)$ does not relate to the indefinite metric, so we didn't need to consider them. And we shall discuss S_1 only. Thus we suppose that $H_+ = H_0$.

Let σ be the spectrum of H_+. For $\lambda \bar{\epsilon} \sigma$, we construct the operator

$$h(\lambda) = \lambda - H_- - \eta\, \Gamma(\lambda - H_+)^{-1}\Gamma^*$$

in $\mathcal{H}_-$. This operator-valued analytic function $h(\cdot)$ is also related to both H and the decomposition $\mathcal{H} = \mathcal{H}_+ \oplus \mathcal{H}_-$ and it is useful.

We consider the following case. Suppose that $\mathcal{H}_+$ is the space $L^2(\sigma, \mathcal{E})$ of all strongly measurable and square-integrable vector-valued functions with values in a separable Hilbert space $\mathcal{E}$ and

$$\| \phi \|^2 = \int_\sigma \| \phi(\omega) \|_{\mathcal{E}}^2 \, d\omega \ , \quad (H_+\phi)(\omega) = \omega\phi(\omega) \ ,$$

for $\phi \epsilon L^2(\sigma, \mathcal{E})$, and there is an operator-valued function $\zeta(\omega)$ such that $\zeta(\omega)$ is an operator from $\mathcal{E}$ to $\mathcal{H}_-$ for $\omega \epsilon \sigma$ and

$$\Gamma\phi = \int_\sigma \zeta(\omega)\phi(\omega)\, d\omega \ , \text{ for } \phi \ \epsilon \mathcal{H}_+ \ .$$

<u>Theorem</u> If $h(\omega \pm i0)$ and $h(\omega \pm i0)^{-1}$ exist for all real ω and

$$\int \| k(\omega \pm i0)^{-1} \zeta(\omega) \|^2 d\omega < \infty \ .$$

Then the indefinite metric $[x,y]$ in $\mathcal{H}'$ becomes positive definite, the space $\mathcal{H}'$ endowed with $[x,y]$ is a Hilbert space, the wave operators $W_\pm$ exist, are unitary operators from $\mathcal{H}_+$ onto $\mathcal{H}'$ and

$$W_\pm\phi = \int h(\omega \pm i0)^{-1} \zeta(\omega)\phi(\omega) d\omega \oplus [\phi - i\eta\zeta^* \int \frac{h(\omega \pm i0)^{-1}\zeta(t)\phi(t)}{t - (\omega \pm i0)} dt]$$

And the scattering operator $S = W_-^{-1}W_+$ is a unitary operator in $\mathcal{H}_+$ which has the form $(S\phi)(\omega) = S(\omega)\phi(\omega)$, where $S(\omega) = I + 2\pi i\eta\zeta(\omega)^* h(\omega + i0)^{-1}\zeta(\omega)$.

For Lee and Wick's model in N $\theta\theta$ sector [1], we determined the concrete form of S by solving a singular integral equation of function of two variables, i.e.

$$H(\omega - \omega_1 - \omega_2 + i0) y(\omega_1, \omega_2) - \int_\mu^\infty \frac{y(\omega_1, \omega', \alpha(\omega_2)\alpha(\omega') + y(\omega', \omega_2)\alpha(\omega')\alpha(\omega_1)}{\omega' - \omega + \omega_1 + \omega_2 - i0} d\omega'$$

$$= \delta(\omega_1 - a_1)\delta(\omega_2 - a_2) \ ,$$

where $y(\cdot,\cdot)$ is the unknown function and α is given,

$H(\lambda) = \lambda - m + \int_{\mu}^{\infty} \alpha(\omega)^2 (\lambda-\omega)^{-1} d\omega$. In this case, the explicit form of the matrix elements of S is

$$S(\omega_1,\omega_2\omega_3;a_1,a_2,a_3) = \prod_{i=1}^{3} e^{-i\arg(H(\omega_j+i0)H(a_j+i0))} \delta(\omega_1,\omega_2,\omega_3;a_1,a_2,a_3)+$$

$$\delta(\omega_1+\omega_2+\omega_3-a_1-a_2-a_3)\frac{2\pi i}{3} \prod_{j=1}^{3} \frac{\alpha(\omega_j)\alpha(a_j)}{H(\omega_j+i0)H(a_j+i0)} \sum_{i,j=1}^{3} \frac{1}{L(a_j)} \times$$

$$\times \frac{|H(a_j+i0)|^2}{\alpha(a_j)^2} \delta(\omega_j-a_j)+f(\omega_j;a_j)$$

where f and L can be expressed by the given function α .

If we change $\eta = -1$ to $\eta = 1$ in the theorem, then the scattering problem concerning indefinite metric becomes the scattering problem with intermediate system similar to the case considered by Lifschitz[3]. We also proved this theorem in the case $\eta = 1$. And then we gave some applications in the case $\eta = 1$ to determine the asymptotic probability of decay of the intermediate system and the corresponding generalization of Wigner-Eisenbud's formula.

REFERENCES

[1] Lee, T.D.; Wick, G.C.:
Nucl. Phys. B9, 2(1969), 209-243; B10, 1(1969), 1-10.

[2] Xia Daoxing:
Scientia Sinica 18(1975), 165-183.

[3] Lifshitz, M.S.:
Dok.Acad. Nauk III (1956), 67-70, 799-802

SOME OPEN PROBLEMS ABOUT COULOMB SYSTEMS

E.H. Lieb*

Departments of Mathematics and Physics
Princeton University
Princeton, N.J. 08544, USA

I. Introduction

While much progress has been made in the rigorous mathematical theory of Coulomb systems, it hardly needs to be said that much remains to be done. This talk is supposed to be a review, but it is not inappropriate to survey the subject by emphasizing some open problems whose solutions seem to be not far beyond the reach of present mathematical technology.

Naturally everyone can produce his own list, but attention should be paid to the following points: (i) Not all the obvious problems that hold for general many-body systems are appropriate now. They may be presently too difficult in the Coulomb case. (ii) There exist special and unusual problems connected with Coulomb systems whose solution seems to be necessary. An example is Problem 9 on the boundedness of the potential. Another is the existence of screening. (iii) While some mathematical physicists appear to believe that the sole aim of their subject is to find rigorous proofs of what is already heuristically known, that is far from the truth. There is also a computational aspect to mathematical physics, namely to find good approximation schemes which yield upper and lower bounds; in the best case these bounds should be capable of improvement to arbitrary accuracy. This goal is especially important for Coulomb problems because there is already a vast literature on uncontrolled approximation schemes for problems in atomic, molecular and solid state physics. As these schemes are widely used to compute physical parameters, sometimes in regimes in which direct experimental verification is difficult or impossible, it is highly desirable to set limits on their accuracy. Some of the problems mentioned here are of this kind.

---

*) Work partially supported by U.S. National Science Foundation grants INT 78-01160 and PHY-7825390.

II. Lower Bounds to Ground State Energies

In this and the next two sections we will be concerned with the non-relativistic Schroedinger equation when the nuclei are held fixed (also called the Born-Oppenheimer approximation). The Hamiltonian for k fixed nuclei of charges $z_1 .., z_k > 0$ located at $R_1, ..., R_k$ and N electrons (in units in which $|e| = 1$, $h^2/2m_e = 1$) is

$$H = -\sum_1^N \Delta_i - \sum_1^N V(x_i) + \sum_{i<j}^N |x_i - x_j|^{-1} + U \tag{2.1}$$

$$\text{where} \quad V(x) = \sum_1^k z_j \, |x - R_j|^{-1} \tag{2.2}$$

$$U = \sum_{i<j}^k z_i z_j \, |R_i - R_j|^{-1} \ . \tag{2.3}$$

One wants to find good estimates for the ground state energy E_0, as well as qualitative properties of the dependence of E_0 on the $\{R_i\}$.

Another important problem is to find corrections to the static nuclei approximation when the nuclear kinetic energy is introduced. Results in this direction have been given by Aventini, Combes, Duclos, Grossman and Seiler [1], for example. We will not say anything more about this here except to note that if the nuclear kinetic energy is introduced, then obviously the nuclei are not fixed. For large nuclear mass, the nuclei will be predominantly around $\{\bar{R}_i\}$, the values of the R_i which minimizes E_o. In the aforementioned problem posed by (2.1), however, all $\{R_i\}$ can be considered. This is not without physical interest. In particular, a widely used approximation to the partition function of matter at moderately high temperatures and pressures is the partition function obtained from the nuclear Hamiltonian in which $E_0(\{R_i\})$ is regarded as a potential, namely $H' = T_{nuc} + E_0(\{R_i\})$, where T_{nuc} is the nuclear kinetic energy.

One time honored and sensible way to approach (2.1) is to try to approximate $(\psi, H\psi)$ in terms of the single particle density ρ_ψ given by

$$\rho_\psi(x) = N \sum_\sigma \int |\psi(x, x_2 ..., x_N; \sigma_1, .., \sigma_N)|^2 dx_2 .. dx_N \tag{2.4}$$

Here, $\{\sigma_i\}$ are the spin coordinates, $\sigma_i \in \{1,\ldots,q\}$, there being q spin states in general. ψ is antisymmetric and $(\psi,\psi) = 1$. The second term in (2.1) is easily evaluated, namely

$$(\psi, V\psi) = \int V(x)\rho_\psi(x)dx. \tag{2.5}$$

Thus we have to worry only about the first and third terms.

Problem 1 (Lower bound to the kinetic energy): We want to find an inequality of the form

$$T_\psi \equiv (\psi, -\Sigma\Delta_i \psi) \geq Kq^{-2/3} \int \rho_\psi(x)^{5/3} dx . \tag{2.6}$$

The choice $\rho^{5/3}$ is made for dimensional reasons. Of course, other possibilities exist, such as $[\int\rho_\psi{}^a]^b$ with $3b(a-1) = 2$, but (2.6) is classic and also the most tractable. The conjecture [2] is that (2.6) holds with

$$K = K^c \equiv 3/5\ (6\pi^2)^{2/3} . \tag{2.7}$$

K^c is the classical value. The proof of (2.7) is important for the following reasons. First, as will be discussed later, (2.6) with K^c essentially gives Thomas - Fermi (TF) theory. There is an enormous literature in which the TF approximation is applied to physical problems. If K^c is correct, these calculations automatically become lower bounds (except for exchange terms which are easily bounded). Second, there is a complementary conjectured upper bound (cf. Problem 4) using Hartree-Fock (HF) theory which also contains (2.6) with K^c, together with an additional term of secondary importance (the Weizsaecker correction). If the HF and TF constants have a common value K^c, then one would have the beginning of a powerful approximation scheme.

What is known [3] is that (2.6) holds with $K \geq K^c\,[1.83/4\pi]^{2/3}$. Analogous bounds also hold in dimensions $\neq 3$.

Problem 2 (The indirect part of the Coulomb repulsion): The third term in (2.1) is usually approximated by the average Coulomb repulsion, D, in the state ρ_ψ and we want a lower bound to the remainder, namely the indirect part of the Coulomb energy, which we call I_ψ. Thus

$$(\psi, \sum_{i<j}^{N} |x_i - x_j|^{-1}\psi) = D(\rho_\psi) + I_\psi \text{ , where} \tag{2.8}$$

$$D(\rho) = 1/2 \iint \rho(x)\rho(y)|x - y|^{-1}\, dxdy \; . \tag{2.9}$$

For dimensional reasons we can seek a bound of the form

$$I_\psi \geq - C \int \rho_\psi(x)^{4/3} dx \; . \tag{2.10}$$

This form was introduced by Dirac [4] as an approximation, with C replaced by $(0.93)\, q^{-1/3}$, and is widely used (TF-Dirac theory). Note that there is no q-dependence in (2.10). It is known that a bound of this form exists [5] with $0.93 \leq C \leq 8.52$. The problem is to find the sharp C , which is probably not far from 1 .

If we now combine the various estimates, TFD theory is obtained, namely

$$E_0 \geq \inf \{E_L(\rho) | \int \rho = N \} \text{ , where}$$

$$E_L(\rho) = \int [q^{-2/3} K^c \rho^{5/3} - V\rho - C\rho^{4/3}] + D(\rho) + U \; . \tag{2.11}$$

Next, we consider upper bounds of the same form as (2.11).

III. Upper Bounds to Ground State Energies

The obvious way to obtain an upper bound is to compute $(\psi, H\psi)$ for some suitable normalized ψ . This, however, is not easy to do effectively in practice. A favorite choice for ψ is a Slater determinant, namely

$$\psi = (N!)^{-1/2} \det [\phi_i(x_j;\sigma_j)] \; , \tag{3.1}$$

where the $\{\phi_i\}$ are N orthonormal functions. Then

$$\rho_\psi(x) = \sum_1^N \rho_i(x) \quad , \quad \rho_i(x) = \sum_\sigma |\phi_i(x;\sigma)|^2 \tag{3.2}$$

$$T_\psi = \sum_1^N T_i \quad , \quad T_i = \sum_\sigma \int |\nabla \phi_i|^2 \tag{3.3}$$

$$I_\psi < 0 . \tag{3.4}$$

(3.4) is especially important and is one good reason for using a HF function. It is a consequence of $|x - y|^{-1}$ being positive definite. (3.4) suggests the following:

<u>Problem 3 (HF Coulomb energy):</u> Find a useful upper bound, other than zero, to I_ψ. The most useful would be $I_\psi \le - C' \int \rho_\psi^{4/3}$.

The most important problem is an upper bound for the kinetic energy, T_ψ, in terms of ρ_ψ. There is none, of course, because T_ψ can be arbitrary large for any given ρ_ψ. If we turn the question around, however, it becomes sensible. That is

<u>Problem 4 (Lower bound to the HF kinetic energy):</u> Let a nonnegative function $\rho(x)$ satisfying (i) $\int \rho = N$ and (ii) $\int [\nabla \rho^{1/2}]^2 < \infty$ be given. Then choose N orthonormal functions $\{\phi_i(x;\sigma)\}$ such that (3.2) holds. How small can T_ψ, given by (3.3), be? The conjecture is that T_ψ can be made to satisfy

$$T_\psi \le q^{-2/3} K^c \int \rho^{5/3} + \int [\nabla \rho^{1/2}]^2 . \tag{3.5}$$

The second term in (3.5) is the Weizsaecker correction.

There are two obvious advantages to (3.5). First, it implies that one can do HF theory by choosing only the single particle density ρ instead of having to choose the N orthonormal functions $\{\phi_i\}$. The result is

$$E_0 \le E_U(\rho)$$

for any $\rho \ge 0$ such that $\int \rho = N$, where

$$E_U(\rho) = \int [q^{-2/3} K^c \rho^{5/3} - V\rho - C'\rho^{4/3} + (\nabla \rho^{1/2})^2] + D(\rho) + U . \tag{3.6}$$

Second, the minimization of (3.6) and (2.11) are similar problems.

While (3.5) is heuristically reasonable, it seems to be a very hard problem, even for $N = 2$, and even if different constants are allowed in (3.5). For $N = 1$ it is trivial (take $\phi = \rho^{1/2}$). The problem was raised by March and Young [6] who purported, incorrectly, to prove it in three dimensions. However, they did correctly prove the analogous conjecture in one dimension, which is

$$T_\psi \leq q^{-2} K_1^c \int \rho^3 + \int [\nabla \rho^{1/2}]^2 \tag{3.7}$$

with $K_1^c = \pi^2/3$. The construction for $q = 1$ is

$$\phi_j(x) = [\rho(x)/N]^{1/2} \exp\{ik_j \int_{-\infty}^x \rho(y)dy\} \tag{3.8}$$

with $k_j = (2\pi/N)[j - (N + 1)/2]$, $j = 1, \ldots, N$.
This is easily generalized to $q > 1$.

IV. Properties of TF and Related Theories

Let us consider (2.11) (TFD theory) or (3.6) (TFDW theory). The properties of these minima are somewhat complicated. Recent progress has been made [7] but much remains to be understood. If the $\rho^{4/3}$ and Weizsaecker terms are omitted, however, the resulting problem (TF theory) is much better understood [3a,8,9,10]. Some interesting open problems remain (cf [3a,8,9,10]).

The TF energy for a neutral system has the property that it decreases under dilation of all nuclear coordinates [9]. This is surely also true for a non-neutral system, but it is not yet proved. Teller's theorem (which holds generally [8]) is implied by this. It states that for N (which need not be an integer) electrons

$$E_0^{TF}(N) \geq \min \sum_{j=1}^{k} E_{atom}^{TF}(z_j, \lambda_j) , \tag{4.1}$$

where $E_{atom}^{TF}(z,\lambda)$ is the TF energy for a single nucleus and electron number λ. The minimum in (4.1) is over all λ_j such that $\Sigma_1^k \lambda_j = N$. The fact (4.1) is one proof of the stability of matter [2], once one has the kinetic energy inequality (2.6). Now consider what happens when k becomes large and N is fixed. For $\lambda << z$,

$E^{TF}_{atom}(z,\lambda) \simeq -z^2\lambda^{1/3}$. This is qualitatively correct when $1 \leq \lambda << z$, but it is nonsense, in comparison with the real atomic situation, if $\lambda < 1$. Take all $z_i = z$. Then for large k (4.1) becomes $-z^2k(N/k)^{1/3}$. In truth, we expect that $E_0(N) \simeq -z^2N$ if $zk << N$; this was in fact proved by Lenard [11] using a modification of the Dyson-Lenard proof [12]. The trouble with TF theory is that it cannot distinguish the situation in which the electron number is less than one.

Problem 5 (Improvement of TF theory for $zk >> N$): Find a simple modification of the TF stability proof [2,3] to obtain $E_0 \geq -z^2N$ when $zk >> N$.

Despite the foregoing problem, TF theory (and also TFD and TFDW theories) have an important feature. If k is fixed and all $z_j \to \infty$, then TF theory gives the correct energy asymptotically [8].

$$\lim_{\text{all } z_j\to\infty} E_0^{TF} / E_0 = 1 \tag{4.2}$$

Note that (4.2) is true only if K^c is used in (2.11); this is another good reason for solving Problem 1. The TF energy for a neutral atom is $-(2.21)\, z^{7/3}$. The next correction is probably order z^2 and is probably that suggested by Scott [13], which we state as

Problem 6 (The Scott correction): Prove that

$$E_0 = E^{TF} + (q/8) \sum_1^k z_j^2 + o(z^2) \tag{4.3}$$

where E^{TF} is given by (2.11) with $C = 0$.

Scott's idea is that the correction to E^{TF} comes from the electrons close to the nuclei and should therefore be additive for each atom in the molecule. Cf. ref. [3a], p. 560.

It is probably not too difficult to get an upper bound to E_0 of the form (4.3) by using a Slater determinant. A lower bound is more difficult to obtain. The following remark may be useful in this connection.

The minimizing TF ρ for a neutral system satisfies the TF equation

$$(5/3)q^{-2/3}K^c\rho(x)^{2/3} = V(x) - \int\rho(y)|x-y|^{-1}\,dy \equiv \Phi(x) \,. \qquad (4.4)$$

Thus $E^{TF} = -(2/3)\,q^{-2/3}K^c\int\rho^{5/3} - D(\rho) + U$. Let h be the single particle Hamiltonian $-\Delta - \Phi(x)$ and let $H_0 = \Sigma_1^N h_i$. Then

$$(\psi, H\psi) = (\psi, H_0\psi) - 2D(\rho,\rho_\psi) + U + (\psi, \Sigma|x_i - x_j|^{-1}\psi) \qquad (4.5)$$

where $D(f,g) = (1/2)\iint f(x)g(y)|x-y|^{-1}\,dxdy$.

As shown in [3a], equ. 45, the last term in (4.5) is bounded below by $-D(\rho) + 2D(\rho,\rho_\psi) - (2.21)Nz^{2/3} - z^{-2/3}\int\rho^{5/3}$. The last two of these terms are $O(z^{5/3})$. Thus, to $O(z^{5/3})$,

$$D(\rho)+E_0 \geqslant E_{00} = \inf \operatorname{spec} H_0 \,. \qquad (4.6)$$

E_{00} is the sum of the lowest N eigenvalues of h (including spin) . Suppose one can show that

$$E_{00} \geq -(2/5)(3/5K^c)^{3/2}\, q\int\Phi^{5/2} + (q/8)\Sigma_1^k z_j^2 \,. \qquad (4.7)$$

The first term in (4.7) is the classical approximation; the second term is the Scott correction for h . Then, recalling (4.4), the desired lower bound is obtained. In other words, basically the Scott correction is the same as the non-classical correction to the single particle Hamiltonian H_0 , in which the potential is the negative of the TF potential.

V. The Boson Problem

Consider the Hamiltonian (2.1) for bosons instead of fermions. As far as the ground state energy is concerned, this is the same as setting $q = N$ in the fermions problem. (4.1) is still true, but the energy of a neutral atom is $-(2.21)q^{2/3}z^{7/3}$. Taking $k = N$ and $z = 1$, we thus obtain

$$E_0 \geq -(\text{const.})N^{5/3} \,. \qquad (5.1)$$

This result was obtained by Dyson and Lenard [12].

(5.1) is correct in that it can be shown [14] that $E_0 \leq -(\text{const.})' N^{5/3}$ for a suitable choice of the R_i (depending on N). However, (5.1) is misleading if the nuclear kinetic energy is included, however large the nuclear mass M_N may be. In the fermion case the nuclear kinetic energy does not alter E_0 very much, if $M_N \gg 1$, but for bosons it does.

With $M_N = 1$ and boson nuclei, Dyson [15] showed that

$$E_0 \leq -(\text{const.}) N^{7/5} \tag{5.2}$$

by a variational calculation with a function ψ that was quite complicated. However, the intuition behind (5.2) is relatively straightforward, as follows.

At high density the Coulomb interaction is weak in some sense. In this regime the Bogolyubov approximation is expected to be valid; the wave function in this approximation describes particles with paired momenta k and $-k$. The Bogolyubov approximation yields an energy [16]

$$E_0' \simeq -(\text{const.})\, N \rho^{1/4}, \quad \text{large } \rho . \tag{5.3}$$

Suppose that the total energy of the bound system is $E_0' + N^{1/3} \rho^{2/3}$, the latter term being the "uncertainty principle" kinetic energy needed to localize the system so that its density is ρ. (5.2) is then obtained by minimizing the foregoing expression with respect to $\rho (\rho \propto N^{8/5})$. The mean spacing between particles is $N^{-1/5}$.

Problem 7 (Boson ground state energy): Find an upper bound of the form (5.2) when the Hamiltonian is (2.1) + (nuclear kinetic energy) and $k=N$, $z_j=1$.

It may be wondered why Problem 7 is presented as an important problem at the frontier of the subject. The reason lies in its conceptual rather than its physical importance. Sections II-IV stressed the fact that fermion systems could be plausibly understood in terms of a fluid of electron density. Quantum mechanics entered through the statement that the self energy of the fluid was $q^{-2/3} K^c \int \rho^{5/3}$. Boson systems, on the other hand, cannot be easily understood in such a simple way. Quantum correlations, of a sort related to superfluidity and superconductivity, play a central role. It is essential that these correlations be understood if progress is to be made with the quantum many-body problem. Problem 7 appears to be the most tractable problem of this genre.

VI. Infinite Systems

Let us first discuss the problem of the thermodynamic limit of the free energy in the canonical ensemble. For the Hamiltonian (2.1) supplemented by the nuclear kinetic energy operator, the limit has been shown to exist [17]. Quantum mechanics does not play much of a role, except that it provides a linear lower bound on E_0 (H-stability), because the difficulty comes from the long range $1/r$ fall off. To control the Coulomb potential, a primitive form of screening was used, namely Newton's theorem which states that two disjoint, neutral, isotropic balls have vanishing interaction energy.

There are several reasons that the use of Newton's theorem is unsatisfactory: (i) It requires that the system be locally isotropic; this property fails for the solid state problem discussed below; (ii) Screening is really the property that the potential of an arbitrary charge inside a surface S can be cancelled outside S by a suitable surface charge on S. Screening holds for the Yukawa potential, but Newton's theorem does not. Thus, the proof in [17] does not hold for the Yukawa potential, although the thermodynamic limit does exist by more standard arguments.

Apart from these criticisms, there is the problem that the use of Newton's theorem obscures the main physical fact required for the thermodynamic limit. This is that two large systems have a weak Coulomb interaction on the average. One can explain this from two alternative points of view: (a) In order for two systems to have a strong interaction they must have large internal charge fluctuations. These fluctuations are rare because they raise the internal energy by a large amount; (b) If large charge fluctuations occur then one can place a suitable charge density σ on a surface S separating the two systems in such a way that the energy of σ will be much more negative than the purported interaction between the two systems. A suitable σ is, in fact, the charge density that screens out the charge fluctuation in one of the two systems.

While it seems to be difficult to construct a proof of the thermodynamic limit using (a), the use of (b) is attractive. One can easily argue rigorously that if large charge fluctuations occur then one can introduce n extra particles into the system and localize them to give a charge density approximating σ. This will lower the free energy by an amount much more than n, thereby giving a contradiction provided the following can be solved.

Problem 8 (The insertion energy of a particle): (α) Consider the Hamiltonian described at the beginning of this section for a neutral collection of N particles in a box of volume V proportional to N. Show that the free energy, or ground state energy, changes by an amount of order unity if a neutral pair of positive and negative particles is introduced into the system. In other words, show that the chemical potential is of order unity. (β) Next, consider the "solid state Hamiltonian" which is given by (2.1) with all $z_j = z$ and the $\{R_j\}$ arranged periodically. Show that the chemical potential is order unity for this system also.

If Problem 8β can be solved then, by the preceding argument, one will be able to prove the thermodynamic limit for the solid state Hamiltonian - something which has not been done so far.

It is absurd that Problem 8 does not have a trivial solution. It is easy to see, using H-stability, that adding O(N) particles changes the energy by O(N). What is needed is the obvious, but unproved fact that adding two particles changes the energy by O(1).

Problem 8 is essentially equivalent to

Problem 9 (Boundedness of the potential): For a neutral system in equilibrium, or in the ground state, show that the average Coulomb potential $\langle\phi(x)\rangle$ is bounded, uniformly in x, by a function of the density and temperature only.

Of course Problem 9 hints at a much more difficult problem, namely to prove the existence of the correlation functions in the thermodynamic limit for all densities and temperatures. In particular screening (i.e. exponential fall off) is expected to occur in the two point function - at least at high temperature. In a remarkable paper [18] Brydges has shown this to be true in certain Coulombic, but classical, lattice models at high temperature.

References

[1] Aventini, P.; Combes,J.M.; Duclos,P.; Grossman,A.; Seiler,R.: On the method of Born and Oppenheimer (to be published), cf. J.M. Combes in The Schroedinger Equation, W. Thirring and P. Urban eds. Springer, Berlin, p. 139 (1976).

[2] Lieb,E.H.; Thirring,W.E.: Phys.Rev. Lett. 35, 687 (1975); errata ibid 35,1116 (1975). See also Studies in Mathematical Physics, E.H. Lieb, B. Simon and A.S. Wightman eds., Princeton Press, p. 269 (1976).

[3] (a) Lieb, E.H.: Rev. Mod. Phys. 48, 553 (1976).
(b) Lieb, E.H.: Bull. Amer. Math. Soc. 82, 751 (1976).
Lieb, E.H.:
The number of bound states of one-body Schroedinger operators. Amer. Math. Soc. Symposium on the Geometry of the Laplace operator, (1979), to appear.

[4] Dirac, P.A.M.: Proc.Camb. Phil. Soc. 26, 376 (1930).

[5] Lieb, E.H.: Phys. Lett. 70A, 444 (1979).

[6] March,N.H.; Young,W.H.: Proc. Phys. Soc. 72, 182 (1958).

[7] Benguria, R.: The von-Weizsäcker and Exchange Corrections in Thomas-Fermi Theory, Ph.D. Thesis, Princeton University, 1979.

[8] Lieb, E.H.; Simon,B.: Adv. in Math. 23, 22 (1977).

[9] Benguria,R.; Lieb,E.H.: Ann. of Phys. (N.Y.) 110, 34 (1978); Commun. Math. Phys. 63, 193 (1978).

[10] Brezis,H.; Lieb,E.H.: Commun. Math. Phys. 65, 231 (1979).

[11] Lenard, A.: Springer Lecture Notes in Physics 20, 114 (1973).

[12] Dyson,F.J.; Lenard,A.: J. Math. Phys. 8, 423 (1967).
Lenard,A.; Dyson,F.J.: ibid. 9, 698 (1968).

[13] Scott, J.M.C.: Phil. Mag. 43, 859 (1952).

[14] Lieb, E.H.: Phys. Lett. 70A, 71 (1979).

[15] Dyson,F.J.: J. Math. Phys. 8, 1538 (1967).

[16] Girardeau,M.; Arnowitt,R.: Phys. Rev. 113, 755 (1959); Foldy, L.L.: ibid. 124, 649 (1961).

[17] Lieb,E.H.; Lebowitz,J.L.: Adv. in Math. 9, 316 (1972).

[18] Brydges, D.C.: Commun. Math. Phys. 58, 313 (1978).

Time Dependent Phenomena in Statistical Mechanics

*Oscar E. Lanford III**

Department of Mathematics
University of California
Berkeley, California 94720

0. Introduction

This lecture will unfortunately not be a systematic review of the subject of rigorous results in non-equilibrium statistical mechanics. A preliminary attempt to outline such a review led me quickly to the conclusion that the field is too diverse to be summarized in a single lecture. I have therefore decided instead to discuss a few related works in depth. The works I have chosen are:

1. The paper of J. Fritz and R. L. Dobrushin[5] on two-dimensional dynamics.
2. The paper of W. Braun and K. Hepp[3] on classical mechanics in the Vlasov limit.
3. A recent preprint by H. van Beijeren, J. L. Lebowitz, H. Spohn, and myself[2] on auto-correlations and fluctuations in the dilute equilibrium hard-sphere gas.

1. Fritz and Dobrushin on Two-Dimensional Dynamics

The problem is to investigate solutions to the Newtonian equations of motion:

$$\frac{d^2\vec{q}_i}{dt^2} = \sum_{i \neq j} \vec{F}(\vec{q}_i - \vec{q}_j); \quad \vec{F} = -\nabla\phi$$

where:

a. The force function $\vec{F}(\vec{q}_i - \vec{q}_j)$ is small for large $|\vec{q}_i - \vec{q}_j|$

but

b. There are infinitely many particles.

In contrast to the existence problem for finite-particle dynamics, this is a singular and delicate problem, in essence because it is difficult to bound the force exerted on any particle and hence difficult to rule out the development of singularities in finite time.

The approach generally taken is to restrict the set of initial phase points. On the other hand, it is undesirable to restrict them too drastically, since one wants to keep enough to form a set of probability one for at least some interesting statistical states of the infinite system. There are two broad approaches to the problem of restricting the initial phase point; I will refer to these two approaches as *mechanical* and *probabilistic*. The mechanical approach is the one which comes to mind first; one seeks explicit verifiable bounds on the initial phase point which are propagated by the equation of motion in such a way as to guarantee that solutions do not "blow up". The work of Fritz and Dobrushin is the most powerful result available in this direction; previously an existence theorem of this sort valid for a large set of initial phase points had been proved only for one-dimensional dynamics.

The probabilistic approach is based on a more subtle idea, which can be traced back to a

*Preparation of this report was supported in part by the National Science Foundation under grant MCS78-06718. It is a pleasure for me to acknowledge also the hospitality of the Seminar für theoretische Physik of the ETH, Zürich, where the first draft was written under ideal conditions.

paper of Sinai[8] on dynamics in one dimension. In this approach, it is assumed that the interparticle potential ϕ is thermodynamically stable, and the idea is to exploit two facts:

a. Any Gibbs state μ_0 for ϕ assigns probability one to a set of initial phase points with very good regularity properties.

b. If the solution flow to the equations of motion exists, it ought to leave μ_0 invariant, and hence almost all solution curves should have good regularity properties not only at time zero but at all other times as well.

To see how these two ideas might be used, imagine that a local existence theorem--up to time δt--can be proved for μ_0-almost all initial phase points and that the solution mapping $T^{\delta t}$ leaves μ_0 invariant. (This is intended only to illustrate the reasoning in a simple case; I know of no system in which the argument I am about to give is actually useful.) The idea is as follows: We are assuming that

$$\{\omega : T^{\delta t}\omega \text{ is defined}\}$$

is a set of μ_0-probability one. But since $T^{\delta t}$ preserves μ_0,

$$\{\omega : T^{\delta t}\omega \text{ is defined and in the domain of } T^{\delta t}\}$$

is also of probability one, *i.e.*,

$$\{\omega : T^{2\delta t}\omega \text{ is defined}\}$$

is of probability one. Iterating this argument,

$$\{\omega : T^{n\delta t}\omega \text{ is defined for all n}\}$$

is of probability one, *i.e.*, a global existence theorem holds for μ_0-almost all initial phase points.

A more circumspect version of this argument avoids even the necessity of having a local existence theorem to start with, and shows that, for a wide class of potentials ϕ, the set of initial phase points for which global solutions exist has probability one with respect to every Gibbs state for ϕ [1],[6],[7],[9]. These results are considerably less powerful than they appear at first glance. Their defect, in comparison to the results of the mechanical approach, is that they do not permit a determination, in any practical way, of whether a given initial phase point admits a solution. This is clear in the illustrative argument given above: Even if the local existence theorem is constructive in the sense that it explicitly identifies the phase points for which $T^{\delta t}$ is defined, it is nevertheless necessary actually to know $T^{(n-1)\delta t}\omega$ in order to determine whether $T^{n\delta t}\omega$ exists. Similarly, such theorems tell us nothing about the existence of solutions with initial conditions describing typical non-equilibrium situations, *e.g.*, which have one well-defined density for large positive q_1 and a different density for large negative q_1.

With these remarks as background, we turn to the work of Fritz and Dobrushin, which establishes the existence of solutions to the equations of motion in two dimensions for a set of initial phase points which is large and which can be described explicitly. We will see, however, that these results do not extend in any obvious way to more than two dimensions.

Let $\omega = (\vec{q}_i, \vec{v}_i)$ be a phase point, let $\vec{\mu} \in \mathbf{R}^2$, and let $\sigma > 0$. Denote by $H(\omega, \vec{\mu}, \sigma)$ the total energy--kinetic plus potential--of those particles in ω which lie inside the disk with center $\vec{\mu}$ and radius σ. Also, let $g(u)$ denote $1+\log(1+u)$, $u \geqslant 0$, and define

$$\overline{H}(\omega) = \sup\left\{\frac{H(\omega, \vec{\mu}, \sigma)}{\sigma^2} : \vec{\mu} \in \mathbf{R}^2,\ \sigma \geqslant g(|\vec{\mu}|)\right\}$$

The condition

$$\overline{H}(\omega) < \infty$$

says in a precise way that local energy fluctuations in the phase point ω grow no faster than logarithmically with distance from the origin.

What Fritz and Dobrushin prove is that, under conditions on the potential to be specified later:

1. If $\overline{H}(\omega) < \infty$, there exists a solution ω_t to the Newtonian equations of motion, defined for all t, with $\omega_0 = \omega$.
2. For this solution, $\overline{H}(\omega_t)$ as a function of t is bounded on bounded intervals. There is no other solution with the specified initial value which has this property.
3. The solution ω_t may be obtained as follows: Delete from ω all particles outside a disk, centered at the origin, of radius M. Evolve the resulting finite system to time t, and let M go to infinity.

The heart of the argument, as might be expected, is an *a priori* estimate:

4. There is a function $\rho(T, h)$ such that, if ω_t is a solution of the Newtonian equations with $\overline{H}(\omega_0) \leqslant h$ and with $\overline{H}(\omega_t)$ bounded for $t \in [-T, T]$, then
$$\overline{H}(\omega_t) \leqslant \rho(T,h) \text{ for } t \in [-T,T].$$

Once the *a priori* estimate has been proved for phase points with a finite number of particles, it as a simple matter to construct, by a compactness argument, at least one solution to the Newtonian equations with a given initial phase point ω as a limit of finite particle solutions, provided only that $\overline{H}(\omega)$ is finite. The constructed solution will have $\overline{H}(\omega_t)$ bounded on bounded intervals. Uniqueness is proved by a contraction argument which, while distinctly non-trivial, contains no real surprises. Finally 3. follows from uniqueness by the general principle that a sequence in a compact space with only one cluster point must converge. The main thing, then, is to prove 4.

The following conditions on the potential are sufficient for the argument to work; the minimal hypotheses are actually somewhat weaker (but more complicated).

1. ϕ is of finite range, even, and twice continuously differentiable away from $\vec{0}$.
2. For some strictly positive β, c,
$$\phi(\vec{\xi}) \geqslant \frac{c}{|\vec{\xi}|^{\beta}} \quad \text{for sufficiently small } |\vec{\xi}|$$
Either ϕ is non-negative or β is at least 4.
3. $$|\vec{\xi}|\,|\nabla\phi(\vec{\xi})| \leqslant \text{const}\,|\phi(\vec{\xi})| \quad \text{for small } \vec{\xi}.$$
4. For some γ,
$$\left|\frac{\partial^2\phi}{\partial\xi_i\partial\xi_j}\right| \leqslant |\vec{\xi}|^{-\gamma} \text{ for small } \xi.$$

(This last condition is used only in the proof of uniqueness.)

Note that hard cores are excluded. We will give a sketch of the proof *assuming ϕ to be non-negative.* The general case is significantly more involved, but the ideas remain the same. The main steps are as follows:

1. Replace the local energy $H(\omega, \vec{\mu}, \sigma)$ by an equivalent smoothly cut-off quantity $W(\omega, \vec{\mu}, \sigma)$. Equivalent will mean:
$$H(\omega, \vec{\mu}, \sigma) \leqslant W(\omega, \vec{\mu}, \sigma) \leqslant \text{const}\,(H(\omega, \vec{\mu}, \sigma) + 1).$$

All estimates will then be made in terms of W rather than H. Thus, we also introduce
$$\overline{W}(\omega) = \sup\left\{\frac{W(\omega, \vec{\mu}, \sigma)}{\sigma^2} : \vec{\mu} \in \mathbf{R}^2, \sigma \geqslant g(|\vec{\mu}|)\right\}$$

Construction of W is relatively straightforward. Choose (with some care) a cut-off function $f(u)$, a smooth positive function on **R** satisfying, among other conditions:

$$f(u) = 0 \text{ if } u \leqslant -3R; \; f(u) = 1 \text{ if } u \geqslant 0,$$

where R denotes the range of the potential.

For each i, define

$$W_i(\omega) = 1 + \vec{v}_i^{\,2} + \sum_{j \neq i} \phi(\vec{q}_i - \vec{q}_j),$$

and put

$$W(\omega, \vec{\mu}, \sigma) = \sum_i f(\sigma - |\vec{q}_i - \vec{\mu}|) \, W_i(\omega).$$

Thus,

$$W(\omega, \vec{\mu}, \sigma) = 2H(\omega, \vec{\mu}, \sigma) + N(\omega, \vec{\mu}, \sigma) + \text{boundary terms}.$$

2. Prove a "partial differential inequality" for $W(\omega_t, \vec{\mu}, \sigma)$, where ω_t denotes a solution to the Newtonian equations:

$$\frac{\partial}{\partial t} W(\omega_t, \vec{\mu}, \sigma) \leqslant K \, g(|\vec{\mu}| + \sigma) \sqrt{W(\omega_t)} \, \frac{\partial}{\partial \sigma} W(\omega_t, \vec{\mu}, \sigma). \tag{1.1}$$

Here, K is to be independent of ω_t, $\vec{\mu}$, and σ.

This inequality seems to me to be the real heart of the proof. It is clever in at least two respects:

a. It exploits effectively the fact that--as is well known-- the energy in a region changes only through boundary effects.

b. It shows how to control those boundary effects in a very clean way; they can be dominated by the effect of changing the size of the region. Thus, if we shrink the region fast enough as ω_t evolves, the energy in the time-dependent region will not increase.

To prove the partial differential inequality (1.1) start by differentiating $W(\omega_t, \vec{\mu}, \sigma)$ with respect to t and using the equations of motion. The resulting expression contains some relatively uninteresting terms coming from differentiating the cut-off functions; these we will not discuss. The remaining terms can be organized as

$$\frac{1}{2} \sum_{i \neq j} (f_i - f_j)(\nabla \phi)(\vec{q}_i - \vec{q}_j) \cdot (\vec{v}_i + \vec{v}_j). \tag{1.2}$$

Here f_i denotes the value of the cut-off function $f(\sigma - |\vec{q} - \vec{\mu}|)$ at $\vec{q} = \vec{q}_i$. Observe that a pair (i, j) contributes nothing to the sum unless $|\vec{q}_i - \vec{q}_j| \leqslant R$ and unless at least one of $\vec{q}_i, \vec{q}_j$ is in the annulus $\{\vec{q} \colon \sigma \leqslant |\vec{q} - \vec{\mu}| \leqslant \sigma + 3R\}$.

Now group the non-zero terms in the sum into those for which $|\vec{q}_i - \vec{q}_j|$ is small enough so that the inequality:

$$|\xi| \, |\nabla \phi(\xi)| \leqslant \text{const} \, \phi(\xi)$$

(valid for small $|\xi|$) can be used to estimate the gradient, and those for which $|\vec{q}_i - \vec{q}_j|$ is not so small. We will deal explicitly only with the former class of terms. (The latter are estimated using the inequality

$$|\nabla \phi| \leqslant \text{const} \sqrt{\phi},$$

valid for any smooth non-negative function of compact support, and then applying the Schwarz inequality to the sum.) One of the respects in which care was exercised in the choicc of the cutoff function f was to guarantee that

$$|f_i - f_j| \leqslant |\vec{q}_i - \vec{q}_j| \, (f_i' + f_j').$$

We also estimate, somewhat crudely,

$$|\vec{v}_i|,\ |\vec{v}_j| \leqslant \text{const}\, g(|\vec{\mu}|+\sigma)\sqrt{\overline{W}(\omega_t)}\,, \tag{1.3}$$

using the fact that the velocity of a particle is bounded by the square root of the energy of any cluster which contains it and that a particle at $\vec{q}$ is contained in a cluster with energy no larger than $g(|\vec{q}|)^2\,\overline{W}(\omega_t)$.

Inserting all these estimates and reorganizing, we find that we can bound the sum of the terms in (1.2) with $|\vec{q}_i-\vec{q}_j|$ small by

$$\begin{aligned}
&\text{const}\, g(|\vec{\mu}|+\sigma)\sqrt{\overline{W}(\omega_t)}\,\sum_i f_i'\sum_{j\neq i}\phi(\vec{q}_i-\vec{q}_j)\\
&\leqslant \text{const}\, g(|\vec{\mu}|+\sigma)\sqrt{\overline{W}(\omega_t)}\,\sum_i f_i'\,W_i(\omega)\\
&=\text{const}\, g(|\vec{\mu}|+\sigma)\sqrt{\overline{W}(\omega_t)}\,\frac{\partial}{\partial\sigma}W(\omega_t,\vec{\mu},\sigma)
\end{aligned}$$

We observe for later use that the only change in these estimates if they are carried out in three rather than two dimensions is that, in the velocity estimate and hence in the final result, $g(|\vec{\mu}|+\sigma)$ is replaced by $[g(|\vec{\mu}|+\sigma)]^{3/2}$.

3. We exploit the fundamental partial differential inequality (1.1) as follows: Let ω_t be a solution of the equations of motion. Choose $\vec{\mu}\in\mathbf{R}^2$, $\sigma\geqslant g(|\vec{\mu}|)$, and a positive time T. We want to estimate

$$W(\omega_T,\vec{\mu},\sigma).$$

To do this, we note that the partial differential inequality implies that if $r(t)$ is any (positive) solution of

$$\frac{dr}{dt}=-K\sqrt{\overline{W}(\omega_t)}\,g(|\vec{\mu}|+r(t))$$

then

$$\frac{d}{dt}W(\omega_t,\vec{\mu},r(t))\leqslant 0.$$

Taking $r(t)$ to be the solution satisfying

$$r(T)=\sigma,$$

we get

$$\frac{W(\omega_T,\vec{\mu},\sigma)}{\sigma^2}\leqslant\frac{W(\omega_0,\vec{\mu},r(0))}{\sigma^2}\leqslant\overline{W}(\omega_0)\left(\frac{r(0)}{\sigma}\right)^2.$$

Taking the supremum over $\vec{\mu}$, σ with $\sigma\geqslant g(|\vec{\mu}|)$:

$$\overline{W}(\omega_T)\leqslant\overline{W}(\omega_0)\sup\left\{\left(\frac{r(0)}{\sigma}\right)^2:\sigma\geqslant g(|\vec{\mu}|)\right\}.$$

4. The problem is now to estimate $r(0)/\sigma$. We have, integrating the differential equation for r,

$$\begin{aligned}
r(0)&=\sigma+K\int_0^T g(|\vec{\mu}|+r(t))\sqrt{\overline{W}(\omega_t)}\,dt\\
&\leqslant\sigma+K\,g(|\vec{\mu}|+r(0))\int_0^T\sqrt{\overline{W}(\omega_t)}\,dt,
\end{aligned}$$

or, writing $z(T)$ for $\int_0^T\sqrt{\overline{W}(\omega_t)}\,dt$,

$$r(0) \leqslant \sigma + K\, g(|\vec{\mu}| + r(0))\, z$$

An elementary (but tricky) analysis of this inequality leads to the explicit bound

$$\frac{r(0)}{\sigma} \leqslant \bar{K}\,(1 + z\, g(z^2))$$

(with $\bar{K}$ another constant) and thus to the inequality:

$$\sqrt{\overline{W}(\omega_T)} \leqslant \sqrt{\overline{W}(\omega_0)}\, \bar{K}\,(1 + z\, g(z^2))$$

This is true for all positive T, so, writing t instead of T and using the fact that

$$\frac{dz}{dt} = \sqrt{\overline{W}(\omega_t)},$$

we get the ordinary differential inequality

$$\frac{dz}{dt} \leqslant \bar{K} \sqrt{\overline{W}(\omega_0)}\,(1 + z\, g(z^2)), \quad z(0) = 0. \tag{1.4}$$

5. We are now essentially done. Solving the differential equation obtained by replacing the inequality (1.4) by an equality gives an *a priori* bound for

$$\frac{dz}{dt} = \sqrt{\overline{W}(\omega_t)}.$$

The only thing which needs to be checked is that the solution to this differential equation does not go to infinity in finite time; this follows from the fact that $\int \frac{du}{1 + u\, g(u^2)}$ diverges at infinity.

What goes wrong in three dimensions is that the methods no longer yield a bound for $r(0)/\sigma$ uniform in $\sigma \geqslant g(|\vec{\mu}|)$. To see this, recall that the differential equation for $r(t)$ becomes, in three dimensions,

$$\frac{dr}{dt} = -K \sqrt{\overline{W}(\omega_t)}\,[g(|\vec{\mu}| + r(t))]^{3/2}$$

Dropping the $r(t)$ inside g gives the *lower* bound

$$r(0) \geqslant \sigma + K\, g(|\vec{\mu}|)^{3/2}\, z(T),$$

or, for $\sigma = g(|\vec{\mu}|)$,

$$\frac{r(0)}{\sigma} \geqslant 1 + K\, g(|\vec{\mu}|)^{1/2}\, z(T).$$

Because the right-hand side becomes arbitrarily large as $|\vec{\mu}|$ goes to infinity, we don't get a uniform bound for $r(0)/\sigma$, and the whole technique of proof fails.

2. Braun and Hepp on the Vlasov Dynamics

We look next at a different kind of infinite particle dynamics, the *Vlasov* or *weak-coupling* or *effective field* limit. In the previous section, we let the system become infinite by letting the volume and the particle number go to infinity at least approximately proportionally, keeping the interparticle potential function $\phi(\vec{\xi})$ fixed. This means that we are looking at a large system on a length scale of the order of the typical distance between particles and on a time scale of the order of the length of time it takes a particle to travel that distance. The limit we now want to look at is one in which the force exerted by any single particle on any other is extremely gentle--no matter how close together the particles are--but in which the forces have a range which is very long compared with the typical interparticle distance, so each particle feels the effects of very many others, *i.e.,* moves in an effective force field which doesn't depend very much on what any individual particle is doing. We should further think of the particles as spread out over the one-particle phase space approximately according to some density. The net force a particle at point $\vec{q}$ feels is therefore proportional to the convolution of this distribution density with the pair force $\vec{F} = -\nabla\phi$ so it changes only over a distance comparable to the range of the potential. It would therefore be quite uninteresting to look at the behavior of the particle over the length of time it requires to travel a typical interparticle distance--it would be subject to a force which is in the limit constant--so we want to look instead on a length scale of the order of the range of the potential. Finally, we should arrange that the overall strength of the interparticle force is such that the length of time it takes a particle's velocity to change significantly is of the same order as the length of time it takes the particle to travel a distance of the order of the range of the potential. We will now proceed to formalize this limit; it will turn out--not surprisingly--that the limiting dynamics is much less singular than either the infinite-particle dynamics discussed in the previous section or the dynamics in the Boltzmann-Grad limit to be discussed in the next.

We consider, then, a system of N particles which we can think of either as in a fixed (*i.e.*, N-independent) container or in free space. The space dimension doesn't matter very much; we take it to be three. We also fix an interparticle potential ϕ which for simplicity we take to be C^∞ and of finite range. The methods to be described work for more general potentials, but it is essential that the gradient of the potential be bounded. We take the force exerted by a particle at $\vec{q}_j$ on a particle at $\vec{q}_i$ to be

$$-\frac{1}{N}\nabla\phi(\vec{q}_i - \vec{q}_j) \equiv \frac{1}{N}\vec{F}(\vec{q}_i - \vec{q}_j)\,,$$

and the particle mass to be unity. (Equivalently, we could take the force to be $\vec{F}$ and the particle mass to be $1/N$, and observe the behavior on a time scale of order $1/N$. The initial velocities should then be of order one in the new time scale, *i.e.*, of order N in the original time scale.)

As long as the system is finite, there is no problem with the dynamics. We will write the solution curves as

$$\omega_t = (\vec{q}_i(t), \vec{v}_i(t))_{1 \leqslant i \leqslant N}$$

To pass to the limit $N \rightarrow \infty$, we re-express the dynamics in a form which is as nearly N-independent as possible. To each phase point ω we associate the discrete probability measure

$$\mu^\omega = \frac{1}{N}\sum_{i=1}^{N} \delta_{\vec{q}_i,\, \vec{v}_i}$$

on the one-particle phase space $\mathbf{R}^3 \times \mathbf{R}^3$. We will write μ_t instead of the more awkward μ^{ω_t}. The N-particle dynamics thus defines a flow $\mathbf{T}_N^t$ on a very sparse subspace of the set M_+^1 of probability measures on $\mathbf{R}^3 \times \mathbf{R}^3$. Having imbedded all the N-particle phase spaces in the single space M_+^1, it is natural to ask the following: Do two initial phase points, perhaps with different numbers of particles but which look similar, *i.e.*, are near to each other in the weak-* topology for measures, have similar time evolutions? The first major result of Braun and Hepp is that they do:

Theorem 2.1. *There is a flow* $\mathbf{T}^t$ *defined on all of* M_+^1 *which reduces on each N-particle subspace to* $\mathbf{T}_N^t$. $\mathbf{T}^t\mu$ *is weak-* continuous in* μ.

We may thus think of $\mathbf{T}^t\mu$ for a continuous measure μ as describing the dynamics of a phase point for a system of infinitely many particles initially distributed over the one-particle phase space according to the measure μ.

Remarks. 1. The fact that this theorem concerns probability measures does not mean that there is any statistical element in it; it is a theorem of classical mechanics, not classical statistical mechanics. Probability measures are simply a convenient representation of the points of an (idealized) infinite system phase space.

2. This theorem describes a situation which is radically different from-- and simpler than--the Boltzmann-Grad limiting regime to be described in the next section. In the latter context,

modulo technicalities, the picture is that the Boltzmann equation generates a semiflow-- *i.e.*, a flow defined only for positive times--on the space of continuous probability measures and *most* discrete measures weak-* approximating a given continuous measure have evolutions which, at least for a short time, approximate the evolution given by the semi-flow. There are, however, necessarily exceptions; a simple way to construct them is to take a "typical" phase point representing approximately some non-equilibrium continuous distribution; let it evolve for a short while; and then reverse all the velocities. Theorem 2.1 says that there are no such exceptions in the Vlasov limit. Their absence means in particular that the limiting dynamics is reversible and cannot describe approach to equilibrium.

To prove Theorem 2.1 we rework the equations of motion for N particles into a form in which it is apparent how to generalize them for initial measures which are not normalized sums of N unit point masses. Let μ_t be a solution curve for the N-particle dynamics. From it, we construct a time- and position-dependent force field on the one-particle position space:

$$\vec{f}(\vec{q}, t) = \int \vec{F}(\vec{q} - \vec{q}_1)\, \mu_t(d\vec{q}_1\, d\vec{v}_1).$$

We then consider the motion of a single particle in this force field, and write $Z^t x = (\vec{Q}^t x, \vec{V}^t x)$ for the solution curve starting at $x = (\vec{q}, \vec{v})$ at time zero. Because the force field is time-dependent, the mappings Z^t do not in general form a group; we could (but won't) consider the two-parameter family of solution mappings $Z(t_1, t_2)$ evolving from time t_2 to t_1 and would then get the multiplication law

$$Z(t_1, t_2) \cdot Z(t_2, t_3) = Z(t_1, t_3).$$

Each Z^t is, nevertheless, a volume-preserving mapping of $\mathbf{R}^3 \times \mathbf{R}^3$ into itself.

The mappings Z^t contain in principle more information than the solution curve $\mu_t = (\vec{q}_i(t), \vec{v}_i(t))$ itself, and the solution curve is reconstructible from Z^t and the initial condition by

$$(\vec{q}_i(t), \vec{v}_i(t)) = Z^t(\vec{q}_i(0), \vec{v}_i(0)).$$

(This uses the fact that $\vec{F}(\vec{0}) = \vec{0}$.) A moment's reflection shows that this last equation can also be written

$$\mu_t = Z^t \mu_0,$$

where the image of a measure μ under a mapping Z means

$$\int g(x)\, (Z\mu)(dx) \equiv \int g(Zx)\, \mu(dx).$$

Thus, we can write the equations defining Z^t directly in terms of μ_0 rather than in terms of μ_t; they are, simply,

$$\frac{d\vec{Q}^t x}{dt} = \vec{V}^t; \quad \frac{d\vec{V}^t x}{dt} = \int \mu_0(dy)\, \vec{F}(\vec{Q}^t x - \vec{Q}^t y).$$

(Reminder about what the notation means: x,y are points of $\mathbf{R}^6$, the one-particle phase space. Z^t is a mapping of $\mathbf{R}^6$ into itself; its value at a point x is a pair of three-vectors which we write $\vec{Q}^t x$, $\vec{V}^t x$.)

Once the equations are written in this form, it is easy to see how to generalize to initial measures μ_0 which are not discrete; simply write and solve the same equations for more general μ_0. Of course, existence and uniqueness have to be proved, but this turns out not to be difficult. It is not necessary that μ_0 be normalized or even positive, but it should have finite total variation. We get in this way, for each μ_0, a family of mappings $Z^t(\mu_0)$ which are weak-* continuous in μ_0 in the sense that the mapping

$$(x, \mu_0) \rightarrow Z^t(\mu_0)x$$

is weak-* continuous into $\mathbf{R}^6$. The time evolution flow acting on measures is then

$$\mu_t = \mathbf{T}^t \mu_0 = Z^t(\mu_0)\, \mu_0 .$$

(Note the double appearance of μ_0.) This time, the group law *does* hold, by virtue of the multiplication law

$$Z^{t_2}(\mu_{t_1})\, Z^{t_1}(\mu_0) = Z^{t_2+t_1}(\mu_0),$$

which follows from the uniqueness of the solution to the equation for $Z^t(\mu_0)$.

Although the flow $\mathbf{T}^t$ on the space of measures is the intuitively natural form in which to view the dynamics (by analogy to the solution flows on the finite-dimensional phase space in the dynamics of finite systems), the $Z^t(\mu)$ have great technical advantages as objects with which to work. For example, Braun and Hepp show that

$$\mu \to Z^t(\mu)$$

is differentiable in the sense that

$$\lim_{h\to 0} \frac{Z^t(\mu + h\nu) - Z^t(\mu)}{h}$$

exists (in the weak-* topology) for all signed measures ν. Heuristically, this ought to mean that

$$\mu \to \mathbf{T}^t \mu$$

is differentiable, but as a technical matter mappings into spaces of measures are almost never differentiable, and

$$\mu \to \mathbf{T}^t \mu$$

is in fact not even continuous (in the norm topology).

If μ_0 is of the form $f_0(\vec{q}, \vec{v})\, d\vec{q} d\vec{v})$, with f_0 smooth, then, since the $Z^t(\mu_0)$ are volume-preserving, μ_t is given by the density

$$f_t = f_0 \circ Z^t(\mu_0)^{-1}.$$

This density is again smooth, and one can write a differential equation for its time development. The differential equation is the *Vlasov equation:*

$$\frac{\partial f}{\partial t} = -\vec{v} \cdot \frac{\partial f}{\partial \vec{q}} - \frac{\partial f}{\partial \vec{v}} \cdot \int \vec{F}(\vec{q} - \vec{q}\,')\, f(\vec{q}\,', \vec{v}\,')\, d\vec{q}\,'\, d\vec{v}\,'.$$

We have been concerned, so far, with the classical mechanics of "pure states". The measures appearing in the theory do not have a probabilistic interpretation; they simply represent phase points. We now introduce some probabilistic considerations. To minimize the confusion between measures with a probabilistic interpretation and measures as phase points, we will generally denote the former by p or P.

Let p be a probability measure on the one-particle phase space $\mathbf{R}^6$; p_N the product of N copies of p as a measure on $\mathbf{R}^{6N}$; and P_N as the measure on M_+^1 which is the image of p_N under our usual representation of N-particle phase points as normalized discrete measures. Here is an alternate description of P_N: Let X denote the mapping

$$x \to \delta_x \text{ from } \mathbf{R}^6 \text{ into } M,$$

which we view as a vector-valued random variable on the probability space $(\mathbf{R}^6, p)$. Then P_N is just the joint distribution of

$$\frac{1}{N}(X_1 + \cdots + X_N)$$

where $X_1, \ldots, X_N$ are independent copies of X. This description of P_N suggests that the Law of Large Numbers and the Central Limit Theorem can be applied in this situation. We are now going to argue that this is indeed the case, and that all we need are the classical versions of these theorems, not refined versions valid for Banach-space valued random variables.

Let g be a function on $\mathbf{R}^6$. Define a function $n(g)$ on $\mathbf{R}^{6N}$ by

$$n(g)(\omega) = \frac{1}{N}\sum_{i=1}^{N} g(x_i) \quad \text{if } \omega = (x_1, \ldots, x_N).$$

If g is the characteristic function of a set, then $n(g)$ measures the fraction of particles which lie in that set. For this reason, we refer to $n(g)$ as an *occupation number observable.* Technically, we will usually want to consider only $n(g)$'s with g a smooth function of compact support.

If the phase point ω corresponds to the measure μ^ω, then

$$n(g)(\omega) = \int g(x)\,\mu^\omega(dx).$$

We use this formula to define $n(g)$ on all of M:

$$n(g)(\mu) = \int g\, d\mu.$$

It is easy to check that the distribution of $n(g)$ with respect to P_N is the same as the distribution of

$$\frac{1}{N}(G_1 + \cdots + G_N)$$

where $G_1, \ldots, G_N$ are independent copies of g regarded as a random variable on the probability space $(\mathbf{R}^6, p)$. Thus, by the Weak Law of Large Numbers, the random variable $n(g)$ on the probability space (M, P_N) converges in probability to the constant $\int g\, dp = n(g)(p)$ as $N \to \infty$. Re-expressed in analytic language, this means that, given any weak-* neighborhood V of p in M, the probability, with respect of P_N, of finding the phase point outside of V goes to zero as $N \to \infty$. Heuristically,

$$P_N \to \delta_p.$$

Now apply the dynamics. Since $\mathbf{T}^t$ is weak-* continuous, we have in the same sense

$$P_N(t) \equiv \mathbf{T}^t P_N \to \delta_{p(t)}\,;\; p(t) = \mathbf{T}^t p.$$

This result can be expressed as "propagation of chaos". If we translate the measure $P_N(t)$ back onto $\mathbf{R}^{6N}$, the resulting measure $p_N(t)$ will generally not be a product measure. What the above argument shows, however, is that if $p_N(t)$ is integrated over all but its first j arguments and N is allowed to go to infinity, the resulting sequence of measures converges to a product of j copies of $\mathbf{T}^t p$.

Braun and Hepp also analyze the more delicate question of fluctuations in the Vlasov limit. Given a function g as above, and given N, define a *fluctuation observable*

$$\Phi_N(g) = \sqrt{N}\,[n(g) - \overline{n(g)}],$$

where the overbar denotes mean value with respect to P_N. From what has already been said about the distribution of $n(g)$, it follows at once from the Central Limit Theorem that $\Phi_N(g)$ converges in distribution (as $N \to \infty$) to a centered Gaussian with variance

$$\int g^2\, dp - \left[\int g\, dp\right]^2.$$

More generally, any finite number of fluctutation observables $\Phi_N(g_1), \ldots, \Phi_N(g_n)$ are asymptotically jointly centered Gaussian with covariance

$$\lim_{N\to\infty} \overline{\Phi_N(g_i)\,\Phi_N(g_j)} = (P_0 g_i | g_j),$$

where $(\ |\ )$ denotes the inner product in $L^2(\mathbf{R}^6, p)$ and P_0 is the projection onto the orthogonal complement of the constant functions in that space. (The reason why P_0 appears is that the *total* particle number does not fluctuate.)

All this is elementary and concerns only time zero. Given the time evolution, however, one can ask about fluctuation observables at other times. Since even the fluctuation observables at time zero converge only in distribution, not in any pointwise sense, we ask only about convergence in distribution. Thus, choose $t_1, \ldots, t_n, g_1, \ldots, g_n$ and ask what happens to the joint distribution of the $\Phi_N(g_i)_{t_i}$ as $N \to \infty$. Braun and Hepp show that, in the limit, the joint distribution is centered Gaussian, and they give a formula for the covariance. We can therefore think of the limiting fluctuations as defining a Gaussian stochastic process with a large and unpleasant instantaneous state space--some space of distributions, say, large enough to contain the typical sample fields for the time-zero fluctuations.

I want to make the point that this stochastic process is of a degenerate type--it is deterministic. I am using the word "deterministic" here in a very simple-minded sense; intuitively, what I mean is that knowing all fluctuations at time zero permits prediction, with certainty, of all fluctuations at other times. Actually, because of the linearity of the evolution of fluctuations, the precise statement is both simpler and more powerful than the intuitive statement: Given any time t and any test function g, there is a test function $\hat{g}$ such that

$$\Phi_N(g)_t - \Phi_N(\hat{g})_0$$

converges to zero in probabiliy, *i.e.*, a fluctuation observable at time t can be predicted with certainty from the value of *one* properly chosen fluctuation observable at time zero. We will see in the next section that this contrasts with what apparently happens in the Boltzmann-Grad limit.

The deterministic character of the fluctuation process is readily checked by examination of the formula for the covariance given by Braun and Hepp. I will not reproduce this formula but will instead describe an analogous but technically simpler situation which seems to me to show clearly what must be going on. Recall that P_N is the distribution of an average of N independent copies of a random variable X with values in M and that time evolution is given by a flow $\mathbf{T}^t$ on M such that

$$\mu \to \mathbf{T}^t\mu$$

is, at least heuristically, differentiable. The model system I want to discuss is one in which P_N is the distribution of the average of N independent copies of a random variable X taking values in a *finite-dimensional* vector space E equipped with a differentiable flow T^t. The analogues of the $n(g)$'s are just the *linear* mappings ψ from E to $\mathbf{R}$, *i.e.*, the elements of the dual space E^* of E. Put

$$\Phi_N(\psi) = \sqrt{N}\ (\psi - \bar{\psi}); \quad \bar{\psi} = E(\psi \circ X).$$

By the Central Limit Theorem, the $\Phi_N(\psi)$ are asymptotically jointly centered Gaussian with covariance

$$E(\psi_1(X)\,\psi_2(X)) - E(\psi_1(X)) \cdot E(\psi_2(X)).$$

Now look at a time t fluctuation observable

$$\Phi_N(\psi)_t \equiv \sqrt{N}\ [\psi \circ T^t - E_N(\psi \circ T^t)],$$

(where E_N denotes the expected value with respect to P_N). Let p denote the expected value of X, and write the general point of E as $x = p + \eta$. By the differentiability of T^t,

$$T^t x = T^t p + DT^t(p)\eta + O(|\eta|^2),$$

and so

$$\psi(T^t x) = \psi(T^t p) + DT^t(p)\,{}^*\psi(\eta) + O(|\eta|^2)$$
$$\Phi_N(\psi)_t(x) = DT^t(p)\,{}^*\psi(\sqrt{N}\,\eta) + O(\sqrt{N}\,|\eta|^2)$$
$$= \Phi_N(DT^t(p)\,{}^*\psi) + O(\sqrt{N}\,|\eta|^2).$$

By the Central Limit Theorem, $\sqrt{N}\,|\eta|^2$ converges to 0 in probability, *i.e.*,

$$\Phi_N(\psi)_t - \Phi_N(DT^t(p)\,{}^*\psi) \to 0 \text{ in probability.}$$

This means in particular that the $\Phi_N(\psi)_t$'s are all asymptotically jointly Gaussian, and we can thus regard them, as above, as defining a Gaussian stochastic process. This time, there are no technical difficulties with the state space; it is just E. If we denote the instantaneous state of this process by $F_t \in E$, so the F_t are all vector-valued random variables on some unspecified probability space, we get

$$F_t = DT^t(p)\, F_0 \text{ with probability one.}$$

I don't recommend the above argument as a pattern for constructing a theory of the time-dependence of fluctuations in the Vlasov limit. The technical difficulties in implementing it directly seem to be formidable. The fact is, nevertheless, that if it could be applied it would lead to exactly the formula for the covariance of the limiting fluctuation observables that Braun and Hepp obtain by a different argument. We can thus think of the fluctuations in the Vlasov limit as evolving linearly, deterministically, and (properly understood) reversibly. Their motion is given simply by linearizing the solution flow $\mathbf{T}^t$ about the particular solution under consideration.

3. Autocorrelations and fluctuations in the equilibrium hard-sphere system

In this section we consider a system of many elastic spheres of diameter ϵ, in a limiting regime in which $\epsilon \to 0$ and in which the number of particles per unit volume goes to infinity like ϵ^2. We refer to this regime as the *Boltzmann-Grad limit;* its essential feature is that the mean free path remains of order one as the limit is taken. The work to be described here exploits the techniques used in my study of the Boltzmann equation [6], and I will begin by sketching these techniques.

Let

$$\mu_t(x_1, \ldots, x_N) = \mu_0(T_{(\epsilon)}^{-t}(x_1, \ldots, x_N))$$

be a time-dependent probability density for a finite system of N hard spheres of diameter ϵ. Here, $T_{(\epsilon)}^t$ denotes the solution mapping for the hard-sphere dynamics, and each x_i denotes a pair $(\vec{q}_i, \vec{v}_i)$ giving the position and velocity of one of the particles. We construct the correlation functions for μ_t and scale the j-th correlation function by ϵ^{2j}; the resulting functions, which we will denote by $r_j(x_1, \ldots, x_j)$ and refer to as the *rescaled correlation functions,* have at least the possibility of remaining bounded and not approaching zero as ϵ approaches zero.

The Liouville equation for the time evolution of μ_t translates into the BBGKY hierarchy for the time evolution of the r_j. This hierarchy can be written schematically as

$$\frac{dr_j(t)}{dt} = H_j^{(\epsilon)}\, r_j(t) + C_{j,j+1}^{(\epsilon)} r_{j+1}(t),$$

and even more schematically as

$$\frac{d\mathbf{r}(t)}{dt} = H^{(\epsilon)}\,\mathbf{r}(t) + C^{(\epsilon)}\,\mathbf{r}(t).$$

Here $H_j^{(\epsilon)}$ denotes the Liouville operator for the system of j hard spheres of diameter ϵ. Writing $S_j^{(\epsilon)}(t)$ for the j-particle streaming operator (*i.e.*, the group generated by $H_j^{(\epsilon)}$), we can formally solve the hierarchy as

$$\mathbf{r}(t) = \sum_{m=0}^{\infty} \int_{0 \leqslant t_1 \leqslant \cdots \leqslant t_m \leqslant t} dt_1 \cdots dt_m \tag{3.1}$$
$$S^{(\epsilon)}(t-t_m)\, C^{(\epsilon)}\, S^{(\epsilon)}(t_m - t_{m-1}) \cdots C^{(\epsilon)}\, S^{(\epsilon)}(t_1)\, \mathbf{r}(0).$$

Having arrived at this point considering a finite--but possibly very large--number of particles, we can now if we like take a thermodynamic limit, letting the particle number and the volume both go to infinity keeping ϵ fixed.

The main step in the argument leading to the Boltzmann equation is to show that if the $r_j(0)$ satisfy some uniform bounds as ϵ goes to zero and converge in a sufficiently strong sense to a sequence of limit functions $r_j^{(0)}(0)$, then the convergence of the series (3.1) is uniform in ϵ, and the series goes over term by term to a series solving the so-called *Boltzmann hierarchy.* Because of limitations on the range of values of t for which the convergence of the series (3.1) can be controlled, we can only show that the $r_j(t)$ converge to a solution of the Boltzmann hierarchy for relatively small values of t--no larger than about one-third of a heuristic estimate of the mean free time. Convergence for larger values of t would follow, however, if bounds on the $r_j(t)$ uniform in ϵ could be established by some other means.

The starting point for the work to be described here is the observation that none of these arguments depends on the fact that $\mu_0(x_1, \ldots, x_N)$ is a probability density. By looking at μ_0's which are not probability densities but which are related in various ways to the hard-sphere equilibrium measure, it is possible to extract information about some time-dependent phenomena in the equilibrium hard-sphere gas. Technically, we will work in the grand-canonical ensemble, so N does not have a definite value, but this causes no problems.

We look first at the velocity autocorrelation of a single particle. Since the particles are treated symmetrically in the equilibrium ensemble, we may as well look at particle 1. We want to analyze the joint distribution of $\vec{v}_1(0)$ and $\vec{v}_1(t)$, for fixed t, as ϵ approaches zero. We can probe this distribution by choosing two continuous functions f, g of compact support on $\mathbf{R}^3$ and studying

$$E^{eq}_{(\epsilon)}\Big(f(\vec{v}_1(0))\, g(\vec{v}_1(t))\Big)$$

where $E^{eq}_{(\epsilon)}(\ \ )$ denotes the mean value in equilibrium. We can obtain a formula for this quantity as follows: Let $\mu_N^{eq}(x_1, \ldots, x_N)$ denote the N-particle component of the probability density for the hard-sphere grand-canonical ensemble, and put

$$\mu_N(0; x_1, \ldots, x_N) = f(\vec{v}_1)\, \mu_N^{eq}(x_1, \ldots, x_N).$$

Then

$$E^{eq}_{(\epsilon)}\Big(f(\vec{v}_1(0))\, g(\vec{v}_1(t))\Big) = \sum_N \int \mu_N(t; x_1, \ldots, x_N) g(\vec{v}_1)\, dx_1 \cdots dx_N.$$

where $\mu_N(t; \cdots)$ is obtained by applying the streaming operator $S_N^{(\epsilon)}(t)$ to $\mu_N(0; \cdots)$. This latter expression can be rewritten as

$$\int g(\vec{v}_1)\, r_1(t; \vec{q}_1, \vec{v}_1)\, d\vec{q}_1\, d\vec{v}_1.$$

(In this formula, we are using a non-standard normalization for the rescaled correlation function r_1. We are also assuming that the thermodynamic limit has not been taken. If it is taken, then r_1 becomes independent of $\vec{q}_1$, and the integral over this variable is suppressed.

Thus, determining the joint distribution of $\vec{v}_1(0)$ and $\vec{v}_1(t)$ reduces to determining $r_1(t)$. To do this in the limit as ϵ goes to zero, we argue as follows:

1. Although $\mu_N(t; x_1, \ldots, x_N)$ is not symmetric in $x_1, \ldots, x_N$, it is symmetric in $x_2, \ldots, x_N$, and this suffices to permit the derivation of the BBGKY hierarchy for the $r_j(t)$.
2. We have

$$|\mu_N(0; x_1, \ldots, x_N)| \leqslant ||f||_\infty\, \mu_N^{eq}(x_1, \ldots, x_N).$$

Since the equilibrium ensemble is invariant under the time evolution, the inequality remains true if we replace $\mu_N(0)$ by $\mu_N(t)$ for any t. By integrating over $x_{j+1}, \ldots, x_N$, we get

$$|r_j(t; x_1, \ldots, x_j)| \leqslant ||f||_\infty \, r_j^{eq}(x_1, \ldots, x_j).$$

This inequality gives us a *time-independent* bound on the size of the $r_j(t)$.

3. $r_j(0; x_1, \ldots, x_j) = f(\vec{v}_1)\, r_j^{eq}(x_1, \ldots, x_j)$.

4. $\lim_{\epsilon \to 0} r_j(0; x_1, \ldots, x_j) = f(\vec{v}_1) \prod_{i=1}^{j} h_\beta(\vec{v}_i)$, where h_β is the Maxwellian with inverse temperature β.

5. The solution of the Boltzmann hierarchy with initial condition as on the right-hand side of 4. is

$$r_j^{(0)}(t; x_1, \ldots, x_j) = e^{At} f(\vec{v}_1) \prod_{i=1}^{j} h_\beta(\vec{v}_i),$$

where A is the *linear Boltzmann operator* or *Rayleigh-Boltzmann operator:*

$$Af(\vec{v}) = \int d\vec{v}_1 \int_{+} d\hat{\omega}\, \hat{\omega}\cdot(\vec{v}-\vec{v}_1)\, h_\beta(\vec{v}_1)\, [f(\vec{v}\,') - f(\vec{v})],$$

where $\vec{v}\,', \vec{v}_1{}'$ are the outgoing velocities for a collision with incoming velocities $\vec{v}, \vec{v}_1$ and momentum transfer direction $\hat{\omega}$, and where $\int_{+} d\hat{\omega}$ means the surface integral over the half of the unit sphere where $\hat{\omega}\cdot(\vec{v}-\vec{v}_1) > 0$.

Now apply the general analysis and the time-independent bound 2. on the $r_j(t)$ to conclude that

$$\lim_{\epsilon \to 0} r_1(t; \vec{v}_1) = e^{At} f(\vec{v}_1)\, h_\beta(\vec{v}_1)$$

for all positive t, and thus

$$\lim_{\epsilon \to 0} E_{(\epsilon)}^{eq} \Big(g(\vec{v}_1(t))\, f(\vec{v}_1(0)) \Big) = (e^{At} f \mid g),$$

where $(\;|\;)$ denotes the inner product in $L^2(h_\beta\, d\vec{v})$. I believe that this result is the first non-trivial case where the validity of a picture derived from the Boltzmann equation can be proved for all times. It has as an immediate consequence that the power-law decay of the velocity autocorrelation function which is believed to occur for all finite ϵ must disappear in favor of exponential decay in the Boltzmann-Grad limit. It also appears that the argument can be extended to show that the stochastic process

$$\vec{v}_1(t), \quad -\infty < t < \infty$$

(with distribution given by the equilibrium ensemble) converges in distribution as ϵ approaches zero to the stationary Markov process constructed from the positivity-preserving semigroup e^{At} as in the Euclidean formulation of quantum mechanics.

The second application of these ideas is to the study of time-dependent fluctuations in the equilibrium hard-sphere gas. If f is a continuous function of compact support on $\mathbf{R}^6$, put

$$n(f)(x_1, \ldots, x_N) = \epsilon^2 \sum_{i=1}^{N} f(x_i).$$

(In contrast to Section 2, we are normalizing by dividing by ϵ^{-2} rather than by the particle number N which is not constant here.) Let

$$\bar{f}_{(\epsilon)} = E_{(\epsilon)}^{eq}(n(f))$$

$$\Phi_{(\epsilon)}(f) = \frac{1}{\epsilon} \Big(n(f) - \bar{f}_{(\epsilon)} \Big).$$

Dobrushin and Tirozzi[4] have shown that the fluctuation observables $\Phi_{(\epsilon)}(f)$ are asymptotically jointly centered Gaussian with covariance given by the inner product in $L^2(h_\beta(\vec{v})\, d\vec{q}\, d\vec{v})$. Our objective is to study the joint distribution of fluctuations at unequal times. Our results are limited in that all we are able to analyze is the covariance

$$\lim_{\epsilon\to 0} E^{eq}_{(\epsilon)}\left[\Phi_{(\epsilon)}(g)_{t_1}, \Phi_{(\epsilon)}(f)_{t_2}\right]$$

and that only for small $|t_1 - t_2|$. Since the equilibrium state is invariant under the time evolution, we may as well take $t_2 = 0$ and $t_1 = t \geqslant 0$.

As before, we proceed by constructing an initial μ adapted to the problem at hand. We take

$$\mu_N(0; x_1, \ldots, x_N) = \left[\sum_{i=0}^{N} f(x_i) - \frac{1}{\epsilon^2}\bar{f}_{(\epsilon)}\right] \mu_N^{eq}(x_1, \ldots, x_N).$$

We let $\mu_N(t; x_1, \ldots, x_N)$ be the corresponding time-dependent density and $r_j(t; x_1, \ldots, x_j)$ the rescaled correlation functions. A straightforward computation shows

$$E^{eq}_{(\epsilon)}\left[\Phi_{(\epsilon)}(g)_t, \Phi_{(\epsilon)}(f)_0\right] = \int g(x_1)\, r_1(t; x_1)\, dx_1,$$

so again all we need to do is to compute $r_1(t)$.

Another straightforward computation shows that

$$r_j(0; x_1, \ldots, x_j) = \sum_{i=1}^{j} f(x_i)\, r_j^{eq}(x_1, \ldots, x_j) + \text{correction},$$

and the correction would vanish if $r_{j+1}^{eq}(x_1, \ldots, x_{j+1})$ were exactly equal to $r_j(x_1, \ldots, x_j)\, r_1(x_{j+1})$. This factorization does not hold for non-zero ϵ, but a detailed computation using a low-density expansion for the equilibrium correlation functions shows that the correction is $O(\epsilon)$ (in an appropriate norm) and so can be ignored in the zero-ϵ limit. Thus,

$$\lim_{\epsilon\to 0} r_j(0; x_1, \ldots, x_j) = \sum_{i=1}^{j} f(x_i) \prod_{i=1}^{j} h_\beta(\vec{v}_i),$$

and what we therefore have to do is to solve the Boltzmann hierarchy with initial condition given by the right-hand side of this equation. The solution turns out to be

$$\sum_{i=1}^{j} f_t(x_i) \prod_{i=1}^{j} h_\beta(\vec{v}_i),$$

where

$$f_t = e^{Lt} f,$$

with L the linear*ized* Boltzmann operator:

$$(Lf)(\vec{q}, \vec{v}) = -\vec{v}\cdot\frac{\partial f}{\partial \vec{q}} + \int d\vec{v}_1 \int_+ d\hat{\omega}\, \hat{\omega}\cdot(\vec{v} - \vec{v}_1)\, h_\beta(\vec{v}_1)$$

$$\left\{ f(\vec{q}, \vec{v}_1{}') + f(\vec{q}, \vec{v}') - f(\vec{q}, \vec{v}_1) - f(\vec{q}, \vec{v}) \right\}$$

We thus get

$$\lim_{\epsilon\to 0} r_1(t; x_1) = (e^{Lt} f)(x_1)\, h_\beta(\vec{v}_1).$$

This time, we have no way of bounding the $r_j(t)$ uniformly in t, ϵ, so the limit is proved only for small (positive) t. For such t, we have

$$\lim_{\epsilon\to 0} E^{eq}_{(\epsilon)}\left(\Phi_{(\epsilon)}(g)_t \Phi_{(\epsilon)}(f)_0\right) = (g|e^{Lt}f).$$

It is attractive to conjecture that the $\Phi_{(\epsilon)}(f)_t$, all f and t, become jointly Gaussian in the zero-ϵ limit. So far as I know, this has not been proved. (A few higher moments can be computed without too much difficulty, and their values are consistent with the conjecture.) I want to note, nevertheless, that *if* the conjecture is correct, it implies that the equilibrium fluctuations in the Boltzmann-Grad limit do not evolve deterministically as in the Vlasov limit. To see this, let

$$\hat{g} = (e^{Lt})^* g.$$

Then

$$\lim_{\epsilon\to 0} E^{eq}_{(\epsilon)}\left([\Phi_{(\epsilon)}(g)_t - \Phi_{(\epsilon)}(\hat{g})_0]\, \Phi_{(\epsilon)}(f)_0\right) = 0$$

for all f. But, for centered Gaussian random variables, orthogonality is equivalent to independence, so the conjecture implies that, asymptotically, the time t fluctuation observable $\Phi_{(\epsilon)}(g)_t$ can be written as the sum of a time zero fluctuation observable $\Phi_{(\epsilon)}(\hat{g})_0$ and a centered Gaussian random variable which is independent of all time zero fluctuation observables. The latter term has variance $||g||^2 - ||\hat{g}||^2$, which is generally non-zero. The presence of a non-zero independent component means that time t fluctuation observables are not predictable with certainty even with complete knowledge of the time zero fluctuation observables.

References

1. Alexander,R.K.: Time evolution for infinitely many hard spheres. Commun. math. Phys. **49,** 217-232(1976)
2. van Beijeren,H.,Lanford,O.E.,Lebowitz,J.L.,and Spohn,H.:Equilibrium time correlation functions in the low density limit. Preprint(1979)
3. Braun,W. and Hepp,K.: The Vlasov dynamics in the 1/N limit of interacting classical particles. Commun. math. Phys. **56,** 101-113(1977)
4. Dobrushin,R.L. and Tirozzi,B.: The central limit theorem and the problem of equivalence of ensembles. Commun. math. Phys. **54,** 173-192(1977)
5. Fritz,J. and Dobrushin,R.L.: Non-equilibrium dynamics of two-dimensional infinite particle systems with a singular interaction. Commun. math. Phys. **57,** 67-81(1977)
6. Lanford,O.E.: Time evolution of large classical systems, in: Dynamical Systems: Theory and Applications, ed. J.Moser. Springer Lecture Notes in Physics **38,** 1-111(1975)
7. Marchioro,C., Pellegrinotti,A. and Presutti,E.: Existence of time evolution for ν-dimensional statistical mechanics. Commun. math. Phys. **40,** 175-185(1975)
8. Sinai,Ya.G.: Construction of the dynamics in one-dimensional systems of statistical mechanics. Teoret. Mat. Fiz. **11,** 248-258(1972) English translation: Theoret. and Math. Phys. **12,** 487-494(1973)
9. Sinai,Ya.G.: Construction of cluster dynamics for dynamical systems of statistical mechanics. Vest. Moscow Univ.,no 1,152-158(1974)

Holonomic Quantum Fields

— The unanticipated link between deformation theory of differential equations and quantum fields —

By

Michio Jimbo, Tetsuji Miwa and Mikio Sato

RIMS, Kyoto University, Kyoto 606, JAPAN

and partly, Yasuko Môri

Department of Mathematics, Ryûkyû University

Naha, Okinawa 903, JAPAN

§1. Introduction

Through recent study of the problems in mathematical physics, a deep, unexpected link has emerged: a link between the monodromy preserving deformation theory for linear (ordinary and partial) differential equations, and a class of quantum field operators ([1] [2] [3]). The aim of this article is to give an overview to the present stage of development in the theory (see also [4]).

The fruit of the above link is multifold. On the one hand it enables one to compute exactly the n point correlation functions of the field in question in a closed form, using solutions to certain non-linear differential equations of specific type (such as the Painlevé equations). On the other hand it provides an effective new tool of describing the deformation theory by means of quantum field operators. Thus it stands as a good example of the fact that not only the pure mathematics is applied to physical problems but also the converse is true.

The following examples will show how the theory actually works.

Example 1. The scaling limit of the 2 dimensional Ising model.

The first example is concerned with the calculation of n point spin correlation functions $\langle\sigma_{j_1k_1}\cdots\sigma_{j_nk_n}\rangle$ of the 2 dimensional Ising model, specifically their scaling limit $\tau_n^{\pm}$ at the critical temperature $T \to T_c \pm 0$. Wu et al. [5] found that the 2 point function $\tau_2^{\pm}$ admits a closed expression

$$(1.1) \qquad \tau_2^{\pm}(\theta) = \text{const.} \begin{Bmatrix} \sinh\frac{\psi}{2} \\ \cosh\frac{\psi}{2} \end{Bmatrix} \exp\left(\frac{1}{2}\int_{\infty}^{\theta} dt\ \frac{t}{2}\left(\left(\frac{d\psi}{dt}\right)^2 - \sinh^2\psi(t)\right)\right)$$

where $2\theta = \text{const.}|T-T_c|\sqrt{(j_1-j_2)^2+(k_1-k_2)^2}$ is the scaled distance, and $\psi = \psi(\theta)$ is a solution of the following non-linear second order differential equation (an equivalent of a Painlevé equation of the third kind ([6]))

$$(1.2)\qquad \frac{d^2\psi}{d\theta^2} + \frac{1}{\theta}\frac{d\psi}{d\theta} = 2\sinh 2\psi.$$

Their result is generalized as follows ([H] III, IV, [S] II-IV). The n point correlation function $\tau_n^{\pm}$ as a function of $2n$ scaled variables $a_\nu = \frac{a_\nu^1+ia_\nu^2}{2}$, $a_\nu^* = \frac{a_\nu^1-ia_\nu^2}{2}$, $(a_\nu^1, a_\nu^2) = \text{const.}|T-T_c|\cdot(j_\nu, k_\nu)$ $(\nu=1,\cdots,n)$, is expressed as

$$(1.3)\qquad \tau_n^{\pm}(a_1,a_1^*,\cdots,a_n,a_n^*) = \text{const.}\left\{\begin{matrix} i\sqrt{\det\sinh H} \\ \\ \sqrt{\det\cosh H}\end{matrix}\right\}\exp\left(\frac{1}{2}\int\omega\right),$$

where $\int\omega$ is a primitive of the closed 1-form

$$(1.4)\qquad \omega = \frac{1}{2}\sum_{\mu\neq\nu} f_{\mu\nu}f_{\nu\mu}\frac{da_\mu - da_\nu}{a_\mu - a_\nu} + m^2\sum_{\mu,\nu=1}^{n}(\delta_{\mu\nu} - g_{\mu\nu}^2)a_\mu^* da_\nu + \text{complex conjugate},$$

and the matrices $F = (f_{\mu\nu})$, $G = (g_{\mu\nu}) = e^{-2H}$ are solutions of a non-linear system of partial differential equations

$$(1.5)\qquad \frac{\partial f_{\mu\nu}}{\partial a_\lambda} = \sum_{\kappa(\neq\mu,\nu)} f_{\mu\kappa}f_{\kappa\nu}\left(-\frac{\delta_{\lambda\kappa}-\delta_{\lambda\mu}}{a_\kappa - a_\mu} + \frac{\delta_{\lambda\kappa}-\delta_{\lambda\nu}}{a_\kappa - a_\nu}\right) + m^2(\delta_{\lambda\mu}-\delta_{\lambda\nu})\sum_{\kappa=1}^{n} a_\kappa^* g_{\kappa\mu}g_{\kappa\nu}$$

$$\frac{\partial f_{\mu\nu}}{\partial a_\lambda^*} = m^2(a_\mu - a_\nu)g_{\lambda\mu}g_{\lambda\nu}$$

$$\frac{\partial g_{\mu\nu}}{\partial a_\lambda} = \sum_{\kappa(\neq\nu)} g_{\mu\kappa}f_{\kappa\nu}\frac{\delta_{\lambda\kappa}-\delta_{\lambda\nu}}{a_\kappa - a_\nu}$$

$$\frac{\partial g_{\mu\nu}}{\partial a_\lambda^*} = \sum_{\kappa(\neq\mu)} (GFG^{-1})_{\mu\kappa}g_{\kappa\nu}\frac{\delta_{\lambda\mu}-\delta_{\lambda\kappa}}{a_\mu^* - a_\kappa^*}$$

subject to the symmetry ${}^tG = G^* = G^{-1}$, ${}^tF = -F$, $F^* = -GFG^{-1}$.

<u>Example 2. The reduced density matrix for the 1 dimentional gas of impenetrable bosons.</u>

The second example deals with a non-relativistic bose gas described by the N body problem in a periodic box $0 \leq x \leq L$

$$(1.6)\qquad \left(-\frac{1}{2}\sum_{i=1}^{N}\left(\frac{\partial}{\partial x_i}\right)^2 + c\sum_{i<j}\delta(x_i - x_j)\right)\Psi(x_1,\cdots,x_N) = E\Psi(x_1,\cdots,x_N),$$

or equivalently by the quantal non-linear Schrödinger equation

$$(1.7)\qquad i\frac{\partial\phi}{\partial t} = -\frac{1}{2}\frac{\partial^2\phi}{\partial x^2} + c\phi^*\phi^2$$

$$[\phi(0,x), \phi^*(0,x')] = \delta(x-x'),$$

with the impenetrability condition $c = +\infty$. The ground state wave function of (1.6) with $c = +\infty$ takes the simple form ([7])

$$(1.8) \qquad \Psi_{N,L}(x_1,\cdots,x_N) = \frac{1}{\sqrt{N!L^N}} \prod_{j<k} |e^{2\pi i x_j/L} - e^{2\pi i x_k/L}|.$$

The subject of interest here is the n particle reduced density matrix

$$(1.9) \qquad \rho_n(x_1',\cdots,x_n';x_1'',\cdots,x_n'') = \langle\Psi_{N,L}|\phi^*(x_1')\cdots\phi^*(x_n')\phi(x_1'')\cdots\phi(x_n'')|\Psi_{N,L}\rangle$$

$$= \frac{N!}{(N-n)!}\int_0^L\cdots\int_0^L dy_{n+1}\cdots dy_N\, \Psi^*_{N,L}(x_1',\cdots,x_n',y_{n+1},\cdots,y_N)$$

$$\times\, \Psi_{N,L}(x_1'',\cdots,x_n'',y_{n+1},\cdots,y_N)$$

in the thermodynamic limit $N,L \to \infty$, with the particle density $\rho_0 = \frac{N}{L}$ kept fixed. A complete parallelism to Example 1 is exhibited in this Example 2. The 1 particle reduced density matrix $\rho_1(x_1';x_1'') = \rho(|x_1'-x_1''|)$ admits a closed expression ([S] XVI, [3]; we choose $\rho_0 = \pi^{-1}$)

$$(1.10) \qquad \rho(x) = \rho_0 \exp\left(\int_0^x dt\, \frac{\sigma(t)}{t}\right),$$

where $\sigma = \sigma(x)$ satisfies a second order non-linear equation

$$(1.11) \qquad \left(x\frac{d^2\sigma}{dx^2}\right)^2 = -4\left(x\frac{d\sigma}{dx} - 1 - \sigma\right)\left(x\frac{d\sigma}{dx} + \left(\frac{d\sigma}{dx}\right)^2 - \sigma\right)$$

which is an equivalent of a Painlevé equation of the fifth kind. For general n, the n particle reduced density matrix $\rho_n(x_1',\cdots,x_n', x_1'',\cdots,x_n'')$ with $x_1'<\cdots<x_n'$, $x_1''<\cdots<x_n''$, is expressible in the form ([3])

$$(1.12) \qquad \rho_n(x_1',\cdots,x_n';x_1'',\cdots,x_n'') = \text{const.}\det(R_I(x_j',x_k''))_{j,k=1,\cdots,n} \times \exp\left(\int\omega\right).$$

Here, denoting by $x_1<\cdots<x_{2n}$ the re-ordering of $x_1',\cdots,x_n'$, $x_1'',\cdots,x_n''$, we have

$$(1.13) \qquad R_I(x_j,x_k) = \frac{1}{2i(x_j-x_k)}(r_{+j}r_{-k} - r_{-j}r_{+k}) \qquad (j\neq k)$$

$$= \frac{1}{4}\sum_{j'(\neq j)} \frac{(r_{+j}r_{-j'} - r_{+j'}r_{-j})^2}{x_{j'}-x_j} + i r_{+j}r_{-j} \qquad (j=k),$$

$$\omega = -\frac{1}{4}\sum_{j<j'}(r_{+j}r_{-j'} - r_{+j'}r_{-j})^2\, \frac{dx_j - dx_{j'}}{x_j - x_{j'}} + i\sum_{j=1}^{2n} r_{+j}r_{-j}dx_j ,$$

and $r_{\pm j}$ $(j=1,\ldots,2n)$ satisfy the system of non-linear partial differential equations

$$(1.14)\quad \frac{\partial r_{\pm j}}{\partial x_k} = \frac{1}{2}(-r_{\pm k}r_{-k}r_{+j} + r_{\pm k}r_{+k}r_{-j})\frac{1}{x_j - x_k} \qquad (k\neq j)$$

$$= \pm i r_{\pm j} - \frac{1}{2}\sum_{j'(\neq j)} (-r_{\pm j'}r_{-j'}r_{+j} + r_{\pm j'}r_{+j'}r_{-j})\frac{1}{x_j - x_{j'}} \qquad (k=j).$$

In both of these examples the relevant non-linear differential equations (1.2), (1.5), (1.11), (1.14) have a common characteristic feature: they arise as monodromy preserving deformation equations of appropriate linear differential equations (§2,§5).

Roughly speaking the monodromy is a notion that measures the character of ramification of a multi-valued function. Take as an example the following system of linear differential equations with rational coefficients

$$(1.15)\quad \frac{dY}{dx} = A(x)Y, \quad A(x):\ m\times m \text{ matrix of rational functions.}$$

Its fundamental solution matrix $Y = Y(x)$ is multi-valued, having as its branch points the poles $a_1,\dots,a_n$ of $A(x)$ and ∞. Thus by analytic continuation along a closed path γ (Figure 1) it undergoes a linear transformation

$$(1.16)\qquad Y(x) \longmapsto Y(x)M_\gamma .$$

The constant matrix M_γ depends only on the homotopy class of γ, and is called the monodromy matrix for $Y(x)$ corresponding to γ. The correspondence

$$(1.17)\qquad \begin{array}{ccc}\pi_1(\mathbb{C}-\{a_1,\cdots,a_n\}) & \longrightarrow & GL(m,\mathbb{C})\\ \gamma & \longmapsto & M_\gamma\end{array}$$

constitutes a representation of the fundamental group $\pi_1(\mathbb{C}-\{a_1,\cdots,a_n\})$, the monodromy representation.

Example 3. The monodromy problem of Riemann-Hilbert.

In contrast with the preceding two examples, the third one is of a purely mathematical character. The problem is this: find a differential equation (1.15) with simple poles, $A(x) = \sum_{\nu=1}^{n} \frac{A_\nu}{x-a_\nu}$, whose solution $Y(x)$ has the prescribed monodromy representation. The answer is given in a particularly simple form in the language of quantum field theory ([H] II, [S] VI). By introducing one dimensional free fermion fields $\psi^{(i)}(x)$, $\psi^{*(i)}(x)$ and a class of field operators $\varphi(a;L)$, we find such a matrix as the analytic continuation of the vacuum expectation value

$$(1.18)\quad Y(x) = 2\pi i(x-x_0)\times\left(\frac{\langle\psi^{*(i)}(x_0)\varphi(a_1;L_1)\cdots\varphi(a_n;L_n)\psi^{(j)}(x)\rangle}{\langle\varphi(a_1;L_1)\cdots\varphi(a_n;L_n)\rangle}\right)_{i,j=1,\cdots,n}$$

with the normalization $Y(x_0) = 1$. As we vary the branch points a_ν, the monodromy representation $Y(x)$ is kept invariant by construction. The classical work of L. Schlesinger [8] states that in this case the coefficient matrices A_ν should satisfy a system of non-linear differential equations (here we choose $x_0=\infty$)

$$(1.19)\qquad \frac{\partial A_\nu}{\partial a_\mu} = \frac{[A_\nu, A_\mu]}{a_\nu - a_\mu} \quad (\mu\neq\nu), \qquad \frac{\partial A_\nu}{\partial a_\nu} = -\sum_{\nu'(\neq\nu)} \frac{[A_\nu, A_{\nu'}]}{a_\nu - a_{\nu'}} .$$

The n point function of $\varphi(a;L)$'s is again expressed in terms of A_ν's as

$$(1.20)\qquad \langle\varphi(a_1;L_1)\cdots\varphi(a_n;L_n)\rangle = \text{const.}\exp\left(\int\omega\right), \quad \omega = \sum_{\mu<\nu} \text{trace}\, A_\mu A_\nu \frac{da_\mu - da_\nu}{a_\mu - a_\nu} .$$

It should be mentioned that these correlation functions (1.3), (1.12), (1.20) in turn play an intrinsic rôle to the non-linear partial differential equations (1.5), (1.14), (1.19) respectively. Namely the latter are rewritten as non-autonomous classical Hamiltonian systems ([9] [10] [3]) involving several "time" variables, whose Hamiltonians are given by the closed 1-form ω ([10] [3]), i.e. essentially the logarithmic derivatives of the correlation functions.

The mathematical theory of isomonodromic deformations goes back to the classical works [8] [11] [12] in the dawn of this century. In recent years another type of deformation theory of linear differential equations has widely attracted people's attention in the branch of non-linear waves: the isospectral deformations, or the soliton theory. The deep interrelationship between these two theories is now becoming apparent ([13] [14] [15]). Our second example stands as an indication of this connection at the quantal level ([3] cf. [16] [17] [18]). We expect that such non-linear completely integrable systems at certain critical coupling ($c = +\infty$ for the present case) should fall within the category of our theory.

All these models described above are those in one space (+ one time) dimension. To generalize our scheme in higher dimensions, it seems inevitable to introduce fields that depend on extended objects (§6). Construction of the model is based on a peculiar commutation relation similar to the one discussed in the literature [19], although its analysis is not worked out except for some special case. The deformation equations are now taken place by a variational formula of Hadamard's type ([20], [S] XIII, XIV).

This paper is organized as follows.

§2 deals with the first of the above examples, the scaling limit of the 2 dimensional Ising model. The mathematical theory of monodromy preserving deformation is formulated in §3, where a system admitting an irregular singularity of rank 1 is discussed. In §4 field theoretical description is given to the Riemann-Hilbert problem (Example 3 above). The second example of impenetrable bose gas is discussed in the next §5. Lastly in §6 we discuss analogous constructions in space-time dimensions more than 2.

An algebraic building block of our whole story is the theory of Clifford group, which we have no room to explain in the text. See [H] I, [3] Appendix.

The authors would like to express their sincere gratitude to Professor E. Brézin who gave them an opportunity to attend and to give a talk at the Lausanne conference.

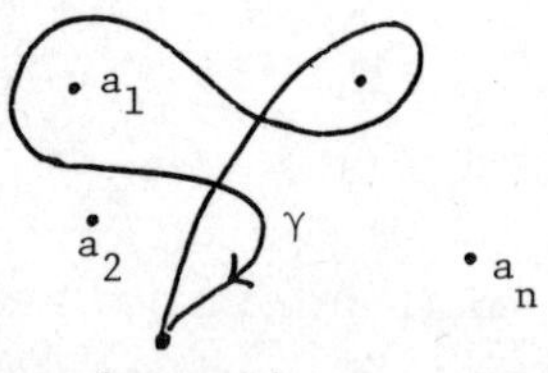

Figure 1

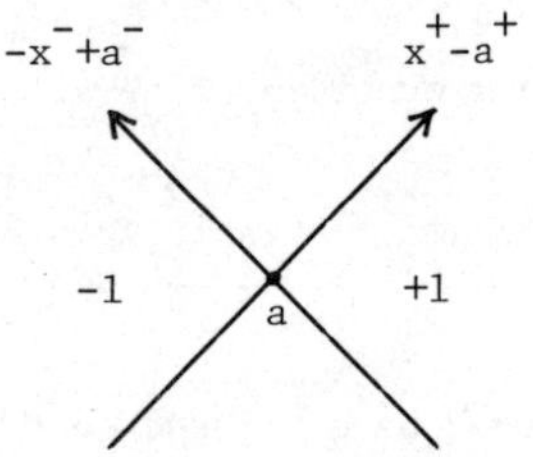

Figure 2

§2. Scaling limit of the two dimensional Ising model

2.1 History. Ever since the monumental work of Onsager [21] on the two dimensional Ising model, calculation of the spin correlation functions has been a longstanding problem. After the exact computation of free energy ([21]) and the spontaneous magnetization ([22]) there appeared an enormous literature dealing with this model from various points of view (see [23]), but it was only quite recently that exact infinite series expressions for two- and multi- spin correlation functions have been derived on the lattice as well as in the continuum limit ([5], [24], [25], [26], [S]I, [H]V, [27]). Among them, however, T. T. Wu, B. M. McCoy, C. A. Tracy and E. Barouch [5] have found the following most remarkable result: the scaling limit of the 2 point correlation function

$$(2.1)\quad \tau_2^{\pm}(\theta)=\lim_{T\to T_c\pm 0}\varepsilon^{-1/4}\langle\sigma_{00}\sigma_{MN}\rangle,\quad 2\theta=\varepsilon\sqrt{M^2+N^2}\;;\;\varepsilon=|T-T_c|\to 0,\ \sqrt{M^2+N^2}\to\infty$$

admits a closed expression

$$(2.2)\quad \frac{d}{d\theta}\log\tau_2^{\pm}(\theta)=\left(-\frac{1}{2\eta}\mp\frac{1}{1\mp\eta}\right)\frac{d\eta}{d\theta}-\frac{1}{4}\frac{\theta}{\eta^2}\left((1-\eta^2)^2-\left(\frac{d\eta}{d\theta}\right)^2\right)$$

in terms of a classical transcendental function $\eta=\eta(\theta)$, known as a Painlevé transcendent of the third. It is defined through the following non-linear second order ordinary differential equation (cf. (3.4))

$$(2.3)\quad \frac{d^2\eta}{d\theta^2}=\frac{1}{\eta}\left(\frac{d\eta}{d\theta}\right)^2-\frac{1}{\theta}\frac{d\eta}{d\theta}-\eta^3+\frac{1}{\eta}\,.$$

The function $\psi(\theta)$ in (1.2) is related to $\eta(\theta)$ by $\eta(\theta)=e^{-\psi(\theta)}$. Their approach was to relate the calculation of $\tau_2^{\pm}(\theta)$ to a linear integral equation

$$(2.4)\qquad \int_{-1}^{1} K_0(\theta|x-x'|)\begin{pmatrix} f(x') \\ g(x') \end{pmatrix} dx' = \begin{pmatrix} \cosh\theta x \\ \sinh\theta x \end{pmatrix}$$

where $K_0(x)$ denotes a modified Bessel function. This type of integral equation has previously been studied by G. Latta [28], who converted (2.4) into a system of linear differential equations

$$(2.5)\qquad \begin{aligned}(1-x^2)\frac{df}{dx} &= (\tfrac{1}{2}-\rho)xf + (\eta^{-1}(\tfrac{1}{2}+\rho) + \theta(1-x^2))g \\ (1-x^2)\frac{dg}{dx} &= (\eta(\tfrac{1}{2}-\rho) + \theta(1-x^2))f + (\tfrac{1}{2}+\rho)xg,\end{aligned}$$

with constants η,ρ left to be determined. The θ-dependence of η,ρ was subsequently determined by J. Myers [29], who found that

$$(2.6)\qquad \rho(\theta) = \frac{\theta}{2\eta}(1-\eta^2 - \frac{d\eta}{d\theta})$$

and that $\eta = \eta(\theta)$ should satisfy the Painlevé equation (2.3).

The fact is: all the Painlevé equations I-VI are known to be monodromy preserving deformation equations (in a due sense) of associated second order <u>linear</u> ordinary differential equations (see §3). The above work of Wu et al. strongly suggests that some kind of monodromy problem should be comprised in the structure of the model itself. We shall see below that this is indeed the case.

<u>2.2 Field theory of the Ising model in the scaling limit</u> ([26], [S] I, [H] VI,V). A key step of Onsager's original calculation of the free energy ([21]) is to introduce free fermion operators p_{mn}, q_{mn} $(m,n\in\mathbb{Z})$, with respect to which the spin operator s_{mn} satisfies the characteristic commutation relations (cf. [30] [31])

$$(2.7)\qquad s_{mn}p_{m'n} = \pm p_{m'n}s_{mn} \quad (m' \lessgtr m), \qquad s_{mn}q_{m'n} = \pm q_{m'n}s_{mn} \quad (m' \lessgtr m).$$

The theory of Clifford group ([H] I, [S] V) tells that a commutation relation of the type (2.7) with <u>free</u> fields uniquely determines s_{mn} up to a constant factor. In the continuum limit (2.1) these free fermion fields are scaled to give 2 dimensional free relativistic Majorana fields $\psi(x) = {}^t(\psi_+(x), \psi_-(x))$ $(x=(x^0,x^1)\in\mathbb{R}^2)$

$$(2.8)\qquad \lim \varepsilon^{-1/2}p_{mn} = \psi_+(x)\pm\psi_-(x), \quad \lim \varepsilon^{-1/2}iq_{mn} = \psi_+(x)\mp\psi_-(x), \quad (T\to T_c\pm 0)$$

whereas the spin operator const. $\varepsilon^{-1/8}s_{mn}$ (with a slight modification for $T < T_c$) gives ([H] V, [26])

$$(2.9)\qquad \varphi^F(x) = :\psi_0(x)e^{\rho_F(x)/2}: \quad (T\to T_c+0), \qquad \varphi_F(x) = :e^{\rho_F(x)/2}: \quad (T\to T_c-0)$$

with

$$(2.10)\qquad \psi_0(x) = \int_{-\infty}^{+\infty} \frac{du}{2\pi|u|}\, e^{-im(x^-u+x^+u^{-1})}\psi(u)$$

$$\rho_F(x)=\int_{-\infty}^{+\infty}\int_{-\infty}^{+\infty} \frac{du}{2\pi|u|}\frac{du'}{2\pi|u'|}\frac{-i(u-u')}{u+u'-i0}\, e^{-im(x^-(u+u')+x^+(u^{-1}u'^{-1}))}\psi(u)\psi(u').$$

Here $x^\pm = (x^0 \pm x^1)/2$, and $\psi(u)$ denotes the creation $(u < 0)$ annihilation $(u > 0)$ operator of the free Majorana field carrying the energy-momentum $(p^0,p^1) = (m\frac{u+u^{-1}}{2}, m\frac{u-u^{-1}}{2})$. By the Kramers-Wannier duality these fields $\varphi^F(x)$, $\varphi_F(x)$ are also regarded as the continuum version of the order and disorder variables introduced by Kadanoff and Ceva [32]. The peculiar commutation relation (2.7) is inherited to the continuum so that

$$(2.11)\qquad \varphi_F(a)\psi_\pm(x) = \varepsilon(x^1-a^1)\psi_\pm(x)\varphi_F(a), \quad \varphi^F(a)\psi_\pm(x) = \varepsilon(a^1-x^1)\psi_\pm(x)\varphi^F(a)$$

holds for x and a spacelike (Figure 2).

The normal product (2.9) of exp (quadratic form in ψ) is a characteristic of the fields in the Clifford group ([H] I). The point is that these fields, or more generally their products $\varphi\cdots\varphi$, are characterized by a kernel function of two variables. In other words the expectation value

$$(2.12)\qquad w_{\pm\pm'}(x,x') = \langle\psi_\pm(x)\psi_{\pm'}(x')\varphi_F(a_1)\cdots\varphi_F(a_n)\rangle \,/\, \langle\varphi_F(a_1)\cdots\varphi_F(a_n)\rangle$$

carries all the informations concerning the product $\varphi(a_1)\cdots\varphi(a_n)$. Viewed as a function of x or x' (2.12) satisfies the free Dirac equation, and has a singularity of a free propagator $\sim \langle\psi_\pm(x)\psi_{\pm'}(x')\rangle$ at $x=x'$. The commutation relation (2.11) now has the following consequence: the Euclidean continuation $w^{Euc}_{\pm\pm'}(x,x')$ of (2.12) is double-valued, changing its sign each time when x or x' makes a circuit around some a_μ $(\mu=1,\cdots,n)$. In this sense the monodromy problem is inherent in the theory, and the commutation relation (2.11) determines the monodromy structure of $w^{Euc}_{\pm\pm'}$.

2.3 Monodromy problem ([H] III, [S] II). Let $z = (x^1+ix^2)/2$ denote the complex coordinate of the Euclidean space-time $\mathbb{R}^2$. Given n points $a_1,\cdots,a_n$ arbitrarily, we consider the following problem. Characterize the functions $w = {}^t(w_+,w_-)$ with the properties

(2.13) (i) Euclidean Dirac equation: $\frac{\partial}{\partial z}w_- = mw_+$, $\frac{\partial}{\partial z^*}w_+ = mw_-$ except at $z=a_1,\cdots,a_n$,

(ii) w is double valued; it changes sign when prolonged once around each a_ν, and

$$|w| = O\left(\frac{1}{\sqrt{|z-a_\nu|}}\right) \qquad (z \to a_\nu),$$

(iii) $|w| = O(e^{-2m|z|})$ as $|z| \to \infty$.

Call $W_{a_1,\cdots,a_n}$ the linear space spanned by these functions. Then it turns out that $\dim W_{a_1,\cdots,a_n} = n$. Moreover a basis $w_1,\cdots,w_n$ of $W_{a_1,\cdots,a_n}$ is shown to satisfy, along with (i), a system of linear differential equations of the form

$$(2.14)\quad \begin{pmatrix} M_F w_1 \\ \vdots \\ M_F w_n \end{pmatrix} = \left(B\frac{\partial}{\partial z} - B'\frac{\partial}{\partial z^*} + F\right)\begin{pmatrix} w_1 \\ \vdots \\ w_n \end{pmatrix}, \quad M_F = z\frac{\partial}{\partial z} - z^*\frac{\partial}{\partial z^*} + \frac{1}{2}\begin{pmatrix} 1 & \\ & -1 \end{pmatrix}$$

where B, B' and F are $n\times n$ constant matrices independent of z. For instance one may choose such a basis $w_{C\nu}$ that

$$(2.15)\quad B = A,\ B' = G^{-1}A^*G \ \text{ with } A = \begin{pmatrix} a_1 & & \\ & \ddots & \\ & & a_n \end{pmatrix},\ A^* = \begin{pmatrix} a_1^* & & \\ & \ddots & \\ & & a_n^* \end{pmatrix}$$

$$^tG = G^* = G^{-1},\ ^tF = -F,\ F^* = -GFG^{-1}.$$

2.4 Deformation theory ([H] III, [S] II). Now we consider the a_ν, a_ν^*-dependence of the above normalized basis $w_{C\nu}$ and the matrices G, F. For notational simplicity we employ the exterior differentiation d with respect to $a_1,\cdots,a_n$, $a_1^*,\cdots,a_n^*$. (In general for a function u of $x_1,\cdots,x_n$, $du = \sum_{j=1}^{n} f_j dx_j$ stands for $\frac{\partial u}{\partial x_j} = f_j$, $j=1,\cdots,n$.) We find the following linear system

$$(2.16)\quad \begin{pmatrix} dw_{C1} \\ \vdots \\ dw_{Cn} \end{pmatrix} = \left(-dA\cdot\frac{\partial}{\partial z} - G^{-1}dA^*\cdot G\frac{\partial}{\partial z^*} + \Theta\right)\begin{pmatrix} w_{C1} \\ \vdots \\ w_{Cn} \end{pmatrix}$$

where $\Theta = (\theta_{\mu\nu})$ is a matrix of 1-forms related to $F = (f_{\mu\nu})$ through $\theta_{\mu\nu} = -f_{\mu\nu} d\log(a_\mu - a_\nu)$ $(\mu\neq\nu)$, $=0$ $(\mu=\nu)$. Equations (2.16) express the necessary and sufficient condition that the system (2.13)-(i)+(2.14) is deformed with respect to a_ν, a_ν^* while preserving its monodromy structure (in our case (2.13)-(ii)). The integrability condition for (2.13)-(i), (2.14) and (2.16) then gives rise to non-linear total differential equations, the "deformation equations":

$$(2.17)\quad dF = [\Theta, F] + m^2[dA, G^{-1}A^*G] + m^2[A, G^{-1}dA^*\cdot G],\quad dG = -G\Theta - {}^t\Theta^* G.$$

In the case of $n=2$, this reduces to the Painlevé equation (2.3). By rescaling $x\theta \longmapsto x$ the linear system (2.5) coincides with a restriction of (2.13)-(i)+(2.14) with $n=2$ on the branch cut between $a_1 = -\theta$ and $a_2 = \theta$.

2.5 Correlation functions ([H] III, IV, [S] IV). Set

$$(2.18)\quad w_{F\nu}(x) = \langle\psi(x)\varphi_F(a_1)\cdots\varphi^F(a_\nu)\cdots\varphi_F(a_n)\rangle \,/\, \langle\varphi_F(a_1)\cdots\varphi_F(a_n)\rangle .$$

These functions arise as coefficients of the local expansion of $w(x,x')$ at $x'=a_\nu$, so that their Euclidean continuations $w_{F\nu}^{Euc}$ share the monodromy property (2.13)-(ii); in fact they are related to $w_{C\nu}$'s through

$$(2.19)\qquad i\begin{pmatrix} w_{F1}^{Euc} \\ \vdots \\ w_{Fn}^{Euc} \end{pmatrix} = G(1+G)^{-1}\begin{pmatrix} w_{C1} \\ \vdots \\ w_{Cn} \end{pmatrix},$$

and hence they also constitute a basis of $W_{a_1,\cdots,a_n}$. The significance of (2.18) is that the logarithmic derivatives $\mp \frac{\partial}{\partial a_\nu^{\mp}} \log\langle \varphi_F(a_1)\cdots\varphi_F(a_n)\rangle$ appear as the second coefficients of the expansion of $w_{F\nu}(x)$ at $x=a_\nu$. This is seen from the following short distance expansion

$$(2.20)\qquad \psi_\pm(x)\varphi^F(a) = \frac{i}{2}\varphi_F(a)\frac{(m(-x^-+a^-))^{\mp 1/2}}{(\mp 1/2)!} + \frac{-i}{m}\frac{\partial\varphi_F}{\partial a^-}\frac{(m(-x^-+a^-))^{1\mp 1/2}}{(1\mp 1/2)!} + \cdots$$

$$- \frac{i}{2}\varphi_F(a)\frac{(m(x^+-a^+))^{\pm 1/2}}{(\pm 1/2)!} - \frac{i}{m}\frac{\partial\varphi_F}{\partial a^+}\frac{(m(x^+-a^+))^{1\pm 1/2}}{(1\pm 1/2)!} - \cdots.$$

By (2.19) it is only a little bit of algebra to relate these coefficients to the coefficients F,G of differential equations. Thus we obtain a closed expression of the Euclidean continuation τ_{Fn} ($=\tau_F^-$ in (1.3)) of the n point function

$$(2.21)\qquad d\log\tau_{Fn} = d\log\sqrt{\det\cosh H} + \frac{1}{2}\omega$$

where $G = e^{-2H}$ and

$$(2.22)\qquad \omega = -\frac{1}{2}\,\mathrm{trace}(F\Theta + \Theta^* GFG^{-1}) + m^2\,\mathrm{trace}(d(AA^*) - G^{-1}A^*GdA - GAG^{-1}dA^*).$$

Mixed correlation functions $\tau_{Fn}^{\nu_1\cdots\nu_m} = \langle\varphi_F(a_1)\cdots\varphi^F(a_{\nu_1})\cdots\varphi^F(a_{\nu_m})\cdots\varphi_F(a_n)\rangle$ are given by

$$(2.23)\qquad \tau_{Fn}^{\nu_1\cdots\nu_m} = \tau_{Fn}\times \mathrm{Pfaffian}\,(i(\tanh H)_{\nu\nu'})_{\nu,\nu'=\nu_1,\cdots,\nu_m}.$$

2.6 Comments. The above construction admits of extensions to several directions.

It is straightforward to generalize the monodromy -1 to an arbitrary phase factor $e^{2\pi i\ell}$ ([H] IV, [S] VII), or even a matrix $e^{2\pi iL}$ ([S] XIV), starting from the free Dirac (instead of Majorana) fields. The deformation theory in 2.3, 2.4 holds without change, the only difference being the algebraic conditions on G and F. The relevant fields are related to the solution of the Federbush model ([H] IV supplement, [33]) whose interaction Lagrangian is

$$(2.24)\qquad \mathcal{L}_{int} = -g\varepsilon_{\mu\nu}J_I^\mu(x)J_{II}^\nu(x)$$

(current-pseudocurrent coupling between two species of fermions). The coupling constant coincides with the exponent of the monodromy: $g=2\pi\ell$.

In the massless case, our construction is directly related to the original problem of Riemann (§4). The corresponding Lagrangian field theory is the massless Thirring model ([H] IV supplement).

Another possibility is to start from free bose fields (neutral or complex) instead of fermions ([H] IV, [S] IX). In the neutral case there results an interacting fermi field $\varphi^B(a) = {}^t(\varphi^B_+(a), \varphi^B_-(a))$, to be compared with the bose field $\varphi^F(a)$ obtained on the basis of free fermions. These two fields $\varphi^B(a)$, $\varphi^F(a)$ share the same S-matrix $(-)^{N(N-1)/2}$ in the N-particle sector ([H] IV, [S] I, IX, [26]), and their n point correlation functions are inverse to each other apart from a simple factor ([H] IV, [S] IX).

§3. Monodromy preserving deformation of linear ordinary differential equations

3.1 History. The most typical example of a linear ordinary differential equation is the Gauss' hypergeometric differential equation $(' = \frac{d}{dx})$

$$x(1-x)y'' + (\gamma-(\alpha+\beta+1)x)y' - \alpha\beta y = 0. \tag{3.1}$$

The coefficients of (3.1) are completely determined by the characteristic indices at the regular singularities 0, 1 and ∞. However this is not the case for a more general equation. Even if all the singularities are regular ones, not only the local indices but also the structure of global monodromy is needed to specify the equation. Here arises naturally the following question, first posed by Riemann [34]: Find a differential equation with regular singularities whose monodromy representation is the prescribed one.

Schlesinger [8] considered the first order system (1.15) with

$$A(x) = \sum_{\nu=1}^{n} \frac{A_\nu}{x-a_\nu} \tag{3.2}$$

and initiated the following approach to the Riemann's problem: Find conditions on $A_1,\cdots,A_n$ so that the monodromy representation is preserved under the variation of the position of the branch points $a_1,\cdots,a_n$. With a suitable normalization of the solution $Y(x)$ at $x=\infty$, he obtained the following system of linear differential equations for $Y(x)$,

$$\frac{\partial Y}{\partial x} = \Big(\sum_{\nu=1}^{n} \frac{A_\nu}{x-a_\nu}\Big)Y, \quad \frac{\partial Y}{\partial a_\nu} = -\frac{A_\nu}{x-a_\nu}\, Y \qquad (\nu=1,\cdots,n), \tag{3.3}$$

and then, as the integrability conditions for (3.3), the system of non-linear differential equations (1.19) for A_ν $(\nu=1,\cdots,n)$.

The simplest nontrivial case of (1.19), $m=2$ and $n=4$, is equivalent to a second order non-linear ordinary differential equation, known as the Painlevé equation of the sixth kind ([11]).

Originally the Painlevé equations emerged in a different context. Painlevé obtained the following six canonical types of equations through the classification of second order non-linear algebraic differential equations whose general solutions have no movable branch points.

(3.4)
(I) $y'' = 6y^2+x$

(II) $y'' = 2y^3+xy+\alpha$

(III) $y'' = \frac{y'^2}{y} - \frac{y'}{x} + \frac{\alpha y^2+\beta}{x} + \gamma y^3 + \frac{\delta}{y}$

(IV) $y'' = \frac{y'^2}{2y} + \frac{3y^3}{2} + 4xy^2 + 2(x^2-\alpha)y + \frac{\beta}{y}$

(V) $y'' = (\frac{1}{2y} + \frac{1}{y-1})y'^2 - \frac{y'}{x} + \frac{(y-1)^2}{x^2}(\alpha y+\frac{\beta}{y}) + \frac{\gamma y}{x} + \frac{\delta y(y+1)}{y-1}$

(VI) $y'' = (\frac{1}{y} + \frac{1}{y-1} + \frac{1}{y-x})\frac{y'^2}{2} - (\frac{1}{x} + \frac{1}{x-1} + \frac{1}{y-x})y'$

$$+ \frac{y(y-1)(y-x)}{x^2(x-1)^2}(\alpha + \frac{\beta x}{y^2} + \frac{\gamma(x-1)}{(y-1)^2} + \frac{\delta x(x-1)}{(y-x)^2}).$$

Subsequently Garnier [12] showed that the Painlevé equations I through V are also obtained as monodromy preserving deformation equations for certain second order linear ordinary differential equations with regular and irregular singularities.

In this section we pursue this approach by Schlesinger and Garnier (see [14] [36] for recent progress), laying stress on the Hamiltonian structure of deformation equations ([9][10][3]). This point of view reveals the mathematical rôle of the "physically" introduced correlation functions appearing in our theory of Holonomic Quantum Fields.

3.2 Monodromy data ([14], [36]). Let us consider (1.15) with

$$(3.5)\qquad A(x) = \sum_{\nu=1}^{n} \frac{A_\nu}{x-a_\nu} + A_\infty, \quad A_\infty = \begin{pmatrix} t_1 & & \\ & \ddots & \\ & & t_m \end{pmatrix}.$$

It has a normalized solution which admits the following asymptotic expansion in a certain sector in the neighborhood of $x=\infty$.

$$(3.6)\qquad Y(x) \sim \hat{Y}_\infty(x)x^D e^{A_\infty x}, \quad D:\ \text{diagonal}, \quad \hat{Y}_\infty(x) = 1 + \sum_{\ell=1}^{\infty} Y_{\infty\ell}x^{-\ell}.$$

We can choose successive sectors $\mathcal{S}_1, \mathcal{S}_2, \cdots$ (see Figure 3.1) and correspondingly the normalized solutions $Y_1(x), Y_2(x), \cdots$ in each sector. However the analytic continuation of $Y_\ell(x)$ into $\mathcal{S}_{\ell+1}$ does not necessarily coincide with $Y_{\ell+1}(x)$, but differs by a constant matrix C_ℓ, called the Stokes multiplier:

$$(3.7)\qquad Y_{\ell+1}(x) = Y_\ell(x)C_\ell \qquad (\ell=1,\cdots,k).$$

These matrices $D, C_1, \cdots, C_k$ constitute a refined notion of the monodromy matrix at the irregular singularity $x=\infty$. In fact the monodromy matrix M_∞ corresponding to the path encircling $x=\infty$ clockwise is given by

$$(3.8) \qquad M_\infty = e^{2\pi i D} C_k^{-1} \cdots C_1^{-1}.$$

We choose a path joining ∞ to a_ν and consider the analytic continuation along this path of the normalized solution $Y_1(x)$ of the first sector. Assuming that the eigenvalues of A_ν are distinct, we conclude that $Y(x)$ has the following local expression at $x=a_\nu$.

$$(3.9) \qquad Y(x) = \hat{Y}_\nu(x)(x-a_\nu)^{L_\nu}, \quad \hat{Y}_\nu(x) = \sum_{\ell=0}^{\infty} Y_{\nu\ell}(x-a_\nu)^\ell, \quad \det Y_{\nu 0} \neq 0,$$

$$(3.10) \qquad A_\nu = Y_{\nu 0} L_\nu Y_{\nu 0}^{-1} .$$

The monodromy matrix M_ν at $x=a_\nu$ is given by

$$(3.11) \qquad M_\nu = e^{2\pi i L_\nu}.$$

Conversely, let $Y(x)$ be an $m\times m$ matrix holomorphic and invertible except for $x = a_1, \cdots, a_n, \infty$. Assume further that $Y(x)$ has local expressions (3.9) and (3.6) at $x = a_\nu$ $(\nu=1,\cdots,n)$ and $x=\infty$, respectively. Then it is shown that $Y(x)$ satisfy a system of the type (1.15)+(3.5). In this sense the following equivalence takes place.

Linear equation		Multi-valued function
$\dfrac{dY(x)}{dx} = A(x)Y(x)$	$\Longleftrightarrow$	$Y(x)$
with singularity data		with monodromy data
$a_1, \cdots, a_n, \infty$		$a_1, \cdots, a_n, \infty$
$A_1, \cdots, A_n, A_\infty$		$L_1, \cdots, L_n, D, C_1, \cdots, C_k$

3.3 Deformation equation ([36]). Now let us deform parameters $a_1, \cdots, a_n$ and $t_1, \cdots, t_m$, keeping the monodromy data $L_1, \cdots, L_n$ and $D, C_1, \cdots, C_k$ fixed. If we denote by d the exterior differentiation with respect to $x, a_1, \cdots, a_n$, and $t_1, \cdots, t_m$, Y satisfies the following system of linear differential equations.

$$(3.12) \qquad dY = \Omega Y, \quad \Omega = \sum_{\nu=1}^{n} A_\nu d\log(x-a_\nu) + d(xA_\infty) + \Theta,$$

$$(3.13) \qquad \Theta_{jj} = 0, \quad \Theta_{jk} = \sum_{\nu=1}^{n} (A_\nu)_{jk} d\log(t_j-t_k).$$

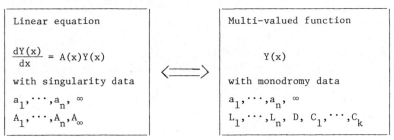

The integrability condition $d\Omega = \Omega\wedge\Omega$ for (3.13) gives rise to the following non linear system of total differential equations.

$$(3.14)\quad dA_\nu = -\sum_{\nu'(\neq\nu)} [A_\nu, A_{\nu'}]d\log(a_\nu - a_{\nu'}) - [A_\nu, d(a_\nu A_\infty) + \Theta].$$

3.4 Hamilton structure. The system (3.14) can be written as a Hamiltonian system. Let us define a Hamiltonian 1-form ω with time variables $a_1,\cdots,a_n$, $t_1,\cdots,t_m$.

$$(3.15)\quad \omega = \frac{1}{2}\sum_{\nu\neq\nu'} \mathrm{trace}\, A_\nu A_{\nu'} d\log(a_\nu - a_{\nu'}) + \sum_\nu \mathrm{trace} A_\nu d(a_\nu A_\infty) + \frac{1}{2}\mathrm{trace}\,\Theta\sum_\nu A_\nu.$$

Then (3.14) is rewritten as

$$(3.16)\quad dA_\nu = \{A_\nu, \omega\}.$$

Here the Poisson bracket is defined by

$$(3.17)\quad \{(A_\nu)_{jk}, (A_{\nu'})_{j'k'}\} = \delta_{\nu\nu'}(\delta_{jk'}(A_\nu)_{j'k} - \delta_{j'k}(A_\nu)_{jk'}).$$

3.5 Painlevé equation of the fifth kind. In the special case when $n=m=2$, the Hamiltonian system (3.16) reduces to the following system.

$$(3.18)\quad H(t,y,z) = z + (-yz(\nu_1 - z) + y^{-1}(\nu_2 - z)(\nu_3 - z) - 2z^2 + (\nu_1+\nu_2+\nu_3)z)\frac{1}{t},$$

$$(3.19)\quad \{z,y\} = y,$$

$$(3.20)\quad \frac{dy}{dt} = \{y, H(t,y,z)\} = -ty - 2z(y-1)^2 - (y-1)(-\nu_1 y + \nu_2 + \nu_3),$$

$$\frac{dz}{dt} = \{z, H(t,y,z)\} = -yz(\nu_1 - z) - y^{-1}(\nu_2 - z)(\nu_3 - z).$$

Eliminating z from (3.20), we obtain the fifth Painlevé equation (3.4)-(V) with parameters

$$(3.21)\quad \alpha = \frac{1}{2}\nu_1^2,\quad \beta = -\frac{1}{2}(\nu_2 - \nu_3)^2,\quad \gamma = -1 + \nu_1 - \nu_2 - \nu_3 \quad\text{and}\quad \delta = -\frac{1}{2}.$$

If we set $\sigma(t) = tH(t,y(t),z(t))$, $\sigma(t)$ itself satisfies the following second order differential equation.

$$(3.22)\quad \left(t\frac{d^2\sigma}{dt^2}\right)^2 = \left(\sigma - t\frac{d\sigma}{dt} + 2\left(\frac{d\sigma}{dt}\right)^2 - (\nu_1+\nu_2+\nu_3)\left(\frac{d\sigma}{dt}\right)\right)^2 + 4\frac{d\sigma}{dt}\left(\nu_1 - \frac{d\sigma}{dt}\right)\left(\nu_2 - \frac{d\sigma}{dt}\right)\left(\nu_3 - \frac{d\sigma}{dt}\right).$$

3.6 Ising model. We show that the deformation theory of Ising model (§2) can be incorporated into the framework of this section. By introducing the generalized

Laplace transformation

$$(3.23)\qquad w_\nu(z,z^*) = \int \frac{du}{2\pi u}\begin{pmatrix}\sqrt{u}\\ \sqrt{u}^{\,-1}\end{pmatrix} e^{m(zu+z^*u^{-1})}\hat{w}_\nu(u)$$

the system (2.14), (2.15) is converted into equations

$$(3.24)\qquad \frac{d}{du}\begin{pmatrix}\hat{w}_1\\ \vdots\\ \hat{w}_n\end{pmatrix} = (-mA - \frac{1}{u}F + \frac{1}{u^2}G^{-1}mA^*G)\begin{pmatrix}\hat{w}_1\\ \vdots\\ \hat{w}_n\end{pmatrix}$$

having two irregular singularities of rank 1 at $u=0$ and $u=\infty$. The monodromy preserving deformation equation for (3.24) is just the same as that for (2.14) ([36]). This suggests the stability of isomonodromic property under the generalized Laplace transformation ([37]).

The algebraic expression for ω (2.22) in terms of F and G gives a Hamiltonian 1 form for (2.17), where the Poisson bracket is defined to be

$$(3.25)\qquad \{G_{jk}, G_{j'k'}\} = 0, \quad \{(FG^{-1})_{jk}, (FG^{-1})_{j'k'}\} = 0, \quad \{G_{jk},(FG^{-1})_{j'k'}\} = \delta_{jk'}\delta_{j'k}.$$

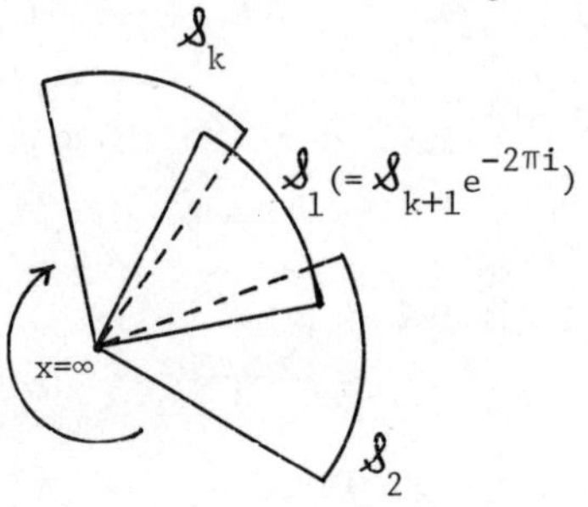

Figure 3 Sectors at $x=\infty$.

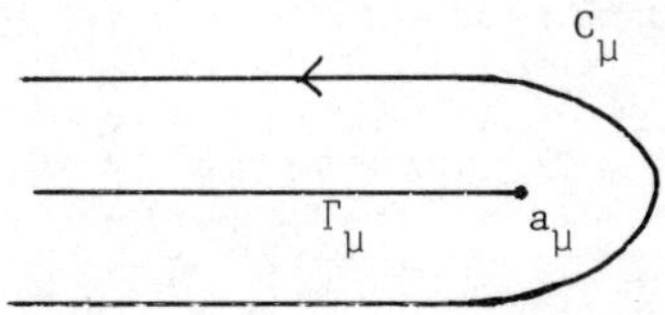

Figure 4

§4. Riemann's monodromy problem

<u>4.1</u> <u>Problem</u>. Let $a_1 < a_2 < \cdots < a_n$ be real numbers. We construct $m\times m$ matrix $Y(x)$ with monodromy data $(a_1,L_1),\cdots,(a_n,L_n)$ in the sense of (3.9), assuming that $L_1,\cdots,L_n$ are sufficiently 'small' (= close to 0) and imposing the following additional conditions (4.1) at $x=\infty$ and $(4.3)_{x_0}$ (or $(4.3)_{a_\nu}$).

$$(4.1)\qquad Y(x) = \hat{Y}_\infty(x)x^D, \quad \hat{Y}_\infty(x) = \sum_{\ell=0}^{\infty} Y_{\infty\ell}x^{-\ell}, \quad \det Y_{\infty 0} \neq 0 .$$

Here D is the 'small' matrix uniquely determined by the following condition.

$$(4.2)\qquad e^{2\pi i L_1}\cdots e^{2\pi i L_n} = e^{2\pi i D} .$$

The meaning of (4.2) is clear: A path encircling $a_n,\cdots,a_1$ successively is homotopic to a path encircling ∞.

The second additional condition is the normalization;

$(4.3)_{x_0}$ $\qquad Y(x_0) = 1$ for $x_0 \neq a_1, \cdots, a_n, \infty,$

or

$(4.3)_{a_\nu}$ $\qquad Y_{\nu 0} = 1.$

By a suitable linear transformation we obtain the solution subject to the Schlesinger's normalization:

$(4.3)_\infty$ $\qquad Y_{\infty 0} = 1.$

4.2 Operator solution ([H] II). Let $\psi^{(j)}(x)$, $\psi^{*(j)}(x)$ $(j=1,\cdots,m)$ denote free fermion operators on $\mathbb{R}$. They satisfy the following anti-commutation relations.

(4.4) $\quad [\psi^{(j)}(x), \psi^{(j')}(x')]_+ = 0,\ [\psi^{*(j)}(x), \psi^{*(j')}(x')]_+ = 0,\ [\psi^{(j)}(x), \psi^{*(j')}(x')]_+ = \delta(x-x').$

The vacuum $|vac\rangle$ is defined by

(4.5) $\quad \psi^{(j)}(p)|vac\rangle = \psi^{*(j)}(-p)|vac\rangle = 0 \qquad \text{for } p > 0$

where

(4.6) $\quad \psi^{(j)}(p) = \int_{-\infty}^{+\infty} dx\, e^{-ixp}\psi^{(j)}(x), \quad \psi^{*(j)}(p) = \int_{-\infty}^{+\infty} dx\, e^{ixp}\psi^{*(j)}(x).$

Then we have the following vacuum expectation values.

$$(4.7) \qquad \langle\psi^{(j)}(x)\psi^{(j')}(x')\rangle = \langle\psi^{*(j)}(x)\psi^{*(j')}(x')\rangle = 0,$$

$$\langle\psi^{(j)}(x)\psi^{*(j')}(x')\rangle = \langle\psi^{*(j)}(x)\psi^{(j')}(x')\rangle = \frac{1}{2\pi}\frac{i}{x-x'+i0}.$$

For an $m \times m$ matrix L we define a field operator $\varphi(a;L)$ by

$$(4.8) \quad \varphi(a;L) = :\exp \rho(a;L):,\ \rho(a;L) = \sum_{j,k=1}^{m} \int_{-\infty}^{+\infty}\int_{-\infty}^{+\infty} dx dx' \psi^{(j)}(x) R(x-a, x'-a; L)_{jk} \psi^{*(k)}(x')$$

$$(4.9) \quad R(x,x';L) = -2i \sin\pi L\ x_-^{-L}\ x_-'^{L}\left(\frac{1}{2\pi}\frac{i}{x-x'+i0}e^{-\pi i L} + \frac{1}{2\pi}\frac{-i}{x-x'-i0}e^{\pi i L}\right).^{(\dagger)}$$

$\varphi(a;L)$ satisfies the following commutation relation with free fields.

$$(4.10) \quad \varphi(a)\psi^{(k)}(x) = \sum_{j=1}^{m} \psi^{(j)}(x)\varphi(a)\, M_0(x-a)_{jk},\ \varphi(a)\psi^{*(k)}(x) = \sum_{j=1}^{m} \psi^{*(j)}(x)\varphi(a)({}^tM_0(x-a)^{-1})_{jk}$$

(†) $x_-^L = 0\ (x > 0), \quad = |x|^L\ (x < 0).$

where $$M_0(x) = \begin{cases} 1 & \text{if } x > 0 \\ e^{-2\pi i L} & \text{if } x < 0 . \end{cases}$$

Let us consider the following matrix of vacuum expectation values.

(4.11)
$$Y_+\begin{pmatrix} x_0, x ; & \begin{matrix} a_1 \cdots a_n \\ L_1 \cdots L_n \end{matrix} \end{pmatrix}_{jk} = 2\pi i (x-x_0) \frac{\langle \psi^{*(j)}(x_0)\psi^{(k)}(x)\boldsymbol{\varphi}(a_1;L_1)\cdots \boldsymbol{\varphi}(a_n;L_n)\rangle}{\langle \boldsymbol{\varphi}(a_1;L_1) \cdots \boldsymbol{\varphi}(a_n;L_n)\rangle},$$

$$Y_-\begin{pmatrix} x_0, x ; & \begin{matrix} a_1 \cdots a_n \\ L_1 \cdots L_n \end{matrix} \end{pmatrix}_{jk} = 2\pi i (x-x_0) \frac{\langle \psi^{*(j)}(x_0)\boldsymbol{\varphi}(a_1;L_1) \cdots \boldsymbol{\varphi}(a_n;L)\psi^{(k)}(x)\rangle}{\langle \boldsymbol{\varphi}(a_1;L_1) \cdots \boldsymbol{\varphi}(a_n;L_n)\rangle}.$$

(4.7) implies that, as a function of x, $Y_+(x)$ (resp. $Y_-(x)$) is the boundary value of a holomorphic function in the upper (resp. lower) half plane. (4.10) implies that $Y_\pm(x)$ satisfy the following relation on the real axis.

(4.12)
$$Y_-(x) = Y_+(x)\, M(x),$$
$$M(x) = M_\nu \cdots M_n \quad (a_{\nu-1} < x < a_\nu),\ M_\nu = e^{-2\pi i L_\nu}.$$

From (4.12) and (4.7) we see easily that the analytic continuations of $Y_\pm(x)$ satisfy the required monodromy property with the normalization condition $(4.3)_{x_0}$. In order to check the stronger requirement (3.9) about the exponent L_ν we should look into the local behavior more closely.

<u>4.3 Infinite series expressions</u>. An application of the Wick's theorem to the explicit formula (4.8), (4.9) and (4.11) leads us to the following infinite series expression for $Y(x_0, x)$

(4.13)
$$Y(x_0, x) = 1 - 2\pi i (x_0 - x) \sum_{\mu,\nu=1}^{n} Z_{\mu\nu}(x_0, x)$$

where $Z_{\mu\nu}(x_0, x)$ is a holomorphic function defined for $(x_0, x) \in (\mathbb{C}-\Gamma_\mu) \times (\mathbb{C}-\Gamma_\nu)$ by the following (see Figure 4 as for Γ_μ):

(4.14)
$$Z_{\mu\nu}(x_0,x) = \delta_{\mu\nu} \int_{-\infty}^{0}\int_{-\infty}^{0} dx_1 dx_2 \frac{1}{2\pi} \frac{i}{x_0 - x_1 - a_\mu} R(x_1, x_2; L_\nu) \frac{1}{2\pi} \frac{i}{x_2 + a_\nu - x}$$

$$+ \sum_{\ell=1}^{\infty} \sum_{\nu_1,\ldots,\nu_{\ell-1}=1}^{n} \int_{-\infty}^{0} \cdots \int_{-\infty}^{0} dx_1 \cdots dx_{2\ell+2} \frac{1}{2\pi} \frac{i}{x_0 - x_1 - a_\mu} R(x_1, x_2; L_\mu) A_{\mu\nu_1}(x_2, x_3)$$

$$\cdots R(x_{2\ell-1},x_{2\ell};L_{\nu_{\ell-1}})A_{\nu_{\ell-1}\nu}(x_{2\ell},x_{2\ell+1})R(x_{2\ell+1},x_{2\ell+2};L_\nu)\frac{1}{2\pi}\frac{i}{x_{2\ell+2}+a_\nu-x}\ .$$

Here we set $A_{\mu\nu}(x,x')=\frac{1}{2\pi}(1-\delta_{\mu\nu})\frac{i}{x+a_\mu-x'-a_\nu-i\varepsilon_{\mu\nu}0}$, $\varepsilon_{\mu\nu}=\mathrm{sgn}(\mu-\nu)$.

Assuming that $\mathrm{Im}\,a_1>\cdots>\mathrm{Im}\,a_n$, we can rewrite (4.13) in a form suitable to check the local behavior.

$$(4.15)\qquad Y(x_0,x)=(x_0-a_\mu)^{-L_\mu}\hat{Y}_{\mu\nu}(x_0,x)(x-a_\nu)^{L_\nu}$$

$$(4.16)\qquad \hat{Y}_{\mu\nu}(x_0,x)=\delta_{\mu\nu}+2\pi i(x_0-x)\int_{C_\mu}dx_1\int_{C_\nu}dx_2\,\frac{1}{2\pi}\frac{i}{x_0-x_1}(x_1-a_\mu)^{L_\mu}$$

$$\times\Big(\frac{1}{2\pi}\frac{i}{x_1-x_2}(1-\delta_{\mu\nu})+\sum_{\rho(\neq\mu)}\sum_{\sigma(\neq\nu)}Z_{\rho\sigma}(x_1,x_2)\Big)(x_2-a_\nu)^{-L_\nu}\frac{1}{2\pi}\frac{i}{x_2-x}\ .$$

Here the contour of integration C_μ is given in Figure 4, and x_0 and x are supposed to be inside of C_μ and C_ν, respectively.

We note that (3.9) follows from (4.15). Moreover, the normalization $(4.3)_{a_\nu}$ is achieved if we take

$$(4.17)\qquad Y_\nu(x)=\lim_{x_0\to a_\nu}(x_0-a_\nu)^{L_\nu}Y(x_0,x).$$

$Y_\nu(x)$ is also expressible in terms of field operators. We set

$$(4.18)\qquad \varphi^{*(k)}(a;L)=:\sum_{j=1}^{n}\int_{-\infty}^{a}dx_1\psi^{*(j)}(x_1)\left(\frac{\sin\pi\,{}^tL}{\pi}|x-a|^{{}^tL-1}\right)_{jk}\exp\rho(a;L):.$$

Then $Y_\nu(x)$ is written as

$$(4.19)\qquad Y_\nu(x)_{jk}=\frac{2\pi i(x-a_\nu)\langle\varphi(a_1;L_1)\cdots\varphi^{*(j)}(a_\nu;L_\nu)\cdots\varphi(a_n;L_n)\psi^{(k)}(x)\rangle}{\langle\varphi(a_1;L_1)\cdots\varphi(a_n;L_n)\rangle}\ .$$

We have already mentioned about $\langle\varphi(a_1;L_1)\cdots\varphi(a_n;L_n)\rangle$ in §1 (1.20). We note that $d\log\langle\varphi(a_1;L_1)\cdots\varphi(a_n;L_n)\rangle$ can be expressed as an infinite series similar to (4.13) ([H] II).

§5. Density matrix for impenetrable bose gas.

5.1 Fredholm theory ([38], [39]). As was pointed out by Schultz [38], the discretized version of (1.7), with $c=+\infty$ and finite particle density ρ_0, is the 1 dimensional XY model defined by the Hamiltonian

$$(5.1)\qquad \mathcal{H}_{XY}(\gamma,h)=-\frac{1}{4}\sum_m\big((1+\gamma)\sigma^x_m\sigma^x_{m+1}+(1-\gamma)\sigma^y_m\sigma^y_{m+1}+2h\sigma^z_m\big)$$

with $\gamma=0$, $h=\cos\pi\rho_0\varepsilon$ (ε=lattice spacing). This is one of typical models solvable by the technique of the Clifford group ([H] V). The structure of (5.1) is quite similar to the Ising model (§2, 2.2). This time, going back to the continuum $p(x) =\lim\varepsilon^{-1/2}p_m$, $q(x) = \lim\varepsilon^{-1/2}q_m$, we obtain the one particle reduced density matrix $\rho_{FF}(x)$ for free fermion (we choose $\rho_0=\pi^{-1}$) as

$$(5.2) \qquad \langle q(x)\, p(x')\rangle - \delta(x-x') = -\frac{2}{\pi}\frac{\sin(x-x')}{x-x'} = -2\rho_{FF}(x-x'),$$

while the scaling limits corresponding to $\langle \varphi_F(a_1)\cdots \varphi_F(a_{2n})\rangle$ and $\langle\psi_\pm(x)\psi_{\pm'}(x')\varphi_F(a_1)\cdots \varphi_F(a_{2n})\rangle$ in (2.12) give the Fredholm determinant and the first Fredholm minor belonging to the kernel (5.2), considered on the union of finite intervals with end-points $a_1,\cdots,a_n$. The n particle reduced density matrix (1.9) itself is the n-th Fredholm minor determinant ([39]). We are to characterize these quantities by deformation theory.

Another approach due to Vaidya-Tracy [40] is to study the double scaling limit of the XY model (5.1). For this we refer to [40] and [S] XV, XVI, [3].

5.2 Deformation theory. Let $a_1 < a_2 < \cdots < a_{2n}$ be real numbers and set $I = [a_1,a_2]\cup[a_3,a_4]\cup\cdots\cup[a_{2n-1},a_{2n}]$. We denote by $\Delta_I(\lambda)$, $\Delta_I\begin{pmatrix} x_1\cdots x_r \\ x'_1\cdots x'_r \end{pmatrix};\lambda)$ and $R_I(x,x';\lambda)$ the Fredholm determinant, the r-th Fredholm minor and the resolvent kernel for the integral kernel $\sin(x-x')/(x-x')$ on I, respectively. Since we have

$$(5.3) \qquad \Delta_I\left(\begin{matrix} x_1\cdots x_r \\ x'_1\cdots x'_r \end{matrix}\;;\;\lambda\right) = (-\lambda)^r\Delta_I(\lambda)\det(R_I(x_j,x'_k;\lambda))_{j,k=1,\cdots,r}\,,$$

$$(5.4) \qquad \partial\log\Delta_I(\lambda)/\partial a_j = (-)^{j+1}\lambda R_I(a_j,a_j;\lambda),$$

$$(5.5) \qquad \partial^2\log\Delta_I(\lambda)/\partial a_j\partial a_k = (-)^{j+k+1}\lambda^2 R_I(a_j,a_k;\lambda)^2 \qquad (j\neq k),$$

the characterization of $d\log\Delta_I(\lambda)$ is sufficient.

We set

$$(5.6) \qquad Y_{I\infty}(x) = \begin{pmatrix} R^+_{+I}(x;\lambda) & R^-_{+I}(x;\lambda) \\ R^+_{-I}(x;\lambda) & R^-_{-I}(x;\lambda) \end{pmatrix},$$

$$(5.7) \qquad R^{\pm'}_{\pm I}(x;\lambda)= \sum_{\ell=0}^{\infty}\lambda^\ell\int_I\cdots\int_I dx_1\cdots dx_\ell \frac{\pm' e^{\pm' i(x_0-x_1)}}{2i}\frac{\sin(x_1-x_2)}{x_1-x_2}\cdots\frac{\sin(x_{\ell-1}-x_\ell)}{x_{\ell-1}-x_\ell}e^{\pm ix_\ell} \qquad (x_0=x).$$

$Y_{I\infty}(x)$ is holomorphic and invertible except for $x=\infty$, $a_1,\cdots, a_{2n}$. The local behavior of $Y_{I\infty}(x)$ at $x=\infty$ and $x=a_j$ $(j=1,\cdots,2n)$ are as follows.

$$(5.8)\qquad Y_{I\infty}(x) = (1+O(\tfrac{1}{x}))\, e^{\begin{pmatrix} i & \\ & -i\end{pmatrix}x},$$

$$(5.9)\qquad Y_{I\infty}(x) = \hat{Y}_I(x) \left(\frac{(x-a_1)\cdots(x-a_{2n-1})}{(x-a_2)\cdots(x-a_{2n})}\right)^{\begin{pmatrix} 0 & \frac{i\lambda}{2} \\ 0 & 0\end{pmatrix}} \begin{pmatrix} 1 & 0 \\ 1 & 1\end{pmatrix}.$$

Here $\hat{Y}_I(x)$ is holomorphic and invertible except for $x=\infty$. We note that the Stokes multipliers are trivial in this case.

These monodromy properties for $Y_{I\infty}(x)$ leads us to the following linear system:

$$(5.10)\qquad dY_{I\infty} = \Omega_I Y_{I\infty},\quad \Omega_I = \sum_{j=1}^{2n} A_j \, d\log(x-a_j) + \begin{pmatrix} i & \\ & -i\end{pmatrix} dx,$$

$$(5.11)\qquad A_j = \begin{pmatrix} r_{+j} \\ r_{-j}\end{pmatrix} \cdot (-r_{-j}\ r_{+j}),\quad r_{\pm j} = \sqrt{(-)^j i\lambda (R^+_{\pm I}(a_j;\lambda)+R^-_{\pm I}(a_j;\lambda))}.$$

As the integrability condition for (5.10) we obtain the Hamiltonian equation, an equivalent of (1.14):

$$(5.12)\qquad dr_{\pm j} = \{r_{\pm j},\ \omega\},$$

with the Hamiltonian ω of (1.13) and the Poisson bracket

$$(5.13)\qquad \{r_{+j},\ r_{+j'}\} = 0,\quad \{r_{-j},\ r_{-j'}\} = 0,\quad \{r_{+j},\ r_{-j'}\} = \delta_{jj'}.$$

5.3 1 particle reduced density matrix. The one particle reduced density matrix $\rho(x)$ for impenetrable bose gas and the level spacing probability distribution function $E_u(t)$ in [41] coincide with $-\frac{1}{2}\Delta_{[0,x]}\begin{pmatrix}0 \\ x\end{pmatrix}; \frac{2}{\pi})$ and $\Delta_{[0,\pi t]}(\frac{1}{\pi})$. If we set

$$(5.14)\qquad \sigma_I^{\det} = t\frac{d}{dt}\log\Delta_I(t),\quad \sigma_I^{\min} = t\frac{d}{dt}\log\Delta_I\left(\begin{matrix}0\\t\end{matrix};\lambda\right),\quad I = [0,t],$$

they satisfy the non-linear equation (3.22) with $(\nu_1, \nu_2, \nu_3) = (0,0,0)$ and $(0,1,1)$ (i.e. (1.11)), respectively. We note that the one particle reduced density matrix $\rho_0^{-1}\rho_{FF}(x) = \sin x/x$ for free fermion also satisfies (1.10), (1.11).

Finally, we give the small and large x behaviors of $\rho(x)$ ($C=1/2\pi$ in (5.15)):

$$(5.15)\qquad \rho_0^{-1}\rho(x) = 1-\frac{1}{3!}x^2+\frac{2C}{3^2}x^3+\frac{1}{5!}x^4-\frac{11C}{3^2\cdot 5^2}x^5-\frac{1}{7!}x^6+\frac{61C}{2^2\cdot 3^3\cdot 5^2\cdot 7^2}x^7+\left(\frac{1}{9!}+\frac{C^2}{3^5\cdot 5^2}\right)x^8+\cdots$$

$$(5.16)\qquad \rho_0^{-1}\rho(x) = \frac{\rho_\infty}{\sqrt{x}}\left(1+\frac{1}{2^3x^2}(\cos 2x-\frac{1}{4})+\frac{3}{2^4x^3}\sin 2x+\frac{3}{2^8x^4}\left(\frac{11}{2^3}-31\cos 2x\right)+\cdots\right).$$

ρ_∞ was determined by [40] to be $\pi e^{1/2}\, 2^{-1/3} A^{-6}$ (A=Glaisher's constant). See [S] XVI, [3] for further expansions up to x^{10} and to x^{-8}.

5.4 Low temperature and large coupling expansions. The above analysis concerns with the case of zero temperature $kT=\beta^{-1}=0$ and infinite coupling $c=\infty$. Density matrix for finite β and c is written down in the form of β^{-1} and c^{-1} expansions. Each coefficient contains a finite number of integrals over the density matrix with $\beta=c=\infty$ and simple elementary functions.

§6. Higher dimensions

6.1 General setting. When the space-time dimension n is greater than 2, analogous construction as in §2 inevitably requires to introduce fields that depend on extended objects. Let us start with two kinds of fields, the main field $\varphi[C]$ and the auxiliary field $\psi[C']$, dependent on a closed spacelike submanifold C (resp. C') of dimensions r (resp r'). Here r, r' are non-negative integers subject to the condition $r+r' = n-2$. When C and C' are in a mutually spacelike position so that they lie on a common spacelike hypersurface, one has the notion of the linking number $\nu(C, C')$. We set the following commutation relation between the main and auxiliary fields for mutually spacelike C and C' (Figure 5):

$$\varphi[C]\,\psi[C'] = (-)^{\nu(C,C')}\,\psi[C']\,\varphi[C]. \tag{6.1}$$

Up to this point (6.1) looks quite similar to the 'tHooft algebra ([19]). However, in order to apply our machinary we require one more point : the auxiliary field $\psi[C']$ should be free in a sense or another, so that (6.1) will completely characterize the main field $\varphi[C]$ (theory of Clifford group, [H] I).

6.2 Construction from local free fields ([S] XII). Unfortunately "free fields depending on extended objects" is not a tractable notion at present. We restrict ourselves to the case $r'=0$, where the auxiliary field is local, e.g. the usual free Dirac field $\psi(x)$. In this case the main field $\varphi[C]$ is given by : exp (quadratic form in ψ):, having

$$w(x, x') = \langle \psi(x)\bar{\psi}(x')\,\varphi[C]\rangle / \langle\varphi[C]\rangle \tag{6.2}$$

as the kernel of the quadratic form. The commutation relation (6.1) now tells that the Euclidean continuation $w^{\mathrm{Euc}}(x,x')$ of (6.2) is ramified around an $n-2$ dimensional submanifold C with the corresponding monodromy -1. More generally it is straightforward to incorporate phase factors (or matrices) $e^{2\pi i \ell}$ in place of -1. The problem is now to construct a deformation theory for w^{Euc}.

6.3 Deformation theory ([S] XIII). To this end we consider a Riemann-Hilbert monodromy problem for Euclidean Dirac equations $(-\partial\!\!\!/ + m)w = 0$. Let Γ be an $n-1$ dimensional closed submanifold of the Euclidean space $\mathbb{R}^n$. Let $M(\xi)$ be an $N\times N$ real analytic matrix on Γ, assumed to be close to the unity. We pose the following problem : find $w(x,x')$ such that

$$(6.3)\quad \begin{array}{lll} \text{(i)} & (-\not\partial_x+m)w(x,x') = \delta^n(x-x') & (x,x'\in \mathbb{R}^n-\Gamma) \\ \text{(ii)} & |w(x,x')| = O(e^{-m|x|}) & (|x|\to\infty,\ x' \text{ fixed}) \\ \text{(iii)} & w(\xi^+,x') = M(\xi)w(\xi^-,x') & (\xi\in\Gamma). \end{array}$$

Here $w(\xi^{\pm},x')$ means the boundary value from inside or outside Γ (Figure 6). Conditions (6.3) uniquely determine $w(x,x') = w(x,x';\Gamma,M)$ as a functional of Γ and M. Consider a small variation $\Gamma' = \{\xi+\delta\rho(\xi)\,|\,\xi\in\Gamma\}$ of Γ by specifying a vector field $\delta\rho(\xi) = (\delta\rho^1(\xi),\cdots,\delta\rho^n(\xi))$ on Γ. We assume the "monodromy" $M'(\xi')$ on Γ' to be preserved in the sense that $M'(\xi+\delta\rho(\xi)) = M(\xi)$ $(\xi\in\Gamma)$. Then the following variational formula is valid:

$$(6.4)\quad \delta w(x,x') = \int_\Gamma d\sigma(\xi) \sum_{\mu=1}^{n} \delta\rho^\mu(\xi)\, \frac{\delta w(x,x')}{\delta\rho^\mu(\xi)},$$

$$\frac{\delta w(x,x')}{\delta\rho^\mu(\xi)} = w(x,\xi^+)\cdot(n_\mu(\xi)\not\partial-\not n(\xi)\partial_\mu)M(\xi)\cdot w(\xi^-,x')$$

where $d\sigma(\xi)$ is the surface element and $n(\xi) = (n_1(\xi),\cdots,n_n(\xi))$ is the unit outer normal.

Our original situation is realized as a limiting case where Γ contains an n-2 dimensional submanifold C and $M(\xi)$ is a step function on Γ (Figure 7). Since (6.4) contains a tangential derivative of $M(\xi)$, $\frac{\delta w}{\delta\rho^\mu(\xi)} = 0$ wherever $M(\xi)$ = constant. This guarantees that the variational equation does not depend on the choice of Γ. The deformation equations in 2 dimensions (§2, §4) are recovered from (6.4) along with (6.3)-(i) and the Euclidean covariance ([S]XIV).

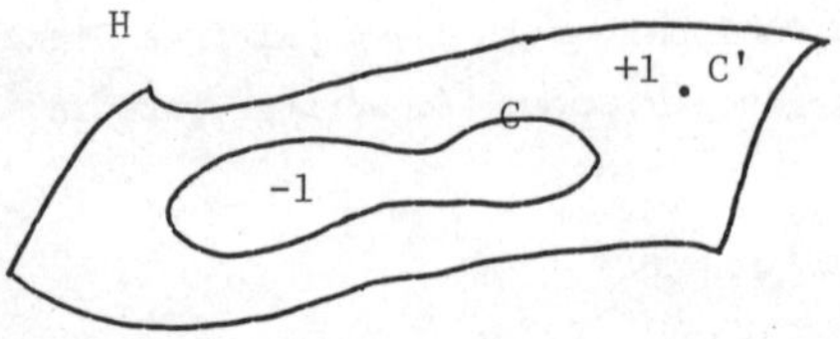

Figure 5
n=3, r=1, r'=0.
H is a spacelike hypersurface that contains C and C'.

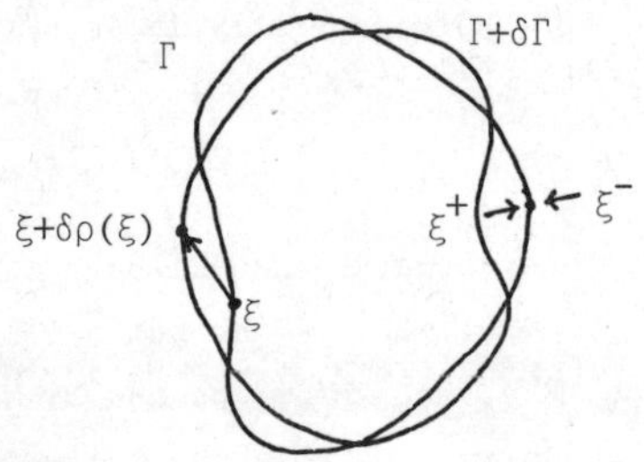

Figure 6

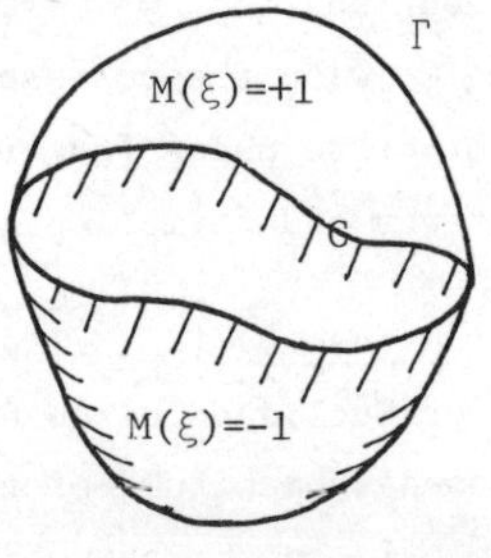

Figure 7

REFERENCES

1. M. Sato, T. Miwa and M. Jimbo, a series of papers entitled "Holonomic quantum fields." I: Publ. RIMS, Kyoto Univ. 14 (1978) 223, II: ibid. 15 (1979) 201, III: ibid., to appear in 15 (1979), IV: RIMS preprint 263 (1978), V: ibid. 267 (1978), IV supplement: ibid. 304 (1979). These papers are referred to in the text as [H] I, etc.

2. ___________ , a series of short notes entitled "Studies on holonomic quantum fields." Proc. Japan Acad. 53A (1977) 6(I), 147(II), 153(III), 183(IV), 219(V), 54A (1978) 1(VI), 36(VII), 221(VIII), 263(IX), 309(X), 55A (1979) 6(XI), 73(XII), 115(XIII), 157(XIV), 267(XV), 317(XVI). Some of them are with different authors (IX: M. Jimbo, XIII, XIV: M. Jimbo and T. Miwa, XV: M. Jimbo, T. Miwa and M. Sato, XVI: M. Jimbo, T. Miwa, Y. Môri and M. Sato). These papers are referred to in the text as [S] I, etc.

3. M. Jimbo, T. Miwa, Y. Môri and M. Sato, Density matrix of impenetrable bose gas and the fifth Painlevé transcendent, RIMS preprint, Kyoto Univ. 303 (1979).

4. M. Jimbo, T. Miwa and M. Sato, Kakuyûgô Kenkyu, Inst. Plasma Phys. Nagoya Univ., 40 Suppl. (1978) 45.
___________ , Proc. of the "International Colloquium on Complex Analysis, Microlocal Calculus and Relativistic Quantum Theory" Les Houches, France (1979), to appear.

5. E. Barouch, B. M. McCoy and T. T. Wu, Phys. Rev. Lett. 31 (1973) 1409.
T. T. Wu, B. M. McCoy, C. A. Tracy and E. Barouch, Phys. Rev. B13 (1976) 316.

6. B. M. McCoy, C.A . Tracy and T. T. Wu, J. Math. Phys. 18 (1977) 1058.

7. M. Girardeau, J. Math. Phys. 1 (1960) 516.

8. L. Schlesinger, J. Reine u. Angew. Math. 141 (1912) 96.

9. J. Malmquist, Arkiv Math. Astro. Phys. 17, No.18 (1922) 1.

10. K. Okamoto, Polynomial Hamiltonians associated to the Painlevé equations, preprint Tokyo Univ. (1979).

11. R. Fuchs, Math. Ann. 63 (1907) 301.

12. R. Garnier, Ann. Ecol. Norm. Sup. 29 (1912) 1.

13. M. J. Ablowitz, A. Ramani and H. Segur, A connection between nonlinear evolution equations and ordinary differential equations of P-type I, II, preprints (1979).

14. H. Flaschka and A.C. Newell, Monodromy and spectrum preserving deformations I, preprint (1979).

15. K. Ueno, Monodromy preserving deformations and its application to soliton theory, RIMS preprint 302, Kyoto Univ. (1979).

16. L. D. Faddeev, Modèles complètement intégrables de la théorie quantique des champs, preprint (French translation by Cambefort in Dijion) (1979).

17. J. Honerkamp, P. Weber and A. Wiesler, Nucl. Phys. B152 (1979) 266.

18. H. B. Thacker and D. Wilkinson, Phys. Rev. D19 (1979) 3660.

19. G.'t Hooft, Nucl. Phys. B138 (1978) 1.

20. J. Hadamard, Œuvre t.II.
P. Lévy, Problèmes concrets d'analyse fonctionelle, Gauthier-Villars, Paris (1951).

21. L. Onsager, Phys. Rev. 65 (1944) 117.

22. C. N. Yang, Phys. Rev. 85 (1952) 808.

23. B. M. McCoy and T. T. Wu, The two dimensional Ising model, Harvard University Press, Cambridge, Mass. (1973).

24. B. M. McCoy, C. A. Tracy and T. T. Wu, Phys. Rev. Lett. 38 (1977) 793.

25. R. Z. Bariev, Phys. Lett. 55A (1976) 456.
Phys. Lett. 64A (1977) 169.

26. M. Sato, T. Miwa and M. Jimbo, Field theory of the two dimensional Ising model in the scaling limit, RIMS preprint, Kyoto Univ. 207 (1976).

27. D. B. Abraham, Comm. Math. Phys. 59 (1978) 17.

28. G. E. Latta, J. Rational Mech. Anal. 5 (1956) 821.

29. J. M. Myers, J. Math. Phys. 6 (1965) 1839.

30. B. Kaufman, Phys. Rev. 76 (1949) 1232.

31. T. D. Schultz, D. C. Mattis and E. H. Lieb, Rev. Mod. Phys. 36 (1964) 856.

32. L. P. Kadanoff and H. Ceva, Phys. Rev. B3 (1971) 3918.

33. B. Schroer, T. T. Truong and P. Weisz, Ann. Phys. 102 (1976) 156.

34. B. Riemann, Werke, p.379.

35. P. Painlevé, Oeuvre t.III, p.187.

36. K. Ueno, Monodromy preserving deformation of linear differential equations with irregular singular points, RIMS preprint, Kyoto Univ. 301 (1979).

37. G. D. Birkhoff, Trans. Amer. Math. Soc. 10 (1909) 436.

38. T. D. Schultz, J. Math. Phys. 4 (1963) 666.

39. A. Lenard, J. Math. Phys. 7 (1966) 1268.

40. H. G. Vaidya and C. A. Tracy, Phys. Lett. 68A (1978) 378.
Phys. Rev. Lett. 42 (1979) 3.

41. M. L. Mehta, Random Matrices, Academic Press, New York (1967).

The authors would like to thank the secretaries in RIMS for carefully typing this manuscript.

INSTABILITY OF PHASE COEXISTENCE AND TRANSLATION INVARIANCE IN TWO DIMENSIONS

Michael Aizenman*

Physics Department, Princeton University
Princeton, N.J. 08540

I. Description of the main result

Collective phenomena which are important features of the physics of bulk systems are already exhibited by deceptively simple lattice models. Among such phenomena are the first order phase transitions, which are associated with the occurance of several distinct phases of a system in a given temperature, magnetic field and other intensive parameters.

The question I would address is the possibility of a stable coexistence of phases at a phase transition. By this I mean the occurance of a state of an extended system which is in a thermodynamic equilibrium and which in different regions of the space

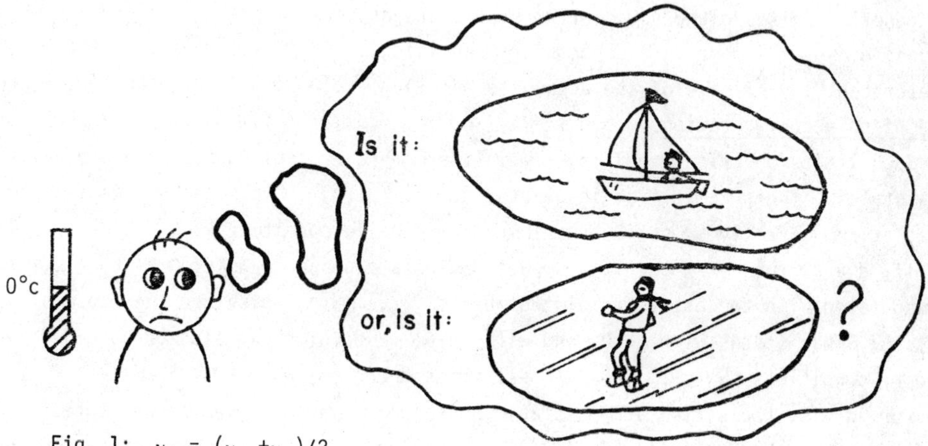

Fig. 1: $\mu_1 = (\mu_+ + \mu_-)/2$

Fig. 2: A would-be stable phase coexistence.

*Supported in part by the National Science Foundation grant PHY - 7825390 locally resembles different phases (fig. 2). This should not be confused with the occurance of mixed ensembles of the pure phases (fig. 1). These represent an uncertainty about the system's phase, rather than stable phase coexistence.

Specifically, I shall describe a recent proof that for the Ising model of ferromagnets there is no stable phase coesixtence in two dimensions, in contrast with its occurance in three and more dimensions.

The Ising model consists of spin variables $\sigma_i = \pm 1$, associated with the sites $i \in \mathbb{Z}^d$ of a d-dimensional lattice, which interact by the Hamiltonian

$$H(\sigma) = -\tfrac{1}{2} \sum_{|i-j| = 1} \sigma_i \sigma_j - h \Sigma \sigma_i \quad \text{(h is the external magnetic field).}$$

For a specified spin configuration on $\partial\Lambda$ - the boundary of a finite volume $\Lambda \subset \mathbb{Z}^d$, the equilibrium state in Λ is given by the probability measure

$$\mu_{\Lambda,\sigma_{\partial\Lambda}}(\{\sigma\}) = \exp.[-\beta\, H(\sigma)] \,/\text{Norm.}$$

on the space of spin configurations with the specified boundary values $\sigma_{\partial\Lambda}$ $(\beta = (hT)^{-1})$. More general finite volume equilibrium states are ensemble averages (≡ convex combinations) of $\mu_{\Lambda,\sigma_{\partial\Lambda}}$ over various boundary conditions. The thermodynamic limit, $\Lambda \uparrow \mathbb{Z}^d$, is described by limits of the above states which yield probability measures on the space of spin configurations, $\Omega = \{\{\sigma_i\}_{i \in \mathbb{Z}^d} \,|\sigma_i = \pm 1\} \equiv \{-1,1\}^{\mathbb{Z}^d}$.

Such limits describe a bulk system in thermodynamic equilibrium, and are called Gibbs states. Their collection is denoted here by $\Delta(\beta)$. Gibbs states can also be characterized directly by the Dobrushin-Lanford-Ruelle condition [1,2].

If $d \geq 2$, for $\beta > \beta_c$ the Ising model exhibits a phase transition at $h = 0$. This is associated with the existence of two phases $\mu_+, \mu_- \in \Delta(\beta)$ which are the two distinct limits of Gibbs states with $h \downarrow 0$ and with $h \uparrow 0$, and which can also be obtained by the boundary conditions $\sigma = +1$, or $\sigma = -1$, imposed on $\partial\Lambda$, as $\Lambda \uparrow \mathbb{Z}^d$.

For $d \geq 3$, the system admits stable phase coexistence at low temperatures. This was first proven by Dobrushin [3] (1972), by analyzing the equilibrium states induced in the volumes $[-n,n]^d = \Lambda_n$ by the boundary conditions:

$$\sigma_i = \text{sgn}(i_d + \tfrac{1}{2}) \quad \text{for} \quad i = (i_1,\ldots,i_d) \in \partial\Lambda_n.$$

At low temperatures the limit $n \to \infty$ yields a Gibbs state which is not invariant under translations and which offers an example of a stable phase coexistence. Additional light was shed on this phenomenon by the work of van Beijeren [4], and the proof of its occurance was recently extended, by Bricmont et. al. [5], to the Widom-Rowlinson model.

The above construction does not produce phase coexistence in two dimensions, as was first proven by Gallavotti [6]. His analysis shows that, at low temperatures, as $n \to \infty$ the long contour, which is a line passing between nearest neighbor pairs of different spins starting and ending between the two such pairs on $\partial \Lambda_n$, has unbounded fluctuations. Specifically, the contour would typically be either far below or far above any given finite region V, provided n is large enough. Consequently, the limiting state (defined on the local observables) is the ensemble average $\mu = (\mu_+ + \mu_-)/2$.

Gallavotti's work, and the further results of Abraham and Reed [7], Lebowitz [8], Messager and Miracle-Sole [9], Russo [10], and Merlini [11] raised and supported the possibility that a stable phase coexistence is not possible in the two dimensional model. That this indeed is the case is implied by the following recent result.

Theorem: Any Gibbs state of the two dimensional ferromagnetic Ising system, $\mu \in \Delta(\beta)$, is of the form: $\mu = \lambda\mu_+ + (1 - \lambda)\mu_-$, for some $\lambda \in [0,1]$.

Following is an outline of the proof, which is given in [12]. The theorem was also proved, independently, by Higuchi [13]. Both derivations incorporate previous results of Russo [10] who proved a similar assertion for states which have one of the main symmetries of the lattice. As remarked in [12], the proof given there can be extended to some other two phase systems at low temperatures.

2. Steps in the proof

The proof can generally be described as stochastic-geometrical. Its starting point is the identification of infinite contours as the interface lines. This permits the characterization of any given spin configuration as a patchwork of several pure-phase components. The fact that locally any spin configuration does occur even in the pure phases has not been overlooked. On general grounds, any choice of interface which satisfies the following three conditions is well justified.

1) For a specified configuration of the interface lines, the distribution of spins in the regions they separate should resemble pure phases.

2) The absense of interface lines in the typical configurations of a given Gibbs state should imply that the state is a combination of pure phases.

3) No interface lines should be found in configurations which are typical for either μ_+ or μ_-.

Of main interest to us is, of course, property 2), which provides a sufficient condition. However this criterion would have been useless had 3) not been satisfied. Condition 3) would presumably prevent the use of contour surfaces as interfaces in higher dimensions.

The verification of the above three conditions for infinite contours is the first step in the proof of the theorem, which is organized as follows:

1) Identification of interface lines

2) Proof of the thermodynamic instability of states with a single interface line, which is assumed to basically flat.

3) Proof that in any Gibbs state the number of interface lines is at most one, and that if such line exists, it has the properties postulated in 2).

3. Brief outline of the arguments

A very useful tool in the study of support properties of Gibbs states is a generalization of the fact that in a finite region, extremal Gibbs states correspond to pure boundary conditions. To formulate it, let $\mathcal{B}_\infty$ be the σ-algebra of subsets of Ω (events) which are measurable at infinity - e.g. the collection of spin configurations with exactly five infintie contours.

Proposition 1:

i) Any Gibbs state, $\mu(\cdot) \in \Delta(\beta)$, can be represented as a convex combination of extremal (non-decomposable) Gibbs states $\nu_\alpha(\cdot)$:

$$\mu(\cdot) = \int \hat{\mu}(d\alpha)\nu_\alpha(\cdot)$$

ii) $\mu \in \Delta(\beta)$ is extremal $\Leftrightarrow \forall A \in \mathcal{B}_\infty$ $\mu(A)$ is either 0 or 1.

iii) If $\mu \in \Delta(\beta)$, $A \in \mathcal{B}_\infty$ then μ conditioned on A is also a Gibbs state, i.e. $\mu(\cdot|A) \in \Delta(\beta)$.

The geometrical elements in our analysis consist, in addition to the infinite contours, of:

+ clusters (- clusters) -- the connected components, in the nearest-neighbor sense, of sites of the lattice on which the spins are +1(or -1)

+ (or -) * clusters -- same as above with the extension of the notion of connectedness to diagonally neighboring sites. (These appear naturally as the boundaries of - clusters).

Two key elements in the first step of the argument are the following support properties.

Lemma 1 (Russo [10]): If in a Gibbs state $\mu \in \Delta(\beta)$ almost surely (a.s.) there are no infinite + clusters then $\mu = \mu_-$ (assuming, of course, $\beta > \beta_c$).

Lemma 2 (Implied by a result of Russo[10]): In two dimensions, configurations with an infinite contour are a-typical (i.e. their occurance has probability zero) in the states μ_+ and μ_-.

Lemma 1 follows (for any d) from the fact that the absense of infinite + contours implies that any finite volume is surrounded by a - * cluster; and from the Markov property of Gibbs states (some care has to be exercised to insure "non-anticipatory" conditioning). Lemma 1 and proposition 1 imply that condition 2) in our criteria for interfaces is satisfied by the infinite contours. Condition 3) is directly implied by lemma 2.

Lemma 2 is proven by showing that (in d = 2) the infinite + clusters of μ_+ form a mesh which completely surrounds any point, preventing the existence of infinite -*clusters, and thus of infinite contours.

The second step in the argument is accomplished by proving the following lemma:

Lemma 3: Let $\Omega_1 = \{\sigma \in \Omega | \sigma$ has exactly one infinite contour, $\gamma(\sigma)$, and γ has finite, $\neq \phi$, intersections with $\{i_1 = \text{const.}\}\}$. Then $\mu(\Omega_1) = 0$ for any Gibbs state $\mu \in \Delta(\beta)$.

Since the level of the lowest intersection of $\gamma(\sigma)$ with $\{i_1 = 0\}$ is a well defined variable on Ω_1, no probability measure supported on Ω_1 can be invariant under T_2 --the translation in the i_2 direction. Using the decomposition to extremal states, one may show that lemma 3 is equivalent to:

Lemma 3′: Let μ be on extremal Gibbs state of the ferromagnetic Ising system, such that $\mu(\Omega_1) = 1$. Then μ is T_2 - invariant.

To prove lemma 3′ one may consider pairs $(\sigma,\hat{\sigma}) \in \Omega_1 \times \Omega_1$, distributed by the product measure $\mu \times T_2\mu$. Since each $\sigma \in \Omega_1$ has exactly one infinite contour, every finite volume may be completely surrounded by a connected set on which the spins are of a constant sign on each side of $\gamma(\sigma)$. The key observation now is that,due to line fluctuations,the two contours $\gamma(\sigma)$ and $\gamma(\hat{\sigma})$ would, $\mu \times T_2\mu$ - almost surely, intersect infinitely often. Thus each volume may both be completely surrounded by a set on which $\sigma \geq \hat{\sigma}$ and by a set on which $\sigma \leq \hat{\sigma}$. This implies that $\mu = T_2\mu$, by an argument similar to the one used in the proof of lemma 1.

The proof of step 3 would not be discussed here.

4. An open problem

One aim of this summary has been to expose some stochastic geometrical methods. It seems conceivable that a further development of these techniques could help to find out whether in three dimensions the "roughening temperature", at which phase coexistence destabilizes, is strictly below T_c.

References:

1. R. L. Dobrushin: Theor. Prob. Applic. 13, 197 (1968).
2. O. E. Lanford III, D. Ruelle: Commun. Math. Phys. 13, 194 (1969).
3. R. L. Dobrushin: Theor. Prob. Applic. 17, 582 (1972).
4. H. van Beijeren: Commun. Math. Phys., 40, 1, (1975).
5. J. Bricmont, J., L. Lebowitz, C. E. Pfister, E. Olivieri: Commun. Math. Phys. 66, 1 (1979).
6. G. Gallavotti: Commun. Math. Phys. 27, 103 (1972).
7. D. B. Abraham, P. Reed: Phys. Rev. Lett. 33, 377 (1979), and D. B. Abraham, P. Reed: Commun. Math. Phys. 49, 35 (1976).
8. J. L. Lebowitz: J. Stat. Phys. 16, 463 (1977).
9. A. Messager, S. Miracle-Sole: J. Stat. Phys. 17, 245 (1977).
10. L. Russo: Comun. Math. Phys. 67, 251 (1979).
11. D. Merlini: Preprint.
12. M. Aizenman: Phys. Rev. Lett., 43, 407 (1979), and M. Aizenman: "Translation Invariance and Instability of Phase Coexistence in the Two Dimensional Ising System", to appear in Commun. Math. Phys.
13. Y. Higuchi: "On the Absense of Non-Translationally Invariant Gibbs States for the Two-Dimensional Ising System". To appear in the Proceedings of the Colloquium on Random Fields (Esztergom, June 1979).

TOPOLOGY AND DEFECTS IN ORDERED MEDIA

G. Toulouse
Laboratoire de Physique de l'Ecole Normale Supérieure
24 rue Lhomond
75231 Paris Cedex 05, France

The topological study of stable configurations is in principal applicable to a vast variety of problems. So far it has been mostly developed for ordinary statistical mechanics problems (this is the so-called theory of defects in ordered media) and for gauge field problems (monopoles and instantons).

An ordered medium is typically a low temperature phase, with a broken symmetry, and characterized by an order parameter (e.g. a ferromagnetic phase). Besides uniform configurations in space of the order parameter, which are absolute ground states, there are distorted configurations which may happen to be stable, for continuous deformations, against return to uniform states. Among these stable distorsions, one makes a distinction between :

i) singular configurations, or defects, with a "core" which may be of dimensionality 0 (points), 1 (lines), 2 (walls),...

ii) non singular configurations, or textures.

There are several motivations for the study of defects. Their practical importance in governing the rigidity and dissipative properties of ordered media (metallurgy : plasticity ; magnetism : hysteresis ; superconductors : high-field magnets, ...). The clue that they provide to the nature of the order (cf the early observations by G. Friedel of defects in liquid crystals and the identification of the smectic order). The possibility of predicting new properties, guiding thereby the search of new materials. The role of the defects in the phase transitions (relations with the concept of disorder parameter, speculations for new types of phase transitions,...).

The study of topological excitations can be divided into the following categories :

- one-defect theory (classification),
- two-defect theory (combinations : coalescence, crossing),
- many-defect theory (phase transitions, hydrodynamics).

Before the introduction of homotopic tools, there used to be a theory of defects, called here the traditional theory, which was an outgrowth of the consideration of crystal dislocations. In brief, within the traditional theory,

i) a defect is characterised, fully and ever, by an integer number (or a set of such numbers),

ii) to every defect corresponds an antidefect,

iii) combinations are ruled by addition of these numbers,

iv) crossing (say, for lines) is permitted.

This traditional theory works well in cases such as translation dislocation lines in crystals, or point defects in ferromagnets. The convincing way to prove that a theory is insufficient, and needs to be improved, consists in showing that it leads to paradoxes. Such a paradox arises here, when one tries to extend this traditional approach to point defects in nematic liquid crystals[1]. The paradox is insoluble within the traditional theory ; its origin lies in a noncommutative feature and its explanation requires homotopy theory (action of π_1 on π_2).

The homotopy theory of defects starts with the consideration of the manifold of internal states $V = G/H$, where G is the symmetry group of the high temperature phase and H is the symmetry group of the low temperature phase. The homotopy group $\pi_r(V)$ classifies the defects of dimensionality $d' = d-r-1$. Thus, in dimension 3,

$\pi_0(V) \rightarrow$ walls,

$\pi_1(V) \rightarrow$ lines,

$\pi_2(V) \rightarrow$ points ;

besides, $\pi_3(V)$ classifies the non singular configurations or textures.

The Whitehead product[2]

$$\pi_p \times \pi_q \rightarrow \pi_{p+q-1}$$

describes the noncommutative features. If $p = q = 1$, it reduces to the commutator product of π_1. If $p = 1$, q arbitrary, it reduces to the action of π_1 on π_q. For $p,q \geq 2$, it gives to the ensemble of homotopy groups, of a manifold, a graded Lie algebra structure.

In the realm of gauge field problems, with G as a gauge group, the homotopy groups of G (or of G/C, where C is the center of G) have been put to use in order to classify the topologically stable objects. Thus in dimension 3 + 1,

$\pi_1 \rightarrow$ monopoles (via the flux strings attached to them)

and

$\pi_3 \rightarrow$ instantons.

In such simple frameworks, (homotopy groups of a Lie group), which is all I have seen so far (in the context of gauge field theories), the Whitehead product is disappointingly vanishing.

Noncommutativity effects for defects in ordered media include coalescence effects (for instance, in 3D uniaxial nematics or in 2D smectics) and crossing effects (for instance, in 3D biaxial nematics). Topological obstructions to crossing bring topological rigidity and entanglement problems. Thus also a mechanism for confinement of linked rings. In all this, theory is now far ahead of experiments.

Some developments of this basic theoretical picture can be listed under the title of double topology. This includes the solitons of G. Volovik and V.P. Mineyev, for systems with a small perturbing energy term. The absolute minima of the free energy define a manifold V. Neglecting the small term, the minima define a larger manifold V'. The Volovik-Mineyev solitons, non singular defects, are classified by the relative homotopy groups $\pi_r(V,V')$. Thus in dimension 3,

$\pi_1(V,V') \rightarrow$ wall solitons,

etc...

Neat examples have been described for the superfluid phases of Helium 3, where the nuclear dipolar energy provides the small energy perturbation.

Double topology includes also the study of cholesteric liquid crystals(4), for which the high energy cost of singularities in the molecular director field leads to a second topological constraint. This leads to the stability, against unlinking, of observed optical patterns of linked rings, a stability characterized by a non zero Hopf index.

In systems with broken translation invariance, complications arise in the topological classification, because of the necessity of including metric considerations (V.P. Mineyev, N.D. Mermin, V. Poénaru).

In conclusion of this rapid survey, we give a list of recent review papers(5).

References

(1) G. Toulouse, Proc. of the XVI Karpacz Winter School (Springer, 1979).

(2) V. Poénaru, G. Toulouse, J. Physique 38, 887 (1977) ; J. Math. Phys., 20, 13 (1979).

(3) G. Volovik, V.P. Mineyev, Zh. E.T.F., 73, 768 (1977).

(4) Y. Bouligand, B. Derrida, V. Poénaru, Y. Pomeau, G. Toulouse, J. Physique 39, 863 (1978).

(5) L. Michel, Proc. of the Tübingen Conf. on Group Theoretical Methods in Physics, 79, (Springer, 1978) ; Rev. Mod. Phys., to appear.
N.D. Mermin, Rev. Mod. Phys., 51, 591 (1979).
V.P. Mineyev, Sov. Phys. Uspekhi, to appear.
V. Poénaru, Proc. of the 1978 Les Houches Summer School on "Ill-condensed Matter", Eds. R. Balian, R. Maynard, G. Toulouse (North-Holland, 1979).

DEBYE SCREENING IN CLASSICAL STATISTICAL MECHANICS

David Brydges, University of Virginia
Paul Federbush, University of Michigan

Two years ago David Brydges, using techniques developed in Constructive Quantum Field Theory, rigorously proved Debye screening for certain Coulomb lattice systems in the low density region [1]. The present talk presents joint work [2] with Brydges generalizing his results from the lattice case, to continuum statistical mechanics; from the pure 1/r interaction, to include the presence of essentially arbitrary short range forces; and from a charge-symmetric situation, to cover arbitrary charge species. The short range forces are required to ensure the stability of the system (the pure 1/r potential is unstable in the classical situation), and also to fall off quickly enough at infinity so as not to interfere with the shielding. Rather then to here detail these two very physical conditions, we present a special case indicative of the general result and of interest in its own right.

We consider a system of charged, $e = \pm 1$, spheres of radius R , so that the potential V is given by

$$V = \begin{cases} \frac{1}{2} \sum\limits_{i \neq j} \frac{e_i e_j}{4\pi |\vec{x}_i - \vec{x}_j|} \; , & |\vec{x}_i - \vec{x}_j| \geq 2R \quad \text{all } i \neq j \\ \infty \; , & \text{otherwise} \end{cases} \tag{1}$$

The grand canonical partition function is then

$$Z = \sum_{N_+, N_-} \frac{1}{N_+!} \frac{1}{N_-!} z^{N_+ + N_-} \int d^3x_1 \ldots \int d^3x_{N_+ + N_-} e^{-\beta V(\vec{x}_i, e_i)} \tag{2}$$

where we are dealing with the charge symmetric situation, $z_+ = z_- = z$. The Debye length, ℓ_D , is defined as equal $(2z\beta)^{-1/2}$. By a limiting procedure to be explained, infinite volume expectation values of products of smeared density operators are defined. For simplicity, we will state our main clustering theorem for the two point function (rather than an arbitrary product).

We introduce dimensionless parameters $\xi_1 = \beta/R$ and $\xi_2 = zR^3$. Δ_i , $i = 1,2$ are arbitrary unit cubes, and $h(\vec{x})$ and $g(\vec{x})$ are functions of absolute value less than 1 .

THEOREM. Given α, $0 < \alpha < 1$, there is c_α and a function $f_\alpha(\xi)$, $0 \leq \xi < \infty$, $f_\alpha(\xi) > 0$, f_α decreasing, such that for $\xi_2 < f_\alpha(\xi_1)$

$$|\int_{\Delta_1} d^3x_1 \int_{\Delta_2} d^3x_2 g(x_1)h(x_2) \ < J(x_1)J(x_2) > | \le c_\alpha e^{-d(\Delta_1,\Delta_2)\alpha/\ell_D} \quad (3)$$

d is the distance between Δ_1 and Δ_2 , and c_α, f_α do not depend on the choice of g,h, or Δ_i . Thus at any fixed temperature, if the density is low enough, there is Debye screening. To specify the limiting procedure we define V_L as

$$V_L = \frac{1}{2} \sum_{i,j} \frac{e_i e_j}{4\pi |\vec{x}_i - \vec{x}_j|} (1 - e^{-|\vec{x}_i - \vec{x}_j|/\lambda \ell_D}) \quad (4)$$

λ is a small parameter. V_s is chosen so that

$$V = V_L + V_s \quad (5)$$

We see that

$$V_L = \frac{1}{2} \int J(x_1) \left(\frac{1}{-\Delta_0} - \frac{1}{-\Delta_0 + 1/\lambda^2 \ell_D^2}\right) J(x_2) \quad (6)$$

where Δ_0 is the free Laplacian on R^3 . For a large box Λ , we define V_L^Λ as (6) with the integrals restricted to Λ , and with Δ_0 replaced by the Laplacian satisfying Dirichlet data on $\partial\Lambda$. With Λ' a box containing Λ , we obtain our infinite volume limit by performing the following three steps in order.

a) Calculating expectation values for the grand canonical ensemble defined in volume Λ' , for potential $V^{\Lambda'} = V_L^\Lambda + V_s$.

b) Letting Λ' become infinite.

c) Letting Λ become infinite.

If we were dealing with a non charge-symmetric system we would impose a condition on the fugacities: that the system described by the same fugacities and V_s alone, should be neutral in the limit. The most important result of the limiting procedure primarily due to the "grounding" on $\partial\Lambda$ and the condition on fugacities, is the absence of a layer of surface charge (whose presence, as one can estimate by simple physical arguments, can lead to effects surviving the infinite volume limit). We do not here discuss how small λ must be chosen.

We feel the treatment of Debye screening in [2] is quite complete, but we mention three problems left open (among many others).

1) Can one generalize to the quantum statistical situation? This completely eludes us. Since in [3] and [4], Brydges and I have handled the short distance difficulties of the Coulomb system, we are particularly frustrated.

2) Can one use a more natural limiting procedure? In particular, can one effectively treat situations in which there is a surface charge?

3) In a system with charges and fugacities (e_i, z_i) we have imposed the condition

$$z_i / \max_i z_i \geq c > 0 \tag{7}$$

Our estimates are not uniform as $c \to 0$, if one species is of extremely low density compared to the other species, and of fractional charge compared to the other species. (e_f is fractional with respect to $e_1, e_2, \ldots, e_r$ if e_f is not a multiple of g.c.d.$(e_1, \ldots, e_r)$.) We have proven that fractional charges are shielded! This <u>may</u> suggest that our limiting condition (7) is merely a weakness of our procedure, and not a physical requirement.

Two of the basic ingredients of our proof are the same as in [1], the use of the Sine Gordon transformation and the Glimm-Jaffe-Spencer cluster expansion. The basic cluster expansion is fully presented in [7], and the particular application to situations where it is combined with a Peierls expansion first developed in [8]. The third basic ingredient is the Mayer expansion used to handle the effects of the short range potential V_s .

Pursuing the decomposition of V given in (4) and the succeeding discussion, we write $e^{-\beta V}$ in (2) as

$$e^{-\beta V_s} \, e^{-\beta V_L^\Lambda} \tag{8}$$

and use

$$e^{-\beta V_L^\Lambda} = \int d\mu_0(\phi) \, e^{i\beta^{1/2} \sum_\alpha e_{i(\alpha)} \phi(x_\alpha)} \tag{9}$$

$$d\mu_0(\phi) \sim d\phi \, e^{-\frac{1}{2} \int \phi [\lambda^2 \ell_D^2 (-\Delta)^2 + (-\Delta)] \phi} \tag{10}$$

μ_0 is a normalized Gaussian measure with inverse covariance a fourth order differential operator as given in (10). Although for this reason some of the techniques of Euclidean Quantum Field Theory are not available to us, we have gained the advantage that there are no renormalization difficulties; no expressions in ϕ need be normal ordered.

Inserting (8) and (9) into (2) we find the following sum (written in a compact notation) still to be integrated over ϕ .

$$\sum \frac{1}{N!} z^N \int e^{-\beta V_s} e^{i\beta^{1/2} \sum e_{i(\alpha)} \phi(x_\alpha)} \tag{11}$$

This sum may be expressed as the exponential of a Mayer series, M, the Mayer series developed for potentials as appearing in V_s together with the one-body potential, $-i\beta^{-1/2} e_i \phi(x)$. Up to a constant term M is given by the sum

$$\sum_i \int \rho_i(x)\, \varepsilon_i(x) + \frac{1}{2!} \sum_{i,j} \int \rho_{i,j}(x,y)\, \varepsilon_i(x)\, \varepsilon_j(y) + \dots. \tag{12}$$

with

$$\varepsilon_i(x) = e^{i\beta^{1/2} e_i \phi(x)} - 1 \tag{13}$$

Here $\rho_{i,j,\dots}(\ \ )$ is a truncated correlation function developed for the potentials in V_s alone. To control the Mayer series expansion, fall-off properties of the ρ's are required. These may be obtained from results in [6], or alternatively, with slightly different conditions on the short range potentials, from results in Appendix 1 of [2] using the formalism of [5].

The following approximation is the motivation behind much of our procedure

$$e^{2\rho \int_\omega [\cos(\beta^{1/2} \phi(x))-1]} \cong \sum_{n=-\infty}^{\infty} e^{-\frac{2\rho\beta}{2} \int_\omega (\phi(x) - n\tau)^2} \tag{14}$$

with ω a cube, and $\tau = 2\pi/\beta^{1/2}$. We write

$$e^{2\rho \int_\Lambda [\cos(\beta^{1/2} \phi(x) - 1]} = \sum_h e^{-\frac{2\rho\beta}{2} \int_\Lambda (\phi(x) - h(x))^2} e^G \tag{15}$$

where $h(x)$ is constant on each cube of a lattice of cubes filling Λ, and assuming values integral multiples of τ. G would equal 0 if (14) were an equality. Equation (15) is our Peierls expansion. Unfortunately we do not have a physical understanding of the Peierls expansion in (15).

Collecting the fruits of our efforts and with some further manipulation we arrive at

$$Z = \sum_h N \int d\mu(\psi) e^E e^G e^R \tag{16}$$

$$d\mu(\psi) \sim d\psi\, e^{-1/2 \int \psi [\lambda^2 \ell_D^2 (-\Delta)^2 + (-\Delta) + \frac{1}{\tilde{\ell}_D^2} + \nu] \psi} \tag{17}$$

$\mu(\psi)$ is a normalized Gaussian measure whose covariance may be read off from (17). $\psi(x)$ is a linear translate of $\phi(x)$, $\psi(x) = \phi(x) - g(x)$, with $g(x)$ chosen to (at least approximately) cancel the terms in the exponent of (15) linear in $\phi(x)$. R arises from the change in variables in the Gaussian integral. $1/\tilde{\ell}_D^2 = 2\rho\beta$; this term in (17) arises from the exponential terms in (15) quadratic in $\phi(x)$. ν is a "small" term arising from terms in the expansion for M quadratic in ϕ. E is the contribution of all terms in M except the first term, whose effect is given by (15), and the ν term. N is a relatively harmless normalization factor. In (16) if E and G were zero, and R a function of h alone, it would be easy to show screening took place. That deviations of E, G, and R from these ideal values are "small" in a suitable range of parameters, that screening occurs, is finally proven by applying the machinery of the cluster expansion to (16). We feel the inexorable ineluctable technical details of this process are treated quite efficiently in [2].

REFERENCES

1. D. Brydges, A rigorous approach to Debye screening in dilute Coulomb systems, Commun. Math. Phys. 58, 313-350 (1978).
2. D. Brydges and P. Federbush, Debye screening, preprint, University of Michigan.
3. D. Brydges and P. Federbush, The cluster expansion in statistical mechanics, Commun. Math. Phys. 49, 233-246 (1976).
4. D. Brydges and P. Federbush, The cluster expansion for potentials with exponential fall-off, Commun. Math. Phys. 53, 19-30 (1977).
5. D. Brydges and P. Federbush, A new form of the Mayer expansion in classical statistical mechanics, J. Math. Phys. 19, 2064-2067 (1978).
6. M. Dunean, D. Iagolnitzer, and B. Souillard, Decay of correlations for infinite-range interactions, J. Math. Phys. 16, 1662-1666 (1975).
7. J. Glimm, A. Jaffe, and T. Spencer, The particle structure of the weakly coupled $P(\phi)_2$ model and other applications. In: Lecture notes in physics, Vol. 25 (eds. G. Velo, A. Wightman), Berlin-Heidelberg-New York, Springer 1973.
8. J. Glimm, A. Jaffe, A convergent expansion about mean field theory I and II, Ann. Phys. 101, 610-630, 631-669 (1975).

EQUILIBRIUM PROPERTIES OF CLASSICAL SYSTEMS WITH LONG RANGE FORCES

Ch. Gruber, Ph. A. Martin
Laboratoire de Physique Théorique, Ecole Polytechnique Fédérale,
CH-1001 Lausanne, Switzerland

We discuss general properties of infinite classical systems of particles interacting with long range forces from the view point of equilibrium equations. Details can be found in [1, 2, 3, 4]. We consider N types of particles in R^ν, denoting by $q = (x, \sigma)$ the position $x \in R^\nu$ and the charge σ of a particle with $\int_V dq \ldots = \int_V dx \sum_\sigma \ldots$, $V \subset R^\nu$.

The particles interact with a two-body force $F(q_1, q_2) = \sigma_1 \sigma_2 F(x_2 - x_1)$ and a uniform background of fixed negative charge $F^{(1)}(q) = -\rho_B \sigma \int F(x-y)\, dy$, $\rho_B \geqslant 0$. $F(x)$ is assumed to be bounded and C^1 everywhere, except possibly for an integrable singularity at the origin; there is no restriction on the decay of $F(x)$ at infinity. Our discussion includes the Coulomb force, $F(x) = \frac{x}{|x|^\nu}$, and jellium models with $\rho_B \neq 0$.

We adopt as a definition of equilibrium states of infinite systems the following BBGKY hierarchy for the correlation functions $\rho^{(n)}(q_1 \ldots q_n)$, $n = 1, 2, \ldots$. ρ is an equilibrium state with respect to the temperature β^{-1}, the sequence of finite space regions $\{V_\lambda\}$ and the effective field E if its correlation functions satisfy:

(1) L¹-clustering: $\int_{R^\nu} dx_1 |\rho_T^{(n)}(x_1\sigma_1 \ldots x_n\sigma_n)| < \infty$ (1)

($\rho_T^{(n)}(q_1 \ldots q_n)$ are the truncated correlation functions)

(2) The BBGKY (equilibrium) hierarchy:

$$\nabla_1 \rho^{(n)}(q_1 \ldots q_n) = \beta \{\sigma_1 E_\rho(x_1) + \sum_{j=2}^{n} F(q_1, q_j)\} \rho^{(n)}(q_1 \ldots q_n)$$
$$+ \beta \int dq\, F(q_1, q) [\rho^{(n+1)}(q_1 \ldots q_n q) - \rho^{(n)}(q_1 \ldots q_n) \rho^{(1)}(q)] \qquad (2)$$

with $E_\rho(x) = E + \lim_{V_\lambda \to R^\nu} \int_{V_\lambda} dy\, (F(x-y) - F(-y))\, C_\rho(y)\, dy$

and $C_\rho(x) = \sum_\sigma \sigma \rho^{(1)}(x\sigma) - \rho_B$ is the charge density in the state ρ.

This definition deserves the following comments:

(a) The hierarchy applies to extremal states of the infinite system having at least the L¹-clustering property (1).
(b) $E_\rho(x)$ is the effective field in the state ρ and the parameter $E = E_\rho(o)$ is its value at $x = o$ (in the Coulomb case $E_\rho(x)$ is the solution of the electrostatic equation $\nabla \cdot E_\rho(x) = C_\rho(x)$ with the condition $E_\rho(o) = E$).
(c) If the force is not integrable in R^ν, the sequence $V_\lambda \to R^\nu$ defines $E_\rho(x)$ as a limit of definite integrals.
(d) If the force is integrable in R^ν, the BBGKY hierarchy reduces to the usual form it takes for short range forces.

We present now the properties of equilibrium states defined as solutions of (1) and (2) (assuming their existence). First, there are explicitely known examples of such states in one space dimension.

A. Examples of equilibrium states [1]

(1) The one dimensional Coulomb-jellium has non trivially periodic eq. states with $E_\rho(x) = 2\int_0^x C_\rho(y)\,dy + E$.

(2) The one dimensional two component Coulomb gas has an eq. state invariant under translations and charge conjugation with $E_\rho(x) = E = 0$ [5].

(3) The one dimensional two component Coulomb gas has also translation invariant eq. states, but not invariant under charge conjugation with non vanishing effective field $E_\rho(x) = E \neq 0$ [6].

B. Shape dependence [2, § 3]

A state verifying (2) may depend genuinely on the sequence of cut offs V_λ entering in (2). This cannot occur if the force decreases faster than the Coulomb force. Precisely, let $\{V_\lambda\} = \{\lambda V_0\}$ sequence of dilated regions and ρ_{EV_0} a solution of (2) with respect to E and $\{\lambda V_0\}$. If $F(x) \le \frac{M}{|x|^\gamma}$ with $\gamma > \nu - 1$, then ρ_{EV_0} is independent of V_0.

C. Local neutrality [2, § 4]

Let τ be a discrete subgroup of the translations and Δ_0 the corresponding fundamental cell. If the force decreases as or slower as the Coulomb force, any τ invariant eq. state is locally neutral. That is, if $\lim_{\lambda\to\infty} \lambda^\gamma F(\lambda\hat{x}) = d(\hat{x}) \neq 0$ ($\hat{x} = \frac{x}{|x|}$) with $\gamma \le \nu - 1$, then the average charge of the fundamental cell must vanish, $\int_{\Delta_0} C_\rho(x)\,dx = 0$.

D. The canonical sum rules [2, § 5]

The canonical sum rules are integral relation that link the n-point to the (n+1)-point correlation function:

$$\left(\sum_{i=1}^{n} \sigma_i\right)\rho^{(n)}(q_1 \ldots q_n) = -\int dq\,\sigma\left(\rho^{(n+1)}(q_1 \ldots q_n\, q) - \rho^{(n)}(q_1 \ldots q_n)\rho^{(1)}(q)\right)$$

One has to note that such relations hold trivially true for a system of N types of particles in a finite volume canonical ensemble, irrespective of the nature of the force. For infinite systems, we have the following theorem: the canonical sum rules hold true if the truncated correlation functions decrease faster than the force, ie if $\rho_T = o(F)$ as $|x| \to \infty$ (for a precise statement see prop. 8 of ref. [2]). We emphasize the following points:

- The sum rules hold for Coulomb systems when screening occurs (in the sense of L^1-clustering).
- Since the sum rules do not hold for infinite free (or short range force)

systems, they have to be seen as an essential characteristic of Gibbs States of long range force system.

- The validity of the sum rules has been noticed by Dyson and Mehta [7, 8] in their analysis of the logarithmic potential.

E. The charge fluctuations [3]

As a consequence of the first sum rule one deduces that the charge fluctuations $\langle (Q_\Lambda - \langle Q_\Lambda \rangle)^2 \rangle$ are non extensive as $\Lambda \to R^\nu$, ie $\lim_{\Lambda \to R^\nu} \frac{1}{|\Lambda|} \langle (Q_\Lambda - \langle Q_\Lambda \rangle)^2 \rangle = 0$.

Furthermore, under the additional clustering assumption $\int dx_1 \, |x_1| \, |\rho_T^{(2)}(x_1\sigma_1, x_2\sigma_2)| < \infty$, $\langle (Q_\Lambda - \langle Q_\Lambda \rangle)^2 \rangle$ is of the order of the surface $|\partial\Lambda|$ of the boundary of Λ [3, § 3].

If the state obeys the sum rules and has suitable clustering properties, one can prove a central limit theorem for the probability distribution $P_{\bar{Q}_1}$ of the normalized charge $\bar{Q}_\Lambda = \frac{1}{|\partial\Lambda|} (Q_\Lambda - \langle Q_\Lambda \rangle)$ with respect to the surface, in dimension $\nu = 2, 3$. In one dimension, $P_{\bar{Q}_1}$ converges to a discrete distribution [3, § 4].

F. Particle fluctuations in one component systems [2, § 6]

When $N = 1$, Q_Λ is identical with the particle number N_Λ. Thus if the first sum rule holds, the bulk compressibility $\chi = \lim_{\Lambda \to R^\nu} \frac{1}{|\Lambda|} \langle (N_\Lambda - \langle N_\Lambda \rangle)^2 \rangle$ vanishes. From this, one deduces that the truncated function $\rho_T^{(2)}$ of a compressible fluid ($\rho_B = 0$) cannot decrease faster than the force. If it did, then the first sum rule would hold implying $\chi = 0$. But $\chi \neq 0$ in the analyticity region, a contradiction! Under the supplementary assumption $|x| |\nabla \rho_T| = O(\rho_T)$ and with Groeneveld result [9], $\rho_T^{(2)}$ decreases exactly as the potential itself. On the other hand, in the one component Coulomb plasma, the sum rules hold (because of screening) and thus, the particle fluctuations are not extensive ($\chi = 0$) in such systems.

G. Equation of state for one component systems [4]

Several definitions of the pressure are a priori possible:

thermal: $p^{(\theta)} = \lim_{\lambda \to \infty} \beta^{-1} \frac{\partial}{\partial V_\lambda} \ln Z(V_\lambda, \beta, N, N_B)$ $\qquad N_B = \rho_B V_\lambda$

mechanical: $p^{(m)} = \lim_{\lambda \to \infty} \beta^{-1} \frac{\partial}{\partial V_\lambda} \ln Z(V_\lambda, \beta, N, \rho_B)$

kinetic: $p^{(k)} = \lim_{\lambda \to \infty} \frac{1}{\nu} \beta^{-1} \frac{1}{|V_\lambda|} \int_{\partial V_\lambda} \rho_{V_\lambda}^{(1)}(x) \, x \, d\sigma$ $\qquad \geqslant 0$

It is $p^{(k)}$ which is equal to the virial pressure and not $p^{(\theta)}$ as it appears often in the literature. For a fluid ($\rho_B = 0$) these definitions are equivalent, but they are not if $\rho_B > 0$; it is argued in [4] that the correct equation of state is defined by $p^{(k)}$ and it is shown that the Coulomb force appears again as the border line where new properties appear.

References

[1] Ch. Gruber, Ch. Lugrin, Ph. A. Martin, Helv. Phys. Acta 51, 829 (1979)

[2] Ch. Gruber, Ch. Lugrin, Ph. A. Martin, Equilibrium properties of classical systems with long range forces, to appear in Journ. of Stat. Phys.

[3] Ph. A. Martin, T. Yalcin, Charge fluctuations in classical Coulomb systems, to appear in Journ. of Stat. Phys.

[4] Ph. Choquard, P. Favre, Ch. Gruber, On the equation of state of classical systems with long range forces (Preprint)

[5] S. F. Edwards, A. Lenard, Journ. of Math. Phys. 3, 778 (1962)

[6] M. Aizenman, Ph. A. Martin, in preparation

[7] F. Dyson, Journ. of Math. Phys. 3, 166 (1962)

[8] F. Dyson, M. Mehta, Journ. of Math. Phys. 4, 701 (1963)

[9] J. Groeneveld, in Statistical Mechanics, foundation and application, Ed. A. Bak, Benjamin, N. Y. (1967)

ON THE LACK OF FRéCHET DIFFERENTIABILITY IN MORE PHASE REGIONS

H.A.M. Daniëls and A.C.D. van Enter
Institute for Theoretical Physics,
University of Groningen, The Netherlands

Abstract: For the version of the Gibbs phase rule proposed by Ruelle [1,2], Fréchet differentiability of the pressure is essential. In more phase regions however the pressure is not Fréchet differentiable in general. We illustrate this on the 2-dimensional Ising model below T_c.

Consider the thermodynamical pressure as a function on interaction spaces. We prove that for the 2-d Ising model below T_c the pressure is not Fréchet differentiable in the space of pair interactions (for the more general case see [3]). This gives a negative answer to the question raised by Ruelle [1,2] whether his condition (R) holds.

Proof: Let ρ^+ and ρ^- be the extremal pure phases and $\rho^+(s_0^z) = m > 0$ (the magnetization). Divide $\mathbb{Z}^2$ in layers L_N^i of thickness $N(L_N^i = \{(n_1,n_2) | Ni \leq n_1 < N(i+1)\})$.

Define R_N by:

$$R_N\, s_j^z = s_j^z \quad \forall\, j \in L_N^{2k+1} \qquad \text{for any } k$$

$$R_N\, s_j^z = -s_j^z \quad \forall\, j \in L_N^{2k} \qquad \text{for any } k$$

Define $\tilde{\rho}_N$ by

$$\tilde{\rho}_N(A) = \frac{1}{2N} \sum_{i=1}^{2N} \rho^+(R_N \circ \tau_i\, A).$$

So $\tilde{\rho}_N$ is ordered in layers and is also translation invariant. We denote the entropy per lattice site of the system in a state ρ by $s(\rho)$. We prove that $s(\tilde{\rho}_N) = s(\rho^+)$ [3]. Denote by Ψ_N the antiferromagnetic interaction of strength one which acts between two spins at distance N.

Take $\varepsilon > 0$. From the variational principle [4] we obtain :

$$P(\Phi_{Is}+\varepsilon\Psi_N) = \sup_{\rho\varepsilon I} s(\rho) - \rho(A_{\Phi_{Is}+\varepsilon\Psi_N}) \geq s(\tilde{\rho}_N) - \tilde{\rho}_N(A_{\Phi_{Is}}) - \varepsilon\tilde{\rho}_N(A_{\Psi_N}) \tag{1}$$

The fact that ρ^+ is an equilibrium state for Φ_{Is} implies [4]:

$$P(\Phi_{Is}) = s(\rho^+) - \rho^+(A_{\Phi_{Is}}) . \tag{2}$$

Combining (1) and (2) we obtain:

$$\frac{P(\Phi_{Is}+\varepsilon\Psi_N) - P(\Phi_{Is}) + \varepsilon\rho^+(A_{\Psi_N})}{\varepsilon} \geq \frac{1}{\varepsilon}\Big\{ s(\tilde{\rho}_N) - s(\rho^+) - \tilde{\rho}_N(A_{\Phi_{Is}}) + \rho^+(A_{\Phi_{Is}}) -$$

$$- \varepsilon\tilde{\rho}_N(A_{\Psi_N}) + \varepsilon\rho^+(A_{\Psi_N})\Big\} \geq 2m^2 - \frac{2}{N\varepsilon}\|\Phi_{Is}\| > m^2 \text{ if N large enough.}$$

Thus P is not Fréchet differentiable and also not analytic (compare [5]) in the space of pair interactions. □

By making use of a theorem by Israel [6] it can be shown that some interaction in the neighbourhoods proposed by Ruelle has an equilibrium state which is ordered in layers. Hence to prove a version of the Gibbs phase rule in infinite dimensional interaction spaces, some conditions on the decrease of the interaction are necessary.

This work is part of the research program of the "Stichting F.O.M.".

1 D. Ruelle: Commun. Math. Physics 53 (1977) pp. 195-208.
2 D. Ruelle: Theor. and Math. Physics 30 (1977) pp. 24-29.
3 H.A.M. Daniëls and A.C.D. van Enter: to appear in Commun. of Math. Physics.
4 D. Ruelle: Statistical Mechanics (1969) Benjamin.
5 D. Iagolnitzer and B. Souillard: Commun. of Math. Physics 60 (1978) pp. 131-152.
6 R.B. Israel: Commun. Math. Physics 43 (1975) pp. 59-68.

LONG TIME TAIL FOR SPACIALLY INHOMOGENEOUS RANDOM WALKS

Herbert Spohn

Theor. Physik, Universität München, Theresienstr. 37, 8 München 2, FRG

The Lorentz gas consists of a single classical particle moving through in R^2 randomly distributed, infinitely heavy hard disk scatterers. The Lorentz particle starts at the origin with a uniform distribution of velocities of magnitude one. The hard disks are in equilibrium conditioned not to overlap the origin. The time integral over the correlation function $\langle v(t)\cdot v(0)\rangle$ of the stationary velocity process $v(t)$ gives the diffusion constant of the Lorentz gas. One deep problem is to show that this diffusion constant is finite, which would follow from a sufficiently fast decay of $\langle v(t)\cdot v(0)\rangle$. Theoretical and numerical results indicate a power law decay as $-t^{-(d/2 + 1)}$ in d dimensions.

To understand qualitatively the origin of the long time tail I use an old recipe in non-equilibrium statistical mechanics which advices to replace the deterministic motion by a stochastic one. Then at each site of the lattice Z^d there is with probability p a scatterer, with probability 1-p no scatterer. The particle performs a random walk: At a site with a scatterer the particle continues with probability 1/2d in one of the 2d directions. At a site with no scatterer the particle continues in the direction it came from. One wants to prove that the velocity autocorrelation function $\langle v(t+1)\cdot v(1)\rangle$ has a long time tail. This seems to be difficult. To simplify: Suppose that at each site except the origin there is a scatterer. Then in one dimension, for the particle starting at the origin, since the only contribution comes from paths where the particle is at the origin at time t, $\langle v(t+1)\cdot v(1)\rangle = -r(t) + (r*r)(t) - \ldots$, where $r(t)$ is the probability of a first return to the origin at time t for the simple random walk. Since $r(t) \simeq t^{-3/2}$ for long times, this implies the expected long time tail for this model. A symmetry argument due to M. Aizenman proves that $\langle v(t+1)\cdot v(1)\rangle \simeq -t^{-(d/2+1)}$ in d dimensions. The long time tail originates from a spacially inhomogeneous distribution of scatterers. A periodic distribution gives an exponential decay. Together with H. van Beijeren I studied the one dimensional random scatterer model. Time is continuous and the distances ℓ_i between scatterers are independently and identically distributed. For small z we prove $\int_0^\infty dt e^{-zt}\langle v(t)\cdot v(0)\rangle = (1/2)\langle \ell_0\rangle + \sqrt{z}\langle(\ell_0-\langle \ell_0\rangle)^2\rangle/\sqrt{8\langle \ell_0\rangle} + O(z^{1/2+\delta})$, $\delta>0$. The negative time tail cannot decrease faster than $t^{-3/2}$.

WHY IS THERE A SOLID STATE?

Charles Radin
Mathematics Department
University of Texas
Austin, TX 78712/USA

We consider the problem (called the "crystal problem") of showing that for a finite system of point particles to have minimum potential energy when interacting through Lennard-Jones type potentials, the particles must lie approximately on the vertices of a lattice with the approximation becoming exact as the particle number diverges.

We report here the first nontrivial progress on this old problem — to be published in the Journal of Statistical Physics in three joint papers, respectively with C.S. Gardner, G.C. Hamrick and R.C. Heitmann.

In the first paper we solve the crystal problem in one space dimension, but only for the Lennard-Jones potential itself.

In the second paper we demonstrate that the crystal problem is highly sensitive to physically negligible perturbation of the potential (at least in one dimension): we exhibit a potential with only lattice ground states, and a family of perturbations physically small of order ϵ, such that the perturbed system has only nonperiodic ground states for all $\epsilon > 0$.

In the last paper we solve the crystal problem in two dimensions but only for the "sticky disk" potential: i.e. hard core, with delta function attraction at the hard core radius. The problem is a geometrical one: to show that those configurations, of a finite number of nonoverlapping disks in the plane with the most number of touching pairs of disks, must have the disks centered on a triangular lattice. Extension of the result to longer range potentials is expected.

Finally we consider the possibility that sticky disks exhibit long range orientational order at nonzero temperature.

SOME REMARKS ON THE SURFACE TENSION.

J. Bricmont * †
J.L. Lebowitz
C.E. Pfister ** ‡

Department of mathematics
Rutgers University
New Brunswick N.J. 08903 USA

* supported in part by N.S.F. Grant Phys 77-22302
† present address: Math. Dept. Fine Hall, Princeton University, Princeton N.J. 08544 USA
** supported by the Swiss Foundation of scientific research
‡ present address: Département de mathématiques, E.P.F.-Lausanne, CH-1007 Lausanne

We consider the surface tension $\beta^{-1}\tau$ for the d-dim. Ising model with nearest neighbour ferromagnetic coupling $J > 0$. τ is defined in e.g. [1] :

$$0 \leqslant \tau = - \lim_{L\to\infty} \frac{1}{(2L+1)^{d-1}} \lim_{M\to\infty} \log \frac{Z_{\Lambda,\pm}}{Z_{\Lambda,+}} \tag{1}$$

where Λ is a box centered at the origin of height $2M$ in the i_1-direction with a base of side $2L+1$. The indices $+$ and $\pm$ refer to the usual $+$ and $\pm$ boundary conditions (b.c.).

We do the following construction: we replace the couplings J by sJ, $0 \leqslant s \leqslant 1$ for all nearest neighbour pairs $\langle ij \rangle$ crossing the plane $i_1 = -\frac{1}{2}$. By symmetry $\tau(s)$ is zero for $s = 0$ and $\tau = \tau(1)$ is equal to

$$\tau = k\int_0^1 ds\ (\rho_s^+ (\sigma_0\sigma_{-1}) - \rho_s^{\pm} (\sigma_0\sigma_{-1})),\ \beta J = k \tag{2}$$

where ρ_s^+ ($\rho_s^{\pm}$) is the corresponding infinite volume Gibbs state with +b.c. ($\pm$b.c.) (2) is justified by correlation inequalities and using such inequalities we prove that $\tau = 0$ whenever there is no spontaneous magnetization. We can also prove that $\tau = 0$ if the spontaneous magnetization is zero for the semi-infinite system defined on $\{i \in \mathbb{Z}^d : i_1 > 0\}$ with free b.c. at $i_1 = 0$ and +b.c. elsewhere. From (2) and correlation inequalities we obtain the lower bound

$$\tau \geqslant 2k \int_0^1 \rho_s^+(\sigma_0)\, \rho_s^{\pm}(\sigma_0)\, ds$$

Using duality and correlation inequalities we prove that for d=2 $\tau(k) = m(k^*)$ where k^* is the dual temperature and m is the inverse correlation length in the state with free b.c. For d=3 $\tau(k) = \alpha(k^*)$ where α is the coefficient of the area law decay of the Wilson loop in the Ising gauge model [2].

For d=3 and k large enough we prove that $\tau - 2k$ and the correlation functions in the state $\rho^{\pm}$ are analytic in $\exp(-2k)$. Furthermore we have the Gibbs formula

$$\frac{d\tau}{dk} = \sum_{\substack{i_1=-\infty \\ i_2=i_3=0}}^{+\infty} \sum_{\substack{j: \\ |i-j|=1}} \left(\rho^+(\sigma_i \sigma_j) - \rho^{\pm}(\sigma_i \sigma_j) \right)$$

see [2] and [3].

References:

1) Gruber C., Hintermann A., Messager A., Miracle Sole S Commun. math. Phys. 56 147(1977)

2) Bricmont J., Lebowitz J.L., Pfister C.E. On the surface tension for lattice systems. To be published in the Proceedings of the 3rd International Conference on Collective Phenomena Moscow 1978. Annals of the New York Academy of Sciences 1979.

3) Bricmont J., Lebowitz J.L., Pfister C.E. Commun. math. Phys. 66 1 and 2 1979 and to appear in Commun. math. Phys. 1979.

ANALYTIC STRUCTURE OF GREEN'S FUNCTIONS IN QUANTUM FIELD THEORY

by
J. Bros
DPhT, CEN Saclay, BP n°2, 91190 Gif-sur-Yvette, France

A tentative review of recent results obtained by several authors will be presented. They mainly concern the derivation of analyticity properties for the Green's functions and collision amplitudes of the $2 \to 2$, $2 \to 3$, $3 \to 3$ particle processes from the general principles of Q.F.T. ; emphasis will be laid on the exploitation of asymptotic completeness which generates the monodromic structure of Green's functions. In this connection, the role played by "generalized Bethe-Salpeter kernels" will be described.

INTRODUCTION

In the present situation of Theoretical Physics where models of Quantum Fields of numerous kinds are under study, bringing with them lots of new ideas and new aspects of Quantum Field Theory, we believe that understanding the analytic structure in complex momentum space of Green's functions of local fields and of the associated n-body relativistic scattering amplitudes remains an important objective of Q.F.T.. At least, this looks justified as far as one believes that fields with asymptotic particles are relevant mathematical concepts and that analyticity remains a basic language for describing certain aspects of the reality of particle physics.

In this lecture, we try to make a review of the methods which have been developed and of the results which have been obtained recently in the axiomatic approach of the analytic structure of Green's functions and scattering amplitudes.

From the various recent developments which we shall present here, we wish to emphasize the following aspects :

i) The use of *Asymptotic Completeness* (in short : A.C.)-considered as an extra-axiom is a powerful tool for introducing and investigating the monodromic structure of Green's functions around Landau singularities in the general analytic framework implied by locality and spectrum. It yields a scheme for an exact (i.e. non perturbative) treatment in large complex regions, of the singularity structure of Green's functions (Note that up to now, such a general treatment had only been proposed in the pure-S-matrix approach [1], and has only been developed consistently in the neighbourhood of the physical regions [2]).

ii) The fact that certain methods of the axiomatic scheme, based on analyticity properties, may implement important steps in the study of models. As an example the analytic continuation of structural equations of the Bethe-Salpeter type (see Part C II below) is a tool which can be used either in working out the consequences of A.C., considered as an axiom, or in studying the validity of A.C. in a class of models of Q.F.T. (in the case of $P(\varphi)_2$ models, see in this connection [45],[17].)

As a guide for understanding the results which we describe here, the reader is invited to refer frequently to Table 1 which represents in short the axiomatic scheme of local Q.F.T. in complex momentum space.

TABLE 1

THE AXIOMATIC SCHEME OF Q.F.T. IN COMPLEX MOMENTUM SPACE

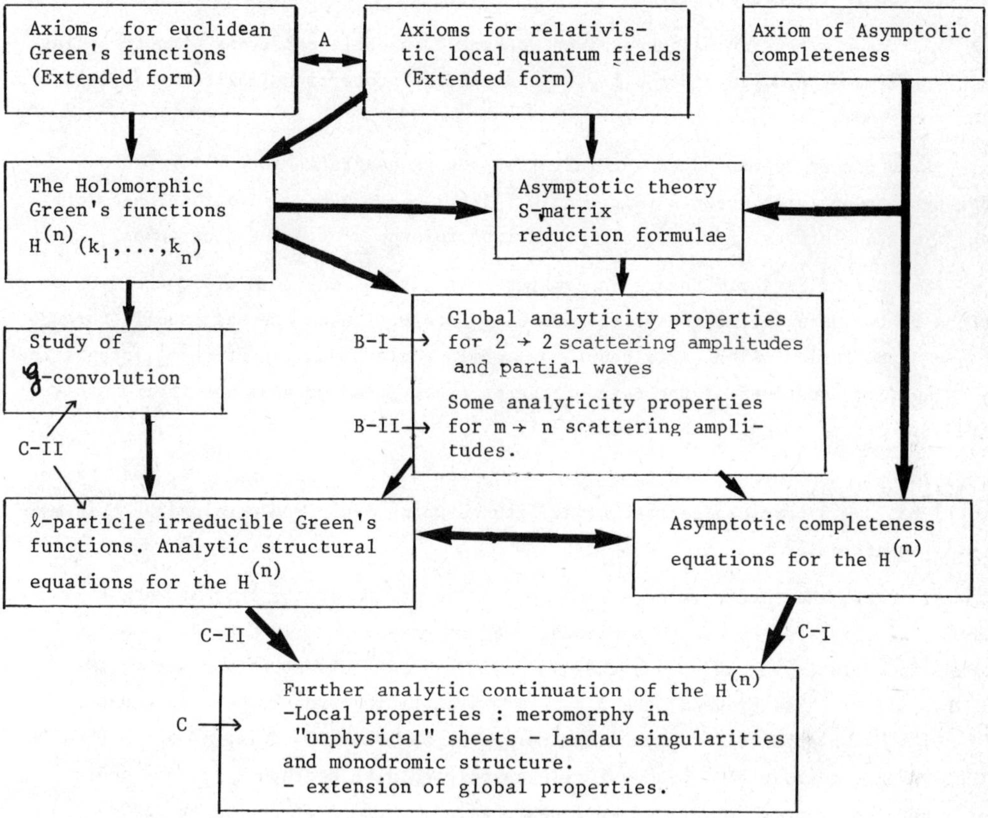

We divide these lecture notes in three parts of unequal length.

Part A is devoted to a work by Eckmann and Epstein [3] on the equivalence of the euclidean formulation of Field Theory in terms of "extended" Schwinger functions and the minkowskian formulation in terms of time-ordered products of fields. This work sets on a firm basis the "bridge" between the euclidean configuration space which is the laboratory of constructive Q.F.T. and the minkowskian momentum space in which the analyticity properties of the scattering amplitudes take place.

Part B deals with some progress in the standard derivation of analyticity properties of the scattering amplitudes without using A.C. as a systematic tool.

In B-I, improved analyticity domains for the 2 → 2 particle scattering amplitude and for corresponding partial waves are presented ; these results are due to Roy and Wanders [4] .

In B-II, some analyticity properties for certain $3 \to 3$ particle scattering amplitudes, obtained by the Russian school of the Steklov Institute [5] are described.

For the reader who is unfamiliar with the subject, we found it useful to precede the results of B-I and B-II with a rather long account of standard results which the specialist is invited to skip.

A larger extension has been given to Part C which is concerned with the joint exploitation of A.C. together with the analyticity properties implied by locality and spectrum.

A.C. can be expressed as an (infinite) set of quadratic integral relations for the n-point analytic Green's functions $H^{(n)}(k_1,\dots,k_n)$, these relations being off-shell extrapolations of the unitarity equations for the scattering amplitudes.

The use of Fredholm theory in complex space allows one to exploit the A.C. equations by two parallel methods : the one uses integration on the mass shell (through *direct* application of the A.C. equations) while the other uses off-shell integration in structural equations of the Bethe-Salpeter type. These methods and their results are separately presented in C-I and C-II.

In C-I, the former, due to V. Glaser, is presented, and some local results for the $2 \to 3$, $3 \to 3$ particle scattering amplitudes which have been obtained by this method [6] are described.

In C-II, the latter is presented ; it is the method of the present author [7], which was suggested to him by a pioneer program proposed in 1960 by K. Symanzik in [8a], and later in [8b] under the name of "Many Particle Structure Analysis" (M.P.S.A.). In this program, the notion of "ℓ-particle irreducible" (in short : "ℓ-p-i") Green's function plays a crucial role ; the starting point of our method is the systematic study of "$\mathcal{G}$-convolution associated with a graph $\mathcal{G}$": this operation on general n-point functions generalizes the construction of perturbative Green's functions.

Some global results for the analytic six-point function $H^{(6)}$ and for the $3 \to 3$ particle scattering amplitude (in the equal mass case for an even field theory) which have been obtained by this method are described ; they can be summarized as follows :

- a study of the equivalence between 3-particle A.C. and 3-particle irreducibility.
- The derivation of a global structural equation for $H^{(6)}$ which exhibits the monodromic structure of the latter in the low energy physical region of the $3 \to 3$ process.
- a proof of the crossing property (on the mass-shell) for the processes

and

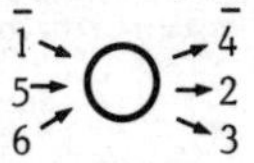

($\bar{j}$ denoting the antiparticle of particle j) in the equal mass case.

In order to help the reader in understanding where the contents of parts A,B,C are situated in the logical organization of the axiomatic scheme, we have indicated the localizations (A,B,C) on table 1.

PART A

THE CONNECTION BETWEEN THE EUCLIDEAN AND MINKOWSKIAN FORMULATIONS OF QUANTUM FIELD THEORY (Q.F.T).

The standard formulation of Q.F.T. in Minkowski space* $\mathbb{R}^{(d+1)}_{(x)} = \mathbb{R}_{(x^{(o)})} \times \mathbb{R}^{d}_{(\vec{x})}$ (d being the dimension of space) is best expressed by the *Wightman axioms* [9] which can be summarized as follows :

-I In the Hilbert space of states $\mathcal{H}$, there exists a unitary representation $U(a,\Lambda)$ of the Poincaré group ("a" denoting a translation in $\mathbb{R}^{d+1}$, and Λ a homogeneous Lorentz transformation of $\mathbb{R}^{d+1}$) whose translation group $U(a,I) = e^{iP^{\mu}a_{\mu}}$ admits the closed light cone $\bar{V}^{+}$ as its spectrum. Moreover there exists a unique state $|\Omega\rangle$ in $\mathcal{H}$, called "*vacuum*", which is invariant under $\{U(a,\Lambda)\}$ $(P^{\mu}|\Omega\rangle = 0)$.

In the following, we shall mostly deal with the case of theories having a minimal mass $m > 0$; in the general developments, we shall assume for simplicity that the spectrum is the set $\{0\} \cup H^{+}_{m} \cup \bar{V}^{+}_{2m}$, where :

$$H^{+}_{m} = \left\{p \in \mathbb{R}^{d+1} \ ; \ p^{2} = m^{2},\ p^{(o)} > 0\right\},\quad V^{+}_{2m} = \left\{p \in \mathbb{R}^{d+1} \ ; \ p^{2} > (2m)^{2}, p^{(o)} > 0\right\}$$

-II There exists a field $A(x)$ (i.e. a set of unbounded operators $A(f) = \int A(x)\, f(x)\, dx$ acting in a common dense domain $\mathcal{D}_{o}$ of $\mathcal{H}$ $\forall f \in \mathcal{S}(\mathbb{R}^{d+1})$) which satisfies the locality condition : $[A(x), A(y)] = 0$ for $(x-y)^{2} < 0$, on $\mathcal{D}_{o}$. $\mathcal{D}_{o}$ can be generated by all vectors of the form $A(f_{1})\ldots A(f_{m})\Omega$. Moreover the field $A(x)$ fulfils a covariance condition under the Poincaré group.

It turns out that all these properties can be fully reexpressed in terms of an infinite set of tempered distributions, called "*Wightman functions*" , namely :

$$W_{n}(x_{1},\ldots,x_{n}) = \langle\Omega, A(x_{1})\ldots A(x_{n})\Omega\rangle \quad .$$

From spectrum and locality, one deduces the important fact that each W_{n} is the boundary value of an analytic functions $\mathcal{W}_{n}(z_{1},\ldots,z_{n})$ $(z_{j} = x_{j} + iy_{j})$ whose analyticity domain contains the whole "euclidean space" :

$$E^{(d+1)n} = \left\{(z_{1},\ldots,z_{n}) \in \mathbb{C}^{(d+1)n} \ ; \ z_{j} = (i\, y^{(o)}_{j},\ \vec{x}_{j})\ ,\ j = 1,2,\ldots,n\right\},$$

except for the set (of measure zero) σ_{n} of all points $(z_{1},\ldots,z_{n})$ for which two time-components *at least* are equal $(y^{(o)}_{i} = y^{(o)}_{j})$.

The infinite set of functions $\{\underline{S}_{n}\}$, called "Schwinger functions" and defined in $E^{(d+1)n} - \sigma_{n}$ by :

---

*the pseudometric being defined by $x^{2} = {x^{(o)}}^{2} - \vec{x}^{2}$, and the light cone V^{+} by $V^{+} = \{x ; x^{2} > 0,\ x^{(o)} > 0\}$

$$\underline{S}_n = \mathcal{W}_n \Big|_{E^{(d+1)n} - \sigma_n}$$

can itself be taken as the starting point of a so-called "*euclidean formulation*" of Q.F.T.

Such an axiomatic formulation in terms of a given set $\{\underline{S}_n\}$ has been proposed by Osterwalder-Schrader [10], and it has been proved in [10] and [11] that the latter entails the existence of a Wightman field whose associated set of Schwinger functions coincide with the given set $\{\underline{S}_n\}$.

This was an important result since it was precisely in euclidean space that the various $P(\varphi)_{2,3}$ models could be most successfully studied [12] and since the Schwinger functions of the latter could be directly constructed and shown to satisfy the O.S. axioms [13].

However a gap had to be filled in order that all the results of the momentum space scheme of Q.F.T. be applicable to euclidean field theory. This gap originates in the following insufficiency of the Wightman axioms : the latter do *not* imply in general the existence of *sharp* time-ordered or retarded operators (since multiplying distributions as A(x) A(y) by the step function $\theta(x^{(o)} - y^{(o)})$ is not licit in general, due to the possible singular behaviour of A(x) A(y) at coinciding time-components) ; a way of avoiding this difficulty consists in working with *regularized* (instead of *sharp*) Green's operators. However the complete momentum-space scheme is definitely more tractable in terms of sharp Green's operators, and in order to set this scheme on a firm basis it was necessary to produce an extended version of the axioms of local fields.

Such an extended version has already been proposed and formulated in terms of retarded operators by O. Steinmann [14]; a more recent version, due to Epstein, Glaser, Stora [15] can be set as follows.

One keeps Wightman axioms I and replaces axioms II by the following set of *postulates for (anti)-time-ordered products*.

T-i) Existence of sets of operators :

$$\left\{\bar{T}(f) = \int \bar{T}(x_1,\dots,x_n)\, f(x_1,\dots,x_n)dx_1\dots dx_n, \quad \forall n \in N,\ \forall f \in \mathcal{S}(\mathbb{R}^{(d+1)n})\right\}$$

such that the distributions $\bar{T}(x_1,\dots,x_n)$ are symmetric with respect to the set of vectors $(x_1,\dots,x_n)$; in short, one puts : $\bar{T}(x_1,\dots,x_n) = \bar{T}(X_n)$, where $X_n = \{1,2,\dots,n\}$.

- Existence of all products : $\bar{T}(I_1).\bar{T}(I_2)\dots\bar{T}(I_p)$ for all partitions $(I_1,I_2,\dots I_p)$ of X_n, in the sense of distributions on a dense domain $\mathcal{D}_o$ of $\mathcal{H}$. (Note that $\bar{T}(x_1)$ coincides with the basic field $A(x_1)$ of Wightman axioms II).

T-ii) *(Anti)-causal factorisation* : for every partition (I,J) of any set X_n, one has : $\bar{T}(X_n) = \bar{T}(I).\bar{T}(J)$ in the following region : $\left\{ (x_1,\dots,x_n) ;\ \forall i \in I,\ \forall j \in J,\ x_i^{(o)} < x_j^{(o)}\right\}$ (Note that this implies the locality of the field $\bar{T}(x) = A(x)$).

-T iii) (resp. T iii')) Time-translation (resp. Poincaré) covariance of $\{\bar{T}(X_n)\}$.

-T iv) If one puts: $T(x) = \sum_{1 \leq p \leq |X|} (-1)^{|X|+p} \sum_{(I_1,\dots,I_p)} \bar{T}(I_1)\dots\bar{T}(I_p)$, then one has the hermiticity property : $T(X)^* = \bar{T}(X)$.

These axioms imply the Wightman axioms and allow the $\mathcal{L}.\mathcal{S}.\mathcal{Z}.$ formalism of scattering theory [16] to be fully justified, as well as the construction of analytic Green's functions $H^{(n)}(k_1\dots k_n)$ in momentum space with slow increase properties in their primitive domains D_n (see part B and [15]).

A corresponding extension of the O.S. axioms was presented in [3] ; the axioms can be set as follows.

Axioms for euclidean Green's functions (Eckmann-Epstein).

-S i) existence of a sequence of tempered distributions $S_n(y_1,\dots,y_n)$ on $E^{(d+1)n}$, which are symmetric with respect to the n points $y_1,\dots,y_n$ $(y_i = (y_i^{(o)},\vec{y}_i))$.

-S ii) *Extended* O.S. positivity :
For every set $\left\{f_m \in \mathcal{S}(\mathbb{R}^{(d+1)n}),\ 1 \leq m \leq N;\ f_o \in C\right\}$ such that :
$\forall m$, SUPP $f_m \subset \left\{(y_1,\dots,y_m)\ ;\ y_1^{(o)} \geq 0,\dots,y_m^{(o)} \geq 0\right\}$, the following inequality holds :

$$\sum_{0 \leq m,n \leq N} \int \overline{f\left((-y_m'^{(o)},\vec{y}_m'),\dots,(-y_1'^{(o)},\vec{y}_1')\right)}\, f_n\left((y_1^{(o)},\vec{y}_1),\dots,(y_n^{(o)},\vec{y}_n)\right)$$

$$\times\ S_{m+n}(y_1',\dots,y_m',y_1,\dots,y_n)dy_1'\dots dy_m'\, dy_1\dots dy_n \geq 0$$

-S iii) (resp. Siii)') Time-translation (resp. euclidean) covariance of $\{S_n\}$.

-S iv) Growth condition :

$$\forall n, |S_n(f_1 \otimes \dots \otimes f_n)| \leq K^n\, n!^L \prod_{j=1}^{n} ||f_j||_s ,$$

for appropriate positive constants K,L,s and a certain Schwartz norm $||f||_s$.

We notice that in these axioms, each distribution S_n is defined on the *whole* space $E^{(d+1)n}$, *including* the set σ_n of coinciding points ; thus S_n represents an extension of the Schwinger function $\underline{S}_n$ (defined on $E^{(d+1)n} - \sigma_n$) and (in view of the following theorem) deserves the name of "euclidean Green's function". We also note that in S ii), the original O.S. positivity condition has been extended from $E^{(d+1)n} - \sigma_n$ to $E^{(d+1)n}$.

Then the following important relationship between the two sets of minkowskian and euclidean axioms has been proved in [3] :

Theorem: There is an equivalence between a set of (anti) time-ordered products satisfying the postulates T i) ... T iv) in the framework defined by Wightman axioms I, and a set of euclidean Green's function satisfying S i) ... S iv). Anyone of these two sets can be reconstructed from the other.

More precisely, if one puts $\bar{\tau}_n(x_1 \dots x_n) = \langle\Omega, \bar{T}(x_1 \dots x_n)\Omega\rangle$ the following relations hold for each couple of distributions $S_n, \bar{\tau}_n)$:

$$\int i^n f_i(x_1,\dots,x_n)\; S_n(x_1,\dots,x_n)dx_1\dots dx_n = \int f(x_1\dots x_n)\; \bar{\tau}_n(x_1\dots x_n)\; dx_1\dots dx_n,$$

where f denotes any test function in $\mathcal{S}(\mathbb{R}^{(d+1)n})$ which is "radially analytic" with respect to the time-components of $(x_1,\dots,x_n)$ in the following sense : the mapping

$$\lambda \in \mathbb{R}_+ \;\to\; f_\lambda(x_1,\dots,x_n) = f\,(\lambda x_1^{(o)}, \vec{x}_1),\dots,(\lambda x_n^{(o)}, \vec{x}_n)$$

extends to an analytic mapping in the domain

$$\left\{ \lambda \in \mathbb{C},\;\; \mathrm{Re}\,\lambda > 0,\;\; \mathrm{Im}\,\lambda > 0 \right\} .$$

Moreover, it has been fully established in [3]that the truncated analytic n-point function $\mathcal{W}_n^{tr}$ in z-space $(z = x+iy)$, the analytic n-point Green's function H_n in k-space $(k = p+iq)$, the various boundary values and the euclidean restrictions S_n^{tr} (resp. S_n^{tr}) of $\mathcal{W}_n^{tr}$ and H_n in their respective spaces are linked together through the following commutative diagram * : this diagram holds for either choice of the basic quantities, $\{S_n\}$ or $\{\bar{T}_n\}$, taken as the starting point of the theory.

Euclidean y-space	Analytic continuation in complex z-space	Boundary values in Minkowskian x-space
S_n, S_n^{tr}	$\longleftarrow\; \mathcal{W}_n^{tr} \;\longrightarrow$	$\{W_n^{tr(\pi)}\}, \tau_n^{tr} = \langle T_n\rangle^{tr},\; \bar{\tau}_n^{tr} = \langle\bar{T}_n\rangle^{tr}$
$\updownarrow \mathcal{F}_E$		$\updownarrow \mathcal{F}_M$
Euclidean q-space	Analytic continuation in complex k-space	Boundary values in Minkowskian p-space
$\tilde{S}_n, \tilde{S}_n^{tr}$	$\longleftarrow\; H^{(n)} \;\longrightarrow$	$\{r^{(n)}\}, \tilde{\tau}_n^{tr}, \tilde{\bar{\tau}}_n^{tr}$

In this diagram, $\mathcal{F}_E$ and $\mathcal{F}_M$ respectively represent the Fourier transformation in the euclidean and minkowskian (d+1)-dimensional spaces ; $\{W_n^{tr(\pi)}\}$ denotes the set of all the "permuted" truncated Wightman functions $W_n^{tr(\pi)}(x_1,\dots,x_n) = W_n^{tr}(x_{\pi 1},\dots,x_{\pi n})$; $\{r_{\mathcal{J}}^{(n)}\}$ denotes the set of all the generalized retarded n-point functions in momentum space (see part B and [15,18]); $\tilde{\tau}_n^{tr}, \tilde{\bar{\tau}}_n^{tr}$ (resp. $\tilde{S}_n^{tr}$) are the $\mathcal{F}_M$-transforms (resp. $\mathcal{F}_E$-transform) of τ_n^{tr}, $\bar{\tau}_n^{tr}$ (resp. S_n^{tr}).

---

*Note however that the facts expressed by this diagram (analyticity properties, algebra of boundary values etc...) were known before and refer to a long historical development produced by numerous authors in this domain [15]. The main point of [3] lies in the rigorous formulation of this rich structure, including the delicate treatment of distributions at "coinciding points" in $x^{(o)}$ (or $y^{(o)}$)-space.

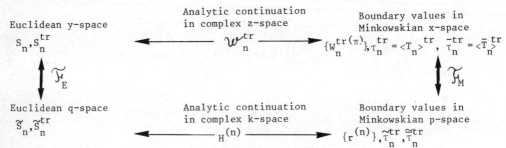

PART B

SOME NEW RESULTS FOR THE 2 → 2 AND 3 → 3 PARTICLE SCATTERING AMPLITUDES, OBTAINED BY STANDARD TECHNIQUES OF ANALYTIC COMPLETION.

We recall the following facts which concern the so-called *primitive analytic structure* of n-point Green's functions in momentum space and are consequences of the (extended) axioms of local fields. For each integer n, there exists an n-point function $H^{(n)}(k)$ of the set of 4(n-1) independent complex variables $k = p + iq \equiv (k_1,\dots,k_n)$, $k_i = p_i + iq_i \in C^4$, $k_1 + \dots + k_n = 0$. The analyticity properties of $H^{(n)}$ are the following :

1) *$H^{(n)}$ is defined and analytic in the following "primitive" domain D_n :*

$$D_n = (\bigcup_{\mathcal{S}} \mathcal{T}_{\mathcal{S}}) \cup (\bigcup_{(\mathcal{S}\mathcal{S}')} \mathcal{V}(\mathcal{R}_{\mathcal{S}\mathcal{S}'}))$$

where : a) $\mathcal{T}_{\mathcal{S}} = \mathbb{R}^{4(n-1)}_{(p)} + i\,\mathcal{C}_{\mathcal{S}}$, and $\mathcal{C}_{\mathcal{S}}$ is a cone in $\mathbb{R}^{4(n-1)}_{(q)}$ which is defined by a *consistent* set of relations

$$\mathcal{C}_{\mathcal{S}} = \left\{q \in \mathbb{R}^{4(n-1)} \;;\; q_J = \sum_{j \in J} q_j \in \varepsilon^{(\mathcal{S})}(J).V^+ \;,\; \forall J \in \mathcal{P}^*(X_n)\right\} ;$$

here $X_n = \{1,2,\dots n\}$ and $\varepsilon^{(\mathcal{S})}$ is a sign-valued function defined on $\mathcal{P}^*(X_n)$. In each tube $\mathcal{T}_{\mathcal{S}}$, $H^{(n)}$ is a function with moderate increase (at infinity and near the boundary of $\mathcal{T}_{\mathcal{S}}$); it thus admits a "distribution boundary value"

$$H^{(n)}_{\mathcal{S}}(p) = \lim_{\substack{\varepsilon \in \mathcal{C}_{\mathcal{S}} \\ \varepsilon \to 0}} H^{(n)}_{\mathcal{S}}(p + i\varepsilon) \text{ on } \mathbb{R}^{4(n-1)}_{(p)} .$$

The link of $H^{(n)}_{\mathcal{S}}$ with the basic field of the theory is the following : the tempered distribution $\tilde{r}^{(n)}_{\mathcal{S}}(p) = \delta_4(p_1 + \dots + p_n)\, H^{(n)}_{\mathcal{S}}(p)$ (defined on $\mathbb{R}^{4n}_{(p)}$) is the Fourier transform of a certain generalized retarded function $r^{(n)}_{\mathcal{S}}(x_1,\dots,x_n)$ such that $r^{(n)}_{\mathcal{S}}(x_1,\dots,x_n) = \langle\Omega, R^{(n)}_{\mathcal{S}}(x_1,\dots,x_n)\Omega\rangle$; the algebra of the generalized retarded operators $R^{(n)}_{\mathcal{S}}$ can be constructed from the set of time-ordered products satisfying the postulates T i)...T iv) (see part A). Locality (or "causal factorization" T iii)) implies support properties for the $R^{(n)}_{\mathcal{S}}$ which entail the analyticity of $H^{(n)}$ in the tubes $\mathcal{T}_{\mathcal{S}}$ through the Laplace-transform theorem.

b) For every couple of tubes $(\mathcal{T}_{\mathcal{S}}, \mathcal{T}_{\mathcal{S}'})$ which are *"adjacent along the face $q_I = 0$"* (i.e. whose associated sign functions $\varepsilon^{(\mathcal{S})}$, $\varepsilon^{(\mathcal{S}')}$ only differ by their values on the subsets I and $X_n - I$), one defines the corresponding *"coincidence region"* $\mathcal{R}_{\mathcal{S}\mathcal{S}'} = \left\{p \in \mathbb{R}^{4(n-1)} ; p_I^2 \neq m^2,\ p_I^2 < M^2\right\}$, which is the region in which $\tilde{r}^{(n)}_{\mathcal{S}}(p) = \tilde{r}^{(n)}_{\mathcal{S}'}(p)$ (the latter is a consequence of the spectral condition).

$\mathcal{V}(\mathcal{R}_{\mathcal{S}\mathcal{S}'})$ represents the intersection of a small complex neighbourhood of $\mathcal{R}_{\mathcal{S}\mathcal{S}'}$ with the convex hull of $\mathcal{T}_{\mathcal{S}} \cup \mathcal{T}_{\mathcal{S}'}$ Note that the set $\left\{\bigcup_{\mathcal{S},\mathcal{S}'} \mathcal{V}(\mathcal{R}_{\mathcal{S}\mathcal{S}'})\right\}$ connects the various tubes $\mathcal{T}_{\mathcal{S}}$ together.

Remarks about D_n : i) For any fixed Lorentz frame, let us put : $k = (k^{(o)}, \vec{k})$, $k^{(o)} = p^{(o)} + i\,q^{(o)}$, $\vec{k} = \vec{p} + i\vec{q}$; then the euclidean space $\mathbb{R}^{3(n-1)}_{(\vec{p})} \times \mathbb{R}^{(n-1)}_{(q^{(o)})}$ is contained in D_n.

ii) Every (n-1)-dimensional section of D_n by $k^{(o)}$-space, at $\vec{k} = \vec{p}$ (real) fixed is a dense domain in $\mathbb{C}^{(n-1)}_{(k^{(o)})}$; it is the exact generalization of a cut-plane in which the upper and lower half-planes ($\mathcal{C}_+, \mathcal{C}_-$) are connected together through a real interval $\mathcal{R}$.

iii) Each manifold $k_I^2 = m^2$ ($\forall I \in \mathcal{P}^*(X_n)$) is a simple polar manifold for $H^{(n)}$.

2) *The discontinuity functions (or "absorptive parts")* $\Delta_I H^{(n)}$ *of* $H^{(n)}$:

With an arbitrary partition (or "channel") (I,J) of X_n, let us associate the following momentum variables :

$$k_I = -k_J = \sum_{i \in I} k_i \quad , \quad \underline{k}_{(I)} = \{ \underline{k}_i = k_i - \frac{k_I}{|I|} \} \quad ,$$

$$\underline{k}_{(J)} = \{ \underline{k}_j = k_j - \frac{k_J}{|J|} \} \quad , \text{ and rewrite : } H^{(n)}(k) = H^{(n)}(k_I ; \underline{k}_{(I)}, \underline{k}_{(J)})$$

We define the corresponding discontinuity or "absorptive part" of $H^{(n)}$:

$$\Delta_I H^{(n)}(p_I ; \underline{k}_{(I)}, \underline{k}_{(J)}) = \lim_{\substack{\varepsilon \to 0 \\ \varepsilon \in V^+}} \left[H^{(n)}(p_I + i\varepsilon ; \underline{k}_{(I)}, \underline{k}_{(J)}) - H^{(n)}(p_I - i\varepsilon ; \underline{k}_{(I)}, \underline{k}_{(J)}) \right]$$

$\Delta_I H^{(n)}$ is : a) a distribution in p_I with support contained in $\{p_I \in H_m^+ \cup \bar{V}_M^+\}$.

b) an analytic function in $(\underline{k}_{(I)}, \underline{k}_{(J)})$ in a domain $D_n^{(I)}(p_I)$ which depends on p_I and is similar to D_n (namely a union of tubes and of neighbourhoods of various "coincidence regions").

c) a *positive kernel* (depending on p_I) in the case $|I| = |J|$ (n even) More precisely, for every suitable test functions χ, φ , the following inequality holds :

$$\int \Delta_I H^{(n)}(p_I ; \underline{p}_{(I)}, \underline{p}_{(J)}) \; \chi(p_I) \; \varphi(\underline{p}_{(I)}) \; \bar{\varphi}(\underline{p}_{(J)}) \; dp_I \; d\underline{p}_{(I)} \; d\underline{p}_{(J)} \geq 0$$

Another basic development in the axiomatic framework of local Q.F.T. is the scattering theory of Haag-Ruelle-Hepp [9b,16], which involves the construction of asymptotic incoming and outgoing states and the derivation of the $\mathcal{L}.\mathcal{S}.\mathcal{Z}.$ reduction formulae which relate the scattering kernels to n-point Green's functions. In the complex momentum space framework, this can be summarized as follows : let $\mathcal{M}_n^{(c)}$ be the complex mass shell, namely :

$$\mathcal{M}_n^{(c)} = \left\{ k \in \mathbb{C}^{4(n-1)} \; ; \; k_1^2 = m^2, \ldots, k_n^2 = m^2 \right\}.$$

The real mass shell $\mathcal{M}_n = \mathcal{M}_n^{(c)} \cap \mathbb{R}^{4(n-1)}_{(p)}$ is composed of the union of all the physical regions

$$\mathcal{M}_n^{(IJ)} = \left\{ p \in \mathbb{R}^{4(n-1)} \; ; \forall i \in I, \; -p_i \in H_m^+ \; ; \forall j \in J, \; p_j \in H_m^+ \right\}$$

of the corresponding reactions $\{A_i\}_{i \in I} \longrightarrow \{A_j\}_{j \in J}$. (the A_i, A_j denoting identical particles whose states are created by the basic field, as asymptotic states).

Let $S_{IJ}^{(n)}(p)$ be the connected scattering amplitude associated with the above reaction. $S_{IJ}^{(n)}\left(\{-p_i\}_{i \in I}, \{p_j\}_{j \in J}\right)$ is a distribution on $\mathcal{M}_n^{(IJ)}$ which is obtained from $H^{(n)}$ $(n = |I| + |J|)$ through the following procedure : consider the "amputated" n-point function

$\hat{H}^{(n)}(k) = \prod_{i=1}^{n} (k_i^2 - m^2)\, H^{(n)}(k)$, which is analytic in a domain $\hat{D}_n$ such that $D_n = \hat{D}_n - \bigcup_{i=1}^{n} \{k ; k_i^2 = m^2\}$. Then in $\hat{D}_n$ take the "time-ordered boundary value" $\hat{\tau}_n$ of $\hat{H}^{(n)}$, namely : $\hat{\tau}_n(p) = \lim_{\substack{\varepsilon > 0 \\ \varepsilon \to 0}} \hat{H}^{(n)}(p^{(o)}(1 + i\varepsilon), \vec{p})$. It can be proved that :

$$S_{IJ}\left(\{-p_i\}_{i \in I} \; ; \{p_j\}_{j \in J}\right) = \hat{\tau}_n(p)\Big|_{\mathcal{M}_n^{(IJ)}} .$$

The derivation of analyticity properties for the scattering amplitudes $S_{IJ}^{(n)}$ on the complex manifold $\mathcal{M}_n^{(c)}$ is based on the following analysis.

$\hat{D}_n \cap \mathcal{M}_n^{(c)} = \emptyset$, but $\hat{D}_n$ is *not* a natural domain of holomorphy. Standard techniques of analytic completion can then be applied to computing parts of the holomorphy envelope of $\hat{D}_n$. Any such extension D'_n of $\hat{D}_n$ which has a non-empty intersection with $\mathcal{M}_n^{(IJ)}$ yields a "*mass shell domain*" from which $S_{IJ}^{(n)}$ is the boundary value of an analytic function. $S^{(n)} = \hat{H}^{(n)}\Big|_{\mathcal{M}_n^{(c)}}$

The analyticity properties of the $S_{IJ}^{(n)}$ can have either a local or a global character.

By "*local analyticity*" , we mean the fact that in the neighbourhood of a point p, $S_{IJ}^{(n)}$ is the boundary value of an analytic function from a certain "local tube" in $\mathcal{M}_n^{(c)}$. It has been proved in [20c], that for every couple (I,J), there exists a subregion of $\mathcal{M}_n^{(IJ)}$ in which $S_{IJ}^{(n)}$ satisfies the property of local analyticity.

Under the name of "*global analyticity*", we mean either the explicit knowledge of a certain analyticity domain in $\mathcal{M}_n^{(c)}$, or the existence of such a domain which connects two distant regions : for example, a "*crossing domain*" on $\mathcal{M}_n^{(c)}$ links together a couple of physical regions $\mathcal{M}_n^{(I_1J_1)}, \mathcal{M}_n^{(I_2J_2)}$. Several global analyticity properties have been proved for the two-particle scattering amplitude $(|I| = |J| = 2)$; among them let us quote : i) the "Lehmann- Martin ellipses"[22] which give an explicit size of the analyticity neighbourhood of each physical region in the momentum transfer variable t (at fixed squared total energy s), ii) the crossing domains which exist in the general case of two initial and two final particles with arbitrary masses [20b], and become the domains of dispersion relations for certain mass configurations (for example, cut-planes in the variable s, at fixed t, with $|t| < 4m^2$ in the case of identical neutral particles with mass m [22]).

The "standard techniques of analytic completion" which have been used in obtaining these results can be summarized as follows :

i) the local techniques which are elaborated forms of the edge-of-the-wedge theorem [23] (see in this connection the notions of "essential support of a distribution" [24] or "singular spectrum of a hyperfunction" [25])

ii) the global techniques which essentially reduce to :

a) the continuity (or "disc") theorem [26] ; in practice it is often applied under the form of an explicit continuation of a suitable Cauchy integral (or "dispersion relation"),

b) the interpolation technique [27] it often amounts to a suitable application of the tube theorem (after appropriate conformal mapping).

c) Martin's use of the positivity of the absorptive parts [21] (this special method of analytic continuation can also be given an alternative, more geometrical, form [24a]

B-I. Extension of the analyticity properties of the π-π scattering amplitude by using dispersion relations on curved manifolds (Roy and Wanders [4]).

Let $\mathfrak{M}_4^{(c)} = \{k = (k_1,k_2,k_3,k_4),\ k_1+k_2+k_3+k_4 = 0,\ k_i^2 = m^2,\ 1 \leq i \leq 4\}$ be the complex mass shell in $\mathbb{C}^{12}$, by taking into account the (complex) Lorentz invariance of this 8-dimensional complex manifold, $\mathfrak{M}_4^{(c)}$ can be parametrized by two independent Lorentz invariant variables ; however one usually considers for convenience the three Mandelstam variables $s = (k_1+k_2)^2$, $t = (k_1+k_3)^2$, $u = (k_2+k_3)^2$ which are linked by the relation

$$s+t+u = \sum_{i=1}^{4} k_i^2 = 4m^2 .$$

In [4], the following analyticity properties in $\mathbb{C}^{(2)}_{(stu)}$ have been taken as a basis to further analytic continuation for the mass-shell restriction, $S^{(4)}$ of $\hat{H}^{(4)}$:

-i) Let $\Omega_t = \{(s,t)\ ;\ |t| < 4m^2,\ s \in \mathbb{C}\}$ and Ω_s, Ω_u the similar domains obtained by permutations of (s,t,u). Then $S^{(4)}$ is analytic in $\Omega_t \cup \Omega_s \cup \Omega_u$ minus the union of the "cuts" $\sigma_s = \{(s,t)\ ;\ s \text{ real} \geq 4m^2\}$, σ_t and σ_u ,

-ii) the absorptive part $\Delta_s S^{(4)}$ (i.e. the restriction to $\mathfrak{M}_4^{(c)}$ of $\Delta_{\{12\}} \hat{H}^{(4)}$) is analytic in the Lehmann-Martin ellipse $E(s)$ of the t-plane [22] , for each $s \geq 4m^2$; an identical result holds for $\Delta_t S^{(4)}$ and $\Delta_u S^{(4)}$.

The method of [4] consists in expressing $S^{(4)}(s,t)$ through a "dispersion relation" on a complex analytic manifold $\pi(a,x_o)$ in $\mathbb{C}^2_{(stu)}$, and in using the analytic dependence of the latter with respect to the complex parameters a, x_o as a tool of analytic continuation for $S^{(4)}$. The family of manifolds $\{\pi(a,x_o); x_o \in \mathbb{C},\ a \in V(x_o)\}$ is required to satisfy the following conditions :

a) $\pi(a,x_o)$ is symmetric under the permutations of s,t,u,

b) for $|a| < \varepsilon(x_o)$ ($\varepsilon(x_o)$ being sufficiently small), $\pi(a,x_o)$ lies in the domain $\Omega_s \cup \Omega_t \cup \Omega_u$,

c) $\forall x_o \in \mathbb{C}$, $V(x_o)$ is a star-shaped domain in $\mathbb{C}$ such that, $\forall a \in V(x_o)$, the section of $\pi(a,x_o)$ by every line $s = s_o \geq 4m^2$ is contained in $E_{\bar{s}_o}$.

If these conditions are fulfilled, then $S^{(4)}(s,t)$ can be analytically continued in the domain $\mathcal{D} = \bigcup_{x_o \in \mathbb{C}} \mathcal{D}_{x_o}$, where :

$$\mathcal{D}_{x_o} = \left[\bigcup_{a \in V(x_o)} \pi(a,x_o)\right] - (\sigma_s \cup \sigma_u \cup \sigma_t)$$

The following choice has been made for $\{\pi(a,x_o)\}$:

Put $x = -\frac{1}{16}(st+tu+us)$, $y = \frac{1}{64}stu$; then $\pi(a,x_o)$ is defined by $y = a(x-x_o)$, or in $\mathbb{C}^2_{(stu)}$:

$$(s+4a)(t+4a)(u+4a) = 64a[a(a+1)-x_o] \ .$$

This family of cubics has been shown to satisfy conditions a),b),c) for suitable star-shaped sets $V(x_o)$ and thus yields a computable analyticity domain of the previous form. Moreover, in $\mathcal{D}$ the function $S^{(4)}$ satisfies the following dispersion relation (valid on each curve $\pi(a,x_o)$ such that $x_o \in \mathbb{C}$, $a \in V(x_o)$) :

$$S^{(4)}(s,t,u) = S^{(4)}(s_o,0,4m^2-s_o) + \ldots$$

$$\ldots + \frac{1}{\pi}\int_{4m^2}^{\infty} ds' \, \Delta_s S^{(4)}(s',\tau(s',a,x_o)) \left[\frac{1}{s'-s} + \frac{1}{s'-t} + \frac{1}{s'-u} - \frac{1}{s'-s_o} - \frac{1}{s'} - \frac{1}{s'+s_o-4m^2}\right]$$

in this formula s_o denotes an arbitrary parameter such that $0 < s_o < 4m^2$.

An investigation of the sections of the domains $\mathcal{D}_{x_o}$ at fixed angle θ ($t(s,\cos\theta) = (2m^2 - \frac{s}{2})(1-\cos\theta)$) has been done. Of special importance are the sections at $\theta = \frac{\pi}{2}$ which have been computed for a sample of values of x_o and yield a good approximation of the section $\mathcal{D}_\perp$ of $\mathcal{D}$ at $\theta = \frac{\pi}{2}$. In the complex s-plane, this domain is represented on fig.1 below ; the two cuts $s \geq 4m^2$ and $s \leq 0$ are parts of its boundary.

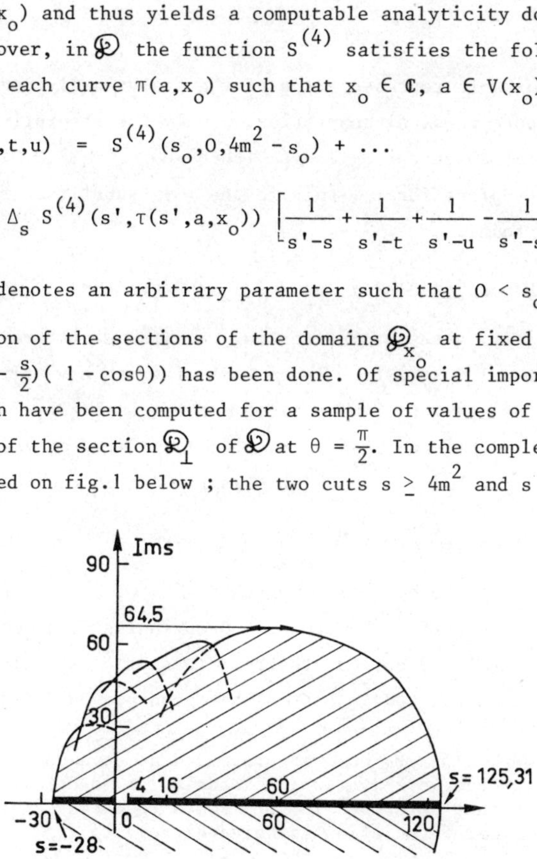

Analyticity of the partial waves

$$f_\ell(s) = \int_0^1 d(\cos\theta)\, S^{(4)}(s,t(s,\cos\theta))\, P_\ell(\cos\theta)$$

The following property has been proved in [4] :

Theorem : i) *for each value of* ℓ , *the function* $f_\ell(s)$ *is analytic in the domain* $\mathcal{D}_\perp$.
ii) f_ℓ *satisfies the following integral representation* :

$$f_\ell(s) = \delta_{\ell o}\, S^{(4)}(4m^2,0,0) + \sum_{\ell'=0}^{\infty} \int_{4m^2}^{\infty} ds'\, K_{\ell\ell'}(s,s')\, a_{\ell'}(s') \ ;$$

Here $a_\ell(s)$ denotes the absorptive part of $f_\ell(s)$ (i.e. its discontinuity on the cut $s \geq 4m^2$), and each kernel $K_{\ell\ell'}(s,s')$ can be explicitly computed, as the integral on a suitable contour of an algebraic expression which contains a certain product of Legendre polynomials $P_\ell \cdot P_{\ell'}$.

Remarks : i) is essentially due to the following fact : for every s in $\mathcal{D}_\perp(x_o)$, one can find a continuous distortion γ_s of [0,1] in the cos θ-plane such that $\bigcup_{\cos\theta\in\gamma_s} (s,\cos\theta) \subset \mathcal{D}_{x_o}$ (this is a simple consequence of the fact that $V(x_o)$ is a star-shaped set).

ii) is a remarkable representation of the analytically continued partial wave f_ℓ in terms of pure physical quantities, namely the absorptive parts of all partial waves $f_{\ell'}$ and the scattering length (through $S^{(4)}(4m^2,0,0)$) ; for want of obtaining a full cut s-plane for $f_{\ell'}$ this is the best substitute for a dispersion relation that can be found.

B-II Some analyticity properties for the 3 → 3 particle scattering amplitudes :

We list very briefly a set of interesting results which have been obtained at the Steklov Mathematical Institute [5] , a part of which have already been presented or announced in the previous Symposium of Mathemetical Physics [28] . The following results of global type have been proved : the validity of one-dimensional dispersion relations for the forward scattering amplitudes of the reactions :

$$\gamma\gamma N \rightarrow \gamma\gamma N \quad \text{in } e^4\text{-order [5a]}, \quad \text{and} \quad \gamma NN \rightarrow \gamma NN \ \text{[5b]}$$

for a mass of the virtual photon satisfying $k^2 \leq a^2 < 0$, a certain absoptive part of the latter being relevant for the study of deep inelastic processes . As a matter of fact, a "generalized optical theorem" (based on the algebra of generalized retarded operators* mentioned in Part A) has been proved for the 3 → 3 particle

---

* Note that all the works [5] are however presented as developments of the Bogoliubov axiomatic framework [29] which leads to an identical analytic and algebraic structure as that described in [18] (see a recent account of it in [5c]). although its conceptual formulation is quite different.

forward scattering amplitude [5a] ; it expresses the latter as a certain sum of absorptive parts ; among these, the discontinuities $\Delta_{\{iji'\}} H^{(6)}$ (represented by the diagram 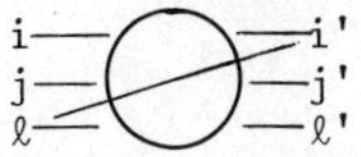

are present and have a physical significance since, due to A.C. or unitary equations (see Part C) the latter can be expressed as inclusive cross sections of the form

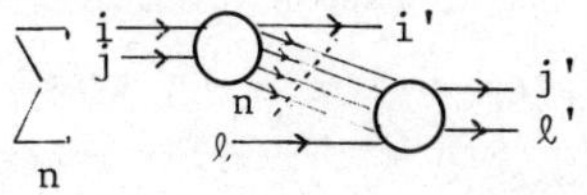

In the same spirit as in [20c], some detailed local analyticity properties for the 3 → 3 particle forward scattering amplitudes with arbitrary masses have been derived in [5e] ; they are based on a straightforward application of the edge-of-the-wedge theorem, which leads to the following description of the situation in the neighbourhood of the physical region Φ_{33} of the forward three-body reaction.

For $H^{(6)}(k_1,\ldots,k_6)$, the set of forward configurations lies in the linear submanifold $\mathfrak{F}$ defined by $k_1+k_4=0$, $k_2+k_5=0$, $k_3+k_6=0$, $k_i \in H^+_{m_i}$ $(i=1,2,3)$, m_i being the mass of particle "i". Then if one takes as variables $s_i = k_j.k_\ell$ $(i,j,\ell)=(1,2,3)$ the physical region Φ_{33} in $\mathfrak{F}$ is the "interior" of the sheet of cubic surface defined in $\mathbb{R}^3_{(s_1 s_2 s_3)}$ by :

$$m_1^2\,s_1^2 + m_2^2\,s_2^2 + m_3^2\,s_3^2 - m_1^2\,m_2^2\,m_3^3 - 2\,s_1 s_2 s_3 = 0 \quad ; \quad s_i \geq m_j m_\ell \quad (i=1,2,3),$$

Φ_{33} is the union of four disjoint parts :

i) the three "inclusive regions" $\Phi_{33}^{(\ell)}$ $(\ell=1,2,3)$ defined by the conditions $(p_i+p_j-p_\ell)^2 \geq M^2_{ij,\ell}$ (for appropriate threshold masses $M_{ij,\ell}$) and the remaining region $\Phi_{33}^{(o)}$ such that $(p_i+p_j-p_\ell)^2 < M^2_{ij\ell}$ for all $(ij\ell)=(123),(231),(312)$.

For $p \in \Phi_{33}^{(\ell)}$, $H^{(6)}$ is analytic in a local tube T_ℓ with basis

$$\left\{q\,;\,q_i+q_\ell \in V^+,\ q_j+q_\ell \in V^+\ ,\ q_i+q_j-q_\ell \in V^+\right\}.$$

For $p \in \Phi_{33}^{(o)}$, $H^{(6)}$ is analytic in a(larger) local tube T_o with basis $\left\{q;\ q_1+q_2 \in V^+,\ q_2+q_3 \in V^+,\ q_3+q_1 \in V^+\right\}$. The intersections of $\{T_o,T_\ell\}$ with the mass shell $\mathcal{M}_6^{(c)}$ have been fully described in [5e]; non-empty local tubes on $\mathcal{M}_6^{(c)}$ are thus obtained for large unbounded subregions of Φ_{33} (for example for p lying in a submanifold of $\Phi_{33}^{(3)}$ of the form : $s_3 = 2s_1$, $s_2 > M_2^2$, $s_3 > f(s_2)$).

Let us finally mention that a certain crossing symmetry property for $H^{(6)}$ restricted to the forward submanifold $\mathfrak{F}$ has been shown in [5f] on the basis of the previous local results and of the Lorentz invariance properties of $H^{(6)}$.

PART C

ANALYTIC STRUCTURE OF GREEN's FUNCTIONS AND SCATTERING AMPLITUDES IMPLIED BY ASYMPTOTIC COMPLETENESS

For each local Q.F.T. whose spectrum contains a discrete part H_m^+ (the case of a single mass m is considered for simplicity), the Haag-Ruelle formalism produces two asymptotic subspaces $\mathcal{H}_{in}$ and $\mathcal{H}_{out}$ of the Hilbert space $\mathcal{H}$ which are both isomorphic to the abstract Fock space of relativistic particles with mass m : these two isomorphisms induce in $\mathcal{H}$ a (partial) isometry from $\mathcal{H}_{out}$ to $\mathcal{H}_{in}$ which is by definition the scattering operator S. However, it is *not* a consequence of the axioms of local Q.F.T. that $\mathcal{H}_{in}$ and $\mathcal{H}_{out}$ span the whole space $\mathcal{H}$.

From now on, we shall consider local field theories in which* $\mathcal{H}_{in} = \mathcal{H}_{out} = \mathcal{H}$. This property, called "*Asymptotic Completeness*" is clearly equivalent to the unitarity of the operator S ($SS^* = S^*S = 1$).

On the basis of Ruelle's discontinuity formulae

$R_{\mathcal{S}} - R_{\mathcal{S}'} = i\,[R_{\mathcal{S}_1}, R_{\mathcal{S}_2}]$ in the algebra of generalized retarded operators** (see [18]), of the reduction formulae and of the analytic structure of the absorptive parts $\Delta_I \hat{H}^{(n)}$ of the functions $\hat{H}^{(n)}$ (see Part B), it can be shown that the latter satisfy the following (infinite) set of relations [30a], which we call "A.C. equations" since they are equivalent to the property of asymptotic completeness : $\forall n$, and for every partition (I,J) of X_n, with $|I| = n_1$, $|J| = n_2$,

$$\text{(A.C.)}\quad \Delta_I \hat{H}^{(n)}(p_I;\underline{k}_{(I)},\underline{k}_{(J)}) = \sum_{\ell=1}^{\infty} \frac{1}{\ell!} \hat{H}_{\alpha}^{(n_1+\ell)}(p_I;\underline{k}_{(I)},\cdot) *_{\ell}^{(I)} \hat{H}_{-\alpha}^{(n_2+\ell)}(p_I;\cdot,k_{(J)})$$

where the operation $*_{\ell}^{(I)}$, called in the following "*mass shell convolution of order ℓ in the channel (I,J)*" is defined by :

$$\hat{H}_{\alpha}^{(n_1+1)} *_1^{(I)} \hat{H}_{-\alpha}^{(n_2+1)} = \frac{2\pi i}{Z}\,\delta(p_I^2 - m^2)\,\theta(p_I^{(o)})\,\hat{H}_{\alpha}^{(n_1+1)}(p_I;\underline{k}_{(I)})\,\hat{H}_{-\alpha}^{(n_2+1)}(p_I;\underline{k}_{(J)})$$

$$\forall \ell \geq 2,\ \hat{H}_{\alpha}^{(n_1+\ell)} *_{\ell}^{(I)} \hat{H}_{-\alpha}^{(n_2+\ell)} = i\left(\frac{2\pi}{Z}\right)^{\ell} \int_{\mathcal{M}_{\ell}(p_I)} \hat{H}_{\alpha}^{(n_1+\ell)}(p_I;\underline{k}_{(I)},\{r_1,\dots,r_{\ell}\}) \dots$$

$$\dots \times \hat{H}_{-\alpha}^{(n_2+\ell)}(p_I;\{-r_1\dots-r_{\ell}\},\underline{k}_{(J)})\,\frac{d\vec{r}_1}{2\omega(r_1)}\cdots\frac{d\vec{r}_{\ell}}{2\omega(r_{\ell})}\,.$$

---

* Such a condition may be less restrictive than it can appear at first sight; for instance one can imagine situations in which the space $\mathcal{H}$ spanned by the states of a certain local field A(x) is not the space of *all* physical states, but a certain sector of physical states which can have its own interest, independently of other types of fields or observables describing other aspects of reality.

** these formulae (which hold for every adjacent couple $(\mathcal{S},\mathcal{S}')$) are generalizations of the standard discontinuity formula for the 2-point "retarded" and "advanced" operators $R_{\pm}$ of a local field A(x). $R_+(x,y) - R_-(x,y) = i\,[A(x),A(y)]$.

in these formulae, $\omega(r_i) = \sqrt{r_i^2 + m^2}$, $Z = -(p^2 - m^2)\, H^{(2)}(p)\Big|_{p^2=m^2}$ and

$$\mathfrak{M}_\ell(p_I) = \left\{(r_1, r_2, \dots, r_\ell);\ r_1 + \dots + r_\ell = p_J = -P_I;\ r_i \in H_m^+,\quad 1 \le i \le \ell \right\}.$$

α denotes the sign of the relevant boundary value of $\hat{H}^{(n_1+\ell)}$ with respect to q_I (i.e. $q_I \in \alpha V^+$, $q_I \to 0$) ; the two cases ($\alpha = \pm$) correspond to either use of "in"-states or "out"-states projections at the right-hand side of Ruelle's formulae ($\langle R\mathcal{G}_1 . R\mathcal{G}_2\rangle = \langle R\mathcal{G}_1|\text{in}\rangle\langle\text{in}|R\mathcal{G}_2\rangle = \langle R\mathcal{G}_1|\text{out}\rangle\langle\text{out}|R\mathcal{G}_2\rangle$) in the derivation of the A.C. equations.

Remarks on A.C. equations and technical assumption :

We first notice that since $\mathfrak{M}_\ell(p_I)$ is empty for $p_I^2 < (\ell m)^2$, the term of order ℓ at the r.h.s. of the A.C. equations has its support contained in $\{k; p_I \text{ real} \in \bar{V}^+_{\ell m}\}$, so that the summation over ℓ always reduces to a finite number of terms if p_I is kept in a bounded region. Strictly speaking, the A.C. equations hold as identities for measures with respect to p_I, taking their values in a space of analytic functions of $\underline{k}_{(I)}, \underline{k}_{(J)}$. The purpose of what follows is to perform some analytic continuation of these equations in the complex variables $s_I = k_I^2$. However this will only be feasible under the following additional assumption*:

Smoothness assumption : the A.C. equations hold as pointwise identities between analytic functions in the relevant domains $\mathcal{D}_n^I(p_I)$, for each fixed value of p_I in $\bar{V}^+_{2m}$.

The analytic continuation procedures which will be used in this Part C are based on the study of the analyticity properties of integrals of analytic functions and specially of the Fredholm resolvent of an analytic kernel. To be more specific, we consider the following general situation : Let $F(k,z,z')$ be defined and analytic in a complex domain $\mathfrak{D} = \bigcup_{k\in\Delta} (k, \mathfrak{D}_k \times \mathfrak{D}'_k)$, $\mathfrak{D}_k$, $\mathfrak{D}'_k$ being domains of complex dimension n ; let Γ_k be an n-dimensional cycle in $\mathfrak{D}_k \cap \mathfrak{D}'_k$ which varies continuously with k (for k in Δ) and let $\omega_k(z)$ be an analytic differential form of order n.

We are interested by the analyticity properties of the Fredholm resolvent $G(k_o, z, z'; \lambda)$ of F corresponding to a given value $k = k_o$ and to the integration space Γ_{k_o}. The following result has been proved in [33] :

Lemma :

The solution G of the resolvent equation :

$$F(k_o, z, z') = G(k_o, z, z'; \lambda) + \lambda \int_{\Gamma_{k_o}} F(k_o, z, z_1)\, G(k_o, z, z'; \lambda)\, \omega_{k_o}(z\,)$$

---

* this property can also be considered as a consequence of the "*smooth spectral condition*" that has been postulated in [30a].

admits a meromorphic continuation of the form :

$$G^{(\dot{\Gamma})}(k,z,z';\lambda) = \frac{N^{(\dot{\Gamma})}(k,z,z';\lambda)}{D^{(\dot{\Gamma})}(k;\lambda)}$$

where $N^{(\dot{\Gamma})}$ (resp. $D^{(\dot{\Gamma})}$) is holomorphic in $\mathfrak{D} \times C_{(\lambda)}$ (resp. $\Delta \times C_{(\lambda)}$).

For each k in Δ, $G^{(\dot{\Gamma})}(k,z,z';\lambda)$ is an analytic continuation of the Fredholm resolvent of $F(k,z,z')$, considered as a kernel on Γ_k. Moreover, $N^{(\dot{\Gamma})}$ and $D^{(\dot{\Gamma})}$ (which can be defined on $\Gamma_k \times \Gamma_k$ by the standard Fredholm formulae) only depend on the set $\dot{\Gamma}$ of (continuously varying) homology classes $\dot{\Gamma}_k$ of Γ_k in $\mathfrak{D}_k \cap \mathfrak{D}'_k$ ($\dot{\Gamma} = \{\dot{\Gamma}_k ; k \in \Delta\}$).

This lemma will be applied in the two schemes of analytic continuation which are developed below in C I and C II, the first scheme being purely based on the analytic continuation of the A.C. equations, while the second scheme uses the analyticity of ℓ-particle irreducible n-point functions; in the latter, the A.C. equations enter as an element of the proof of the irreducibility properties. In a pedagogical spirit, we shall now illustrate the ideas of these two schemes by treating the case of the two-particle structure of $H^{(4)}$ and of the two-body scattering amplitude of an even* Q.F.T. with a single mass m.

We consider the channel $(I,J) = (\{1,2\},\{3,4\})$ and the corresponding low energy region : $s = k_I^2 = (k_1 + k_2)^2 < (4m)^2$; the relative four momenta associated with this channel are :

$$\underline{k}_{(I)} = \frac{k_1 - k_2}{2} = z \ , \quad \underline{k}_{(J)} = \frac{k_3 - k_4}{2} = z'$$

In the considered region, the A.C. equations for $\hat{H}^{(4)}$ can be written as follows :

$$(\text{A.C.})\begin{cases} \Delta_I \hat{H}^{(4)}(p_I;z,z') \equiv \hat{H}_+^{(4)}(p_I;z,z') - \hat{H}_-^{(4)}(p_I;z,z') = \\ = i(\frac{2\pi}{Z})^2 \times \frac{1}{2} \displaystyle\int_{u \in S^2} \hat{H}_{\pm}^{(4)}(p_I;z,z_1(u,s))\, \hat{H}_{\mp}^{(4)}(p_I;z_1(u,s)z') \frac{2}{\sqrt{s(s-4m^2)}}\, d\mu(u) \end{cases}$$

To derive this formula, we have assumed that $k_I = (k_I^{(o)},0)$, with $k_I^{(o)} = \sqrt{s}$ (this is not a restriction since $\hat{H}^{(4)}$ is Lorentz invariant) ; the mass shell integration manifold

$$\mathcal{M}_2(p_I) = \left\{ z_1 \in \mathbb{R}^4 ;\ (z_1 + \frac{p_I}{2})^2 = (z_1 - \frac{p_I}{2})^2 = m^2 \right\}$$

is then a sphere which can be parametrized as follows :

$z_1 = z_1(u,s) = (0, \frac{\sqrt{s-4m^2}}{2}\, u)$, with $u \in S^2$; $d\mu(u)$ is the surface element on the unit sphere S^2

Let us now introduce the following Fredholm resolvent equation, depending on

---

* the results which we present have however been proved in the general case.

the complex parameter s (z(u,s) being the analytic continuation for complex values of s of its definition given for s real) :

$$
(F)\begin{cases} \hat{H}_+^{(4)'}(k_I; z(u,s)\ ,\ z'(u',s)) - \hat{H}_-^{(4)}(\ldots) = \\ = \frac{i}{2}\left(\frac{2\pi}{Z}\right)^2 \int_{u_1\in S^2} \hat{H}_+^{(4)'}(k_I; z(u,s), z_1(u_1,s))\ \hat{H}_-^{(4)}(k_I; z_1(u_1,s), z'(u',s))\ d\mu_s(u), \end{cases}
$$

where $d\mu_s(u) = \frac{2}{\sqrt{s(s-4m^2)}}\ d\mu(u)$.

In this equation, $\hat{H}_-^{(4)}$ represents the function $\hat{H}^{(4)}\big|_{\mathcal{M}_4^{(c)}} = S^{(4)}$ considered as given in a part of its analyticity domain* of the following form :

$$
D_- = \Big\{k \in \mathcal{M}_4^{(c)};\ -\varepsilon < \mathrm{Im}\ s < 0\ ;\ (2m)^2 < \mathrm{Re}\ s < (4m)^2;\ u,u' \in S_c^2: \\ |\mathrm{Im}\ u| < a, |\mathrm{Im}\ u'| < a\Big\}\ ,
$$

where ε and a are certain positive numbers.

According to the previous lemma, the Fredholm resolvent $\hat{H}_+^{(4)'}$ of $\hat{H}_-^{(4)}$ is defined as a meromorphic function $\frac{N(s,u,u')}{D(s)}$ in the domain D_- ; but since equation (F) is identical for s real $((2m)^2 < s < (4m)^2)$ with the mass shell restriction of equation (A.C.) (they both coincide with the standard "elastic unitarity equation" for $S_{IJ}^{(4)}$), it is a straightforward consequence of the edge-of-the-wedge theorem that $\hat{H}_+^{(4)'}$ is the analytic continuation in D_- of $\hat{H}_+^{(4)}$, the latter being the determination of $\hat{H}^{(4)}$ defined in the part

$$
D_+ = \Big\{k \in \mathcal{M}_4^{c};\ 0 < \mathrm{Im}\ s < \varepsilon\ ;\ (2m)^2 < \mathrm{Re}\ s < (4m)^2\ ; u,u' \in S_c^2, \\ |\mathrm{Im}\ u| < a, |\mathrm{Im}\ u'| < a\Big\}
$$

of its analyticity domain.

So far, a meromorphic continuation of $\hat{H}_+^{(4)}$ through the cut : $(2m)^2 < s < (4m)^2$ has been performed in the complex mass shell $\mathcal{M}_4^{(c)}$; however it is easy to get rid of the constraints $k_i^2 = m^2$ (i = 1,2,3,4), by a simple argument of analytic continuation for both sides of equation (F).

Let us add the following remarks concerning this meromorphic continuation of $\hat{H}^{(4)}$ and $S_{IJ}^{(4)}$ in a second sheet :

i) the mechanism of analytic continuation of the partial waves across the cut $(2m)^2 < s < (4m)^2$, through using the "elastic unitarity relations", has been explained long ago in [31] and refined in [32]; however the present method is more general and

*Strictly speaking, the domains D_+, D_- are contained in the holomorphy envelope of the primitive domain D_4 (see [20a]).

more suitable for treating the scattering amplitude $S_{IJ}^{(4)}$.

ii) The pathology which was discovered by A. Martin [32] in the partial wave approach and concerns the possible accumulation of poles from the second sheet on the real interval $(2m)^2 < s < (4m)^2$, corresponds in the present approach to the singular case of Fredholm theory in which the denominator $D(s,\lambda)$ happens to have a fixed zero at $\lambda = \frac{i}{2}(\frac{2\pi}{Z})^2$ (the "physical value" of the Fredholm parameter λ); this case is here discarded thanks to the "smoothness assumption" which has been postulated above ; the latter may be compared with analogous regularity assumptions postulated in [32] .

iii) A local meromorphic continuation $\hat{H}_-^{(4)'}$ of $\hat{H}_-^{(4)}$ in D_+ can be derived similarly ; it can also be shown that $\hat{H}_\pm^{(4)'}$ admits an analytic continuation in a domain of the form $\{|s-4m^2| < \varepsilon \ , \ s \neq 4m^2 + \rho\}$ and that it coincides with $\hat{H}_-^{(4)'}$ in the intersection of the latter with D_+ ; the reason for this "two-sheeted monodromic structure" of $\hat{H}^{(4)}$ around the manifold $s = 4m^2$ lies in the change of sign which is produced in the integral at the r.h.s of equation (F) when s describes a small loop enclosing the point $s = 4m^2$.

iv) Extensions of the second sheet analyticity domain of $S^{(4)}$ outside the local regions described above can be obtained in principle by applying more thoroughly the previous lemma on Fredholm equation. Indeed, for each continuous distortion $\Gamma(s)$ of $S^{(2)}$ in the complex sphere $S_c^{(2)}$ such that the set $\{z = z(u,s),\ u \in \Gamma(s)\}$ is contained in the (first sheet) analyticity domain of $\hat{H}^{(4)}$, there exists a second sheet meromorphic continuation of $\hat{H}^{(4)}$. We notice for example that such a distortion has been produced in the cos θ-plane (see B-I) for the construction of the analyticity domain of the partial waves proposed in [4] . One can assert that any domain of the partial waves obtained by such a distortion technique is the projection onto s-plane of a second-sheet domain of $S^{(4)}$.

An alternative method [7] for deriving these analytic continuation properties of $\hat{H}^{(4)}$ in the two-particle unphysical sheet consists in using the Bethe-Salpeter equation :

$$(\text{B.S.})\quad \hat{H}_{in}^{(4)}(k_I;z,z') = \hat{L}_{in}^{(4)}(k_I;z,z) + \frac{1}{2}\int_{\Gamma(k_I)} \hat{H}_{in}^{(4)}(k_I;z,z_1)\ \hat{L}_{in}^{(4)}(k_I;z_1,z')d_4z_1$$

in which $\hat{H}_{in}^{(4)}$ denotes the "left-amputated" form of $H^{(4)}$, namely :

$\hat{H}_{in}^{(4)}(k) = H^{(4)}(k).(k_1^2-m^2)(k_2^2-m^2)$. Equation (B.S.) defines $\hat{L}_{in}^{(4)}$ (namely the "Bethe-Salpeter kernel") as a Fredholm resolvent of $\hat{H}_{in}^{(4)}$, depending analytically on the complex parameter k_I. In (B.S.), $\Gamma(k_I)$ denotes a four-dimensional contour which coincides with the euclidean space $\{z_1 = (iy_1^{(o)},x_1)\}$ when $k_I = (iq_I^{(o)},\vec{p}_I)$; when k_I varies in $\mathbb{C}^4 - \{k_I;k_I^2 = 4m^2 + \rho,\ \forall\rho \geq 0\}$, $\Gamma(k_I)$ has to be distorted inside the primitive analyticity domain of $\hat{H}_{in}^{(4)}$. The required conditions for applying the C.F. lemma are then fulfilled and the following results can be obtained :

a) the function $\hat{L}^{(4)}(k) = \hat{L}^{(4)}_{in}(k).(k_3^2-m^2)(k_4^2-m^2)$ is analytic in $\hat{D}_4$, except for possible "C.D.D. polar manifolds" of the form : $s = k_I^2 =$ constant,

b) the absorptive part $\Delta_I \hat{L}^{(4)}_{in}$ of $\hat{L}^{(4)}_{in}$ satisfies the following relation (Δ) in the region $(2m)^2 < s < (4m^2)$:

$$(\Delta)\ \Delta_I \hat{L}^{(4)}_{in}(p_I;z,z') = (\mathbb{1} - \tfrac{1}{2}\hat{L}^{(4)}_{in}) \odot_{\Gamma_+(p_I)} (\Delta_I \hat{H}^{(4)}_{in} - \tfrac{1}{2}\hat{H}^{(4)} *_2^{(I)} \hat{H}^{(4)}_{in}) \odot_{\Gamma_-(p_I)} (\mathbb{1} - \tfrac{1}{2}\hat{L}^{(4)}_{in})$$

In this relation, the following operator notation has been used :

$$(F \odot_{\Gamma(k_I)} G)(k_I;z,z') = \int_{\Gamma(k_I)} F(k_I;z,z_1)\, G(k_I;z_1,z')\, d_4 z_1 \ ;$$

$\mathbb{1}$ denotes the identity kernel on the space $\Gamma(k_I)$; $\Gamma_+(p_I)$ and $\Gamma_-(p_I)$ denote limiting positions of $\Gamma(k_I)$ from the respective sides $q_I^{(o)} > 0$ and $q_I^{(o)} < 0$, obtained when $q_I^{(o)} \to 0$. With these notations, equation (B.S.) can be rewritten :

$$(\mathbb{1} + \tfrac{1}{2}\hat{H}^{(4)}_{in}) \odot_{\Gamma(k_I)} (\mathbb{1} - \tfrac{1}{2}\hat{L}^{(4)}_{in}) = \mathbb{1} \quad \text{(for each value of } k_I\text{)},$$

and this shows the "invertibility" of equation (Δ) under the following form :

$$(\Delta')\ \Delta_I \hat{H}^{(4)}_{in} - \tfrac{1}{2}\hat{H}^{(4)} *_2^{(I)} \hat{H}^{(4)}_{in} = (\mathbb{1} + \tfrac{1}{2}\hat{H}^{(4)}_{in}) \odot_{\Gamma_+(p_I)} \Delta_I \hat{L}^{(4)}_{in} \odot_{\Gamma_-(p_I)} (1 + \tfrac{1}{2}\hat{H}^{(4)}_{in})$$

From (Δ) and (Δ') one readily concludes the equivalence of the A.C. equation : $\Delta_I \hat{H}^4 - \frac{1}{2}\hat{H}^{(4)} *_2^{(I)} \hat{H}^{(4)} = 0$ with the equation : $\Delta_I \hat{L}^{(4)}_{in} = \Delta_I \hat{L}^{(4)} = 0$ in the region $(2m)^2 < s < (4m)^2$; if the B.S. kernel $\hat{L}^{(4)}_{in}$ (or $\hat{L}^{(4)}$) satisfies this condition, one says that it is "two-particle irreducible" (see C.II for the perturbative interpretation of this notion).

Now, if $\Delta_I \hat{L}^{(4)} = 0$, it results from a standard application of the edge-of-the-wedge theorem to $\hat{L}^{(4)}$, that the latter is *analytic in a complex neighbourhood of the region* $\{k = (k_1 \ldots k_4) \in \mathcal{M}_4^{(I,J)} ;\ s = p_I^2 < (4m)^2\}$. Then by exploiting this information in equation (B.S.), namely by applying the C.F. lemma to $\hat{H}^{(4)}_{in}$ considered as the Fredholm resolvent of $\hat{L}^{(4)}_{in}$, it is possible to recover the analytic continuation of $\hat{H}^{(4)}_{in}$ across the cut $(2m)^2 < s < (4m)^2$, as well as the two-sheeted monodromic structure of $\hat{H}^{(4)}_{in}$ around the singularity $s = 4m^2$; the threshold behaviour of $\hat{H}^{(4)}_{in}$ in $(s - 4m^2)^{1/2}$ can also be derived from a general study of the Landau singularities produced in complex Fredholm theory (see [33] theorems 3 and 4) ; let us simply notice here that the singularity $s = 4m^2$ appears as being produced by the pinch of the two polar singularities $(z_1 \pm \frac{k_I}{2})^2 = m^2$ which are present for the analytic function $\hat{L}^{(4)}_{in}$: this mechanism is a generalization of the production of pinch singularities in the Feynman integrals.

Finally, the geometrical equivalence between the two previous procedures of analytic continuation in the second sheet can be explained as follows ; when p_I is such that $(2m)^2 < p_I^2 < (4m)^2$, $\Gamma_+(p_I)$ is homologous to $\Gamma_-(p_I) + \tilde{e}(p_I)$ in the z_1-section of the primitive analyticity domain of $\hat{L}_{in}^{(4)}(p_I,z,z_1)$: here $\tilde{e}(p_I)$ denotes a closed four dimensional contour which can always be supposed to be contained in an *arbitrary small* complex neighbourhood in $\mathbb{C}^4$ of the mass shell sphere $\mathcal{M}_4^{(IJ)}(P_I)$ (see [34]). Therefore the region of the half-plane Im $s \leq 0$ for which $\Gamma_+(p_I)$ can be distorted, with the constraint of staying inside the domain of $\hat{L}_{in}^{(4)}$ is identical with the region for which $\tilde{e}(p_I)$, and therefore $\mathcal{M}_4^{(IJ)}(p_I)$, can be distorted inside the same domain ; but this region of the s-plane is precisely that which can be reached through the first procedure (see remark iv) above).

However, differences between the two methods may appear in more general cases at a more global level of analytic continuation (see C II).

C-I. *Local analytic continuation of the 2 → 3 and 3 → 3 particle scattering amplitudes obtained by a direct use of A.C. equations.*

a) The 2 → 3 case : The scattering amplitude associated with the reaction channel (I,J) = ({4,5},{1 2 3}) is a function of the five Lorentz invariants, namely : the three subenergy variables $s_{12} = (k_1+k_2)^2$, $s_{13} = (k_1+k_3)^2$, $s_{23} = (k_2+k_3)^2$ (the squared total energy $s = (k_4+k_5)^2 = (k_1+k_2+k_3)^2$ being such that $s_{12}+s_{23}+s_{31} = s+3m^2$) and two independent momentum transfer variables such as $t_{14} = (k_1+k_4)^2$ and $t_{35} = (k_3+k_5)^2$. The local analyticity of $S_{IJ}^{(5)} = H^{(5)}\big|_{\mathcal{M}_5^{(IJ)}}$ in t_{14} and t_{35} in a complex neighbourhood of all the physical points is an old result [35] which is the analogue of the Lehmann ellipses of $S^{(4)}$. In the variables (s_{12},s_{23},s_{31}), $S_{IJ}^{(5)}$ had been shown (in [20c]) to be analytic in a certain local tube with conical basis contained in $\{(s_{ij});\ \text{Im}\ s_{12} > 0,\ \text{Im}\ s_{23} > 0,\ \text{Im}\ s_{31} > 0\}$, the size of the latter being very "thin" in the low energy strip $(3m)^2 \leq s < (4m)^2$. A much more satisfactory result has been obtained in [6] by carrying out Glaser's method : in the low energy strip, the A.C. equations for $\hat{H}^{(5)}$ can be written :

$$\Delta_{\{jk\}}\hat{H}^{(5)} = \hat{H}_+^{(5)} - \hat{H}_{i-}^{(5)} = \frac{1}{2}\hat{H}_+^{(4)} \underset{2}{*}^{\{jk\}} \hat{H}_{i-}^{(5)} = \frac{1}{2}\hat{H}_-^{(4)} \underset{2}{*}^{\{jk\}} \hat{H}_+^{(5)},$$

where $(i,j,k) = $ p.c. (1,2,3), and $\hat{H}_+^{5}$, $\hat{H}_{i-}^{(5)}$ denote the boundary values of $\hat{H}^{(5)}$ from the respective tubes :

$$\left\{q;q_{12} \in V^+,\ q_{13} \in V^+,\ q_{23} \in V^+\right\} \quad \text{and} \quad \left\{q;q_{ij} \in V^+, q_{ik} \in V^+, -q_{jk} \in V^+,\ q_{123} \in V^+\right\}$$

Then by plugging into the above equation a local decomposition of $\{\hat{H}_+^{(5)}, \hat{H}_{i-}^{(5)}\}$ involving an auxiliary set of analytic functions F_i introduced in [20c] (namely : $\hat{H}_+^{(5)} = F_{1+} + F_{2+} + F_{3+}$, $\hat{H}_{i-}^{(5)} = F_{i-} + F_{j+} + F_{k+}$ with $F_{i\pm}$ analytic in "wide" local tubes with bases $\Theta_i^{\pm} = \{q;\ \pm q_{jk} \in V^+,\ q_{123} \in V^+\ \}$, and by performing the analytic continuation of $\hat{H}_+^{(4)} \underset{2}{*}^{\{jk\}} F_{i-}$ in $\Theta_i^{\underline{i}}$ under the form $\hat{H}_+^{4'} \underset{2}{*}^{\{jk\}} F_{i-}$, one obtains a certain analytic continuation of each F_{i+} across the two-particle cut $s_{jk} = 4m^2 + \rho$ $(\rho \geq 0)$. By iterating twice this procedure and then coming back to $\hat{H}_+^{(5)}$, the authors of [6] have obtained the following result :

Theorem : "*At almost*" *every point* $p = (p_1 \ldots p_5)$ *of the low energy physical region* $(s = (p_1 + p_2 + p_3)^2 < (4m)^2)$ *of the* $2 \to 3$ *particle scattering* $S_{IJ}^{(5)}$ *(equal mass case* ; $I = \{45\}$, $J = \{123\}$) *the latter is the boundary value of an analytic function from a local tube whose basis in the space of the invariants is* $\mathrm{Im}\, s > 0$.

Remarks : i) the exceptional points at which this result fails are the submanifolds with equations $(p_i + p_j - p_k)^2 = m^2$, with $\{i,j,k\} = \{1,2,3\}$; the latter include in particular the two-particle thresholds $s_{ij} = (2m)^2$ $(\{i,j\} \subset \{1,2,3\})$. The microlocal study of [6] does not yield the analytic structure of $S_{IJ}^{(5)}$ at these exceptional points, although the latter could in principle be investigated by a more global use of Glaser' method.

ii) An aspect of the obtained result is that $S_{IJ}^{(5)}$ has been analytically continued across the three two-particle cuts $s_{ij} = (2m)^2 + \rho$ $(0 < \rho < 5m^2)$; this can be visualized with the help of fig. 2, in which hatchings represent the regions of $\mathrm{Im}\, s_{ij}$-space corresponding to an "unphysical sheet" analytic continuation.

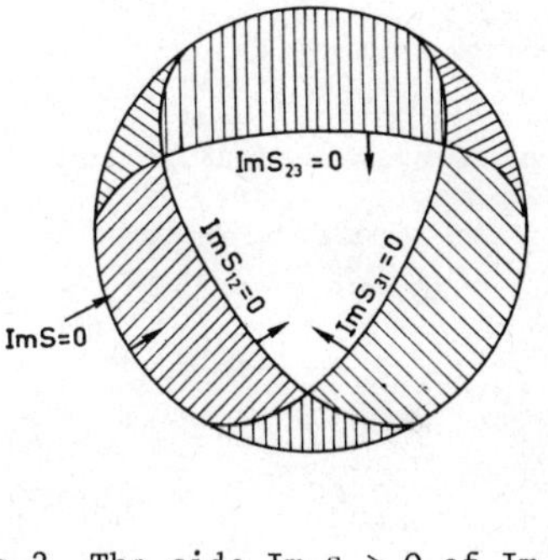

Fig.2. The side $\mathrm{Im}\, s > 0$ of $\mathrm{Im}\, s_{ij}$-space (a section of it by a sphere centered at the origin has been drawn).

b) The $3 \to 3$ *case* : The scattering amplitude associated with the reaction channel $(I,J) = (\{456\},\{123\})$ is a function of eight Lorentz-invariants, namely :

i) the six (incoming and outgoing) subenergy variables $s_{12} = (k_1 + k_2)^2$, $s_{23} = (k_2 + k_3)^2$, $s_{31} = (k_3 + k_1)^2$, $s_{45} = (k_4 + k_5)^2$, $s_{56} = (k_5 + k_6)^2$, $s_{64} = (k_6 + k_4)^2$ which are linked by the relation : $s_{12} + s_{23} + s_{31} = s_{45} + s_{56} + s_{64} = s + 3m^2$; s denotes the total energy variable $s = (k_1 + k_2 + k_3)^2 = (k_4 + k_5 + k_6)^2$.

ii) three independent momentum-transfer variables such as :

$$t_{14} = (k_1 + k_4)^2 \;,\; t_{25} = (k_2 + k_5)^2 \;,\; t_{36} = (k_3 + k_6)^2 \;.$$

By using the A.C. equations for the six absorptive parts $\Delta_{\{jk\}} \hat{H}^{(6)}$ $(\{j,k\} \subset \{1,2,3\})$ and $\Delta_{\{\ell,1,2,3\}} \hat{H}^{(6)}$ $(\ell \in \{456\})$, the authors of [6] have performed a local analytic continuation of $S_{IJ}^{(6)} = \hat{H}^{(6)}\big|_{\mathcal{M}_6^{(IJ)}}$ which is similar to that of $S_{IJ}^{(5)}$ described above. Here again, their starting point is the "weak" local analyticity of $S^{(6)}_{IJ}$ derived from locality and spectrum which (in its refined form [20c]) implies a certain decomposability property of $S^{(6)}_{IJ}$ in terms of auxiliary analytic functions.

From considerations of perturbation theorv and of S-matrix theory, one can expect that the Landau varieties associated with the following tree-graphs and triangle graphs will be obstacles to analytic continuation, since these varieties have real "α_+-branches" in the low-energy physical region ($s < (4m)^2$) ; these varieties are :

$$\theta_{h\ell} \equiv (k_i + k_j + k_\ell)^2 - m^2 \equiv s_{ij} + t_{i\ell} + t_{j\ell} - 4m^2 = 0 \quad \text{for}$$

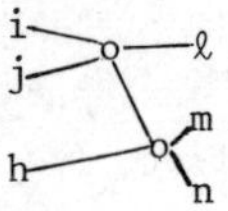

$$\lambda_{h\ell} \equiv \frac{s_{ij}\, s_{mn}\, t_{h\ell}}{m^2} + (s_{ij}^2 + s_{mn}^2 + t_{h\ell}^2) - 2(s_{ij}s_{mn} + s_{mn}t_{h\ell} + s_{ij}t_{h\ell}) = 0$$

$$\text{and} \quad s_{ij} - 4m^2 = 0 \ , \ s_{mn} - 4m^2 = 0$$

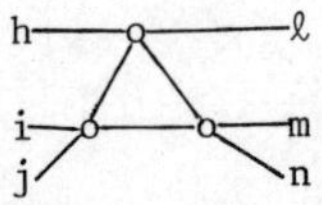

The α^+-branch of $\lambda_{h\ell} = 0$ is defined as follows (see fig. 3) :

$$\left\{ \lambda^+_{h\ell} = 0 \right\} = \left\{ (p_1 \ldots p_6) \in \mathfrak{M}_6^{(I,J)} ;\ \lambda_{h\ell} = 0,\ s_{ij} + t_{h\ell} < 4m^2,\ s_{mn} + t_{h\ell} < 4m^2 \right\}$$

Indeed, the varieties $\theta_{h\ell} = 0$, $\lambda^+_{h\ell} = 0$ are involved in the microlocal result of , which can be described as follows :

Theorem : *At almost every point $p = (p_1 \ldots p_6)$ of the low energy physical region ($s < (4m)^2$) of the $3 \to 3$ particle scattering amplitude $S^{(6)}_{IJ}$ (equal mass case ; $I = \{4,5,6\}$, $J = \{1,2,3\}$), the latter is the boundary value of an analytic function from a local tube whose basis in the space of the invariants is :*

i) Im $s > 0$ is p does not belong to any variety $\theta_{h\ell} = 0$, $\lambda^+_{h\ell} = 0$,

ii) Im $s > 0$, Im $\theta_{h\ell} > 0$ (resp. Im $\lambda_{h\ell} > 0$), if p belongs to a single variety $\theta^+_{h\ell} = 0$ (resp. $\lambda^+_{h\ell} = 0$),

iii) Im $s > 0$, {Im $\theta_{h\ell} > 0$, Im $\lambda_{h'\ell'} > 0$} , if p belongs to the intersection of the corresponding varieties $\{\theta_{h\ell} = 0,\ \lambda^+_{h'\ell'} = 0\}$.

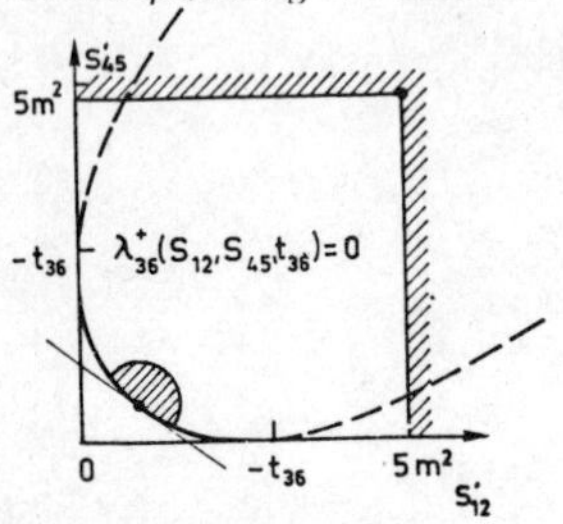

Fig.3. A section of the $3 \to 3$ physical region at t_{36} fixed ; the latter exhibits the relevant branch $\lambda^+_{36} = 0$ of $\lambda_{36} = 0$, and a typical local analyticity domain (Im $\lambda_{36} > 0$) for $p \in \{\lambda^+_{36} = 0\}$ ($s'_{12} = s_{12} - 4m^2$; $s'_{45} = s_{45} - 4m^2$).

Remark : the exceptional set at which this result fails is the union of all the submanifolds $(p_i + p_j - p_h)^2 = m^2$, ($\{i,j,h\} = \{4,5,6\}$) and $(p_\ell + p_m - p_n)^2 = m^2$ ($\{\ell,m,n\} = \{1,2,3\}$). (see also the remark after the analogous result in $2 \to 3$ case).

C-II. *The method of ℓ-particle irreducible kernels ; some local and global results for $H^{(6)}$ and for the 3 → 3 particle scattering amplitudes.*

The ℓ.p.i. kernels have a "natural definition" in the perturbative expansion of lagrangian Q.F.T. : the *ℓ.p.i.-part* $I \equiv (\ell) \equiv J$ *of* $H^{(n)}$ *with respect to a given channel* (I,J) is defined there as the formal series of all Feynman amplitudes which correspond to ℓ.p.i. graphs in the following "topological" sense : a graph $\mathcal{G}$ is ℓ-particle irreducible with respect to (I,J) if $(\ell+1)$ internal lines of $\mathcal{G}$ at least must be cut in order to yield two disjoint connected graphs, carrying respectively the sets of external lines labelled by I and by J.

Equations of the B.S. type were primitively introduced in this context, as identities of formal series which relate the expansions of the $H^{(n)}$ with those of their ℓ.p.i. parts ; an intuitive graphical notation was used for writing the latter; for example for equation (B.S.) :

$(\text{B.S.})_{\text{graph}}$

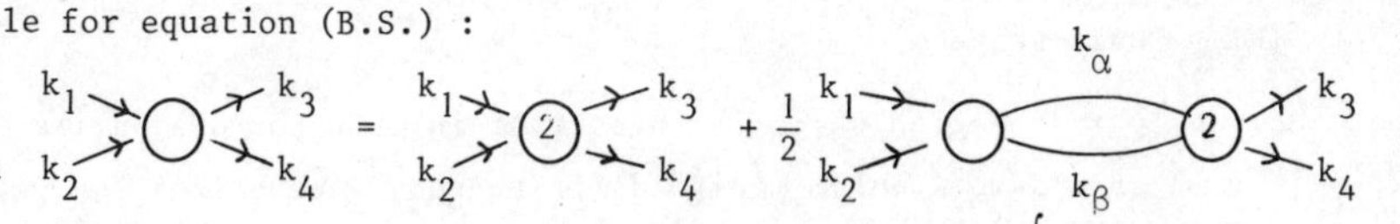

where the notation of the last term indicates the operation $\int dk_\alpha dk_\beta \, \delta(k_1+k_2-k_\alpha-k_\beta)$, as a Feynman amplitudes ; the link with the previous notation for (B.S.) is :

$$z = \frac{k_1-k_2}{2}, \quad z_1 = \frac{k_\alpha-k_\beta}{2}, \quad z' = \frac{k_3-k_4}{2}.$$

In [8] , Symanzik proposed to use these equations for defining the ℓ.p.i. parts of $H^{(n)}$ in axiomatic Q.F.T. and outlined a program for applying these equations to the Many-Particle Structure Analysis (M.P.S.A.) of Green's functions.

A systematic and rigorous exploitation of these ideas in the complex momentum space framework was initiated in [7] and developed in collaboration with M. Lassalle and more recently A. Katz in [36,37] as a first step, we had to give a mathematical status to the generalized graphical Feynman notation (as used in $(\text{B.S.})_{\text{graph}}$); this leads us to introduce and study *$\mathcal{G}$-convolution products* ; we now summarize the (old and recent) results of this study.

$\mathcal{G}$-convolution :

Let us define a *general n-point function* $F^{(n)}(k_1,\dots,k_n)$ as an analytic function in the domain D_n described in part B, and whose absorptive parts $\Delta_I F^{(n)}$ satisfy the properties a) and b) described there, except that the support of $\Delta_I F^{(n)}$ with respect to p_I can be more general and depend on the channel (I,J). We say that $F^{(n)}$ *is ℓ.p.i. with respect to the channel (I.J.) if the support of* $\Delta_I F^{(n)}$ *is contained in* $\bar{V}^+_{(\ell+1)m}$ (in the single mass case).

Let $\mathcal{G}$ denote a connected m-looped graph with n external lines, set of vertices $\{v \in \mathcal{V}\}$, set of internal lines $\{i \in \mathcal{I}\}$: n_v denotes the number of lines incident to vertex v, and $k_v \in \mathbb{C}^{4(n_v-1)}$ the momentum variables attached to v, which are linear functions $k_v(k,k_{int})$ of $k \in \mathbb{C}^{4(n-1)}$ and $k_{int} \in \mathbb{C}^{4(m-1)}$ (the external and internal momentum variables of $\mathcal{G}$); the momenta $k_i = k_i(k,k_{int}) \in \mathbb{C}^4$ are similarly introduced for all internal $i \in \mathcal{I}$. $\mathcal{G}$ will be represented graphically with "fat vertices" (or "bubbles") in place of the pointwise vertices of Feynman graphs.

The following "conservation" theorem has been proved (see [38] for a more précise statement) :

<u>*Theorem*</u> : *Being given a set of general n_v-point functions $\{F^{(n_v)}; v \in \mathcal{V}\}$ and a set of general 2-point functions $\{F_i^{(2)}; i \in \mathcal{I}\}$ which satisfy appropriate decrease properties at infinity in euclidean space, there exists a general n-point function $F_{\mathcal{G}}^{(n)}$, called "$\mathcal{G}$-convolution product of the functions $F^{(n_v)}$, $F_i^{(2)}$", whose restriction to the euclidean space is defined by :*

$$F^{(n)}(k) = \int_{\{k_{int} \in \text{Eucl-space}\}} \prod_{v \in \mathcal{V}} F^{(n_v)}(k_v(k,k_{int})) \prod_{i \in \mathcal{I}} \left[F_i^{(2)}(k_i(k,k_{int}))\right]^{-1} .dk_{int}$$

Note that the analytic function which is integrated in this formula admits $[k_i(k,k_{int})]^2 = m^2$ as a *simple* polar manifold (it thereby generalizes the Feynman integrands).

We call $\odot_\ell$*-product* $F_1 \odot_\ell F_2$ the $\mathcal{G}$-convolution product of* $F_1, F_2, H^{(2)}$ associated with $\mathcal{G}$ = 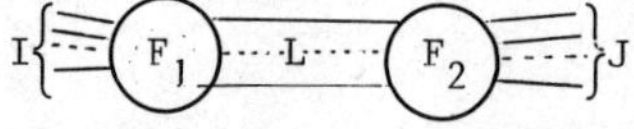 ; $\ell = |L|$. Of special interest is the case $|I| = |J| = \ell$, for which one can consider the *iterated* $\odot_\ell$*-product* of a general 2 -point function $F^{(2\ell)}$, as well as the *Bethe-Salpeter inverse* $G^{(2\ell)}$ of $F^{(2\ell)}$; $G^{(2\ell)}$ is the solution of the Fredholm resolvent equation : $F^{(2\ell)} = G^{(2\ell)} + \frac{1}{\ell!} F^{(2\ell)} \odot G^{(2\ell)}$, and on the basis of the C.F. lemma stated above, one can prove :

<u>*Theorem*</u> : *$G^{(2\ell)}(k_I,z,z')$ (where $z = \underline{k}_{(I)}, z' = \underline{k}_{(J)}$) is a meromorphic function of the form $N^{(2\ell)}(k_I,z,z')/D^{(2\ell)}(k_I)$, where $N^{(2\ell)}$ (resp. $D^{(2\ell)}$) is a general 2ℓ-point (resp. 2-point) function which is analytic at least in the somain $D_{2\ell}$ of $F^{(2\ell)}$.*

Note that N and D can be expressed as convergent series which involve all the iterated $\odot_\ell$-products of $F^{(2\ell)}$

The computation of the *absorptive parts of $\mathcal{G}$-convolution products* has a crucial importance for M.P.S.A.; here we shall briefly present some results for the absorptive parts of $\odot_\ell$-products.

a) absorptive part of an iterated $\odot_\ell$-product in a crossed channel.

Let $[F^{(2\ell)}]^{\odot_\ell r}$ the iterated $\odot_\ell$-product of order r of $F^{(2\ell)}$: it is associated with the graph $\mathcal{G}$ = 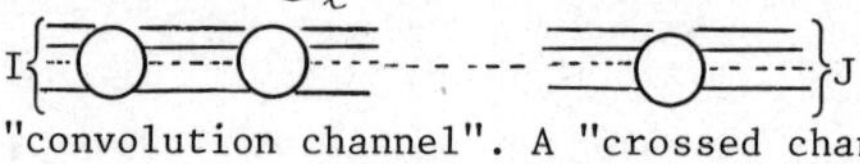, with (r+1) "bubbles", and (I,J) is called the "convolution channel". A "crossed channel" (I',J') is by

---

* In certain developments, one prefers to replace $H^{(2)}$ by $(p^2-m^2)^{-1}$ which has no C.D.D. zeros.

definition such that $I' \cap I$, $J' \cap I$, $I' \cap J$, $J' \cap J$ are all $\neq \emptyset$. The following general statement has an easy graphical interpretation in perturbation theory :

lemma : $[F^{(2\ell)}]^{\odot_\ell r}$ *is r-particle irreducible with respect to all the crossed channels.*

b) *absorptive part of a* $\odot_\ell$*-product in the convolution channel* (I,J) .

We give the expression of the absorptive part of $F \odot_\ell G$ in terms of those of F and G in the cases $\ell = 1,2,3$ (for arbitrary values of $|I|$ and $|J|$) :

i) $\ell = 1$ [30a]: $\Delta_I(F \odot_1 G) = F^+ . \Delta_I G^{amp.} + \Delta_I F . G^{amp-}$

ii) $\ell = 2$ [30b]: $\Delta_I(F \odot_2 G) = F^+ \odot_{\Gamma_+} \Delta_I G + \Delta_I F \odot_{\Gamma_-} G^- + \hat{F}_{out} *_2 \hat{G}_{in}$,

valid for $(2m)^2 < p_I^2 < (3m)^2$

iii) $\ell = 3$ [37] : $\Delta_I(F \odot_3 G) = F^+ \odot_{\Gamma_+} \Delta_I G + \Delta_I F \odot_{\Gamma_-} G^- + \hat{F}_{out} *_3 \hat{G}_{in} + \ldots$

$$+ \sum_{(hij)} \left[F^+ \odot_{\gamma_+}^{ij} *^h \Delta_{\{ij\}} G^- + \Delta_{\{ij\}} F^+ \odot_{\gamma_-}^{ij} *^h G^- \right] ,$$

valid for $(3m)^2 < p_I^2 < (5m)^2$, under suitable assumptions of local analyticity for F and G.

In i), $G^{amp} = G \times [H^{(2)}(k_I)]^{-1}$ (or $G \times (k_I^2 - m^2)$). In ii) and iii) the notations $\odot_{\Gamma_+}$ (in place of $\odot_2, \odot_3$) refer to appropriate distortions Γ_+, Γ_- of the integration cycle in the limits $k_I = p_I \pm i\varepsilon$, $\varepsilon \in V^+$, $\varepsilon \to 0$; $\hat{F}_{out} = F \odot \prod_{i \in L} (k_i^2 - m^2)$ $\hat{G}_{in} = G . \prod_{i \in L} (k_i^2 - m^2)$ (as in equation (B.S.)). In iii) the signs $\odot_{\gamma_\pm}^{ij} *^h$ correspond to an integration prescription on the momenta k_i, k_j, k_h ($\{i,j,h\} = L$) such that $k_h^2 = m^2$ and $\frac{k_i - k_j}{2} \in \gamma_+$ or γ_- (suitable off-shell contours).

These discontinuity formulae (obtained by contour distortion and residue technique) are basic for proving the equivalence between ℓ-particle A.C. and the ℓ-particle irreducibility of an appropriate 2ℓ-point Bethe-Salpeter kernel (see above for the case $\ell = 2$, and below for the case $\ell = 3$).

A remark concerning the applications of $\mathcal{G}$*-convolution* :

Apart from its basic role in M.P.S.A., which will be illustrated below by our analysis of $H^{(6)}$, the formalism of $\mathcal{G}$-convolution provides a "natural" framework for writing and studying the equations that Green's functions of a given lagrangian Q.F.T. must satisfy. However this immediately sets the question of divergences at infinity for $\mathcal{G}$-convolution products which involve the physical n-point functions $H^{(n)}$.

In M.P.S.A. one gets rid of this difficulty as follows : in the $\mathcal{G}$-convolution products and equations of the B.S. type, a regularized version of the $H^{(n)}$ is used, namely, $H^{(n)}(k_1 \ldots k_n) \times \prod_{i=1}^{n} (m^2 - \rho)^{\alpha}(k_i^2 - \rho)^{-\alpha}$ with ρ and α sufficiently large (note that H_n is thus unmodified on the mass shell) ; thereby the analyticity properties of the $H^{(n)}$ which can be derived in M.P.S.A. are shown to be independent of the asymptotic properties of the latter at infinity and purely determined by the geometry.

However, for setting the $\mathcal{G}$-convolution equations of a lagrangian field theory, it seems necessary to use a *renormalized* form of $\mathcal{G}$-convolution products. In [39], a generalization of Zimmermann's renormalization procedure [40] has been presented; it defines a *renormalized integrand* for $\mathcal{G}$-convolution products involving functions $F^{(n_v)}$ which belong to Weinberg classes [41] of a special type. The convergence theorem and the asymptotic properties of the integrals which are stated in [39] furnish a starting point for developing rigorously the approach of Q.F.T.-models based on the system of equations satisfied by Green's functions in momentum space : this study has been undertaken by M. Grammaticou.

Results of Many-Particle Structure Analysis :

the main points in this scheme are the following :

i) show on the basis of A.C. that appropriate n-point functions, introduced by B.S. inversion and $\mathcal{G}$-convolution operations involving the physical n-point functions $H^{(n)}$, have ℓ-particle irreducibility properties,

ii) express each $H^{(n)}$ through $\mathcal{G}$-convolution structural equations which may involve other $H^{(n')}$ with $n' < n$, together with various ℓ.p.i. parts of $H^{(n)}$. These equations should.

a) be valid in well-defined regions of complex momentum space, whose size increase with the rank ℓ of the ℓ-particle structure which is exploited.

b) exhibit for each of their terms an analytic and monodromic interpretation : this amounts to analyze how the Landau singularities attached to various graphs contribute to Green's functions and scattering amplitudes and thus to give a global description of the Riemann surfaces of the latter above the regions mentioned in a).

This analysis can be done through a recursion over ℓ.

The case $\ell = 1$: has been completely treated in [30a]: the content of this preliminary case is that the residue factorization property holds* at all the poles of the $H^{(n)}$ which are poles of $H^{(2)}$ (in all channels and for theories with an arbitrary mass spectrum).

---

* a generalization of this property to the case of bound states and unstable particles can also be established [42] .

The case $\ell = 2$: has been partially presented above : the 2-p.i.part $L_s^{(4)}$ of $H^{(4)}$ in a given channel (I,J) ($k_I^2 = s$) has been introduced through equation (B.S.); it is a general 4-point function meromorphic in $\hat{D}_4$; moreover the two-particle A.C. assumption implies that $L_s^{(4)}$ is 2.p.i. and this implies in turn the two-sheeted monodromic structure of $H^{(4)}$ around $s = 4m^2$. A similar result has been derived for all the $H^{(n)}$ without new B.S. inversion [30b].

Moreover, by using the property (a) of the absorptive parts of $\odot_2$-products in crossed channels, it has been possible to construct and recognize "parts" of $H^{(4)}$ which are simultaneously 2.p.i. with respect to two channels s and u ($L_{su}^{(4)} = L_s^{(4)} + L_u^{(4)} - H^{(4)}$) or with respect to the *three* channels s,t,u ($L_{stu}^{(4)} = L_s^{(4)} + L_t^{(4)} + L_u^{(4)} - 2\,H^{(4)}$). Such kernels, which satisfy crossing symmetry, may be used for global problems ; in this connection, let us mention Sommer's recent work [43] in which the problem of reconstructing $H^{(4)}$ from $L_{su}^{(4)}$, considered as given in a certain domain, has been solved by a fixed point technique ; this work also contains an interesting off-shell extrapolation of Martin's domain obtained by the exploitation of positivity properties of $\Delta_s H^{(4)}$ (see part B : absorptive parts $\Delta_I H^{(n)}$, property c)).

The rest of this section is devoted to a further step :

One-two-and three-particle structure analysis of $H^{(6)}$ (even Q.F.T. with a single mass)

Notations: We put $H^{(6)}(k_1,\dots,k_6) = H^{(6)}(K,Z,Z')$, where $K = k_1 + k_2 + k_3 = -(k_4 + k_5 + k_6)$, $Z = (\underline{k}_4,\underline{k}_5,\underline{k}_6)$ with : $\underline{k}_i = k_i + \frac{K}{3}$, $i = 4,5,6$; $Z' = (\underline{k}_1,\underline{k}_2,\underline{k}_3)$ with $\underline{k}_i = k_i - \frac{K}{3}$, $i = 1,2,3$. K is kept in the manifold $K = (K^{(o)},\vec{0})$, and $K^{(o)}$ varies in a domain $\Delta_M^{cut} = \Delta_M - \{K^o \geq 3m\}$, where $\Delta_M = \{K^{(o)};0 < \mathrm{Re}\,K^o < M$, with $3m < M \leq 5m\}$

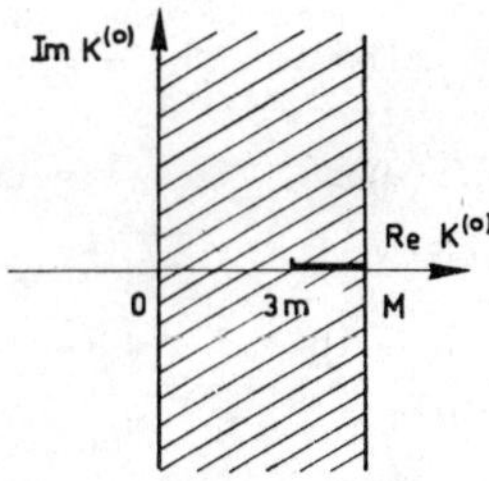

Fig.4: The domain Δ_M^{cut}

Fig. 5. (below) The situation in Re $Z'^{(o)}$ plane ; ω_h is the polar manifold $k_h^2 = m^2$, σ_h is the branch manifold $s_{ij} = (k_i + k_j)^2 = 4m^2$ ($\{i,j,h\} = \{1,2,3\}$) ; the dark disk at the center of fig.5a) represents the physical region $\mathcal{M}_6^{(IJ)}$ ($I = \{4,5,6\}$, $J = \{1,2,3\}$), in projection.

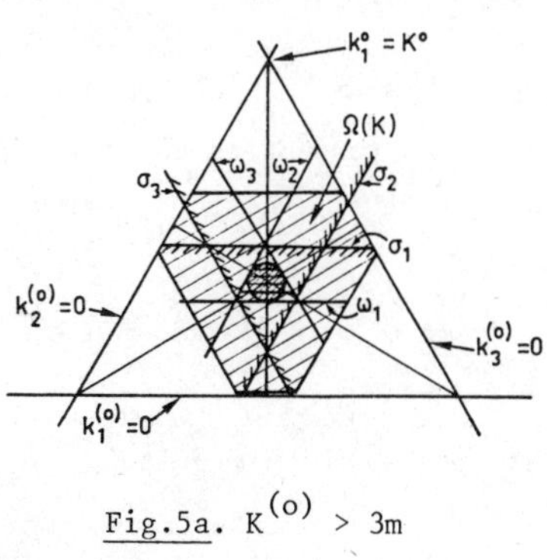

Fig.5a. $K^{(o)} > 3m$

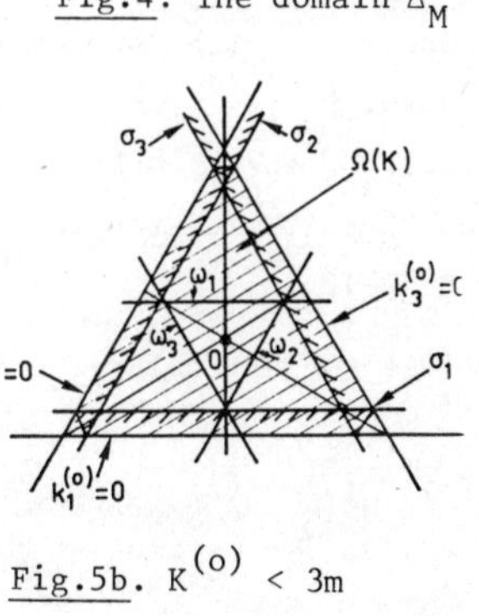

Fig.5b. $K^{(o)} < 3m$

For every $K^{(o)}$ in Δ_M, one defines the following domain D(K) in Z-space :

$$D_K = \left\{ Z = (Z^{(o)}, \vec{Z});\ \text{Im } Z^{(o)} \in \mathbb{R}^2,\ \text{Re } Z^{(o)} \in \Omega(K),\ \text{Re } \vec{Z} \in \mathbb{R}^6\ ; |\text{Im } \vec{Z}| < \varepsilon \right\}$$

where the domain $\Omega(K)$ is represented in light hatchings on fig.5.

The following domains $\mathcal{D}_M, \mathcal{D}_M^{cut}$ are used in the results below :

$$\mathcal{D}_M(\text{resp.} \mathcal{D}_M^{cut}) = \left\{ (K^{(o)}, Z, Z');\ K^{(o)} \in \Delta_M(\text{resp. } \Delta_M^{cut})\ ; Z \in D(K)\ ; Z' \in D(K) \right\}$$

One-and-two-particle dressing of $H^{(6)}$ in the channel (I,J).

The aim of this step is to express* $H^{(6)}$ in terms of a suitable irreducible part $G^{(6)}$ and of auxiliary kernels which only involve $H^{(4)}$ and $L^{(4)}$. We first define the latter, namely :

T_{in} is the tree-graph product $\sum_{h,n}$ 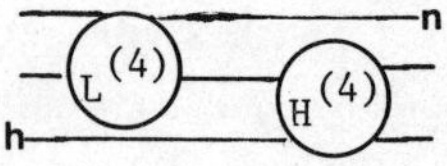 , and similarly

$T_{out} = \sum_{hn}$ 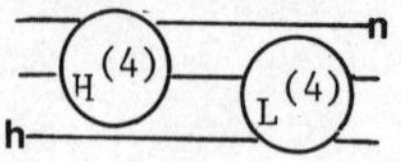 ; they admit B.S. inverses (for $\odot_3$-product)

which we call respectively U_{in} and U_{out}. We also introduce the operators

$\Lambda = \mathbb{1} - \frac{1}{6} \sum_{hn}$ 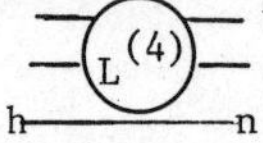 and $V = \mathbb{1} + \frac{1}{6} \sum_{hn}$ 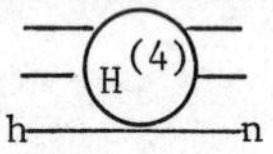 which satisfy :

$$\Lambda V = \mathbb{1} - \frac{1}{6} T_{in},\quad V \Lambda = \mathbb{1} - \frac{1}{6} T_{out}.$$

The B.S. inversion of T_{in}, T_{out} allows to define the inverse of Λ as :

$$\Lambda^{-1} = V.(\mathbb{1} + \frac{1}{6} U_{in}) = (\mathbb{1} + \frac{1}{6} U_{out}).V.$$

We now introduce the six-point function $\mathbb{G}^{(6)}$ through the $\mathcal{G}$-convolution equation :

$\mathbb{G}^{(6)} = \Lambda H_1^{(6)} \Lambda - \Lambda T_{out}$ (note that $\Lambda T_{out} = T_{in} \Lambda$), where

$H_1^{(6)} = H^{(6)} -$ 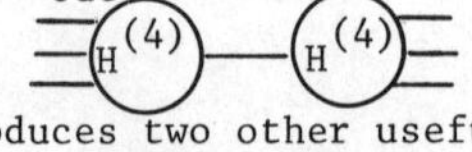 is the 1-p.i. part of $H^{(6)}$ in the channel (I,J). One also introduces two other useful six-point functions $\mathbb{G}_{in}^{(6)}$, $\mathbb{G}_{out}^{(6)}$ which are such that : $\mathbb{G}^{(6)} = \Lambda \mathbb{G}_{out}^{(6)} = \mathbb{G}_{in}^{(6)} \Lambda$.

The following irreducibility properties can be proved for $\mathbb{G}^{(6)}$, $\mathbb{G}_{in}^{(6)}$, $\mathbb{G}_{out}^{(6)}$:

a) They are 1-p.i with respect to *all* channels,

b) $\mathbb{G}_{in}^{(6)}$ (resp. $\mathbb{G}_{out}^{(6)}$) is 2-p.i. with respect to the triplet {4,5,6} (resp.{1,2,3}); $\mathbb{G}^{(6)}$ is 2-p.i with respect to *both* triplets. ("2-p.i. with respect to the triplet {4,5,6}" means "2-p.i. with respect to the three channels ({ij},{h,1,2,3}) with {h,i,j} = {4,5,6}").

This implies in particular the following property of $\hat{\mathbb{G}}^{(6)} = \prod_{i=1}^{6} (k_i^2 - m^2)\ \mathbb{G}^{(6)}$

---

*Actually, $H^{(6)}$, $H^{(4)}$, $H^{(2)}$ are replaced here by the corresponding regularized forms (see our remark on $\mathcal{G}$-convolution products).

Lemma : *$\hat{\mathbb{G}}^{(6)}$ is analytic in the domain $\mathfrak{D}_M^{cut}$, such that $M = 11m/3$.*

Now the equation which defines $G^{(6)}$ can be inverted and yields the following "two-particle dressing equation"

$$H_1^{(6)} = (\mathbb{1} + \frac{1}{6} U_{out}) V \mathbb{G}^{(6)} V(\mathbb{1} + \frac{1}{6} U_{in}) + U_{out} V$$

By using the analyticity properties of $H^{(4)}$ and $\mathbb{G}^{(6)}$, and by exploiting in complex space some standard combinatorics of Fredholm theory, one can prove

Theorem : *In $\mathfrak{D}_M^{cut}$ (with $M = 11m/3$) the following decomposition of $\hat{H}^{(6)}$ holds :*

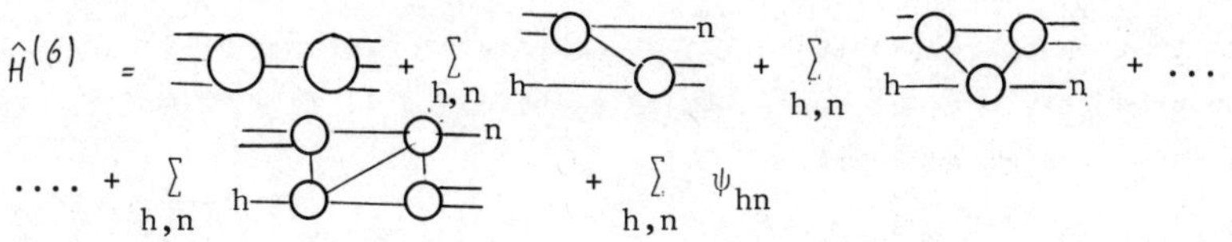

in the latter, all the "bubbles" of the $\mathcal{G}$-convolution products represent $H^{(4)}$ and each analytic function ψ_{hn} of the residual sum ($h \in \{4,5,6\}$, $n \in \{1,2,3\}$) has the following properties :

i) ψ_{hn} is analytic in a "first-sheet domain" which is $\mathfrak{D}_M^{(cut)} - (\bar{\sigma}_h \cup \bar{\sigma}_n)$; here the sets $\bar{\sigma}_h$, $\bar{\sigma}_n$ denote the following "cuts" :
$\bar{\sigma}_h = \{(K,Z,Z'); s_{ij}(K,Z) = 4m^2 + \rho;\ \rho \geq 0\}$ *and* $\bar{\sigma}_n = \{(K,Z,Z'); s_{\ell m}(K,Z') = 4m^2 + \rho;\ \rho \geq 0\}$.

ii) ψ_{hn} admits a local analytic continuation across the sets σ_h and $\bar{\sigma}_n$ on the Riemann surface associated with the Feynman graph

Besides, each $\mathcal{G}$-convolution term of this decomposition is analytic in $\mathfrak{D}_M^{cut}$ minus the Landau singular set associated with the corresponding graph $\mathcal{G}$; it moreover admits a local analytic continuation on the Riemann surface associated with $\mathcal{G}$.

Remark : The only singularities of $\hat{H}^{(6)}$ which are produced in $\mathfrak{D}_M^{cut} - \bigcup_{h,n} (\bar{\sigma}_h \cup \bar{\sigma}_n)$, considered as a first sheet defined by the extension of the primitive analyticity domain, are the leading Landau singularities of the 0-loop, 1-loop and two-loop "truss-bridge graphs"; the ℓ-loop graphs ($\ell > 2$)

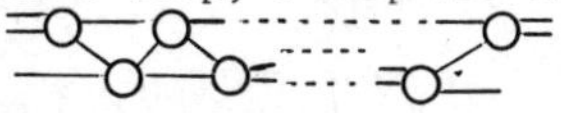

are potentially present in the Neumann expansions of U_{in}, U_{out} and illustrate the two-particle dressing of $H_1^{(6)}$ from both sides of $\mathbb{G}^{(6)}$ in the expression given above ; however, it is a consequence of the previous theorem that their Landau singularities are *effective only in other sheets*. This analysis has not yet been done, but our method should make it feasible.

By taking the restriction of $\hat{H}^{(6)}$ to the complex mass-shell $\mathcal{M}_6^{(c)}$, one obtains the following corollary of the previous theorem :

Corollary : *In $\mathfrak{D}_M^{cut} \cap \mathcal{M}_6^{(c)}$, the $3 \to 3$ particle scattering amplitude admits the following decomposition as a sum of analytic functions :*

$S_{IJ}^{(6)} = \sum_{h,n}$ 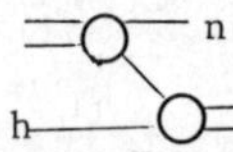 $+ \sum_{h,n}$ 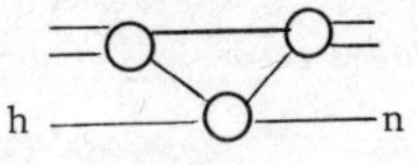 $+ \sum_{h,n} \psi_{hn}$,

where each function $\hat{\psi}_{hn}$ is analytic in $\mathcal{D}_M^{cut} \cap \mathcal{M}_6^{(c)} - (\bar{\sigma}_h \cup \bar{\sigma}_n)$; across each cut $\bar{\sigma}_h, \bar{\sigma}_n$ $\hat{\psi}_{hn}$ admits a local analytic continuation which spreads in a two sheeted Riemann surface around each threshold σ_h, σ_n.

Remark :One reobtains as a subproduct, and for $K^{(o)} < M$, the result of [6] described in C.I. ; in this region, one obtains actually here a more global description of the analytic structure of $S_{IJ}^{(6)}$ and $\hat{H}^{(6)}$; however our limitation $M = 11m/3$ should be improved (it is due to the apparent obstruction of certain crossed channel singularities in $\mathcal{G}$-convolution products, but may hopefully be expected to be fictitious : this point deserves a more careful analysis).

The three-particle irreducible kernel $L^{(6)}$.

Let $L_{out}^{(6)}$ (resp. $L_{in}^{(6)}$) be the B.S. inverse of $\mathbb{G}_{out}^{(6)}$ (resp. $\mathbb{G}_{in}^{(6)}$) with respect to the $\odot_3$-product, and let us put : $L^{(6)} = \Lambda\, L_{out}^{(6)}$ ($= L_{in}^{(6)} \Lambda$). The following property of $L^{(6)}$ has been proved :

Theorem : There is an equivalence between the three-particle A.C. equation : $\Delta_I \hat{H}^{(6)} = \frac{1}{3!} \hat{H}^{(6)} *_3^{(I)} \hat{H}^{(6)}$ and the fact that $L^{(6)}$ is 3.p.i., i.e. $\Delta_I L^{(6)} = 0$, in the region $(p_I)^2 = (K^{(o)})^2 < M^2$ (the present limitation $M = 11m/3$ being hopefully provisional).

The proof of the latter amounts to showing the following relation for appropriate integration cycles $\Gamma_+(p_I)$, $\Gamma_-(p_I)$:

$$\Delta_I L^{(6)} = (\mathbb{1} - \frac{1}{3!} L_{in}^{(6)}) \Lambda \odot_{\Gamma_+} \left[\Delta_I \hat{H}_1^{(6)} - \frac{1}{3!} \hat{H}_1^{(6)} *_3 \hat{H}_1^{(6)}\right] \odot_{\Gamma_-} \Lambda (\mathbb{1} - \frac{1}{3!} L_{out}^{(6)}) ;$$

in fact, as it has been proved in [30a], the bracket at the r.h.s. of this relation vanishes if and only if the A.C. equation for $H^{(6)}$ holds.

Thanks to this equivalence property, the following consequence of three-particle A.C. can be derived : $\hat{L}^{(6)} = \prod_{i=1}^{6} (k_i^2 - m^2)\, L^{(6)}$ is analytic in the whole domain $\mathcal{D}_M$, and therefore in a complex neighbourhood of the three-particle physical region $s = p_I^2 < M^2$. Then by plugging this new information into the equation : $\mathbb{G}^{(6)} = L^{(6)} + \frac{1}{6} \mathbb{G}^{(6)} \Lambda^{-1} L^{(6)}$, one can show that $\hat{G}^{(6)}$ admits *local* analytic continuation across the three-particle cut $\bar{\sigma} = \{(K,Z,Z') ; (3m)^2 < s = K^2 < M^2\}$ from both sides Im s > 0, Im s < 0. As a consequence, the same property can be established for $\hat{H}^{(6)}$; more precisely, this property holds for each term at the r.h.s. formula of the decomposition theorem.

Remark :This result contains as a subproduct the local analyticity properties which are sufficient to ensure macrocausality conditions* for the 3 → 3 scattering process

*in the sense of Stapp-Iagolnitzer's S.matrix theory

in the considered region $s < M^2$.

We shall finally present a joint application of the above local analyticity properties based on A.C. and of a typical extension of $\hat{D}_6$, obtained by methods of analytic completion.

A crossing domain for the 3π → 3π scattering amplitude.

We consider the crossing problem for the couple of channels $(I,J) = (\{456\},\{123\})$ and $(I'J') = (\{156\},\{423\})$, namely we seek an analyticity domain Θ on $\mathcal{M}_6^{(c)}$ for $S^{(6)} = \hat{H}^{(6)}\big|_{\mathcal{M}_6^{(c)}}$ whose boundary contains parts of both physical regions $\mathcal{M}_6^{IJ}$ and $\mathcal{M}_6^{I'J'}$, and such that it touches $\mathcal{M}_6^{IJ}$ from the side $\mathrm{Im}\, s > 0$ $(s = k_I^2)$ and $\mathcal{M}_6^{I'J'}$ from the side $\mathrm{Im}\, s' > 0$ $(s' = k_{I'}^2)$: the corresponding boundary values of $S^{(6)}$ will then be respectively the scattering amplitudes $S_{IJ}^{(6)}$ and $S_{I'J'}^{(6)}$.

We shall define Θ as a neighbourhood in $\mathcal{M}_6^{(c)}$ of a certain domain Θ_α in $\mathcal{M}_6^{(c)} \cap \mathcal{V}_\alpha$, $\mathcal{V}_\alpha$ being the following one-dimensional submanifold of "forward configurations" :

$$\mathcal{V}_\alpha = \Big\{(k_1,\ldots,k_6)\ ;\ k_1 + k_4 = 0,\ k_2 + k_5 = 0,\ k_3 + k_6 = 0\ ;\ k_1^2 = m^2\ ,$$

$$k_1 = (k_1^{(o)}, k_1^{(1)}, 0, 0),\ k_2 = (m\,\mathrm{ch}\,\alpha, 0, m\,\mathrm{sh}\,\alpha, 0),\ k_3 = (m\,\mathrm{ch}\,\alpha, 0, -m\,\mathrm{sh}\,\alpha, 0)\Big\}$$

α will be chosen sufficiently small. On this manifold, $S^{(6)}$ is singular since the pole $(k_1 + k_4 + k_5)^2 = m^2$ of $\hat{H}^{(6)}$ reduces to $k_5^2 = m^2$ on the mass shell. Therefore on $\mathcal{V}_\alpha$, it is necessary to consider the crossing problem for the *forward part** of $\hat{H}^{(6)}$ which we define by :

$$\hat{H}_F^{(6)} = \hat{H}^{(6)} - \sum_{hn}{}'$$

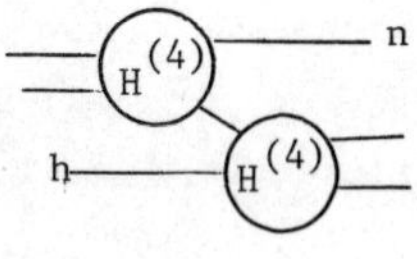

where the summation $\sum'$ runs over the set $\{(h,n);\ h \in \{4,5,6\},\ n \in \{1,2,3\};\ h \neq n+3\}$.

The following lemma can be proved for $\hat{H}_F^{(6)}$.

Lemma : there exists a crossing domain Θ_α on $\mathcal{V}_\alpha$ $(0 < \alpha < \alpha_0)$ for the "forward scattering amplitude" $S_F^{(6)} = \hat{H}_F^{(6)}\big|_{\mathcal{M}_6^{(c)}}$. This domain, whose shape is indicated on fig. 6 tends to a cut-plane in the variable s when α tends to 0.

We note that on $\mathcal{V}_\alpha$, the physical region of the channels (I,J), (I',J') are symmetric with respect to the origin if one takes $k_1^{(o)}$ as the variable, and that :

$$k_1^{(o)} = \frac{s - m^2(1 + 4\mathrm{ch}^2\alpha)}{4m\,\mathrm{ch}\,\alpha} = -\frac{s' - m^2(1 + 4\,\mathrm{ch}^2\alpha)}{4m\,\mathrm{ch}\,\alpha}$$

* The situation was similar in the works which have been presented or mentioned in Part B II.

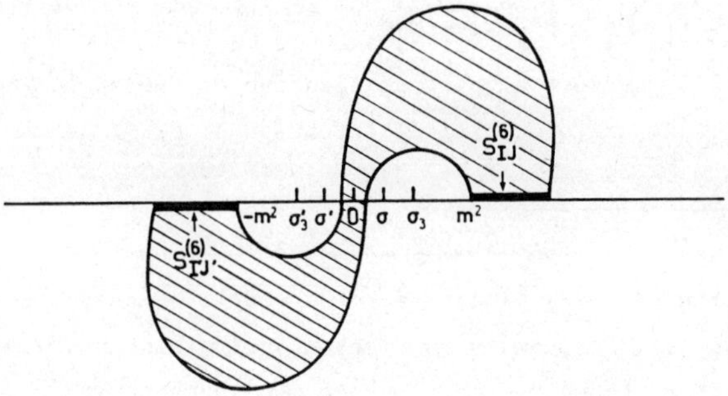

Fig.6. The crossing domain Θ_α in the $k_1^{(o)}$-plane

The proof is obtained by performing the analytic completion of the union of the two following regions in the space of two complex variables $k_1^{(o)}$ and $\zeta = k_1^2$.

a) For ζ real $< \zeta_\alpha = \dfrac{(2-\mathrm{ch}^2\alpha)^2}{\mathrm{ch}^2\alpha}\, m^2$, $\hat{H}_F^{(4)}$ is analytic in the $k_1^{(o)}$ cut-plane (the cuts being given by $s \geq 9m^2$, $s' > 9m^2$). This results from the Jost-Lehman-Dyson completion [44] in k_1-space. Note that $\zeta_\alpha < m^2$ and that $\zeta_\alpha \to m^2$ when $\alpha \to 0$.

b) When $k_1^{(o)}$ varies in a neighbourhood of the low energy physical region considered above (limited by $s < M^2$), $\hat{H}_F^{(4)}$ is analytic in ζ in a neighbourhood $\mathcal{N}_\alpha$ of ζ_α *whose size is independent of* $k_1^{(o)}$ *and* α, when α tends to zero. This results from the fact that such a domain is contained in the analyticity domain obtained in the decomposition theorem for $\hat{H}^{(6)}$.

The standard interpolation technique in two complex variables then yields the crossing domain of fig. 6, if $\zeta_\alpha - m^2$ is small compared with the size of $\mathcal{N}_\alpha$: this is satisfied if α is chosen sufficientltly small.

To obtain the crossing property for $\hat{H}^{(6)}$ itself, it is sufficient to notice that $\hat{H}_F^{(6)}$ is analytic in a neighbourhood Θ of Θ_α in $\mathcal{M}_6^{(c)}$, and that in Θ , the tree-graph terms

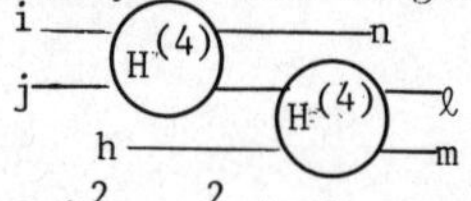

are analytic except on the corresponding poles $(k_i + k_j - k_n)^2 = m^2$ (this follows from the analyticity of the $2 \to 2$ forward scattering).

REFERENCES

[1] G.F. CHEW, The Analytic S-Matrix, W.A. Benjamin, New York (1966) and references quoted therein.
[2] D. IAGOLNITZER, The S-matrix, North Holland, Amsterdam (1978) and references quoted therein.
[3] J.P. ECKMANN, H. EPSTEIN, Comm. Math. Phys., 64, 95 (1979).
[4] S.M. ROY, G. WANDERS, Physics Letters, 74B, 347 (1978) and Nucl. Phys. B141, 220 (1978).
[5a] A.A. LOGUNOV et al., Theor. Math. Phys. 33, 149 (1977).

[5b] L.M. MUSAFAROV, Theor. Math. Phys. 38, 36 (1979).
[5c] V.P. PAVLOV, Theor. Math. Phys. 37, 154 (1978).
[5d] V.P. PAVLOV, Theor. Math. Phys. 35, 3 (1978).
[5e] L.M. MUSAFAROV, V.P. PAVLOV, Theor. Math. Phys. 35, 151 (1978).
[5f] L.M. MUSAFAROV, Theor. Math. Phys. 35, 291 (1978).
[6] H. EPSTEIN, V. GLASER, D. IAGOLNITZER, in preparation.
[7] J. BROS in "Analytic Methods in Math. Physics", p.85, Gordon and Breach, New York (1970).
[8a] K. SYMANZIK, J. Math. Phys. 1, 249 (1960).
[8b] K. SYMANZIK, in Symposium on Theoretical Physics, 3, New York, Plenum Press (1967).
[9a] R.F. STREATER and A.S. WIGHTMAN, "PCT, Spin & Statistics and all that", Benjamin, New York (1964).
[9b] R. JOST, "The general theory of quantized fields", Ann. Math. Soc. Providence R.I. (1965).
[10] K. OSTERWALDER, R. SCHRADER, Comm. Math. Phys. 33, 83 (1973) and 42, 281 (1975).
[11] V. GLASER, Comm. Math. Phys. 37, 257 (1974).
[12] J. GLIMM, A. JAFFE, T. SPENCER, Ann. Math. 100, 585 (1974).
[13] See [3] and references quoted therein.
[14] O. STEINMANN, Comm. Math. Phys. 10, 245 (1968).
[15] H. EPSTEIN, V. GLASER, R. STORA, "Structural Analysis of Collision Amplitudes", p.7, North-Holland, Amsterdam (1976) and references quoted therein.
[16] K. HEPP in : "Axiomatic Field Theory", Gordon and Breach, New-York (1966).
[17] T. SPENCER and F. ZIRILLI, Comm. Math. Phys. 49, 1 (1975).
[18] D. RUELLE, Nuovo Cimento 19, 356 (1961) and Thesis, Zurich (1959).
[19] See R. JOST [9 b] and references quoted therein.
[20] J. BROS, H. EPSTEIN, V. GLASER,
a) Nuovo Cim. 31, 1265 (1964),
b) Comm. Math. Phys. 1, 240 (1965),
c) Helv. Phys. Acta. 45, 149 (1972).
[21] A. MARTIN : Nuovo Cimento, 42 A 930 (1966) and 44 1219 (1966).
[22] A. MARTIN, "Scattering theory : unitarity analyticity and crossing", Springer-Verlag (1970).
[23] H. EPSTEIN , J. Math. Phys. 1 254 (1960).
[24] a) J. BROS, b) D. IAGOLNITZER in Publ. R.I.M.S., Kyoto Univ. 12 Suppl. (1976).
[25] M. SATO, T. KAWAI, M. KASHIWARA, Lect. Notes in Math., Springer Verlag (1972).
[26] A.S.WIGHTMAN, in "Relations de dispersion ..." Hermann, Paris (1960), ref.therein.
[27] J. BROS, V. GLASER, l'enveloppe d'holomorphie de l'union de deux polycercles. (1961).
[28] M.C. POLIVANOV, "Math.Pb. in Math.Phys.",375 Lect. Notes in Phys. Springer (1977).
[29] N.N. BOGOLIUBOV, D.V. SHIRKOV, "Intr. to the theory of Quantized Fields", Moscow (1957).
[30a] J. BROS, M. LASSALLE, Comm. Math. Phys. 43, 279 (1975).
[30b] J. BROS, M. LASSALLE, Comm. Math. Phys. 54, 33 (1977).
[31] W. ZIMMERMANN, Nuovo Cimento, 21 249 (1961)
[32] A. MARTIN, in "Problems of Theoretical Physics", Moscow, Bauka (1969).
[33] J. BROS and D. PESENTI , "Fredholm theory in complex manifolds" to be published in "Journal de Math. Pures et Appliquées". (preprint Orsay 1979).
[34] J. BROS, Cours de 3ème cycle, Lausanne, Mai 1979.
[35] R. ASCOLI, Nuovo Cimento, 18 754 (1960)
[36] J. BROS, M. LASSALLE, "Structural Analysis of Collision Amplitudes", p.97, North-Holland, Amsterdam (1976).
[37] A. KATZ, Thèse de 3ème cycle Paris, Juin 1979.
[38] M. LASSALLE, Comm. Math. Phys. 34, 185 (1974).
[39] J. BROS, M. GRAMMATICOU, "Renormalized $\mathcal{G}$-convolution I", Saclay (1978) to be published in Comm. Math. Phys ;
M. GRAMMATICOU, "Renormalized $\mathcal{G}$-convolution II", Ecole Polytechnique, Paris 1979.
[40] W. ZIMMERMANN, Comm. Math. Phys. 15, 208 (1969).
[41] S. WEINBERG, Phys. Rev. 118 (1960).
[42] J. BROS and D. PESENTI, in preparation.
[43] G. SOMMER, in preparation.
[44] R. OMNES, in "Relations de dispersion ..." Hermann, Paris (1960), ref. therein.
[45] F. DUNLOP, M. COMBESCURE, n-particle irreducible functions in euclidean Q.F.T. Preprint IHES (1978), and H. KOCH, Thesis, Genève (1979).

CONSTRUCTIVE FIELD THEORY

Arthur Jaffe[1]
Harvard University

1. Osterwalder-Schrader Quantization

The cornerstone of constructive quantum field theory is the existence of non-linear quantum fields compatible with both special relativity and quantum mechanics. The solution to this problem requires construction of the Hilbert space $\mathcal{H}$ of quantum mechanics, a unitary representation $U(a,\Lambda)$ of the Lorentz group on $\mathcal{H}$ and the quantum field $\Phi(f)$ itself, which acts as a linear operator on $\mathcal{H}$. There are two standard (and closely related) constructions to solve the problem: The first method is based on Hilbert space methods, approximate Hamiltonians H_n defined on Fock space. With this method one constructs the local field $\Phi(f)$, satisfying

$$[\Phi(f), \Phi(g)] = 0$$

when f, g have space-like separated supports. (One can also obtain C*-algebras of bounded functions $\mathfrak{A}(B)$ of $\Phi(f)$ for suppt $f \subset B$, which satisfy the Haag-Kastler axioms.) Then one constructs the vacuum representation (yielding $\mathcal{H}$ and $U(a,\Lambda)$) by taking limits of the ground states ρ_n of H_n.

The second standard construction method is to verify Euclidean axioms for a field theory at imaginary time, i.e. with time t analytically continued to $-it$. Thus Euclidean symmetry replaces Lorentz symmetry. In the case of bosons, locality analytically continues to commutativity of $\phi(\vec{x}, t) \equiv \Phi(\vec{x}, -it)$. The Wightman functions of Φ analytically continue to moments of a probability measure $d\mu(\phi)$ on the space of classical Euclidean field configurations. For convenience, we take this configuration space to be $\mathcal{S}'(R^d)$ where d is the space-time dimension. (In the case of fermions, the classical fields are elements of a Grassmann algebra, rather than ordinary random variables.)

The inverse problem of recovering $(\Phi,U,\mathcal{H})$ from $(\phi,d\mu(\phi))$ requires a statement of Euclidean axioms. One such set of axioms for bosons was given by Nelson and involves the assumption of a Markov property for $(\phi,d\mu)$. We present a simple version of the axioms of Osterwalder and Schrader, which we refer to as "Osterwalder-Schrader quantization." The three axioms are

O.-S. 1 (Regularity). The regularity assumption is a technical restriction on

$$S\{f\} = \int_{\mathcal{S}'} e^{i\phi(f)}\, d\mu(\phi) \quad .$$

Assume for some $p < 2$,

$$|S\{f\}| \leq \exp \text{const} \left(\|f\|_{L_1} + \|f\|_{L_p}^p \right) \quad .$$

O.-S. 2 (Invariance). For E, a Euclidean transformation on R^d (rotation, translation, reflection)

$$S\{Ef\} = S\{f\} \quad .$$

O.-S. 3 (Reflection Positivity). Let Π be a hyperplane in R^d, θ the reflection in Π and $\Pi_\pm$ the connected components of $R^d \setminus \Pi$. For $\{f_j\}$, any finite sequence of functions supported in Π_+, assume

$$S\{f_i - \theta \bar{f}_j\} = M_{ij}$$

is a nonnegative matrix.

---

[1]Supported in part by National Science Foundation under Grant PHY79-16812.

Theorem 1 [1]. If $S\{f\}$ satisfies O.-S. 1-3, then there exists (Φ,U,H,Ω) satisfying the Wightman axioms, with the possible exception of uniqueness of the vacuum Ω, and such that vacuum expectation values $<\Omega,\Phi(x_1)\ldots\Phi(x_n)\Omega>$ continue analytically to

$$\int \phi(x_1)\ldots\phi(x_n)\, d\mu(\phi) = <\Omega,\Phi(\vec{x}_1,-it_1)\ldots\Phi(\vec{x}_1,-it_n)\Omega>$$

Uniqueness of the vacuum Ω is equivalent to ergodicity of $d\mu$.

Pictorially, the theorem is represented in Figure 1, which illustrates the connection between $d\mu$ and H. The vertical projection from L_2 to H is given by orthogonal projection of L_2 onto L_2^+ (generated by $\exp(i\phi(f))$, suppt $f \subset \Pi_+$) followed by identifying L_2^+ as a subspace of H with the scalar product on L_2^+ given by O.-S. 3.

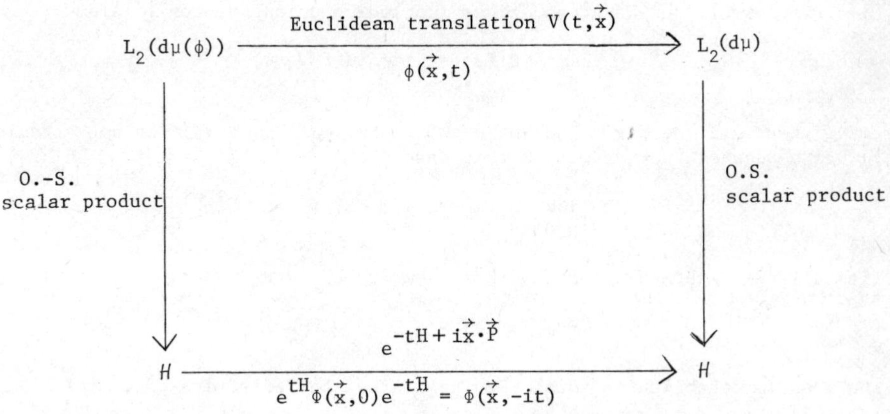

Figure 1. Commutative diagram describing Osterwalder-Schrader quantization. The time translation on $L_2(d\mu)$ is mapped into the semigroup e^{-tH} on H. The field ϕ is mapped into the imaginary time field $\Phi(\vec{x},-it)$.

2. The ϕ_3^4 Model.

To be specific, we present the ϕ^4 model in $d=3$ space-time dimensions. In particular, let

$$V_\kappa(\phi_\kappa(x)) = \lambda\phi_\kappa(x)^4 + a_\kappa\phi_\kappa(x)^2 - \mu\phi_\kappa(x) \quad ,$$

where $\phi_\kappa(x)$ denotes a lattice field on a lattice with spacing κ^{-1}. Also let Λ denote a ball of radius Λ, centered at the origin. Let $\int dx$ denote $\varepsilon^3\Sigma_{\text{lattice}}$ and define

$$d\mu_{\kappa,\Lambda} = Z^{-1}_{\kappa,\Lambda} \exp\left(-\int_\Lambda V_\kappa(\phi_\kappa(x))\,dx\right) d\phi$$

where $d\phi$ is Gaussian measure with mean zero and covariance the lattice version of

$$(-\Delta+I)^{-1}(x,y) = C(x,y) = \frac{1}{4\pi|x-y|}e^{-|x-y|} \quad .$$

Also, $Z_{\kappa,\Lambda}$ is chosen so that $\int d\mu_{\kappa,\Lambda} = 1$. By taking $\kappa = 2^{-n}$, we can imbed each lattice in a fixed continuum space of functions.

Theorem 2 [2]. (Existence). There exist constants $\alpha,\beta > 0$ such that for all $\lambda > 0$, for μ,σ real and for

$$a_\kappa = -\alpha\lambda\kappa + \beta\lambda^2 \ln\kappa + \sigma \quad ,$$

$$S\{f\} = \int e^{i\phi(f)}\, d\mu(\phi)$$
$$= \lim_{\substack{\kappa\to\infty \\ \Lambda\to\infty}} S\{f\}_{\kappa,\Lambda} = \lim_{\substack{\kappa\to\infty \\ \Lambda\to\infty}} \int e^{i\phi(f)}\, d\mu(\phi)_{\kappa,\Lambda}$$

exists, satisfies O.-S. 1-3, and is asymptotic to the standard ϕ^4 theory.

Remark [3]. Existence theorems also exist for $P(\phi)_2$ models with P bounded from below, for Yukawa$_2$ (scalar and pseudoscalar) and for the Abelian Higgs$_2$ model.

3. The Phase Transition Picture

Theorem 3. Let μ,λ be fixed and $\sigma \gg 0$. Then the measure $d\mu(\phi)$ is ergodic and (Φ,U,H) satisfies all the Wightman axioms (including unique vacuum). Also

$$\lim_{\mu\to 0+} S\{f\} = \lim_{\mu\to 0-} S\{f\} .$$

Theorem 4. Let μ,λ be fixed and let $\sigma \ll 0$. If $\mu=0$, then $d\mu(\phi)$ is not ergodic and (Φ,U,H) has a degenerate vacuum. In this case,

$$S\{f\}_+ = \lim_{\mu\to 0+} S\{f\} = \lim_{\mu\to 0-} S\{-f\} = S\{-f\}_-$$

satisfies all the Wightman axioms and has symmetry breaking

$$\langle\Phi\rangle_{\Omega_+} = -\langle\Phi\rangle_{\Omega_-} \neq 0 .$$

The idea in establishing these results is that the effective potential

$$V_{eff}(\phi) = \lambda\phi^4 + (\sigma-\sigma_c)\phi^2 - \mu\phi + \text{small higher order remainders}$$

is minimized at $\langle\phi\rangle$. Here we have the various potentials illustrated in Figure 2.

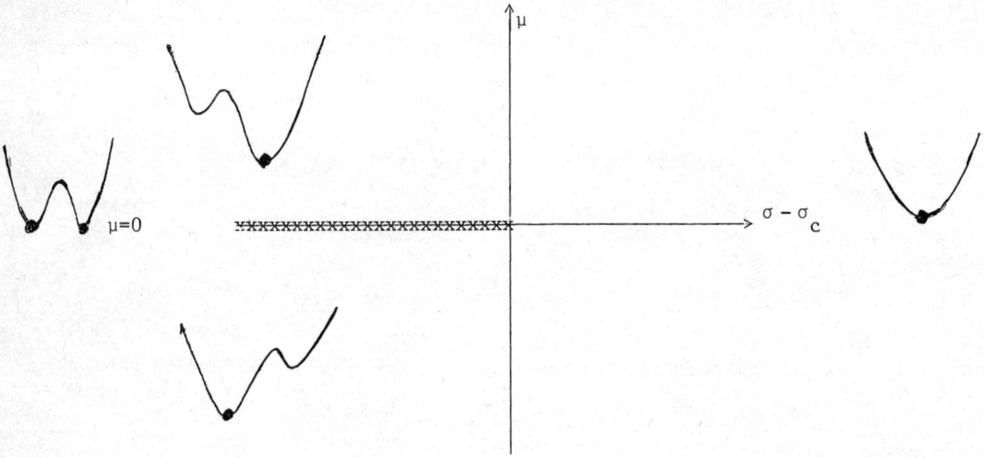

Figure 2. $V_{eff}(\phi)$ in the four quadrants of the $(\mu,\sigma-\sigma_c)$ parameter space for fixed λ. Here $\sigma_c = \sigma_c(\lambda)$ and the symmetry for $\sigma < \sigma_c$ and $\mu = 0$ is clear.

4. Renormalization

The divergent constant a_κ is an infinite mass renormalization. The constant σ is the finite part of the mass counterterm after performing the cancellation of infinities. To complete the mass renormalization, we wish to choose $\sigma > \sigma_c$ in such a way that the physical mass

$$m = \inf(\text{spectrum } H \setminus \{0\})$$

has a predetermined value. This is justified by

Theorem 5 [4]. For λ fixed, $m(\sigma) \searrow 0$ as $\sigma \searrow \sigma_c$. Hence given m there exists $\sigma(m)$ with this mass gap.

It is believed (though not proved) that $m(\sigma)$ obeys an asymptotic power law as $\sigma \searrow \sigma_c$, namely

$$m(\sigma) \simeq \left(\frac{\sigma - \sigma_c}{\sigma_c}\right)^\nu .$$

Here ν is the critical index for the mass (or inverse correlation length). We introduce the dimensionless variable

$$\tau = \frac{\sigma - \sigma_c}{\sigma_c} ,$$

to write

$$m(\sigma) \simeq \tau^\nu , \qquad \tau \simeq 0+ .$$

In addition to parameterizing H by the physical mass m, we would like to choose λ so that the long range force between two particles is given by a Yukawa potential of the form

$$\frac{g}{4\pi r} e^{-mr} ,$$

where g is the dimensionless coupling constant. That such a renormalization is in general impossible follows from the nonlinear nature of the theory: In fact, consider the truncated four point function

$$\langle x_1 x_2 x_3 x_4 \rangle_T = \langle x_1 x_2 x_3 x_4 \rangle - \langle x_1 x_2 \rangle \langle x_3 x_4 \rangle - \langle x_1 x_3 \rangle \langle x_2 x_4 \rangle - \langle x_1 x_4 \rangle \langle x_2 x_3 \rangle ,$$

where we use "x" to abbreviate $\phi(x)$. Define

$$\bar{g} = -m^3 \int \langle x_1 x_2 x_3 x_4 \rangle_T dx_2 dx_3 dx_4 \Big/ \left(\int \langle xy \rangle dy\right)^2 .$$

Theorem 6 [5]. For $\sigma > \sigma_c(\lambda)$,

$$0 \leq \bar{g} \leq \text{const.}$$

where the constant is independent of m, λ.

As a consequence of Theorem 6, and the fact that $g = \text{const.}\ \bar{g}$, we see that g is bounded for $\sigma > \sigma_c$. In particular, $g \to 0$ as $\sigma \to \infty$. The critical behavior of g and of $\bar{g}$ is described by exponents: For $d = 3$,

$$g \sim \tau^{3\nu + 2\gamma - (2\Delta + \gamma)}$$

$$= \tau^{3\nu + \gamma - 2\Delta}$$

where γ is the exponent for the susceptibility and Δ is the "gap" exponent relating the four point to the two point function. From Theorem 6 we conclude

(1) $$3\nu + \gamma - 2\Delta \geq 0 \quad .$$

On the other hand, if $g \neq 0$ at $\sigma = \sigma_c$, it is necessary that

(2) $$3\nu + \gamma - 2\Delta = 0 \quad .$$

The relations (2) is called hyperscaling. By the scaling relations $\Delta = \beta + \gamma$, and $\alpha + 2\beta + \gamma = 2$, we rewrite (2) as

(3) $$3\nu - \gamma - 2\beta = 0$$

or

(4) $$3\nu + \alpha - 2 = 0 \quad .$$

5. Ising Model

The ϕ^4 model is closely related to the Ising model. Let us fix κ in the cutoff action V_κ and take the $\Lambda \to \infty$ limit. We can repeat the O.-S. construction, using invariance under lattice translation for $d\mu_\kappa$. In place of e^{-H}, we obtain a transfer matrix κ with eigenvalues $1 \geq \lambda_1 \geq \lambda_2 \geq \ldots \geq 0$. Define $m = \ln \lambda_1$ as the inverse correlation length, $m = \xi^{-1}$. Then $m = m(\lambda, \sigma, \kappa)$ is a function of the remaining parameters λ, σ, κ.

The Ising limit of the ϕ_3^4 model is a limit with $\lambda \to \infty$ in such a way that m remains fixed and $\phi^2 \to \text{const.} \neq 0$, i.e. $\phi \to \pm \text{const.}$ It is clear that for fixed κ, we can choose σ so that m is fixed and $\phi^2 \to \text{const}$ as $\lambda \to \infty$. Because of the $\lambda^2 \ln \kappa$ term in the mass renormalization, however, it is not clear whether the $\lambda \to \infty$ and $\kappa \to \infty$ limits can be interchanged. We can also express the hyperscaling relations (2-4) in the Ising model.

6. Numerical Calculations

There are two frameworks [6] within which ν and the hyperscaling relations have been calculated: high temperature series, and Borel summation. High temperature series have been used by Wortis et al. in the Ising case, and by Baker and Kincaid in the ϕ^4 case. Borel methods have been used by LeGuillou and Zinn-Justin in the ϕ^4 case. Neither method has matehmatically justified error bounds. The results are

$\underline{\text{Ising}_3 \text{ H.T.:}}$	$3\nu + \alpha - 2 = .039 {}^{+.02}_{-.03}$	Wortis et al.
$\underline{\phi_3^4 \text{ H.T.:}}$	$3\nu + \gamma - 2\Delta = .028 \pm .003$	Baker-Kincaid
$\underline{\phi_3^4 \text{ Borel:}}$	$3\nu - \gamma - 2\beta = 0.000 \pm .003$	LeGuillou Zinn-Justin

Thus the high temperature series suggest a breakdown of hyperscaling. It is possible to argue that the difference in these calculations can be attributed to the exponent ν. In particular,

$$\underline{\text{Ising}_3 \text{ H.T.:}} \qquad \nu = .638 {}^{+.002}_{-.001}$$

$$\underline{\phi_3^4 \text{ Borel :}} \qquad \nu = .6300 \pm .0008$$

Thus

(5) $$3\left(\nu_{\text{H.T.}} - \nu_{\text{Borel}}\right) = .024 {}^{+.006}_{-.004}$$

which appears to account for the discrepancy between Baker-Kincaid and LeGuillou-Zinn-Justin of .028 above. Subtracting (5) from .028 is compatible with zero (hyperscaling).

Only time will resolve these discrepancies, when the high temperature series are taken to higher order. We ask: (Q1.) Are the high temperature exponents different from the Borel exponents for ${\phi_3}^4$? (Q2.) Are the high temperature ν_{I_3} and $\nu_{{\phi_3}^4}$ equal? (Q3.) Do the Ising and ${\phi_3}^4$ scaling limits differ? (Taking the limit $\sigma \to \sigma_c$ with m fixed.) If this is ture the universality hypothesis breaks down. In that case, ν_{I_3} and $\nu_{{\phi_3}^4}$ could be different. If this is the case, its presumed origin lies in the $O(\lambda^2)$ mass renormalization term alluded to above.

7. Statistical Physics Models

Note that the equivalence of field thoery with statistical mechanics has led to new methods to study problems in statistical physics. In particular the equivalence of the grand partition function for the Coulomb gas to the Euclidean Sine-Gordon field theory allowed the application of constructive field theory methods to the Coulomb gas. Brydges and Federbush [7] have established Debye screening, namely a finite inverse correlation length

$$m = m_D(1+O(z)) \quad .$$

m_D is the Debye mass $m_D = \sqrt{2\beta z}$, $\beta = 1/kT$ and z is the activity. The Sine-Gordon transformation exhibits this explicitly, namely, using Fourier transformation,

$$Z_{\text{Coulomb}} = \int e^{2z \int :\cos\left(\beta^{\frac{1}{2}}\phi(x)\right) :dx} \, d\phi \quad ,$$

where $d\phi$ is the zero mass free field measure. Then m_D arises from the quadratic term in the cosine, namely $-\frac{1}{2}(2z\beta{:}\phi^2{:}) = -\frac{1}{2}{m_D}^2{:}\phi^2{:}$, a mass term.

In $d = 2$ dimensions, the question of whether the Coulomb gas has a condensation into a dipole phase has special significance. In fact the xy model ($n = 2$ nonlinear σ-model) has been analyzed by Kosterlitz and Thouless [8] as approximately factorizing

$$Z_{xy} \simeq Z_{\text{Spin Wave}} \, Z_{\text{Vortex Gas}} = Z_{\text{Spin Wave}} \, Z_{\text{Coulomb Gas}}$$

The spin wave (free field) part yields polynomial decay of correlations, while the Coulomb gas has exponential decay (screening) at high temperatures. If the Coulomb gas condenses to a dipole phase, it does not screen. This can be seen from the Sine-Gordon transformation for dipoles:

$$Z_{\text{Dipole}} = \int e^{2z \int :\cos(\beta^{\frac{1}{2}}\nabla\phi) :dx} \, d\phi \quad .$$

Here the mass in Z_{Coulomb} is replaced by a change in the coefficient of the kinetic energy term in Z_{Dipole} , i.e. by a dielectric constant. Thus screening does not occur in the dipole phase and the above approximation suggests that for $T < T_c$, $m \equiv 0$ in the xy model. This explanation of the line of critical points ($m \equiv 0$ for $T < T_c$) supposed to exist in the xy model is not mathematically rigorous at this time. Fröhlich and Spencer hope to show the condensation for the $d = 2$ Coulomb system. This would be a major step in establishing the xy behavior, c.f. Figure 3.

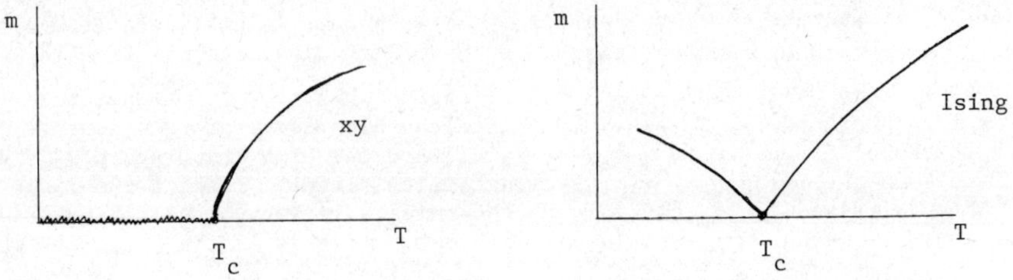

Figure 3. The line of critical points for the xy model, compared with the Ising model (d=2,n=1).

Another fascinating and related problem is the roughening phase transition. Here we consider a $d = 3$ Ising model with + boundary conditions for $x_1 > 0$ and - boundary conditions for $x_1 < 0$, c.f. Figure 4.

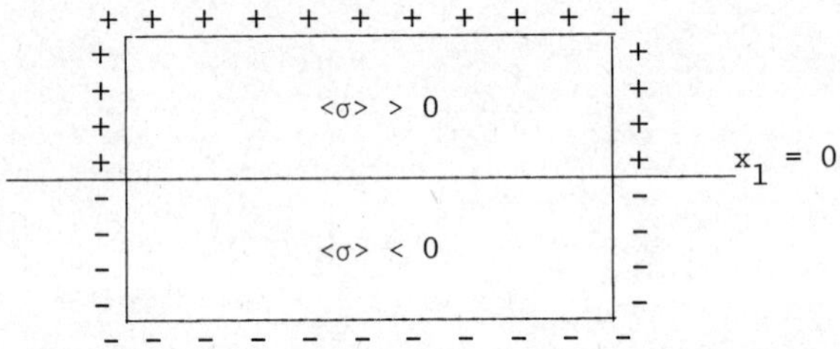

Figure 4. Phase separation in Ising_3, for $T < T_{c_{\text{Ising}_2}}$.

It is known that for $T < T_{c_{\text{Ising}_2}}$, a sharp interface exists. Also for $T > T_{c_{\text{Ising}_3}}$, we have a translation invariant state (no interface). It is of interest to understand the roughening (disappearance) of the interface as a model for melting.

In a special case, where $J_{x_1} \to \infty$ (the solid on solid model), the roughening transition is again described by a $d = 2$ Coulomb gas [9]. It is of great interest to study this problem mathematically. Again, the roughening transition is related to dipole condensation for the Coulomb gas. In addition, we expect the roughening transition does not occur in the continuum ϕ_3^4 field theory, because of the logarithmically divergent height-height correlations in the s.o.s. model.

I have given these examples to show a wealth of mathematical-physics problems associated with statistical physics and quantum fields . If these questions seem too simple, tell us at the next M ∪ Φ conference how to construct a non-Abelian (asymptotically free) gauge theory in $d = 4$ dimensions, or a nonlinear σ-model without cutoffs in $d = 2$ dimensions!

References

1. The original Osterwalder-Schrader theorem is in Commun. Math. Phys. 31, 83(1973), 42, 281 (1975). A theorem similar to this can be found in J. Glimm and A. Jaffe in "New Developments in Quantum Field Theory and Statisticl Mechanics", edited by P. Mitter and M. Lévy, 1976 Cargèse Lectures. The version here is contained in our forthcoming book. The case $p = 2$ (e.g. free field) can be handled with a separate assumption.

2. The ϕ_3^4 existence theorem combines J. Glimm and A. Jaffe, Fort. d. Physik 21, 327 (1973) with J. Feldman and K. Osterwalder, Ann. Phys. 97 80 (1976) and with J. Magnen and R. Sénéor, Ann. de l'Inst. H. Poincaré 24, 95 (1976). See Commun. Math. Phys. 56, 237 (1977). The strong coupling theory is given in J. Fröhlich, B. Simon and T. Spencer, Commun. Math. Phys. 50, 79 (1976). Other references can be found in these proceedings, e.g. to the work of Gallavotti et al.

3. References to the construction of $P(\phi)_2$ models can be found in "Statistical Mechanics and Quantum Field Theory" 1970 Les Houches Lectures, C. DeWitt and R. Stora Editors, Gordon and Breach; in "Mathematics of Contemporary Physics," R. Streater Editor, 1971 London Mathematical Society Symposium, Academic Press; and in "Constructive Quantum Field Theory" edited by G. Velo and A. Wightman, 1973 Erice Summer School, Springer Lecture Notes in Physics V25. The construction of the Yukawa_2 model is given in R. Schrader, Ann. Phys. 70, 412 (1972), E. Seiler, Commun. Math. Phys. 42, 163 (1975), J. Magnen and R. Sénéor, Commun. Math. Phys. 51, 297 (1976) and A. Cooper and L. Rosen, Trans. Am. Math. Soc. 234, 1(1977). The recent work of D. Brydges, J. Fröhlich and E. Seiler on the Higgs_2 model is in press in Commun. Math. Phys. and in these proceedings. The $1/n$ expansion of Kupiainen is also in these proceedings, as is the work in progress of Balaban. Multiphase ϕ_2^6 models are studied by K. Gawedzki, Commun. Math. Phys. 59, 117 (1978) and S. Sommers, preprint.

4. See 1973 Erice and 1976 Cargèse Lectures.

5. J. Glimm and A. Jaffe, Ann. l'Inst. H. Poincaré 22, 97 (1975) and in the 1979 Cargèse Lectures, Plenum Press, in press.

6. M. A. Moore, D. Jasnow and M. Wortis, Phys. Rev. Lett. 22, 940 (1969); J. Kincaid, G. Baker and W. Fullerton, LA-UR-79-1575; G. Baker and J. Kincaid, LA-UR-79-2655; J. Le Guillou and J. Zinn Justin, Phys. Rev. Lett. 39, 95 (1977).

7. D. Brydges, Commun. Math. Phys. 58, 313 (1978); D. Brydges and P. Federbush, Commun. Math. Phys., in press; and P. Federbush, these proceedings.

8. M. Kosterlitz and D. Thouless, J. Phys. C5, 1124 (1972).

9. See John D. Weeks, "The Roughening Transition" in 1979 Geilo (Norway) School, to appear.

1/N EXPANSION-SOME RIGOROUS RESULTS

A. Kupiainen

Department of Physics, Harvard University

1. Introduction

It was noted by Stanley [1] in 1967 that certain n-component lattice spin systems exhibit considerable simplification as n becomes large. In fact, formally these models become the exactly soluble spherical model when $n=\infty$. Later Wilson [2] and others developed systematic expansions in powers of 1/n. These 1/n expansions have several interesting features: they are non-perturbative, supposedly valid near the critical point and for scale invariant theories essentially the only expansions available.

The only rigorous result on large n limit was that of Kac and Thompson [3], who proved the convergence of the free energy as $n\to\infty$ to that of the spherical model for the models considered by Stanley. In this note we present some new results proving that the 1/n expansion is asymptotic and establishing a mass gap arbitrary near the $n=\infty$ critical temperature for n sufficiently large.

2. The Model and Results

We consider the n component nonlinear σ-model (classical Heisenberg model) on lattice. Let Λ be a torus obtained from a cube in Z^d (d arbitrary) and $\phi:\Lambda\to R^n$ be the field with the probability distribution

$$d\mu(\phi) = Z^{-1} e^{\frac{1}{2}(\phi,(\Delta-m^2)\phi)} \prod_{i\varepsilon\Lambda} \delta(\phi_i^2-n\beta)d^n\phi_i$$

where β is the inverse temperature and m^2 is a constant to be chosen later [Note that $d\mu$ is <u>independent</u> on m^2 since $e^{-\frac{1}{2}m^2(\phi,\phi)} = e^{-\frac{1}{2}m^2 n\beta}$]. By inserting the Fourier expansion of δ-function and performing the gaussian integrals one gets e.g. for the generating functional of μ:

$$\langle e^{(g,\phi)}\rangle = \langle e^{\frac{1}{2}(g,(-\Delta+m^2-2ia)^{-1}g)}\rangle_a \tag{1}$$

with $\langle\cdot\rangle_a$ the expectation in the "dual" measure

$$d\hat{\mu}(a) = Z_1^{-1}\det(-\Delta+m^2-2ia)^{-n/2} e^{-in\beta tra} \prod_{i\varepsilon\Lambda} da_i \tag{2}$$

Choosing m^2 now such that a=0 is the saddlepoint of (2) i.e.

$$(-\Delta+m^2(\beta))^{-1}_{00} = \beta \tag{3}$$

one can obtain the formal 1/n expansion as a loop expansion about a=0. As is well known, (3) gives the mass gap of the spherical model, which is positive above some critical temperature T_S. We consider only $T > T_S$.

Let now $\langle\phi_A^\alpha\rangle$ be the correlation function $\langle \prod_{i\varepsilon A} \phi_i^{\alpha_i}\rangle$ Our main results are summarized in the following two theorems:

<u>Theorem 1.</u> Let $\sum S_k(A,\alpha)n^{-k}$ be the formal 1/n expansion for $\langle\phi_A^\alpha\rangle$. Then, for all $T > T_S$

$$|\langle\phi_A^\alpha\rangle - \sum_{k=0}^{r-1} S_k(A,\alpha)n^{-k}| \le n^{-r}R_r(A,\alpha,\beta)$$ □

The expansion is thus asymptotic. R_r has an explicit formula and bound. In particular we get exponential falloff; the following theorem can be stated also for more general correlations.

Theorem 2. There exist constants α_1 and α_2, only depending on the dimension d, such that for all $T > T_S$ and $n > \alpha_1 m(\beta)^{-\alpha_2}$ there is a mass gap:

$$<\phi^1_0 \phi^1_x> \leq (-\Delta+\mu(\beta,n)^2)^{-1}_{ox}$$

where (the physical mass) $\mu>0$. □

α_1 and α_2 are explicit and one gets bounds for the critical temperature of the form $T_C < T_S + \frac{a}{n^b}$.

We have also proved similar results for the 1/n expansion for the continuum $(\vec{\phi}^2)^2$ Quantum Field Theory. For further details see [4].

3. Methods

We will now briefly discuss some methods involved. We start from Theorem 2. The dual transformation (1) gives

$$<\phi^1_0 \phi^1_x> = <(-\Delta+m^2-2ia)^{-1}_{ox}>_a \tag{4}$$

The difficulty with (4) is that $d\mu$ is a complex measure and $(-\Delta+m^2-2ia)^{-1}_{ox}$ is non local in a. The second problem is solved by using a convergent Neuman expansion due to Brydges and Federbush [5]:

$$(-\Delta+m^2-2ia)^{-1}_{ox} = \sum_{\omega:o\to x} \prod_{k\in\Lambda} (2d+m^2-2ia_k)^{-n(\omega,k)} \tag{5}$$

Here ω are random walks from o to x and $n(\omega,k)$ the number of times ω hits k. Although complex, $<\cdot>_a$ turns out to be reflection positive. One can apply chessboard estimates to (4) and (5):

$$<\phi^1_0 \phi^1_x> \leq \sum_{\omega} \prod_{k\in\Lambda} \frac{z(n(\omega,k)}{z(0)}$$

where $z(t) = <\prod_{k\in\Lambda}(2d+m^2-2ia_k)^{-t}>^{1/|\Lambda|}$. Theorem (2) is thus a consequence of a pressure estimate $\left|\frac{z(t)}{z(0)}\right| \leq (2d+\mu^2)^{-t}, \forall\Lambda$. We need sharp upper bounds for $z(t)$ and lower bound for $z(0)$. The idea is to introduce a cutoff in a-integrals thus confining the integration near the saddle point, which is easier to study. Dependence on the cutoff is shown to be small using chessboard estimates.

The expansion itself is generated by integrating by parts with respect to the quadratic form $\left.\frac{\partial^2 F}{\partial a_i \partial a_j}\right|_{a=0}$ where F is defined by $d\hat{\mu} = Z_1^{-1} e^F[da]$.

One gets an explicit formula for the reminder, involving expectations of the form

$$< \prod_i a_i \prod_{<kl>} (-\Delta+m^2-2ia)^{-1}_{kl} > \qquad (6)$$

(6) is estimated by inserting (5) and chessboarding the resulting products of local observables thus reducing again to pressure estimates.

References:

[1] Stanley, H.E., Phys. Rev. 176, 718 (1968).
[2] Wilson, K.G., Phys. Rev. D7, 2911 (1973).
[3] Kac, M.; Thompson, C.J., Phys. Norv. 5, 163 (1971).
[4] Kupiainen, A., Thesis, Princeton University (1979).
[5] Brydges, D; Federbush, P., Comm. Math. Phys. 62, 5 (1978).

This work is supported in part by the National Science Foundation under Grant No. PHY-77-18762.

THE RENORMALIZATION GROUP IN THE EUCLIDEAN SCALAR SUPERRENORMALIZABLE FIELD THEORIES

G. Benfatto+

Istituto di Matematica
Università di Roma
00100 ROMA - Italy

1. Introduction.

In this talk I will refer about an approach to the ultraviolet stability problem in the scalar euclidean field theories, which has been developed in Roma in the last two years, following some initial ideas of G. Gallavotti(1,2,3).

The heart of the method is the rigorous analysis (in 2 or 3 space-time dimensions) of a recursion relation, which can be thought as a representation of the renormalization group (in the definition of K.G. Wilson(5)).

The techniques which allow to study this recursion relation are typical of the Statistical Mechanics and the theory of Gaussian Processes associated to Elliptic Operators(4). I will not refer about these technical aspects of the method, but I will only sketch the main ideas.

2. Notation.

Let $\varphi(\xi)$ be the gaussian random field with covariance $C = (1 - D)^{-1}$, where D is the Laplacian in R^d, $d = 2,3$. Formally

$$\begin{aligned} C &= (1 - D)^{-1} = \sum_{0}^{\infty}{}_K C_K \\ C_K &= (\gamma^{2K} - D)^{-1} - (\gamma^{2K+2} - D)^{-1}, \quad \gamma > 1 \end{aligned} \tag{1}$$

+ Supported in part by C.N.R., G.N.F.M.

Then $\varphi(\xi)$ can be thought as a sum of independent random fields $\varphi_k(\xi)$ of covariance C_k. Observe the scaling relation

$$C_k(\xi-\eta) = \gamma^{(d-2)k} C_0(\gamma^k(\xi-\eta)) \tag{2}$$

which implies that the fields $\varphi_k(\xi)$ are identically distributed up to scale factors. $\varphi_k(\xi)$ will be called the field of frequency k and $d\hat{P}_k$ will denote its probability distribution.

We now define the cut-off field (which is essentially the field regularized à la Pauli-Villars)

$$\varphi^{(\leq N)}(\xi) = \sum_0^N {}_k \varphi_k(\xi) \tag{3}$$

and the corresponding measure

$$d P_N = \prod_{k=0}^{N} d\hat{P}_k \tag{4}$$

We want to study the measure

$$d\mu_N = \exp\{V_I^{(N)}\} d P_N \tag{5}$$

where

$$V_I^{(N)} = -\lambda \int_I :\varphi^{(\leq N)4}(\xi): d\xi + \{\text{counter terms}\} \tag{6}$$

The counter terms are those needed in the formal perturbation theory (then they are absent for d = 2).

3. The recursion relation.

The ultraviolet stability problem is essentially the following: to prove the existence of $E_+(\lambda)$, $E_-(\lambda)$, independent of N, such that

$$\exp\{-E_-(\lambda)|I|\} \leq \int \exp\{V_I^{(N)}\} dP_N \leq \exp\{E_+(\lambda)|I|\} \tag{7}$$

The main idea is to calculate the partition function iteratively.Then, if we define the effective potential of order k,$\overline{V}_I^{(N,k)}$,by

$$\exp\{\overline{V}_I^{(N,k)}\} = \int \exp\{V_I^{(N)}\}\prod_{i=k+1}^{N} d\hat{P}_i \tag{8}$$

we have to study the recursion relation

$$\exp\{\overline{V}_I^{(N,k)}\} = \int \exp\{\overline{V}_I^{(N,k+1)}\} d\hat{P}_{k+1} \tag{9}$$

Eq. (9) can be thought as a representation of the renormalization group in the definition of K.G. Wilson[5].Its analysis is made easier by the scaling properties of the field $\varphi^{(\leq N)}(\xi)$,which imply that the structure of $\overline{V}_I^{(N,k)}$ is essentially independent of k.Roughly:

$$\overline{V}_I^{(N,k)} = V_I^{(k)} + W_I^{(N,k,t)} + \Delta_I^{(N,k,t)} \tag{10}$$

where $W_I^{(N,k,t)}$,which is the contribution of the terms up to order t not contained in $V_I^{(k)}$ in the λ-expansion of $\overline{V}_I^{(N,k)}$,has a finite limit as $N \to \infty$ and,for some $\rho > 0$

$$|\Delta_I^{(N,k,t)}| \leq \delta_{k,t}\, \lambda^{t+1} (\log\lambda)^{\rho} |I| \tag{11}$$

with

$$\sum_{0}^{\infty}{}_{k}\, \delta_{k,t} < \infty \tag{12}$$

if t is large enough ($t \geq 3$ for d=3,$t \geq 1$ for d=2).

In the proof of eq. (10) there are two main technical problems:

1) The fields $\varphi_k(\xi)$ are unbounded,so that it is not possible to simply use the Taylor formula in order to control the "remainder" $\Delta_I^{(N,k,t)}$ in eq. (10).Furthermore,in order to bound the contribution to $\Delta_I^{(N,k,t)}$ coming from $W_I^{(N,k+1,t)}$,it is necessary to control the Hölder norm of $\varphi_{k+1}(\xi)$. We overcome these difficulties by using:

a)for the upper bound,the positivity properties of the interaction and $W_I^{(N,k,t)}$;

b)both for the upper and lower bound,the fact that the points where the fields $\varphi_k(\xi)$ are "large" or "rough" give a negligible contribution to the integral in eq. (9), thanks to the support properties of the free field of covariance $C_0(\xi-\eta)$[4].

2) The solution of problem 1) could give the rough estimate

$$\Delta_I^{(N,K,t)} \sim |I|^{t+1} \qquad (13)$$

To obtain the right dependence on I, we use again the support properties of the gaussian field of covariance $C_0(\xi-\eta)$ and the exponential decay of $C_0(\xi-\eta)$. I will briefly sketch the idea.

We divide I in "large" boxes $\Box_i$ (of side ℓ independent of I), separated by a corridor Γ of width $\delta \ll \ell$ (see Fig. 1 for the case d=2).

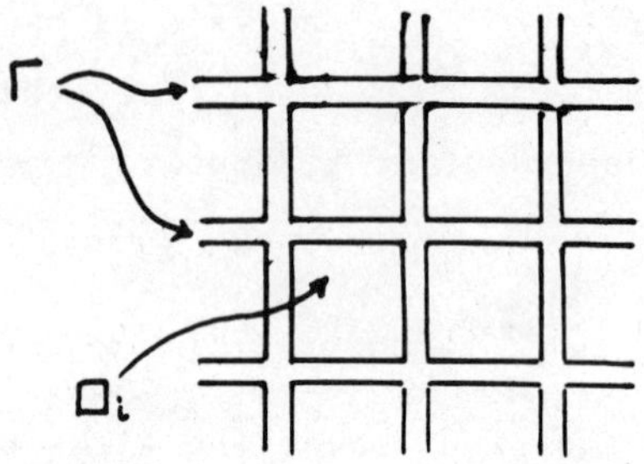

Fig. 1

By the Markov property of the measure $d\hat{P}_k$, we can write the integral in eq. (9) (suppressing all the indices) in the following way

$$\int e^{\bar{V}_I} P(d\varphi) \simeq \int e^{\bar{V}_\Gamma} P(d\varphi_\Gamma) \prod_i \int e^{\bar{V}_{\Box_i}} P(d\varphi_{\Box_i} | \varphi_\Gamma) \qquad (14)$$

where φ_Γ and $\varphi_{\Box_i}$ denote the restriction to Γ and $\Box_i$ of φ and $P(d\varphi_{\Box_i} | \varphi_\Gamma)$ is the probability distribution of $\varphi_{\Box_i}$, conditioned to φ_Γ. The approximate equality in eq. (14) comes from the fact that $\bar{V}_I$ is not local; however the terms connecting different regions give a negligible contribution, as it can be shown using the exponential decay of $C_0(\xi-\eta)$. Using the results of 1) it is now possible to apply the Taylor formula to $\int e^{\bar{V}_{\Box_i}} P(d\varphi_{\Box_i}|\varphi_\Gamma)$, uniformly in φ_Γ up to a small error. There remains to calculate $\int \exp(\bar{V}_\Gamma) P(d\varphi_\Gamma)$. This can be done by iterating the procedure described before many times with new grids obtained from the one of Fig. 1 by suitable translations, chosen so that any point will eventually be in the interior of some $\Box_i$. In this way we construct successively the contributions to the r.h.s. of eq. (10) coming from the different series of boxes. The estimate of $\Delta_I^{(N,K,t)}$ will obviously be proportional to the number of boxes $\square_i$ contained in I, that is it will be proportional to $|I|$.

I want now to give an idea of the main result, which is of course eq. (10). Let us define the normalized fields

$$Z_K(\xi) = \varphi_K(\xi) \Big/ \sqrt{2 C_0(0)\, \gamma^{(d-2)K}} \tag{15}$$

$$X_K(\xi) = \varphi^{(\le K)}(\xi) \Big/ \sqrt{2 C_0(0) \sum_{i=0}^{K} \gamma^{(d-2)i}} \tag{16}$$

It is easy to verify the following recursion relation

$$X_K(\xi) = \left[Z_K(\xi) + \sqrt{\Gamma_K}\, X_{K-1}(\xi) \right] \Big/ \sqrt{1+\Gamma_K} \tag{17}$$

$$\Gamma_K = \begin{cases} K \ , & d=2 \\ (1-\gamma^{-K})/(\gamma-1)\ , & d=3 \end{cases} \tag{18}$$

By eq. (2) the field $\tilde{Z}_K(\xi) = Z_K(\gamma^{-K}\xi)$ is a gaussian field with covariance $C_0(\xi-\eta)/2C_0(0)$, independent of k. Then $V_I^{(k)}$ (see eq. (6)) has an expression in terms of the field $\tilde{Z}_K(\xi)$ as

$$V_I^{(K)} = -\lambda \gamma^{2(d-2)K} \left[2C_0(0)\right]^2 \gamma^{-dK} \int_{\gamma^{dK} I} d\xi \, :\left[\tilde{Z}_K(\xi) + \sqrt{\Gamma_K}\, X_{K-1}(\gamma^{-K}\xi)\right]^4: \\ + \{ \text{counter terms} \} \tag{19}$$

If we can show that $X_{K-1}(\xi)$ is "essentially" bounded, then the contribution of the terms of order t in the λ-expansion of $\log \int \exp\{V_I^{(K)}\} d\hat{P}_K$ is proportional to

$$\left[\lambda \gamma^{-(4-d)K}\right]^t \gamma^{dK} \, |I| \tag{20}$$

Some straightforward algebra shows that this property is not destroied by $W_I^{(N,K,t)}$, thanks to the counter terms. Indeed it is possible to show that the Taylor series in the variable λ of the effective potential of order k, $\bar{V}_I^{(N,k)}$, can be thought as a power series in the "effective coupling constant" $\lambda\gamma^{-(4-d)}$, with coefficients uniformly bounded with respect to N. Eq. (20) immediately implies eq. (12).

References

1. G. Gallavotti, Memorie dell'Accademia dei Lincei, XV, 23, 1978
G. Gallavotti, Annali di Matematica Pura e Applicata, 1979.

2. G. Benfatto, M. Cassandro, G. Gallavotti, F. Nicolò, E. Olivieri, E. Presutti and E. Scacciatelli, Comm. Math. Phys. 59, 143, 1978.

3. G. Benfatto, M. Cassandro, G. Gallavotti, F. Nicolò, E. Olivieri, E. Presutti and E. Scacciatelli, "On the ultraviolet stability in the eucledean scalar field theories", Preprint IHES/P/79/251 (in print on Comm. Math. Phys.).

4. G. Benfatto, G. Gallavotti and F. Nicolò, "Elliptic equations and Gaussian processes", Preprint IHES/P/79/251 (in print on J. Funct. Anal.).

5. K.G. Wilson, Phys. Rev. B4, 3174, 1971.

EXPANSION AND SUMMABILITY METHODS

IN CONSTRUCTIVE FIELD THEORY

J. Magnen and R. Sénéor
Centre de Physique Théorique
Ecole Polytechnique, 91128 Palaiseau

I. Introduction

The following models : $\lambda P(\Phi)$ in $D=2$ dimensions [N,GJ1] (P semi bounded from below), Φ^4 in $D=3$ [GJ2], Yukawa in $D=2$ [Sc,S] and $D=3$ [FMS, MS1], Sine Gordon or massive QED in $D=2$ [FS], Higgs model in $D=2$ [BFS]... have been studied with various degree of achievement in the framework of constructive field theory.

Λ being a space cutoff and K a momentum cutoff, the construction of the Euclidean version of these models consists in 2 steps

1) Ultraviolet stability of the free energy per unit volume

This is the removing of the momentum cutoff. More precisely, $\mathcal{L}_I^{(K)}$ being the cutoff interaction with suitable renormalization counterterms, $\exists$ a, $A>0$ independent of K s.t.

$$e^{-a|\Lambda|} \leq Z_\Lambda^{(K)} = \int e^{-\int_\Lambda \mathcal{L}_I^{(K)}} \leq e^{+A|\Lambda|}$$

and $Z_\Lambda = \lim_{K\to\infty} Z_\Lambda^{(K)}$ exists.

2) Thermodynamic limit of Schwinger functions

$$\exists \lim_{\Lambda\nearrow\infty} < \pi\,\Phi(f) >_\Lambda = \lim_{\Lambda\nearrow\infty} Z_\Lambda^{-1} \int \pi\,\Phi(f)\, e^{-\int_\Lambda \mathcal{L}_I^{(K)}}$$

the limit satisfying at least the Osterwalder Schrader axioms (O - S).

Expansion methods are used among other ones (correlation inequalities - functional space analysis - renormalization group...) to prove 1) and 2).

In principle, these combined methods are enough general to control all superrenormalizable local models far from critical regions (i.e weak and strong coupling regions). Generally expansions give detailed information on the physical structure.

Example of expansions

1. Convergent perturbation expansion with respect to the coupling constant
   Sine Gordon $D = 2$

2. Weak coupling cluster expansion (GJS1)
It is the quantum field analogous of the high temperature expansion in statistical mechanics. It solves 2) knowing 1). The following expansions are related

a) n - particle expansion (GJS2, GJ3)
It is adapted to the study of the Hamiltonian spectrum up to the $(n+1)$ - particle threshold. Roughly speaking it is as n - successive cluster expansions.

a') Spencer's expansion [Sp]
Nice to study n - particle irreducible parts (e.g. Bethe - Salpeter kernels)

b) mean field expansion (GJS3)
See A. Jaffe contribution.

3) <u>Phase space cells expansion</u> (GJ2)

It solves 1) for superrenormalizable theories.
We will now explain how to prove 1) in some simple cases and what are the main ideas of phase space expansions.

II. Ultraviolet stability

It will be a consequence of superrenormalizability and of the positivity of the effective potential.

1) Superrenormalizability
For each vertex, using power counting argument, one gains an extrapower of convergence ρ :

$$\rho = 2n - (n-1)D \quad \text{for } \Phi^{2n}$$

$$\rho = \frac{4-D}{2} \quad \text{for Yukawa } \Phi\bar{\Psi}\Psi \text{ and QED}$$

D	2	3	4
Φ^4	2	1	0
$\Phi\bar{\Psi}\Psi$	1	1/2	0

A vertex, such that one of the momenta of its lines is bigger than ξ is then "convergent like $\xi^{-\rho}$. This is the asymptotic freedom of superrenormalizable theories $(\rho > 0)$.

2) <u>Positivity of the effective potential</u>*

By this we mean for a scalar field theory as e.g. Φ^4 :

$$\Phi^4_{(x)} \geq 0 \qquad \text{and} \qquad :\Phi^4_K:(x) \geq \begin{cases} -\text{cst}(\log K)^2 & D=2 \\ -\text{cst}\,K^2 & D=3 \end{cases}$$

with Φ_K the cutoff field.

For a theory with fermions like e.g. Yukawa or QED, we perform the integration over the fermion leaving the boson field constant, then

$$V^{\text{eff}}(\Phi) = \lim_{\Lambda\nearrow\infty} -\frac{1}{|\Lambda|}\,\ell n \quad e^{-\int_\Lambda \mathcal{L}_I}$$

One can define similarly an effective potential in a "cubic" box with periodic boundary condition in which Φ is constant.
One gets

- for Yukawa and Φ large $\quad V^{\text{eff}} \simeq \begin{cases} \Phi^2\,\ell n|\Phi| & D=2 \\ |\Phi|^3 & D=3 \end{cases}$

Introducing a fermion upper cutoff K, $V^{\text{eff}}_{(K)} \geq -\log K$ for $D=2$ and $\geq -K^3$ for $D=3$

- for QED

V^{eff} is zero by gauge invariance. Breaking it with a momentum cutoff on the fermions gives in $D=3$

$$V^{\text{eff}}_{(K)} \sim \begin{cases} c^t\,\frac{A^4}{K} & |A| << K \\ c^t\,A^2K & |A| >> K \end{cases}$$

We now briefly describe the phase space cell expansion and the use of points 1) and 2) in order to prove I.). For simplicity we treat the case of Φ^4 in $D=3$.

3) <u>The phase space expansion</u>

Following Nelson (N), one wants to show that the measure of space where $-V^{\text{eff}}$ is big is very small, proving thus that $\quad e^{-\int \mathcal{L}_I} < \infty$.
To do that one performs a truncated perturbation expansion, where each vertex produced is localized in space (in cubes Δ) and in momentum in the sense that at least one the fields of the vertex has a momentum bigger than $|\Lambda|^{-1/D}$ (well localized field). As a consequence of 1) each vertex (properly renormalized) has a convergent factor $|\Delta|^{\bar{\rho}/D}$**. The expansion is then as follow (with Λ a unit cube).

* The idea of using effective potential in constructive field theory is due to Balaban and Gawędzki [BG]

** Note that $\bar{\rho}$ has to be smaller than ρ for technical reasons (in particular translation invariance).

Let $M >> 1$, then define $\mathcal{D}_1, \ldots, \mathcal{D}_i, \ldots$ covers of Λ with cubes of equal sizes. The cubes of $\mathcal{D}_i$ are obtained by subdivision of cubes of $\mathcal{D}_{i-1}$, with $|\Delta_i| = |\Delta_{i-1}|^{1+\epsilon}$ for some $\epsilon > 0$.

A perturbation step in Δ can be symbolized by

$$Z = \int e^{-\int \mathcal{L}_I} = \int e^{-\int \mathcal{L}_I}\Big|_{\substack{\text{momentum cutoff} \\ |\Delta|^{-1/D} \text{ in } \Delta}} + \int P\, e^{-\int \mathcal{L}_I}$$

$$= (I_\Delta + P_\Delta)\, Z$$

The expansion is then

$$Z = \Big[\prod_{\Delta_1 \in \mathcal{D}_1} (I_{\Delta_1} + P_{\Delta_1} \prod_{\substack{\Delta_2 \subset \Delta_1 \\ \Delta_2 \in \mathcal{D}_2}} (I_{\Delta_2} + P_{\Delta_2} \prod_{\substack{\Delta_3 \subset \Delta_2 \\ \Delta_3 \in \mathcal{D}_3}} (I_{\Delta_3} + P_{\Delta_3} \prod \ldots\ldots \Big]\Big]\, Z$$

Supposing that each produced "vertex P_Δ" gives a convergent factor $|\Delta|^{\bar{\rho}/D}$ (as expected) and ϵ is small enough, the expansion is convergent and for each term of the sum

$$|\pi I_\Delta \, \pi P_\Delta \, Z| \leq \text{cst}\ \pi\ \Delta^{+\bar{\rho}/D \frac{1}{2}}$$

one half of the convergent factors being used to bound $e^{-\int \mathcal{L}_I}$ since

$$|\pi I_\Delta \, e^{-\int_\Delta \mathcal{L}_I}| \leq e^{-\sum_\Delta |\Delta|\, |\text{lower bound of } \mathcal{L}_I \text{ in } \Delta|} \leq \prod_\Delta \text{cst}$$

We now show that $\quad |\int \pi P_\Delta \, e^{-\int \mathcal{L}_I}| \leq \text{cst}\ \pi\, |\Delta|^{\bar{\rho}/D}$

This bound is obtained via gaussian integration (e^{-x^2} domination) and use of the positivity of the effective potential (e^{-x^4} domination).

Each P_Δ vertex is roughly $\int_\Delta \Phi^4 = (\Delta^{1/4}\, \bar{\Phi}_\Delta)^4$ with $\bar{\Phi}_\Delta = \left(\frac{1}{\Delta} \int_\Delta \Phi^4 \right)^{1/4}$.

At least one of the 4 fields is well localized and can be integrate (gaussian integration). In fact for fields in $\Delta \in \mathcal{D}_i$ well localized (i.e. with lower cutoff at $|\Delta|^{-1/D}$) one has

$$|\int \prod_{\Delta \in \mathcal{D}_i} \bar{\Phi}_\Delta| \sim \prod_{\Delta \in \mathcal{D}_i} |\Delta|^{-1/2 + 1/D}$$

which is the best bound compatible with power counting. The other $|\Delta|^{1/4}\, \bar{\Phi}_\Delta$ are bounded using (e^{-x^4} domination)

$$\left| \int \pi\, \Delta^{1/4}\, \bar{\Phi}_{\Delta}\; e^{-\sum_{\Delta} |\Delta| \bar{\Phi}_{\Delta}^{4}} \right| \sim \prod_{\Delta} x\, e^{-x^4} \sim \prod_{\Delta} \text{cst}$$

It follows that for each P_{Δ} vertex one gets $|\Delta|^{1/4 - 1/2 + 1/D} = |\Delta|^{\bar{\rho}/D}$ as convergent factor.

4) The phase space expansion for Yukawa and QED.

The Yukawa theory can be treated similarly as Φ^4 after the fermions have been integrated out (Matthews Salam formulation), the effective potential V^{eff} being used as Φ^4 for non gaussian domination.
The procedure is however different for QED since by gauge invariance $V^{eff} = 0$. The method is to break the gauge invariance by introducing a (a priori non necessary) cutoff on the fermion fields. As a result $V_K^{eff} \neq 0$ and can be used for domination.

5) Improvement of the convergence : superlocalization

In the proof of the convergence of the expansion in section 3) only $|\Delta|^{\epsilon_1}$ per P_{Δ} vertex is necessary. Superrenormalizability insuring $|\Delta|^{1/D - 1/4}$ it remains at our disposal $|\Delta|^{1/D - 1/4 - \epsilon_1}$. This extra power can be used to define the expansion i.e. to perturb in smaller cubes, increasing the localization.
For example, for Φ^4 in $D = 2$, one can consider as well localized not only fields with lower cutoff at $|\Delta|^{-1/2}$ but at $|\Delta|^{-1/4 - \epsilon_2}$ for some $\epsilon_2 > 0$. This last remark can probably be used to cure the defects of the n-particle expansion (see I 2) and IV).

III Analyticity in the coupling constant and Borel summability

In all the models discussed above the analyticity domain is given by the condition Re $V^{eff} > 0$ (regardless of the possibilities of extending it by complex scaling [EMS]).In the same way, the bounds on the nth derivatives are essentially the ones given by domination with V^{eff} (comparable to what one gets using Lipatov's ideas [P]).

As a consequence Φ^4 in $D = 1$, $D = 2$ or $D = 3$[GGS, EMS, MS2],Yukawa in $D = 2$ and 3 [R,MS1], and probably QED in $D = 3$ are Borel summable at $\lambda = 0$ showing the complete determination of these theories by the perturbation series.

IV Questions

We now present a list of open questions related to the above topics.

On expansions :

1) In the present status of the n-particle expansion, its convergence required that the coupling constant go to zero as $n \to \infty$. Remove this constraint.

2) Define an expansion for strictly renormalizable asymptotically free theories

3) Define a lattice version of the phase space expansion (applicable to gauge fields), (see also T. Balaban report)).

On analyticity and Borel summability

1) If Borel summability holds, how to extend known properties of the perturbation series to the global Green functions?

2) What is the largest analyticity domain in λ for ϕ^4 in $D = 2$ or 3? Can we see the left singularities of the Borel transform (instanton contributions)?

3) Are the perturbation series Borel summable in each pure phase of ϕ^4 in $D = 2$, as it is expected?

4) Is the 1/N expansion Borel summable for O(N) symmetric $(\vec{\phi}^{\,2})^2$ in $D = 1$, 2 or 3?

5) Using Feynman diagrams estimates can we reconstruct the Borel transform, perform its analytic continuation and define a global theory (see for ϕ^4 in $D < 4$ [S - R])?

What to think about Khuri result [K] showing a contradiction between Borel summability of perturbation series of ϕ^4 in $D = 4$ and the existence of a non trivial ultra violet fixed point?

REFERENCES

[BFS] D.C. Brydges, J. Frölich, E. Seiler : Construction of quantized gauge fields II IHES preprint (1979).

[BG] T. Balaban, C. Gawędzki : private communication

[EMS] J.P. Eckmann, J. Magnen, R. Sénéor : Comm. Math. Phys. 39, 251-271(1975)

[FMS] J. Fröhlich, J. Magnen, R. Sénéor : Cargèse, unpublished (1976).

[FS] J. Fröhlich, E. Seiler, Helv. Phys. Acta, Vol. 49, 1976.

[GGS] S. Graffi, V. Grecchi, B. Simon PRL. 32 B 631(1970)

[GJ 1] J. Glimm, A. Jaffe, in les Houches 70, Gordon and Breach 1972 1 - 108.

[GJ 2] J. Glimm, A. Jaffe : Fortchr. Physik 21, 327 - 376 (1973).

[GJ 3] J. Glimm, A. Jaffe : Comm. Math. Phys. 67, 267 - 293 (1979).

[GJS 1] J. Glimm, A. Jaffe, T. Spencer in Erice 1973, Springer Lecture Notes in Phys. No 25.

[GJS 2] J. Glimm, A. Jaffe, T. Spencer, Ann Math. 100, 585 - 632 (1974).

[GJS 3] J. Glimm, A. Jaffe, T. Spencer, Comm. in Math. Phys. 45, 203 - 216 (1975)

[K] N. Khuri, Phys. Review D, Vol. 14, No 10, 2665 (1976)

[MS 1] J. Magnen, R. Sénéor in preparation

[MS 2] J. Magnen, R. Sénéor, Comm. Math. Phys. 56, 237 - 276 (1977).

[N] E. Nelson, in Mathematical theory of elementary particles MIT Press Cambridge (1966).

[P] G. Parisi P.L. Vol. 66 B No 4 382

[R] P. Renouard, Analyticité et Sommabilité de Borel pour Y_2, to be published in Ann. Inst. H. Poincaré.

[Sch] R. Schrader Ann. of Phys. 70, 412 - 457 (1972).

[S] E. Seiler, Comm. Math. Phys. 42, 163 - 182 (1975).

[Sp] T. Spencer, Comm. Math. Phys. 44, 143 - 164 (1975).

[SR] E. Speer, V. Rivasseau, Borel transfrom in Euclidean φ^4_ν ... Preprint Ecole Polytechnique (1979)

THE PHYSICAL-REGION MULTIPARTICLE S MATRIX

D. IAGOLNITZER
DPhT, CEN Saclay, BP 2, 91190 Gif-sur-Yvette, France

1. INTRODUCTION. This text is a short guide to several recent works on the physical-region analytic structure of the S matrix in relativistic quantum theory, for sets of external particles with strictly positive masses. In the absence so far, in the multiparticle case, of a satisfactory understanding and derivation of more general properties, the study of the physical region has in fact appeared to have its own interest both from a conceptual viewpoint and for applications and further developments. For details on the general physical and mathematical background, see[1].

We first recall in Sect.2 the physical region properties that are expected, and are equivalent to basic properties of macroscopic causality and macrocausal factorization (= asymptotic properties of transition amplitudes between sets of initial and final wave functions corresponding respectively to non causal, and causal, configurations of displaced particles).

In Sect.3, we briefly explain[2] the particular phenomena that occur in two-dimensional space-time, and yield factorization of the multiparticle S matrix itself, for a class of models, into a product of two-body S-matrices.

In Sect.4, we return to four-dimensional space-time and present recent works[3-8] on the derivation, in S-matrix theory, of the physical-region structure described in Sect.2. The subject was presented[9] in the sixties as successfully achieved on the basis of unitarity and "maximal analyticity". However, the proofs of [9] rely on some assertions which are crucially non correct, even in the simplest cases. On the other hand, the related approach[10] based on macrocausality and unitarity made use of an *assumption* of "separation of singularities" in unitarity equations (see example in Sect.4), which is an ad hoc way of eliminating crucial problems that appear otherwise. The results obtained in [3-8] in the framework of essential support theory, or of hyperfunction theory, will be illustrated here for $3 \to 3$ processes.

Related results have also been obtained in axiomatic field theory, in two recent approaches[11,12]. Information was derived previously from microcausality and the spectral condition. It is, however, much weaker than macrocausality. The new results are based on a further exploitation of the off-shell structure of the theory, together with the axiom of asymptotic completeness and regularity assumptions : see the lecture by Bros.

2. PHYSICAL-REGION STRUCTURE[1]. The "scattering function" f of a given process is the connected momentum-space S-matrix kernel, after factorization of its energy-momentum conservation δ-function. Expected physical-region properties of f are :

2.1. Analyticity outside $+\alpha$-Landau surfaces $L^+(G)$ of connected graphs G, and plus $i\varepsilon$ rule = f is, in the neighborhood of $+\alpha$-Landau points, the boundary value of an analytic function from the "plus $i\varepsilon$" directions dual, at each point $P=\{P_k\}$ of $L^+(G)$, to the causal direction(**s**) $\hat{u}_+(P;G)$.

The plus $i\varepsilon$ rule gives no information at points that lie at the intersection of several $+\alpha$-Landau surfaces, or at the "M_o points" such that two initial, or two final, P_k are colinear. Property 2.1 is then stated in general in terms of an *essential support*, or equivalently *microanalyticity*, property, which is a natural and precise mathematical formulation of a general property of *macroscopic causality* in terms of particles. Recent works[5,6] study the situation at M_o points, where the analysis is more delicate.

2.2. Discontinuity formulae around the $+\alpha$-Landau singularities, which give detailed information on the structure at $+\alpha$-Landau points. Their general statement is again in terms of an essential support property, which is now equivalent to a general property of *macrocausal factorization* and yields decompositions of f, in appropriate parts of the physical region, as a sum of terms that generalize Feynman integrals and are associated with the various $+\alpha$-Landau surfaces $L^+(G)$ encountered. These terms are well defined, modulo analytic backgrounds, in terms of on-mass-shell S-matrix elements associated with each vertex of G. Recent works[4] concern the nature of the $+\alpha$-Landau singularities in refined situations.

3. TWO-DIMENSIONAL SPACE-TIME[2]. We consider for simplicity a theory with equal-mass particles, possibly of different types, and a $3 \to 3$ process. In each sector of the physical region corresponding to a given relative position of the initial 2-momenta, and of the final 2-momenta, one encounters a number of $+\alpha$-Landau surfaces associated with graphs with one internal line and triangle graphs, which, in view of two-body kinematics in two-dimensional space-time, all coincide with the codimension one surface Σ defined by requiring that the set of initial 2-momenta should coincide with the set of final 2-momenta. They are the only ones in models in which there is no particle production. The plus $i\varepsilon$ directions associated at any point P of Σ with two different subsets of these surfaces are opposite, and properties 2.1, 2.2 of Sect.2 then entail that the (non connected) scattering function is along Σ a sum of two well defined boundary values of analytic functions from opposite directions. It is analytic outside Σ. In models in which it is strictly zero outside Σ, this implies consistency relations between sums of products of two-body functions, and the two above boundary values add up in a way which exhibits factorization of the 3-body S matrix into a product of two-body S-matrices : see[2], where the extension to the m-body S-matrix, $m > 3$, is also given.

3. S-MATRIX THEORY. We shall here consider, for simplicity, a theory with only one type of particle and a $3 \to 3$ process below the 4-particle threshold. In that region,

and away from M_o points, the only $+\alpha$-Landau surfaces $L^+(G_\beta)$ correspond to the 9 graphs with one internal line, such as G_1 = 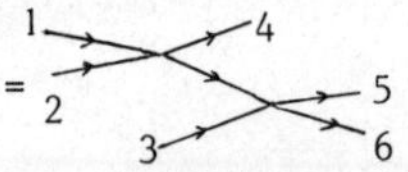 , and 9 triangle graphs. *Macro-causality*[1] entails that f can be decomposed correspondingly as a sum of 18 terms f_β, each of which is analytic outside $L^+(G_\beta)$ and is along $L^+(G_\beta)$ a plus $i\varepsilon$ boundary value. The aim is then to show that f_β is necessarily a generalized Feynman integral associated with G_β (see Sect.2.2).

We start from the "unitarity equation" that directly follows from $SS^{-1}=SS^\dagger=1$ and the decomposition of S into connected components. In order to first study f_1, we write it, following[8], in the form :

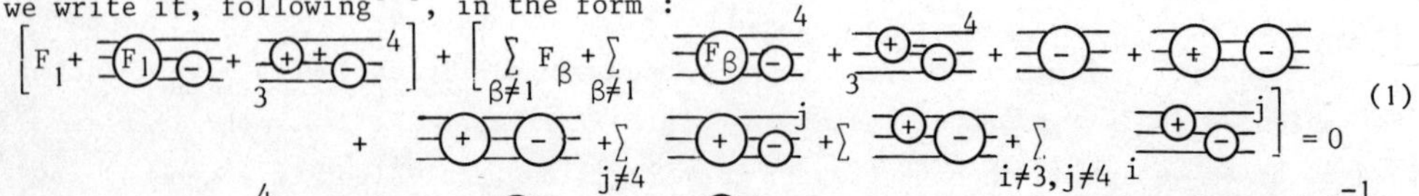

where $F_\beta = f_\beta \times \delta^4(\Sigma p_i - \Sigma p_j)$, 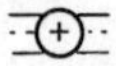 and 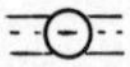 are connected kernels of S and S^{-1}, terms such as 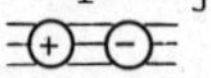 are integrals over *on-mass-shell* internal 4-momenta, and 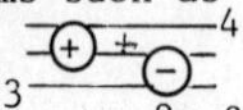 is the plus $i\varepsilon$-pole part of 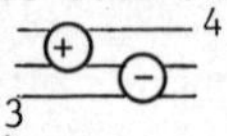 . (This last term has a singularity $\delta(k^2-\mu^2)$, $k=p_1+p_2-p_4$, along $L^+(G_1)$).

The first three terms, after factorization of their conservation δ-function, are, like f_1, analytic outside $L^+(G_1)$ and plus $i\varepsilon$ boundary values along $L^+(G_1)$. Moreover, following the idea of "separation of singularities in unitarity equations" according to a certain common underlying topological structure[1,4], their singularities along $L^+(G_1)$ should cancel among themselves (independently of the singularities of the terms in the second bracket of (1)) and their *sum* a should thus be analytic. If this is established, (see (i)(ii) below), one shows easily, by using two-particle unitarity, that F_1 = 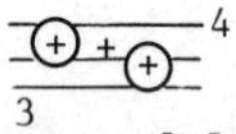 . Q.E.D. The analyticity of a is derived as follows :

(i) The detailed study[3] of the terms in the second bracket of (1) shows that they do *not* contain $\hat{u}_+(P;G_1)$ in their ess-support at P, if $P \in L^+(G_1)$ lies outside a certain subset $\Omega_+ \cup N$ of $L^+(G_1)$. The equality (1) and previous information an a then ensure that a is analytic *outside* this subset.

Property 2.1. (= macrocausality) is *not* sufficient to obtain this result, because one encounters, for a term such as  , the so called u=0 problem. It is a general aspect, in terms of essential supports, of the fact that no information is a priori obtained on a product of boundary values of analytic functions, if they cannot be obtained from *common* directions. It is solved in [6], in the framework of essential support theory, on the basis of a refined version of macrocausality, which gives information on the way rates of exponential fall-off tend to zero when causal directions are approached. A slightly different, somewhat more refined u=0 conjecture is on the other hand proposed in [5] on the basis of general results on "phase-space integrals". It is supported also by recent results[7] on products of holonomic functions with regular singularities.

(ii) Individual terms such as  ,..., do have $\hat{u}_+(P;G_1)$ in their ess-support at points of $\Omega_+ \cup N$. If one cannot show it is absent from the ess-support of a, this singularity will propagate all along $L^+(G_1)$ in the final step, when two-particle unitarity is used, and *no* result on f_1 will be obtained anywhere.

This is solved by making use of a weak "no sprout" assumption on f, which is related to refined macrocausality, and entails a corresponding no sprout property for a itself. Then by Theorem 2 of [3], a , being analytic outside $L^+(G_1)$ and a plus $i\varepsilon$ boundary value along $L^+(G_1)$, cannot be analytic at some points, and singular at other points of $L^+(G_1)$. Hence, it is indeed analytic all along $L^+(G_1)$.

For details, see [3,8]. The method presented above is that of [8], in which triangle graphs are also treated and where some extensions are given.

REFERENCES

[1] - D. IAGOLNITZER, *The S Matrix,* North-Holland (1978) and references therein.

[2] - D. IAGOLNITZER, Phys. Rev. D18, 1275 (1978), Phys. Lett. 76B, 207 (1978).

[3] - D. IAGOLNITZER, H.P. STAPP, Comm. Math. Phys. 57, 1 (1977).

[4] - T. KAWAI, H.P. STAPP in Publ. R.I.M.S. Kyoto Univ. 12, Suppl.,(1977).

[5] - M. KASHIWARA, T. KAWAI, H.P. STAPP, Comm. Math. Phys. 66, 95 (1979).

[6] - D. IAGOLNITZER, Comm. Math. Phys. 63, 49 (1978).

[7] - M. KASHIWARA, T. KAWAI, preprint.

[8] - D. IAGOLNITZER, preprint.

[9] - D. OLIVE, in *Hyperfunctions and Theoretical Physics,* Lecture Notes in Mathematics 449, Springer-Verlag (Heidelberg), 1975, p.133, and references therein.

[10]- H.P. STAPP, in *Structural Analysis of Collision Amplitudes,* ed. R. Balian, D. Iagolnitzer, North-Holland (1976), p.191 and references therein.

[11]- J. BROS, Cours de Lausanne, in preparation.

[12]- H. EPSTEIN, V. GLASER, D. IAGOLNITZER, in preparation.

Quantized Gauge Fields:
Results and Problems

Erhard Seiler
Max-Planck-Institut
für Physik und Astrophysik
München, Fed.Rep.Germany

I. Introduction

At the previous $M \cap \Phi$ conference in Rome a number of talks dealt with the subject of lattice gauge theories[1]. At least for me one main motivation for these investigations was the fact that the lattice provides the only known gauge invariant cutoff for the continuum theory and therefore should be a good starting point for an attempt to construct these continuum gauge theories. Here I want to report some results that have been obtained in that line, in particular on work by D. Brydges, J. Fröhlich and myself[2-5] in which we construct the first (presumably) nontrivial example of a quantized gauge field theory - the abelian Higgs model - obeying all of Wightman's axioms with the possible exception of clustering. I also want to draw your attention to those aspects of that construction that are unsatisfactory in the sense that one cannot hope to generalize them to more interesting and realistic models; I think that we are facing not so much a technical problem as a conceptual one: we do not really understand what phase space localization means for intrinsically nonlinear fields.

II. Some Facts on Lattice Gauge Theories

The lattice approximation "approximates" Euclidean space-time $\mathbb{R}^\nu$ by the simple cubic lattice $\varepsilon\mathbb{Z}^\nu$; a general formalism for lattice gauge theories has been proposed by Wilson[6] and studied by numerous authors. Formalism and many references may be found for instance in[2] ; here I just want to recall the most basic facts:

A lattice gauge field is a map from the links (nearest neighbor pairs) of the lattice into the gauge group G (typically a compact Lie group):

$$(x, y) \longmapsto g_{xy} \in G$$

such that $g_{yx} = g_{xy}^{-1}$. Wilson defined a lattice Yang-Mills action and a corresponding prabability measure for these fields. In our treatment the lattice gauge field arises from a continuum gauge field $A \equiv A_\mu dx_\mu$ (a 1-form with values in the Lie-algebra of G): $g_{xy} = \mathrm{Pexp}\, i \int_x^y A$ where the integral is along the link (x,y) and P indicates path ordering. For the abelian case G = U(1) we will consider in particular gauge fields A_μ that are Gaussian random fields with a covariance $\langle A_\mu(x)\, A_\mu(y)\rangle_t = D^t_{\mu\nu}(x-y)$ where $D^t_{\mu\nu}$ is the Fourier transform of $\left(\delta_{\mu\nu} - \frac{p_\mu p_\nu}{p^2+\mu^2}\right)\frac{1}{p^2+\mu^2}\, e^{-tp_1^2}$; $t > 0$ serves as ultraviolet and $\mu^2 > 0$ as infrared cutoff.

By a general result of Garsia[7] these Gaussian random fields may be assumed to be Hölder continuous of any index $\alpha < \frac{1}{2}$, so $\int A$ makes sense.

A Bose matter field (Higgs field) is a map Φ from the sites of the lattice into some finite dimensional Hilbert space carrying a unitary representation U of G.

The action for the Higgs field coupled minimally to the gauge field is

$$S_{M,\Lambda} = \frac{1}{2}\sum_{\substack{\text{links}\\ \text{in } \Lambda}} \varepsilon^{\nu-2}\,|\Phi(x) - U(g_{xy})\,\Phi(y)|^2 + \frac{1}{2}\sum_{\substack{\text{sites}\\ \text{in } \Lambda}} \varepsilon^{\nu}\left(m^2\,|\Phi(x)|^2 + V(|\Phi(x)|)\right) \tag{1}$$

where V is some polynomial that is bounded from below and grows at least as $|\Phi|^4$, Λ is some reasonable bounded set in $\mathbb{R}^\nu$.

There is an associated joint probability measure $d\mu$ for the Higgs and gauge fields

$$d\mu_{\Lambda,\varepsilon}(A,\Phi) = \frac{1}{Z_{\Lambda,\varepsilon}}\, dm(A)\, e^{-S_{M,\Lambda}} \prod_{x\in\Lambda} d\Phi(x) \tag{2}$$

$$Z_{\Lambda,\varepsilon} = \int \tilde{Z}_{\Lambda,\varepsilon}\, dm(A) \tag{3}$$

$$\tilde{Z}_{\Lambda,\varepsilon}(A) = \int e^{-S_{M,\Lambda}} \prod_{x\in\Lambda} d\Phi(x) \tag{4}$$

dm (A) denotes the probability measure for the gauge fields. The task is now to send ε, t and μ^2 to zero (which requires of course some renormalization), Λ to $\mathbb{R}^\nu$ and verify the Osterwalder-Schrader axioms for the correlation functions of gauge invariant fields in that limit. This has been done in[2-4] (with the exception of clustering).

Among the many results on lattice gauge theories there are two which play an important rôle there (see [2]):

(1) Universal Diamagnetism:

$$|\tilde{Z}(A)| \leq \tilde{Z}(0) \tag{5}$$

This is true for arbitrary group G and even for Fermi matter and expresses a physical property clearly related to the diamagnetism in nonrelativistic quantum mechanics that was discussed by Hunziker[8] at this conference.

(2) Correlation Inequalities:

$$\langle F_1F_2\rangle - \langle F_1\rangle\langle F_2\rangle \geq 0 \tag{6}$$

for $G = U(1)$, if F_1, F_2 are functions in the multiplicative cone generated by $|\Phi|$, cos A, cos (arg Φ). These inequalities have a number of interesting physical consequences[5]; in the constructive program they are used to control the limits $\Lambda \nearrow \mathbb{R}^{\nu}$ and $\mu^2 \searrow 0$.

III. Continuum Limit: External Yang-Mills Fields

In [3] it is shown that the probability measures

$$1/\tilde{Z}_{\Lambda,\varepsilon}(A)\, e^{-S_{M,\Lambda}} \prod_{x\in\Lambda} d\Phi(x)$$

converge weakly as $\varepsilon \searrow 0$ provided $\nu = 2$ and A is Hölder continuous. The limit is independent of the orientation of the lattice; this crucial fact is responsible for the Lorentz invariance of the final field theory.

The proof of convergence is rather involved; it requires a lot of subresults some of which might be of independent interest and not limited to two dimensions, for example we prove L^P convergence of the kernels

$$(-\Delta_A^{\varepsilon} + m^2)^{-1}(x,y)$$

for $p < \frac{\nu}{\nu-2}$ (Δ_A^{ε} is the finite difference covariant Laplacean) and convergence of $\det((\Delta^{\varepsilon})^{-1}\Delta_A^{\varepsilon})$ for $\nu = 2$ (the proof could be extended to $\nu = 3$ and, after renormalization, even $\nu = 4$ with some effort). This last result contains the statement that the sum of the graphs  and  converges even though they diverge individually as $\varepsilon \searrow 0$. A similar result for the Fermi case, i.e. $\det\left[(i\partial^{\varepsilon}+m)^{-1}(i\partial_A^{\varepsilon}+m)\right]$ has been obtained by Challifour and Weingarten[9].

IV. Removal of All Cutoffs for Abelian Higgs$_2$

The results mentioned under (III) are sufficient also to obtain a weak limit for the joint measure $d\mu_{\varepsilon,\Lambda,t}$ as $\varepsilon \searrow 0$ as long as $t > 0$; this follows simply from the diamagnetic bound and the dominated convergence theorem, using the Hölder continuity of the sample fields A_{μ} for $t > 0$.

To remove the t-cutoff one has to do a "stability expansion"[4] involving cancellation of some divergent graphs ( +  ,  + 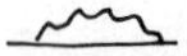) against appropriate counterterms. The stability expansion expresses an "unnormalized expectation" as a telescopic sum:

$$Z_{t_N} \langle F \rangle_{t_N} = \sum_{n=1}^{N} \left(Z_{t_{n+1}} \langle F \rangle_{t_{n+1}} - Z_{t_n} \langle F \rangle_{t_n} \right) \qquad (7)$$

and bounds each term in the sum by an expression of the form

$$\text{const} \prod_{i=1}^{n} t_i^{\delta} \; e^{c(\log t_n)^2} (n!)^r (\log t_n)^n \qquad (8)$$

which insures convergence of (7) as $N \rightarrow \infty$ if the sequence $\{t_n\}$ is chosen appropriately (for instance $t_n = \exp(-n^{\gamma})$, $0 < \gamma < 1$). The proof of the bound (8) uses integration by parts to cancel divergencies, a procedure to estimate large Feynman graphs in terms of small ones and a power counting lemma for convergent graphs.

Finally the limits $\Lambda \nearrow \mathbb{R}^2$ and $\mu^2 \searrow 0$ are taken using the correlation inequalities (6) (which give monotonicity) and upper bounds that are not too hard. The existence of the limit $\mu^2 \searrow 0$ (no infrared divergencies) is a signal of the Higgs mechanism for dynamical mass generation.

In the end one obtains a Wightman field theory for the fields $F_{\mu\nu} = \partial_\mu A_\nu - \partial_\nu A_\mu$ and $:|\Phi|^2:$; expectations of "string" and "loop" observables

$$\bar{\Phi}(x)\, e^{i \int_x^y A}\, \Phi(y) \quad , \quad e^{i \oint A}$$

may also be constructed and fulfill some "axioms" described in [10] and to be discussed in more detail elsewhere.

V. Problems

The application of the techniques of constructive quantum field theory to gauge fields turns out to be rather tricky. I think that the technical difficulties that are coming up here are really the top of an iceberg consisting of unresolved conceptual problems.

The first one is the following: Gauge fields are really intrinsically nonlinear: One can consider them as fields taking values in an infinite dimensional fibre bundle with the gauge group as structure group[11,12] or, alternatively, as maps from a space of loops into the Lie group G. So far constructive methods have not even been successfully applied to the much simpler nonlinear σ-models that share intrinsic nonlinearity with gauge theories. In our treatment we simply force a linear structure on these fields in order to be able to apply the mathematics of random fields. It is perhaps interesting to note that one meets a related difficulty in trying to prove existence theorems for the classical Yang-Mills equations, both in the elliptic and hyperbolic cases. There one would like to dominate some suitable norm by the classical action resp. energy, but of course the concept of norm requires a linear structure.

It is also obvious that in our construction of Higgs_2 we did not pay much attention to the geometrical meaning of gauge fields which should be interpreted as connections in some fibre bundle. Of course we know that (euclidean) quantized fields will be distributions, so one would have to develop a concept of "distribution valued connections".

Furthermore, if we dreamed of ever constructing a theory like QCD_4 we would have to be able to see and exploit "asymptotic freedom" in the construction.

The central problem behind all this is the following: Because of the lack of a linear structure we do not really know what it means to decompose the fields into parts living in different momentum ranges (or belonging to different scale sizes). This momentum or phase space localization plays an essential rôle in constructive field theory both for stability expansions[13] and for mean field (= semiclassical) expansions[14]; this rôle becomes particularly obvious in the approach of Gallavotti and coworkers (cf. [15]) that is inspired by Wilson's renormalization group.

VI. How to Proceed

Of course I don't know the final answer.

What I outlined in this talk is a rather pragmatic approach: It fixes a gauge and proceeds as if the gauge fields were not fundamentally different from other kinds of fields; the penalty seem to be the tremendous technical complications that discourage us from attacking higher dimensional and nonabelian models with these methods. It should also not be forgotten that gauge fixing will run into the Gribov ambiguities[11] that will certainly make things more difficult even though they do not seem to force any formal modifications of the Faddeev-Popov prescription[16].

A more gauge invariant way to decompose fields might try to imitate polyhedral approximations to a curved surface or the finite element method of the numerical treatment of PDE's. A polyhedron is piecewise flat; so one might try to define "slowly varying" gauge fields to be piecewise "as flat as possible", i.e. one might require them to obey the classical equations over the elementary cells of a certain scale size. The matter fields one would require to be piecewise covariantly harmonic (an analogous approach to ordinary scalar fields is equivalent to the standard lattice approximation). Obviously such an approach would require a lot of knowledge of the classical boundary value problems, more than we know at the moment.

A different approach will be discussed by T. Bałaban[17] at this conference; it is inspired by the work of Gallavotti et al. on Φ_3^4 and according to the announced results deserves great attention even though (or because ?) it avoids dealing with the conceptual problems mentioned here.

References

[1] Mathematical Problems in Theor. Phys., Proc. Rome 1977, G. dell'Antonio et al., Springer-Verlag Berlin, Heidelberg, New York 1978.

[2] Brydges, D., Fröhlich, J. and Seiler, E., Construction of Quantized Gauge Fields I, General Results, to appear in Ann. Phys.

[3] ... II, Convergence of the Lattice Approximation, to appear in Comm.Math.Phys.

[4] ... III, The Two-Dimensional Abelian Higgs Model Without Cutoffs, preprint in prep.

[5] ..., Nucl.Phys. B152 (1979) 521.

[6] Wilson, K., Phys.Rev. D10 (1975) 2445.

[7] Garsia, A.M., Proc. of the 6th Berkeley Symp. on Math. Statistics 2, p. 369.

[8] Hunziker, W., these proceedings.

[9] Challifour, J. and Weingarten, D., Indiana preprint 1979.

[10] Fröhlich, J., Lecture given at the Coll. on "Random Fields", Esztergom 1979.

[11] Singer, I.M., Commun. Math.Phys. 60 (1978) 7.

[12] Narasimhan, M.S., Ramadas, T.R., Commun.Math.Phys. 67 (1979) 121.

[13] Glimm, J., Jaffe, A., Fortschr.Phys. 21 (1973) 327.

[14] Glimm, J., Jaffe, A., Spencer, T., Ann.Phys. 101 (1976) 631.

[15] Benfatto, G., these proceedings.

[16] Hirschfeld, P., Nucl.Phys. B157 (1979) 37.

[17] Bałaban, T., these proceedings.

CLUSTERING, CHARGE-SCREENING AND THE MASS-SPECTRUM IN LOCAL QUANTUM FIELD THEORY

Detlev Buchholz and Klaus Fredenhagen
II. Institut für Theoretische Physik, Universität Hamburg
F.R.G.

Many fundamental results in quantum field theory (such as the Haag-Ruelle collision theory, the dispersion relations, the TCP-theorem and the spin-statistics theorem) are based on the assumption that physical states are local excitations of some vacuum state. The conventional way of expressing this hypothesis is to assume that there exist sufficiently many local field operators which generate the physical states from the vacuum [1]. A less technical input is the assumption that physical states cannot be distinguished from the vacuum by measurements made in the causal complement of some sufficiently large, bounded region [2]. It is well known that neither one of these conceptions is adequate in gauge quantum field theory. There one is faced with the problem that the charged physical states cannot be described in terms of local fields [3] or as strictly localized excitations of the vacuum [2], and it is not quite clear what the actual localisation properties of such states are. Therefore the above mentioned canon of fundamental results is not directly applicable in gauge field theory, which is quite annoying, because the only models which seem to be relevant in physics belong to this category.

In view of this situation it is of considerable interest to derive localisation properties of physical states from other assumptions, which are general enough to cover gauge field theories and which clearly display the physical situation under consideration. The adequate framework for such an analysis is the theory of local observables [2]. There the basic object is a net $\mathcal{O} \rightarrow \mathfrak{A}(\mathcal{O})$ of $*$- algebras associated with the bounded regions $\mathcal{O}$ of Minkowski space. The algebras $\mathfrak{A}(\mathcal{O})$ are regarded as being generated by the local (gauge invariant) observables which can be measured in $\mathcal{O}$, and they are therefore subject to the principles of locality and translational covariance. One also considers the algebra $\mathfrak{A}$ which is generated by all local observables.

The physical states of a given charge are represented by vectors in some irreducible representation $(\pi, \mathcal{H})$ of $\mathfrak{A}$, and it is assumed that the translations are unitarily implemented on $\mathcal{H}$ with generators P_μ (energy-momenmentum) satisfying the spectrum condition. This general characterisation of physical states should apply to all field theoretic models of elementary particles [4].

In this note we present the results of an analysis of the localisation properties of physical states for the restricted class of models with a complete particle interpretation and a minimal mass (mass gap). So we disregard models like quantum electrodynamics, but our starting point is general enough to cover massive gauge field theories like the Higgs model and quantum chromodynamics. The above restrictions on the models can be easily expressed in terms of the mass-operator $P_\mu P^\mu$. We assume that to each particle type there corresponds an irreducible representation $(\pi_p, \mathcal{H}_p)$ of $\mathfrak{A}$, called a particle representation, in which $P_\mu P^\mu$ has an isolated eigenvalue $m^2 > 0$. The corresponding eigenvectors represent the one-particle states and we assume that the energy-momentum spectrum of these states has finite multiplicity. The fact that m^2 is isolated guarantees that there are no massless excitations in the theory.

If the algebra of local observables $\mathfrak{A}$ has a particle representation, it turns out that it has also a vacuum representation $(\pi_0, \mathcal{H}_0)$, i.e. an irreducible representation, in which the energy-momentum spectrum contains the isolated point 0; the corresponding (unique) eigenvector Ω is called the vacuum. Moreover it follows that for a dense set of vectors Φ, Ψ in $\mathcal{H}_p$, arbitrary local operators $A \in \mathfrak{A}$, and arbitrary $n \in \mathbb{N}$ [5]

$$\left|(\Phi, \pi_p(\alpha_{\underline{x}}(A))\Psi) - (\Phi, \Psi)\cdot(\Omega, \pi_0(A)\Omega)\right| \leq c_n \cdot (1 + |\underline{x}|)^{-n}. \tag{1}$$

(Here $\alpha_{\underline{x}}$ is the automorphism implementing the spacelike translation $\underline{x}$.) This estimate, which resembles the well known clustering theorem in conventional field theory [6], shows that the vectors in $\mathcal{H}_p$ cannot be distinguished from the vacuum Ω by <u>local</u> measurements in the causal complement of some sufficiently large, bounded region. However, this does not exclude the existence of sequences of observables Q_n in the spacelike complement of an increasing set of bounded regions $\mathcal{O}_n$, $\mathcal{O}_n \to \mathbb{R}^4$, which discriminate these vectors. The limit points Q of these observables commute with all local observables and therefore also with energy and momentum [7]. So they may be interpreted as charges.

Such charges, which can be measured at spacelike infinity, are characteristic for gauge field theories, the standard example being the electric charge. It follows immediately from (1) that in a massive theory all particles carry zero electric charge, because the electric field (which is a local observable) decreases rapidly at infinity [5]. The first general proof of this charge-screening phenomenon in massive abelian gauge field theories was given by Swieca [8]. It is an open and difficult question whether a generalisation of this result holds also for non-abelian gauge field theories.

So far we did not need to specify whether $\mathfrak{A}$ is an algebra of bounded or unbounded operators. We restrict now our attention to bounded observables (which can be obtained by spectral resolution of unbounded oberservables) and assume that the local algebras $\mathfrak{A}(\mathcal{O})$ are C^*-algebras. We must also consider algebras associated with certain unbounded regions of Minkowski space. If $\mathcal{S}$ is any pointed, spacelike cone with arbitrarily small aperture and $\mathcal{S}'$ its causal complement we denote by $\mathfrak{A}(\mathcal{S}')$ the C^*-algebra which is generated by all local algebras $\mathfrak{A}(\mathcal{O})$ with $\mathcal{O} \subset \mathcal{S}'$. Our main result is then that the restrictions of the representations $(\pi_p, \mathcal{H}_p)$ and $(\pi_o, \mathcal{H}_o)$ to the algebra $\mathfrak{A}(\mathcal{S}')$ are unitarily equivalent. That is, there exists a unitary operator V from $\mathcal{H}_p$ onto $\mathcal{H}_o$ such that

$$V \pi_p(A) = \pi_o(A) V \ , \quad A \in \mathfrak{A}(\mathcal{S}'). \tag{2}$$

This relation shows that it is not possible to distinguish the states in $\mathcal{H}_p$ and $\mathcal{H}_o$ by measurements in the unbounded regions $\mathcal{S}'$. As a consequence, the above mentioned charges Q, which can be determined in the causal complement of an arbitrarily large bounded region $\mathcal{O}$, i.e. on the infinite sphere, cannot be approximated by observables on segments of this sphere. Because of the topological differences between a sphere, and a sphere with a point removed one may reasonably argue that in a massive theory Q can only be a topological charge.

In conclusion we want to stress that relation (2) can also be used to establish in our general framework some of the results mentioned at the beginning. The reasoning is very similar to that expounded in [2]. In particular we found that if $\mathfrak{A}$ has particle representations then there exist also representations carrying the composite charges. If these representations are irreducible then the particles can only be Bosons or Fermions, otherwise they must obey para-statistics. Moreover, to each particle there corresponds an antiparticle with the same mass and one can construct incoming and outgoing collision states for these particles. So starting only from assumptions on the mass-spectrum and the causality-requirement for local observables we get a very satisfactory account of qualitative features found in relativistic particle physics. A detailed elaboration on these results will be published elswhere.

References:

1. R.F.Streater and A.S.Wightman: PCT, Spin and statistics and all that (Benjamin, New York, 1964)

2. S.Doplicher, R.Haag, and J.E.Roberts: Commun.Math.Phys. 23 (1971) 199; 35 (1974) 49

3. R.Ferrari, L.E.Picasso and F.Strocchi: Commun.Math.Phys. 35 (1974) 25

4. H.J.Borchers: Commun.Math.Phys. 1, (1965), 281

5. D.Buchholz and K.Fredenhagen: Nucl.Phys. B 154 (1979) 26

6. D.Ruelle: Helv.Physica Acta 35 (1962) 147

7. H.Araki: Progr.Theor.Physics 32 (1964) 844

8. J.A.Swieca: Phys.Rev. D 13 (1976) 312

THE ULTRAVIOLET STABILITY BOUNDS FOR SOME LATTICE σ - MODELS AND LATTICE HIGGS-KIBBLE MODELS

T. Bałaban
Warsaw University, Inst. of Math.
00-901 Warszawa, PKiN IX p.

A.) Let us consider a σ-model with an action

$$S(\Lambda,g_0,\varepsilon) = \sum_{x\in\Lambda}\sum_{\mu=1}^{d} \frac{\varepsilon^{d-4}}{2g_0^2}\left[1-\mathrm{Re}\ \mathrm{tr}(U^*(x+\varepsilon e_\mu)U(x))\right] + E(\Lambda,g_0,\varepsilon)$$

$U(x)\in G$, $G = U(N)$, $SU(N)$ (or any Lie subgroup of $U(N)$), $\Lambda\subset\varepsilon\mathbb{Z}^d$ is a parallelepiped and periodic boundary conditions are assumed. This model is considered for methodological reasons - it is easier than Higgs-Kibble model and the methods were worked out for it at first. It is also much simpler to describe them in this case.

<u>A result</u>: if $E(\Lambda,g_0,\varepsilon)$ is property chosen (it is defined as a sum of normalization constants and a counterterm computed from perturbation expansion), $d = 2,3$, then there exist $E_\pm$ independent of Λ, ε such that

$$\exp(-E_-|\Lambda|) \leqslant Z(\Lambda,g_0,\varepsilon) \leqslant \exp(E_+|\Lambda|)\ ,$$

$$Z(\Lambda,g_0,\varepsilon) = \int\prod_{x\in\Lambda} dU(x)\ \exp(-S(\Lambda,g_0,\varepsilon))\ ,$$

dU is a Haar measure on the group G ($|\Lambda| = \varepsilon^d$ (number of sites in Λ)).

<u>A method</u>: iteration of renormalization group transformations in Kadanoff-Wilson form (3), (4). Domain Λ is divided into blocks $B(y)$, y - a site of a block, each block $B(y)$ contains L^d sites ($L = 2,3,\ldots,$ the estimates are the best for $L = 2$).

Let $\underset{\sim}{V} : \Lambda' \ni y \longrightarrow V(y)\in G$ denote a configuration defined on the set Λ' of sites of the blocks. We define integral transformation T_a :

$$T_a[F] = \int\prod_{x\in\Lambda} dU(x)\ t_a(\underset{\sim}{V},\underset{\sim}{U})F(\underset{\sim}{U})\ ,$$

$$t_a(\underset{\sim}{V},\underset{\sim}{U}) = \prod_{y\in\Lambda'}\exp\Big(-a\frac{\varepsilon^{d-4}}{2g_0^2}\Big|V(y) - L^{-d}\sum_{x\in B(y)}U(x)\Big|^2\Big)\ \cdot$$
$$\cdot\left[\int dV'\exp\Big(-a\frac{\varepsilon^{d-4}}{2g_0^2}\Big|V' - L^{-d}\sum_{x\in B(y)}U(x)\Big|^2\Big)\right]^{-1}$$

where $|A|^2 = \mathrm{tr}(A^*A)$. Of course a normalization condition holds:

$$\int\prod_{y\in\Lambda'} dV(y)\ t_a(\underset{\sim}{V},\underset{\sim}{U}) = 1\ .$$

The use of renormalization group transformation T_a is based on a crucial observation that it provides a natural "slicing" of momentum range, an effective integration in $T_a[\exp(-S)]$ is over momenta $\approx \frac{2\pi}{L\varepsilon} < p < \frac{2\pi}{\varepsilon}$. Thus it implies a kind of "phase space cell expansion" as in (2). This is done by method of Benfatto, Cassandro, Gallavotti et al. (1).

Let us now describe the essential points of the method.

1.) The integration over each U(x), $x \in \Lambda$, is divided into subintegrations over U "small" : $g(\varepsilon)^{-1}|1 - U(x)| \leq O(1)(1+\log g(\varepsilon)^{-1})^p$, and U "large", $g(\varepsilon) = g_0\, \varepsilon^{\frac{4-d}{2}}$ and Λ is next divided into two subdomains $\Lambda = M \cup M^c$, where M is contained in the set of sites x for which the corresponding U(x) are "small".

$T_a[\exp(-S)]$ is estimated by different ways in M^c and M.

2.) In M^c the integral is estimated using positivity properties of the action. It is a difficult part of the proof because the action is changed after each application of renormalization group transformation and the positivity properties are not evident.

3.) In M it is calculated and estimated by cumulant expansion. To do this we expand each U(x), $x \in M$, in power series

$$U(x) = \exp(i\varepsilon g_0 \tfrac{1}{2} \tau_a \phi^a(x)) = 1 + i\varepsilon g_0 \tfrac{1}{2} \tau_a \phi^a(x) + \dots ,$$

the integral is expressed in terms of ϕ -variables and after rescaling $\phi(\varepsilon n) = \varepsilon^{-\frac{d-2}{2}} \phi'(n)$ and other transformations it can be written in a form

$$\text{const.} \int \prod_{n \in M_s} d\phi'(n) \exp\left(-\tfrac{1}{2}\langle \underset{\sim}{\phi}', \rho^M_{eff}\, \underset{\sim}{\phi}'\rangle + \mathcal{P}^M(g(\varepsilon), \underset{\sim}{\phi}', \underset{\sim}{V})\right) \chi(\underset{\sim}{\phi}').$$

Now cumulant expansion method of Benfatto et al. is used for this integral. It is a laborious part of the proof.

4.) The very important ingredients in the considerations of the points 2) and 3) are the properties of the quadratic part $\frac{1}{2}\langle \underset{\sim}{\phi}', \rho_{eff}\, \underset{\sim}{\phi}'\rangle$ of the action (exponential decay of covariances ρ_{eff}^{-1} , $(\underset{\sim}{\rho}^M_{eff})^{-1}$ and so on). In fact a lot of uniform estimates are needed for a sequence $\{\rho^{(k)}_{eff}\}$ obtained by iteration of renormalization group transformations.

The procedure described in the points 1.) - 4.) is iterated until $L^k\varepsilon = O(1)$, and then the required bounds are obtained.

B.) A really interesting model is Higgs-Kibble model on a domain Λ with action

$$S(\Lambda,\lambda,g_o,\varepsilon) = \sum_{x\in\Lambda}\sum_{\mu=1}^{d} \varepsilon^{d-2}\left|U(x,x+\varepsilon e_\mu)\,\phi(x+\varepsilon e_\mu) - \phi(x)\right|^2 + \sum_{x\in\Lambda} \varepsilon^{d}\left[\lambda|\phi(x)|^4 + (m^2 + \delta m^2(\lambda,g_o,\varepsilon))\,|\phi(x)|^2\right] + \sum_{P\subset\Lambda,\, P\text{-plaquettes}} \frac{\varepsilon^{d-4}}{2g_o^2}\left[1 - \mathrm{Re}\;\mathrm{tr}U(P)\right] + E(\Lambda,\lambda,g_o,\varepsilon)$$

and partition function

$$Z(\Lambda,\lambda,g_o,\varepsilon) = \int \prod_{x\in\Lambda} d\,\phi(x) \prod_{\langle x,x'\rangle\subset\Lambda} dU(x,x')\exp(-S(\Lambda,\lambda,g_o,\varepsilon)) ,$$

where $U(x,x')\in G$ (only the cases $G = U(1)$, $SU(2)$ were considered), $\phi(x) \in$ corresponding vector space ($= \mathbb{C}^1, \mathbb{C}^2$).

A result: if $\delta m^2(\lambda,g_o,\varepsilon)$ and $E(\Lambda,\lambda,g_o,\varepsilon)$ are properly chosen, d=2,3, then there exist $E_\pm$ independent of Λ , ε such that

$$\exp(-E_-|\Lambda|) \leqslant Z(\Lambda,\lambda,g_o,\varepsilon) \leqslant \exp(E_+|\Lambda|) .$$

A method: renormalization group transformations in a Wilson form (4) are used again. We define

$$T_a\left[F\right] = \int \prod_{x\in\Lambda} d\,\phi(x) \prod_{\langle x,x'\rangle\subset\Lambda} dU(x,x')\; t_a(\underset{\sim}{\psi}, \underset{\sim}{V} ; \underset{\sim}{\phi}, \underset{\sim}{U})F(\underset{\sim}{\phi}, \underset{\sim}{U})$$

$$t_a(\underset{\sim}{\psi}, \underset{\sim}{V} ; \underset{\sim}{\phi}, \underset{\sim}{U}) = \prod_{y\in\Lambda'} \mathrm{const.}\exp\Big(-a\,\varepsilon^{d-2}\Big|\psi(y) - L^{-d}\sum_{x\in B(y)} U(\Gamma_{y,x})\phi(x)\Big|^2\Big) \cdot \prod_{\langle y,y'\rangle\subset\Lambda'} \exp\Big(-a\,\frac{\varepsilon^{d-4}}{2g_o^2}\Big|V(y,y') - L^{-r(d)}\sum_{\Gamma_{y,y'}} U(\Gamma_{y,y'})\Big|^2\Big) \cdot \left[\int dV' \exp\Big(-a\,\frac{\varepsilon^{d-4}}{2g_o^2}\Big|V' - L^{-r(d)}\sum_{\Gamma_{y,y'}} U(\Gamma_{y,y'})\Big|^2\Big)\right]^{-1} ,$$

where $\Gamma_{y,x}$, $\Gamma_{y,y'}$ are properly chosen contours connecting points y,x or y,y', and for a contour Γ passing through the points $x_1, x_2, \ldots$ $\ldots, x_n$ $U(\Gamma) = U(x_1,x_2)U(x_2,x_3) \ldots U(x_{k-1},x_k)$.

The above transformation T_a transforms a gauge-invariant function into a gauge-invariant one.

Proof of stability bounds proceeds in the same manner as the proof in A.), with one additional step: a gauge fixing procedure is necessary in order to reduce the gauge invariance and to make possible the introduction of "large" and "small" fields and the estimations of points 2.) and 3.).

It is the author's duty to mention here, that for non-abelian case the point 3.) is not completed yet.

References.

(1) G. Benfatto, M. Cassandro, G. Gallavotti, F. Nicolò, E. Olivieri, E. Presutti and E. Scacciatelli "Some Probabilistic Techniques in Field Theory" Commun. Math. Phys. 59, 143-166 (1978).

(2) J. Glimm, A. Jaffe "Positivity of the ϕ_3^4 Hamiltonian" Fortschritte der Physik 21, 327 (1973).

(3) L. Kadanoff "The application of renormalization group techniques to quarks and strings" Rev. Mod. Phys., Vol.49, No.2, 267-296 (1977).

(4) K. Wilson "Quantum Chromodynamics on a Lattice" in "New Developments in Quantum Field Theory and Statistical Mechanics," Cargèse 1976, ed. by M. Lévy and P. Mitter.

MATHEMATICAL ASPECTS OF QUANTUM FIELD THEORY IN CURVED SPACE-TIME

Bernard S. Kay

Blackett Laboratory, Imperial College, London SW7 2BZ, England

Institute for Theoretical Physics, University of Berne, Switzerland

In this short paper, I discuss the question:-

To what extent does the basic structure of flat space-time quantum field theory generalize when we curve the space-time?

The discussion is meant to introduce and complement that of two recent papers [1,2] to which the reader is referred for further details, motivation and references. For completeness, mention should also be made of closely related work by Ashtekar and Magnon [3], Hajicek [4], Isham [5] and Moreno [6]. There has also been interesting work by Dimock [7] and Wald [8] on the existence of an S-matrix for space-times which are Minkowskian off a compact set.

One may divide the basic structure of flat space-time quantum field theory into:

A) What might be called the algebraic structure - i.e. local algebras, states, etc., and

B) What might be called the vacuum structure - i.e. the whole complex of ideas arising from Poincaré (actually time-translation) invariance: The existence of a preferred vacuum state, its associated Hilbert space representation, the existence of a positive Hamiltonian generating the dynamics etc.

I shall illustrate the problem by considering the covariant generalization of the Klein Gordon equation

$$(g^{\mu\nu}\nabla_\mu \partial_\nu + m^2)\phi = 0$$

Such a simple linear system can be understood quite completely. (See especially the work of Segal and co-workers e.g. [9]) and thus one can answer our question quite explicitly. I shall outline this answer in four parts:-

1) The algebraic structure A generalizes for a class of space-times known as globally hyperbolic space-times. These are space-times (M,g) where M has the form $\mathbb{R} \times C$ where $\mathbb{R}$ constitutes a global time - coordinate on M in such a way that the equal-time surfaces $\{t\} \times C$; $t \in \mathbb{R}$ are Cauchy surfaces. For such a space-time, we have Leray's theorem that C_0^∞ Cauchy data on any Cauchy surface define a unique solution $\phi \in C^\infty(M)$ which has compact support on all other Cauchy surfaces. Given a classical linear system with such a well-posed Cauchy problem, it is then straightforward to construct the algebraic aspects of quantization [2]. One equips the space of Cauchy data $C_0^\infty(C) + C_0^\infty(C)$ (or equivalently the space of solutions ϕ) with the natural

symplectic form arising from the conserved current $\phi_1 \overleftrightarrow{\partial}_\mu \phi_2$, and constructs the Weyl algebra over the resulting symplectic space. Thus one obtains a complete theory in the sense of the C^*algebra approach to quantum field theory.

2) However, there is, in general, nothing that could be called a vacuum state and no natural Hilbert space-based theory. I shall illustrate the physical meaning of this with a simple analogy: namely, a time-dependent system in flat-space-time. Take the flat space-time Klein Gordon equation with a variable mass: $(\Box + m^2(t))\phi = 0$ where $m^2(t)$ takes some constant value $m_1{}^2$ before some "early" time t_1; a different constant value $m_2{}^2$ after some "late" time t_2 and varies smoothly in between. Clearly, the "vacuum representation" for mass m_1 will be appropriate for $t < t_1$ and that for a mass m_2 will describe appropriately dressed particles at late times. But these two representations are unitarily inequivalent, and, in general, the dynamics between times t_1 and t_2 will not be implemented. So it makes neither mathematical nor physical sense to have a representation that will do for all times.

3) If one specializes further to (globally hyperbolic) stationary space-times, one has a time-translation invariance and thus one expects to have a vacuum structure B. Standard theorems [10, 11, 12] essentially reduce the problem to the construction of a "first quantized system": One seeks a symplectic map K from $C_0^\infty(C) + C_0^\infty(C)$ with dense range in some complex Hilbert space H which intertwines the classical dynamics with a positive energy unitary group on H: One says that K simultaneously Hilbertizes the classical Cauchy data, and unitarizes the dynamics. The construction of the vacuum then proceeds via second quantization.

We have the:

Theorem [1] If, in a stationary space-time (i) there is a Cauchy surface C (ii) In stationary coordinate systems suitably adapted to C the metric satisfies $g_{00}\sqrt{g^{00}} > \epsilon > 0$. Then a first quantized system exists, has a mass gap and the complexified range of K is a core for the one-particle Hamiltonian.

The proof depends on results on the causal structure of space-times and results relating the Cauchy problem in n + 1 dimensions to essential self-adjointness properties in n dimensions. Conditions (i) and (ii) essentially specify what constitutes "good behaviour at infinity". In fact, they are always satisfied when C is compact.

4) Finally, I shall illustrate further the physical meaning of the conditions of the theorem by giving an example where they fail: Take the so-called "Rindler wedge". That is one of the wedges in two-dimensional Minkowski space-time which is space-like separated from some origin. This is a globally hyperbolic stationary space-time, the time translation symmetry being generated by Lorentz boosts. Now the space-time contains a Cauchy surface, but the positivity condition (ii) is violated - because the

generator of the symmetry gets more and more light-like as we approach the null lines bounding our space-time. This is known to relativists as an ergosurface. The result of this is that, even though we're quantizing a massive quantum field, we end up with a quantum theory with no mass gap [13].

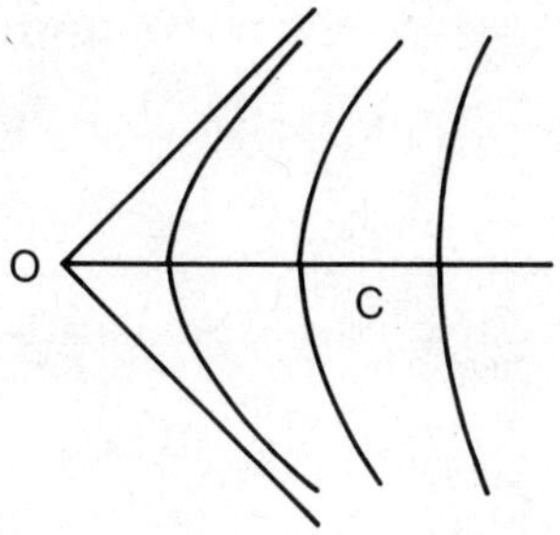

FIG 1 : THE RINDLER WEDGE

REFERENCES

1) B.S. Kay: Linear Spin-Zero Quantum Fields in External Gravitational and Scalar Fields I. A One Particle Structure for the Stationary Case, Commun. math. Phys. 62, 55 (1978)

2) B.S. Kay: Generally Covariant Perturbation Theory (Linear Spin-Zero Quantum Fields in External Gravitational and Scalar Fields II). Commun. math. Phys. (in press).

3) A. Ashtekar, A. Magnon: Proc. Roy. Soc. Lond. A346, 375 (1975).

4) P. Hajicek: In Differential geometric methods in mathematical physics II. Proceedings Bonn, 1977. Ed. K. Bleuler, H.R. Petry, A. Reetz: Berlin: Springer 1978.

5) C.J. Isham: In Bonn Proceedings (as Ref. 4).

6) C. Moreno: J. Math. Phys. 18, 2153 (1977), J. Math. Phys. 19, 92 (1978)

7) J. Dimock: Scalar Quantum Field in an External Gravitational Field, SUNY Buffalo preprint (1979)

8) R.M. Wald: Annals of Physics 118, 490 (1979).

9) I.E. Segal: In: Cargese lectures on theoretical physics, New York: Gordon and Breach 1967.

10) I.E. Segal: Ill. J. Math. 6, 506 (1962)

11) M. Weinless: J. Funct. Anal. 4, 350 (1969).

12) B.S. Kay: A Uniqueness Result in the Segal-Weinless Approach to Linear Bose Fields, J. Math. Phys. (in press).

13) S.A. Fulling: Phys. Rev. D7, 2850 (1973)

CLASSICAL AND QUANTUM ASPECTS OF
THE INVERSE SCATTERING METHOD

V.S. Gerdjikov*), P.P. Kulish**) and M.I. Ivanov*)

*) Joint Institute for Nuclear Research, Dubna, USSR

**) Leningrad Department of the Steklov Mathematical Institute,
Leningrad D-11, USSR

1. The inverse scattering method (ISM) [see the review paper of Ablowitz[1)]] enabled us to solve exactly a number of physically important non-linear evolution equations (NLEE). Starting from an arbitrary linear operator (or an operator bundle), it allows us to construct the exact solutions of a whole class of NLEE; this class is generated by a certain integro-differential operator Λ, closely related to the operator L [1-6)].

For the particular case[5,7,8)]

$$L = \left[i\,\sigma_3 \frac{d}{dx} + \lambda \begin{pmatrix} 0 & q(x,t) \\ p(x,t) & 0 \end{pmatrix} - \lambda^2 \right] \Psi(x,\lambda,t) = 0 \tag{1}$$

$$\Lambda = \frac{i}{2}\,(1 + iI)\,\sigma_3 \frac{d}{dx}\,, \qquad I = \begin{pmatrix} q \\ p \end{pmatrix}(x,t) \int_x^\infty dy\,(p,-q)\,(y,t) \tag{2}$$

it is possible to prove[5)], that the vectors $\begin{pmatrix} q \\ p \end{pmatrix}$ and $(1 + iI)\begin{pmatrix} q_t \\ -p_t \end{pmatrix}$ may be expanded over the eigenfunctions $\Psi^{\pm} = \begin{pmatrix} (\psi_1^{\pm})^2 \\ (\psi_2^{\pm})^2 \end{pmatrix}$ of the operator Λ, $\psi^{\pm}$ being the Jost solutions of L defined by their asymptotics at $x \to \infty$. The coefficients in the expansion for $\begin{pmatrix} q \\ p \end{pmatrix}$ are determined by the scattering data of L, and those for $(1 + iI)\begin{pmatrix} q_t \\ -p_t \end{pmatrix}$ by the t-derivatives of the scattering data of L. Thus one may consider the ISM as a generalized Fourier transform[2,5,9,10)], where instead of $e^{i\lambda^2 x}$ we use $\Psi^{\pm}(x,\lambda)$, and instead of $(1/i)(d/dx)$ there naturally appears the integro-differential operator Λ.

Using the uniqueness and invertibility of the expansions over $\{\Psi^{\pm}\}$, we obtain the most general form of the NLEE related to L [Eq. (1)]:

$$f(\Lambda)\,(1 + iI)\begin{pmatrix} q_t \\ -p_t \end{pmatrix} + g(\Lambda)\begin{pmatrix} q \\ p \end{pmatrix} = 0\,, \tag{3}$$

where f and g are arbitrary functions. For example, if $f(z) = 1/z$, $g(z) = 4iz$, and $q = -p^*$, we obtain the modified non-linear Schrödinger equation[5,8]),
$iq_t + q_{xx} + i(q^2q^*)_x = 0$.

The spectral theory of the operator Λ [5,11]) allows us to derive, in a uniform manner, the main properties of these NLEE[5,9,10]), and in particular their complete integrability[9,12-18]) and the explicit form of the action-angle variables. We also see that to each NLEE we can relate an infinite number (a hierarchy) of symplectic structures[5,9,19]), generated by the symplectic two-forms:

$$\Omega_k = \frac{i}{2} \int_{-\infty}^{\infty} dx\ (\delta p, \delta q) \wedge \Lambda^k (1 + iI) \begin{pmatrix} \delta q \\ -\delta p \end{pmatrix} .$$

Using the important notion of gauge equivalence[20,21]), we obtain that the class of NLEE (3) contains the massive Thirring model and other physically interesting equations.

2. Now let us briefly discuss the quite recent development of the quantum version of the ISM[6,22-26]). This method has been used to quantize the sine-Gordon equation[25]), different versions of the non-linear Schrödinger equation[6,23]), and some spin models on a chain[24]) (for a review, see Ref. 26).

Here we consider the vector non-linear Schrödinger equation:

$$i \frac{\partial q_\alpha}{\partial_t} = - \frac{\partial^2 q_\alpha}{x^2} - 2\varepsilon \left(\sum_{\beta=1}^{n} q_\beta^\dagger q_\beta \right) q_\alpha(x,t) , \qquad \varepsilon = \pm 1 , \tag{4}$$

in the case $n = 2$; the generalization to $n > 2$ is trivial. The operator-valued functions $q_\alpha(x,t)$ satisfy the following commutation relations:

$$\left[q_\alpha(x) ,\ q_\beta^\dagger(y) \right] = \delta_{\alpha\beta}\, \delta(x-y) .$$

As usual, when we start from a given classical theory to construct the appropriate quantum one, there arises the infinitesimal transition (monodromy) matrix $\tau_n(\lambda)$; in our case[6]):

$$\tau_n(\lambda) = \begin{pmatrix} \left(1 - \frac{i\lambda}{2}\Delta\right) e & i \int_{x_{n-1}}^{x_n} q_\alpha(y)\, dy \\ -i\varepsilon \int_{x_{n-1}}^{x_n} dy\ q_\alpha^\dagger(y) & 1 + \frac{i\lambda}{2}\Delta \end{pmatrix} , \qquad e = \begin{pmatrix} 1 & 0 \\ 0 & 1 \end{pmatrix} , \tag{5}$$

where the points x_n divide the interval (-L,L) into $K = 2L/\Delta$ equal parts.

The main relation in the quantum ISM is

$$R(\lambda-\mu)\; T_L(\lambda) \otimes T_L(\mu) = T_L(\mu) \otimes T_L(\lambda)\; R(\lambda-\mu) \;, \tag{6}$$

where $R(\lambda-\mu)$ is a 9×9 (c-number!) matrix

$$R(\lambda) = \begin{pmatrix} r(\lambda) & 0 & 0 & 0 \\ 0 & be & ce & 0 \\ 0 & ce & be & 0 \\ 0 & 0 & 0 & 1 \end{pmatrix}, \quad r(\lambda) = \begin{pmatrix} 1 & 0 & 0 & 0 \\ 0 & c & b & 0 \\ 0 & b & c & 0 \\ 0 & 0 & 0 & 1 \end{pmatrix} \tag{7}$$

$$T_L(\lambda) = \prod_{\overleftarrow{n=1}}^{K} \tau_n(\lambda) = \begin{pmatrix} A_{\alpha\beta} & B_\beta \\ C_\alpha & D \end{pmatrix}, \quad b(\lambda) = \frac{i}{\lambda + i}, \quad b + c = 1 \,.$$

From (5) we obtain the commutation relations between the matrix elements of $T_L(\lambda)$; in particular we obtain that $[\mathrm{tr}\, T_L(\lambda), \mathrm{tr}\, T_L(\mu)] = 0$, i.e. $\mathrm{tr}\, T_L(\lambda)$, is the generating functional of commuting quantum integrals of motion.

Now we construct the states in the Hilbert space $\mathcal{H} = \bigotimes_{n=1}^{K} \mathcal{H}_f^{(n)}$, on which $\mathrm{tr}\, T_L(\lambda)$ is diagonal. First we note that the vacuum $|0\rangle \in \mathcal{H}$, for the operators $q_\alpha(x)$, satisfies

$$\left(B_\beta(\lambda)|0\rangle = 0\right) \mathcal{A}_{\alpha\beta}(\lambda)|0\rangle = \delta_{\alpha\beta}\, e^{-i\lambda L}\, |0\rangle \,, \quad D(\lambda)|0\rangle = e^{i\lambda L}\, |0\rangle \;.$$

The eigenstates of $\mathrm{tr}\, T_L(\lambda)$ in $\mathcal{H}$ are generated from the vacuum $|0\rangle$ by $C_\alpha(\lambda)$. Using the commutation relations (6) we see[6] that these states are labelled by two sets of quasi-momenta $\lambda_1, \ldots, \lambda_N$, $\Lambda_1, \ldots, \Lambda_M$, satisfying the generalized Bethe ansatz[27]:

$$\prod_{j=1}^{N} \frac{\Lambda_k - \lambda_j}{\Lambda_k - \lambda_j + i} = \prod_{\substack{j=1 \\ j\neq k}}^{M} \frac{\Lambda_k - \Lambda_j - i}{\Lambda_k - \Lambda_j + i}$$

$$e^{2iL\lambda_j} \prod_{\substack{k=1 \\ k\neq j}}^{N} \frac{\lambda_j - \lambda_k - i}{\lambda_j - \lambda_k + i} = \prod_{k=1}^{M} \frac{\Lambda_k - \lambda_j + i}{\Lambda_k - \lambda_j} \;.$$

The corresponding eigenvalue of tr $T_L(\lambda)$ equals

$$e^{-i\lambda L}\prod_{i=1}^{N}\frac{1}{c(\lambda-\lambda_i)}\left[\prod_{i=1}^{N}c(\lambda-\lambda_i)\prod_{k=1}^{M}\frac{1}{c(\lambda-\lambda_k)}+\prod_{k=1}^{M}\frac{1}{c(\Lambda_k-\lambda)}\right]+e^{i\lambda L}\prod_{i=1}^{N}\frac{1}{c(\lambda_i-\lambda)},$$

where $c(\lambda-\mu) = (\lambda - \mu)/(\lambda - \mu + i)$.

*

* *

REFERENCES

1) M.J. Ablowitz, Studies in Appl. Math. 58, No. 1 (1978).

2) M.J. Ablowitz, D.J. Kaup, A.C. Newell and H. Segur, Studies in Appl. Math. 53, 249 (1974).

3) F. Calogero and A. Degasperis, Nuovo Cimento 32B, 201 (1976).

4) A.C. Newell, *in* Non-linear evolution equations solvable by the spectral transform (ed. F. Calogero), Research Notes in Mathematics 26, 127 (1978).

5) V.S. Gerdjikov, M.I. Ivanov and P.P. Kulish, JINR preprint E2-12590 (1979).

6) P.P. Kulish, LOMI preprint R-3-79 (1979), Leningrad.

7) E.A. Kuznetzov and A.V. Mikhailov, Teor. Mat. Fiz. 30, 303 (1977).

8) D.J. Kaup and A.C. Newell, J. Math. Phys. 19, 798 (1978).

9) V.S. Gerdjikov and E.Kh. Khristov, JINR preprint E2-12742 (1979).

10) D.J. Kaup and A.C. Newell, Advances in Math. 31, 67 (1979).

11) V.S. Gerdjikov and E.Kh. Khristov, JINR preprints E5-11668 and E2-12731 (1979).

12) L.D. Faddeev, *in* Solitons (ed. R. Bullough) (Springer, 1977).

13) H. Flashka and A.C. Newell, Integrable systems of non-linear evolution equations, preprint (1974).

14) M. Adler, Inventiones Math. 50, 219 (1979).

15) B. Kostant, Inventiones Math. 48, 101 (1978).

16) A.G. Reiman and M. Semenov Tjan'-Shan'skii, Funktzional'nii analiz i ego prilozhenia (in press).

17) Yu.I. Manin and D.R. Lebedev, Preprint ITEP-155 (1978).

18) I.M. Gel'fand and L.A. Dikii, Institute of Applied Math., Preprint 136/78, Moscow (1978).

19) P.P. Kulish and A.G. Reiman, Zapiski nauchnikh seminarov LOMI 77, 134 (1978).

20) V.E. Zakharov and L.A. Takhtadjan, Teor. Mat. Fiz. 38, 26 (1979).

21) V.E. Zakharov and A.V. Mikhailov, JETP 74, 1953 (1978).

22) E.K. Skljanin and L.D. Faddeev, Dokl. Akad. Nauk USSR 243, 1430 (1978).

23) E.K. Skljanin, Dokl. Akad. Nauk USSR 244, 1338 (1979).

24) P.P. Kulish and E.K. Skljanin, Phys. Lett. 70A, 461 (1979).

25) E.K. Skljanin, L.A. Takhtadjan and L.D. Faddeev, Teor. Mat. Fiz. 40, No. 1 (1979).

26) L.D. Faddeev, LOMI preprint (1979).

27) C.N. Yang, Phys. Rev. Lett. 19, 1312 (1967).

MAGNETIC MONOPOLES AND NON-ABELIAN GAUGE THEORIES

D.I. Olive

Blackett Laboratory
Imperial College, London

Magnetic monopoles arise naturally as soliton solutions in spontaneously broken gauge theories in D = 3+1 space time. The possible values of the magnetic charge are restricted by a generalized Dirac quantization condition, and still further by stability requirements. Key questions concern the quantum field theory of these objects and the possibility of a dual symmetry between electric and magnetic charges. These could have a bearing on quark confinement. In one class of model a universal mass formula hints at a rôle played by hidden extra dimensions of space time.

MAGNETIC MONOPOLES AND NON-ABELIAN GAUGE THEORIES

Experiments in particle physics have revealed that apparently disparate phenomena can all be explained in terms of gauge theories. Mathematically these are principal bundles over a base manifold which, in physics, is Minkowski space time with three space and one time dimension (D=3+1). This base leads to special features, the dimensionlessness of the gauge coupling constant and the possible existence of isolated sources of magnetic field, namely magnetic monopoles. These features could bear upon the still unsolved problems of particle physics, namely quark confinement and the unification of all the particle interactions.

There has therefore been a renewed interest in the theoretical study of magnetic monopoles. The fact that we can pose unanswered questions concerning them certainly means we have a lot to learn about gauge theories. In any case recent progress in understanding monopoles has synthesized many hitherto unrelated ideas, some of them important in other branches of theoretical physics; solid state physics, statistical physics, general relativity etc. Thus we may well be on the way to seeing a grander mathematical structure, underlying much diverse physics.

In this talk I shall attempt to explain some of this.

Gauge theories are specified by the relevant Lie group and the simplest choice, namely U(1), governs Maxwell theory as we shall see.

Maxwell's equations in vacuo exhibit a peculiar duality invariance with respect to the substitution [1]

$$\underline{E} + i\underline{B} \rightarrow e^{i\theta}\,(\underline{E} + i\underline{B}) \qquad (1)$$

rotating between the electric and magnetic fields (if D = 3+1!). To maintain this symmetry in the presence of matter one must introduce magnetic as well as electric charges. Classical equations of motion involving $\underline{E}$, $\underline{B}$ and the particle coordinates can easily be written down with the desired symmetry (for a review giving more details of this and many other points see [2]. But if a particle carrying electronic charge q is to be described by a quantum mechanical wave function there is a problem since the minimal coupling principle $\underline{p} \rightarrow \underline{p} - q\underline{A}$ introduces the gauge potential $\underline{A}$ and so the operator $-i\hbar\nabla - q\underline{A} = -i\hbar(\nabla - iqA/\hbar)$ must enter the Schrodinger equation. Now $\underline{B} = \nabla \wedge \underline{A}$, and so $\oint d\underline{S}.\underline{B} = 0$ by Stokes theorem if $\underline{A}$ is defined globally, and so magnetic charge cannot exist.

But $\underline{A}$ is not uniquely defined: we have the gauge ambiguity $A^\mu \rightarrow A^\mu + \partial^\mu \chi$ and $\psi(x) \rightarrow \exp(iq\chi(x)/\hbar)\psi(x)$, where $\chi(x)$ is usually a single valued function of position. Notice that the phase factor $\exp(iq\chi/\hbar)$ lies in a U(1). This is the gauge group.

Surround a magnetic charge g by a sphere split into northern and southern hemispheres N and S, with gauge potentials A^N and A^S defined in each hemisphere but not indefinitely extendable into the opposite hemisphere. For consistency we demand the existence of a gauge transformation connecting A^N and A^S on the equator where the hemispheres overlap. So $\underline{A}^N = \underline{A}^S + \nabla\chi$ and $\psi^N(x) = \exp(iq\chi(x)/\hbar)\psi^S(x)$.

Since ψ^N and ψ^S are each single valued, and we suppose, non zero, exp $(iq\chi(x)/\hbar)$ must be single valued as the equator is encircled (but not necessarily χ itself). The change in phase factor (which should be 1) is formally $\exp(iq\oint dl\nabla\chi/\hbar)$ = exp. $(iq\oint \underline{dl}\,(\underline{A}^N-\underline{A}^S)/\hbar = \exp(iq\oint \underline{dS}.\underline{B}(x)/\hbar)$, using Stokes theorem in the appropriate hemispheres for $\underline{A}^N$ and $\underline{A}^S$. The surface integral $\oint \underline{dS}\ \underline{B}(x)$ extends over the whole sphere and therefore equals the magnetic charge g enclosed. So

$$\exp(i\, q\, g/\hbar) = 1 \tag{2}$$

This is Dirac's argument [3], as refined by Wu and Yang [4]. It is close to the mathematical treatment in fibre bundle theory.

It follows that for any electron charge q and magnetic charge g in nature

$$qg = 2n\pi\hbar,\ n = 0,\ \pm1,\ \pm2\ \ldots \tag{3}$$

and that if there is a $g \neq 0$, $q = nq_0$ as observed in nature.

Notice also that as the equator is encircled, the phase factor exp. $(i\, q\chi/\hbar)$ provides a map from the circle S' (the equator) to the circle S' (the gauge group U(1)), whose "winding number" is $g/2q\hbar$, i.e. the n in equation (3). Thus g is a "topological quantum number", and we shall see further examples yet it is quantized in the same way as the electrical charge q which could not be said to be topological.

Thus there are two ways of looking at a similar object, a quantized charge. This is connected to the fact that although quantum field theory is believed to provide the correct theoretical framework for elementary particle physics there are two distinct ways of associating particles with fields:

1) As a quantum excitation; particles are created by the fourier components of <u>any</u> field satisfying canonical commutation relations

$$\left[\phi\,(x,t),\ \pi(y,t)\right] = i\hbar\delta(x-y) \tag{4}$$

irrespective of the field equations.

2) As solitons; particles with structure occur as smooth localized finite energy classical solutions to special non-linear equations - loosely speaking as "waves at rest".

The simplest example of a soliton occurs in Sine Gordon theory in one space and one time dimension, (D=1+1), with Lagrangian density $\mathcal{L} = \frac{1}{2}\,(\dot\phi^2-\phi'^2\ -V(\phi))$ and self interaction $V(\phi) = \frac{\alpha}{\beta}\,(1-\cos\beta\phi)$. $\mathcal{L}$ has a symmetry $\phi\to\phi+2\pi/\beta$ which generates the group of integers. The vacuum corresponds to $\phi = 0,\ \pm2\pi/\beta,\ \pm4\pi/\beta,\ \ldots$ etc, the minima of $V(\phi)$, which are equivalent because of the symmetry. The soliton solution arises when ϕ changes by $2\pi/\beta$ between adjacent minima as x runs from $-\infty$ to ∞. To see that there is such a solution consider the energy:

$$H = \tfrac{1}{2}\int_{-\infty}^{\infty} dx\, [\dot{\phi}^2 + \phi'^2 + V(\phi)] = \tfrac{1}{2}\int dx\, [\dot{\phi}^2 + \left(\phi' \mp \sqrt{V(\phi)}\right)^2] \pm \int dx\, \phi'\sqrt{V(\phi)}$$

$$\geq \left| \int d\phi \sqrt{V(\phi)} \right| \qquad (5)$$

This lower bound[5] on H is saturated exactly if $\dot{\phi} = 0$ which means the soliton is stationary, and if $\phi' = \pm\sqrt{V(\phi)}$, for soliton/antisoliton. So the soliton satisfies first rather than second order equations and its mass is given by a simple integral. Time dependent multi-soliton solutions also exist. All these solutions possess a "quantum number", proportional to $\phi(x=\infty) - \phi(x = -\infty)$, related to the topological way the scalar field ϕ realizes its boundary condition at large distances. This charge corresponds to the current $\varepsilon_{\mu\nu}\partial^\nu\phi$ which is conserved independently of the equations of motion.

The most remarkable feature of the Sine Gordon model is that it is possible to construct a quantum field operator[6] ψ (x) which creates the soliton in the sense (1) above, and satisfies the equations of motion of the massive Thirring model[7]. This model has a phase convariance $\psi \to \exp.(iq\chi)\ \psi$ whose Noether current $\bar{\psi}\gamma^\mu\psi$ coincides up to a factor with the $\varepsilon^{\mu\nu}\partial_\nu\phi$ already mentioned.

At this stage one may say that two space time dimensions are very special, but there are also other special cases where "miracles" may happen, amongst them our Minkowski space time D = 3 +1, when the aforementioned duality symmetry[1] can occur, and in which there is the special twistor structure[8] relating it to CP3.

When one tries to generalize the concept of a soliton with a topological quantum number supplied by non trivial boundary conditions upon a scalar field at large distances, to D = 3 + 1 one is lead to consider a very interesting type of theory namely a Yang-Mills gauge theory[9] based on a "big" gauge group G (the symmetry of $\mathcal{L}$) together with a Higgs field $\phi(x)$[10]. Like the scalar field above, this has a set of non vanishing vacuum expectation values, but now this set is a continuous manifold, the coset space G/H of points related by the action of G on ϕ. H is the subgroup of G which leaves invariant a $\phi(x)$ of the vacuum set and is observed physically as the exact gauge symmetry group whose fields persist at long range.

The reason for all this is that without the gauge field one of the positive contributions to the energy is $\tfrac{1}{2}\int d^3x\,(\nabla\phi)^2$, which, since $\nabla\phi = O(1/x)$, diverges unless ϕ is constant at large distances, which is uninteresting. If we can gauge G, $\nabla_i\phi$ is replaced by the covariant derivative $\mathfrak{D}_i\phi = \nabla_i\phi + ie\, W_i\phi$. W_i is the Lie algebra valued gauge or connection field for G. Now the energy integral can converge if W_i has a term inversely proportional to e x, and there is a delicate cancellation between the two terms in $\mathfrak{D}_i\phi$. Then the space-space (or magnetic) component of the gauge field strength (or curvature) $B_i = \tfrac{1}{2}\varepsilon_{ijk}F_{jk}$, is inversely proportional to ex^2. This inverse square law magnetic field suggests that the soliton is a magnetic monopole with magnetic charge inversely proportional to e in agreement with[3].

't Hooft and Polyakov were the first to realize this[11]. They considered the simplest example in which G = SO(3) and the Higgs field formed an isotriplet

(ϕ_1, ϕ_2, ϕ_3) so that the vacuum expectation values m_0, i.e.the minima of the SO(3) invariant self interaction, formed a sphere S^2. The special point about spacetime D = 3 + 1 is that the set of points at large distances also form a sphere S^2 so that the boundary condition on the Higgs field provides a map from one S^2 (far away) to another S^2 (m_0 in isospace), if we have finite energy. This map can again be classified by an integer winding number labelling homotopic equivalence classes [12]. By theorems on exact homotopy sequences [13] this is

$$\begin{array}{ccccc} \pi_2(m_0) = \pi_2(S^2) = \pi_2 & (SO(3)/U(1)) & \cong & \pi_1(U(1)) \\ \| & & & \| \\ Z & & & Z \end{array}$$

showing the relation to the previous topological way of classifying monopoles.

For a general (simply connected) G the same theorem yields

$$\pi_2(m_0) \equiv \pi_2(G/H) \cong \pi_1(H) \tag{6}$$

Physically the left hand side is defined by the boundary conditions on the scalar Higgs field while the right hand side concerns the long range magnetic fields [14].

The soliton monopoles look like Dirac monopoles at long distances but now have a detailed finite internal structure and mass, and so provide an interesting model for elementary particles whose structure usually results from renormalized quantum corrections.

The mass of the soliton monopole can be estimated, rather as in the bound (5), whenever the Higgs field ϕ lies in the adjoint representation of G, irrespective of G. Define the Lie algebra generator

$$Q = e\hbar \sum_{i=1}^{\dim G} \phi_i T_i / a \tag{7}$$

In the vacuum ϕ is covariantly constant

$$\mathcal{D}^\alpha_\phi \equiv (\nabla^\alpha - i e W_i^\alpha T_i)\phi = 0$$

and so has constant length, a say. In a representation corresponding to a field ψ, Q has constant eigenvalues q which can be thought of as the Q charges carried by ψ. The generators of the exact symmetry group H are precisely those generators of G which commute with Q. It follows that Q generates an invariant U(1) subgroup of H, and that H has the structure [15]

$$H = \text{" } U(1) \times K \text{ "}$$

at least locally near the unit element. So Q is K singlet and so if G = SU(4), K = SU(3) this would be a good model of QCD + QED (quantum chromodynamics and Maxwell theory). Other currently fashionable gauge theories with a major Higgs field in the adjoint representation are the "grand unified theories" of strong and electroweak interactions, based on G = SU(5) [16], SO(10) [17] or E6 [18]. Then Q is the weak hypercharge.

The Higgs field kinetic energy $\frac{1}{2}(\mathcal{D}^{\mu}\phi)^2$ has a term quadratic in the gauge particles which can be interpreted as a mass term[10] for the gauge fields. The mass matrix M is then, by equation (7).

$$(M^2)_{\alpha\beta} = e^2\hbar^2\ \phi_o\ T^{\alpha}\ T^{\beta}\ \ \phi_o = a^2(Q^2)_{\alpha\beta}$$

remembering that when ϕ is in the adjoint representation, the T's are totally antisymmetric structure constants. Diagonalizing and equating eigenvalues, we have for gauge particle masses[9, 10]

$$M = a\ |q|$$

To evaluate the mass of the monopole soliton, consider the classical Hamiltonian

$$H = \tfrac{1}{2}\int d^3x\ [\mathcal{E}^2 + \mathcal{B}^2 + (\mathcal{D}^i\phi)^2 + (\mathcal{D}^o\phi)^2 + V(\phi)\ ]$$

writing $(\mathcal{B}^i)^2 + (\mathcal{D}^i\phi)^2 = (\mathcal{B}^i \mp \mathcal{D}^i\phi)^2 \pm 2\mathcal{D}^i\phi\mathcal{B}_i$ we see

$$H \geqslant |\int d^3x\, \mathcal{D}^i\phi\mathcal{B}^i| = |\int d^3x\ \nabla^i(\phi\mathcal{B}^i)\ | = |\int dS^i\ \phi\mathcal{B}^i\ |$$

using the Bianchi identity $\mathcal{D}_i\mathcal{B}_i = 0$. Now far away, $\mathcal{D}^i\phi = 0$ and $\phi\,\mathcal{B}^i/a$ is the magnetic field associated with the invariant U(1) subgroup. So $H \geqslant a|g|$ where g is the U(1) magnetic charge of the monopole soliton[5,21]. The bound is saturated if the terms dropped vanish. $V(\phi) = 0$ means the self interaction coupling constant must vanish, even though ϕ "remembers" its length a in vacua.[22] This leaves $\mathcal{E}^i = \mathcal{D}^i\ \Phi = 0, \mathcal{B}^i = \pm\mathcal{D}^i\phi$, which mean that the soliton is at rest and satisfies first order "self duality" equations. Similar arguments applied to a dyon soliton[23] carrying both U(1) electric charge q and magnetic charge g yields.[5,21]

$$M = a\sqrt{q^2 + g^2} \tag{8}$$

When $V(\phi) = 0$ the Higgs particles are massless (and chargeless) so the remarkable conclusion is that all the particles of the theory, whether gauge, Higgs or solitonic satisfy the universal mass formula (8), irrespective of their origin, or the gauge group G [19,20]. The formula (8) manifests the electromagnetic dual symmetry (1) with respect to $q + ig \rightarrow e^{i\theta}(q + ig)$. 't Hooft [24] has found another manifestation of this symmetry (1).

When G = SU(2), so H = U(1), the 'self dual' equations can be solved analytically [22]

$$\Phi_a(r) = \frac{r^a}{e\,r^2}\ (\xi\ \coth\xi - 1) \qquad \xi = aer$$

$$W^i{}_a(r) = -\varepsilon_{aij}\ \frac{rj}{er^2}(1 - \xi/\sinh\xi)$$

This "Bogomolny-Prasad-Sommerfield" monopole is not a Dirac monopole since the Higgs field contributes equally with the magnetic field to the long range energy density. This also has the effect [25] that forces between like charged BPS monopoles vanish while forces between unlike ones are doubled, at least to an accuracy of $(\text{separation})^{-N}$.

This raises the question as to whether like-like forces vanish altogether, or equivalently whether there exist time independent solutions describing N(>1) like BPS monopoles (as happens [26] for vortices in superconductors at the transition point between types I and II). Nahm [27] and Weinberg [28] have shown, using index theorems, that if such a solution exists it lies on a manifold of 4N dimensions. 3N are spatial displacements but the physical meaning of the remaining N dimension is as yet unclear.

The charge quantization condition satisfied by two dyons with charges (q_i, g_i), $i=1,2$ is $q_1g_2-q_2g_1 = 2\pi hn_{12}$ where n_{12} is an integer [29]. If we accept the existence of an electron (q,o) this condition implies that dyons with $g=\frac{2\pi}{q}$ have $q_i=q(n+\delta)$ where n is an integer and δ is undetermined, but the same for all dyons with the same g. CP invariance implies that if (q,g) exists so does (-q,g). This then implies $\delta = 0$ or $\frac{1}{2}$ so that we see that the departure of δ from these values is a measure of CP non invariance. Another measure of this is the θ angle of instanton theory [30]. The action A of a CP conserving non-Abelian gauge theory is replaced by $A+\hbar\,\theta t$ where t is the topological (Pontryagin) integer, associated with the gauge field. Witten [31] has shown that these two parameters δ and θ, measuring CP violation are related, $\delta = -\theta/2\pi$ in such a way that their periods coincide.

Now let us return to the mass formula (8) and notice that if we let aq and ag constitute two extra components of momentum so that we have a six momentum $\mathbb{P}^\mu = (p^0, p^1, p^2, p^3, aq, ag)$ then (8) states that for a particle state $\mathbb{P}^2 = 0$. Where do these extra dimensions come from? First we shall justify $p^5 = aq$ by noting that since the Higgs field lies in the adjoint representation, like the gauge potential W^μ, it can be thought of as a fifth component of the latter: [32] $\phi = W^5$, providing we agree that all fields are x^5 independent; $\partial_5 = 0$. Since $\mathfrak{D}^\mu\phi = F^{\mu 5}$ the gauge + Higgs Lagrangian density in D = 3+ 1 forms a gauge Lagrangian density in D = 4+1.

Now the ordinary, gauge invariant, momentum operator P^u must generate covariant displacements

$$[P_\mu, \psi(x)] = i\hbar\mathfrak{D}_\mu\psi \equiv i\hbar\,(\partial_\mu - i\,e\,W_\mu^i\,T_i)\,\psi$$

Hence if the five dimensional theory is $g(x^5)$ invariant, while $\partial_5 = 0$

$$[P_5, \psi(x)] = e\hbar W_5^i\,T^i\psi = aQ\psi$$

by (7). This shows that P_5 acts as the charge operator [20]. If ψ couples minimally with no explicit mass term, the corresponding particle must be five dimensional light-like, so satisfying (8).

At the price of an x_5 dependent gauge transformation we can gauge the Higgs field ϕ to zero sincethis is an "axial gauge" via[20].

$$g(x^5, x^\mu) = T \exp i e \int_0^{x_5} W_i^{\,5} T_i \, dx_5 = \exp i(a\, x_5\, Q/\hbar)$$

Gauge non invariant fields now depend on x^5, periodically if Q has quantized eigenvalues, and so now the theory is very reminiscent of the Kaluza-Klein theory[33].

The clue to understanding $p_6 = ag$ is the question as to how to quantize the theory in D = 3+ 1. There is an ambiguity which we shall seek to fix by demanding as much symmetry as possible, by fitting ϕ and W^μ, which carry different D = 3 +1 spins into a single symmetry multiplet [34]. The only known symmetry generator with spin is a supercharge, in fact an anticommuting generator of SO(2) or SO(4) extended supersymmetry. These symmetric theories are nice in D = 5 + 1 or 9 + 1 spacetime dimensions respectively and consist of a fermion in the adjoint representation of the gauge group G coupling minimally. In D = 5 + 1 the fermion is Weyl and in 9 + 1 Weyl and Majorana [35]. In D = 5 + 1 the supercharge Q_α carries an 8 component Dirac index and satisfies the fundamental anticommutator

$$\{Q_\alpha\,,\, Q_\beta^\dagger\} = [\,(1 + \Gamma_7)\;\; \Gamma\cdot\mathbb{P}\Gamma^0\,]_{\alpha\beta} \tag{9}$$

Taking $\alpha=\beta$ and summing, we find $P^0 = H = \frac{1}{8}\{Q_\alpha, Q_\alpha^\dagger\} \geqslant 0$ so that the Hamiltonian H is guaranteed to be positive, and to vanish only for supersymmetric states with $Q_\alpha = 0$ [36]. In particular if the theory is quantized in a supersymmetric way, there are <u>no</u> quantum corrections to the energy of the vacuum. Now equation (9) also implies

$$\mathbb{P}^0\, \mathbb{P}^2 = \frac{1}{8}\{(\Gamma\cdot\mathbb{P}Q)_\alpha\; (\Gamma\cdot\mathbb{P}\,Q)_\alpha^\dagger\} \geqslant 0$$

so $\mathbb{P}^2 \geqslant 0$ and vanishes if and only if $\Gamma\cdot\mathbb{P}Q = 0$. Now when we put $\partial_5 = \partial_6 = 0$ we find [37] that indeed $\mathbb{P}_5 = aq$, $\mathbb{P}_6 = ag$ and that $\Gamma\cdot\mathbb{P}Q$ does vanish, at least classically for the solitons. Hence if (9) and the vanishing of the supersymmetry condition $\Gamma\mathbb{P}Q$ holds in the quantum theory, so does $\mathbb{P}^2 = 0$ and hence the mass formula (8). The analogous statements for the supersymmetric Sine-gordon theory in D = 1+1 have been checked partially, [38]although there is some disagreement as yet [39].

The charges q and g we have discussed so far are Abelian, but non- Abelian gauge groups H seem to occur as exact symmetries in nature. Let H be simple and compact with generator T_i satisfying

$$[T_i, T_j] = i f_{ijk}\, T_k$$

where f_{ijk} are totally antisymmetric structure constants. The field strength tensor $F_{\alpha\beta} = \sum_{i=1}^{\dim H} F^i_{\alpha\beta} T_i$ can be divided into "electric" components $\mathcal{E}_i = F_{io}$ (i=1,2,3) and "magnetic" components $\mathcal{B}_i = \frac{1}{2}\varepsilon_{ijk}F_{jk}$. A reasonable way of recognizing generalized electric and magnetic charges is to look at the components of E_i and B_i at large distances and seek a generalized inverse square law form

$$\begin{rcases} \mathcal{E}_i(\underline{r}) = (\frac{1}{4\pi}) \quad (r_i/r^3)\ Q(\underline{r})\,;\ \mathcal{D}_i Q = 0 \\ \mathcal{B}_i(\underline{r}) = (\frac{1}{4\pi}) \quad (r_i/r^3)\ G(\underline{r})\,;\ \mathcal{D}_i G = 0 \end{rcases} r \text{ large}$$

This means we are considering configurations with no "radiation". This is important since the distinctive feature of a non Abelian theory is that gauge particles do carry charge which can therefore be radiated. The Q's and G's are gauge covariant objects and their equivalence classes modulo gauge transformations will yield the macroscopic structure of these configurations.

Let us examine Q in quantum field theory, assuming for the time being that G = 0, so that there exists a non-singular gauge transformation putting $W_i = 0$ (for large r) so that $Q = \text{const} = \int dS\,\mathcal{E} = \int d^3x \nabla\mathcal{E} \approx \int d^3x\ j^0 = \bar{Q}$, where j_μ is the non-covariant current weakly equal to $\partial^\nu F^{\nu\mu}$ by the Gauss law constraint equations, and including contributions of the form $i\ e\ \Sigma_i \Pi\ D\,(T_i)\Phi\ T_i$ from all the fields including the gauge field itself. By the canonical commutation relations (4), we find

$$[\bar{Q}_i, \bar{Q}_j] = i\ e\ \hbar\ f_{ijk}\ \bar{Q}_k$$

We "measure" the $\bar{Q}$'s via the eigenvalues of a maximum mutually commuting set called a Cartan subalgebra, I say. Then by the theory of Lie algebras, if $|\Lambda>$ denotes a simultaneous eigenstates of the $\bar{Q}_i$, $i \in I$

$$<\Lambda \mid \bar{Q}_i \mid \Lambda> = e\,\hbar \begin{cases} \Lambda_i & i \in I \\ 0 & i \notin I \end{cases}$$

where $\Lambda_i \in \Lambda\,(H)$, is a "weight" of the group H, i.e. a set of simultaneous eigenvalues of T_i, $i \in I$, in a single valued representation of H. Since $|\Lambda>$ is constructed from the Fock space of matter fields, it is their transformation laws which determine the set of weights and hence the global structure of H. Here it is important to distinguish the different groups with the same Lie algebra but different global properties e.g. SU(2) with weights $(0, \pm\frac{1}{2}, \pm 1, \pm\frac{3}{2}, \pm 2\ \dots$, and SO(3) with weights $(0, \pm 1, \pm 2, \dots)$. In general if H is simply connected $(\Pi_1(H) = 1)$, $\Lambda(H) = \{\lambda_i,\ 2\alpha\lambda/\alpha^2 = \text{integer, any root } \alpha\}$ while instead if H has a trivial centre $(Z(H) = 1)$, $\Lambda(H) = \Lambda_{root}(H) = \{\lambda,\ \lambda = \Sigma n_i\,\alpha_i,\ n_i \text{ integers}, \alpha \text{ root}\}$. There are intermediate cases we shall not discuss here. If $|\Lambda>$ is a physical state, the expectation values of Q and $\bar{Q}$ coincide, and we have found a specific quantum structure for Q, rather analogous to the Noether charge in the massive Thirring model.

Now let us turn to the magnetic charge G, treating it by analogy with the Sine Gordon theory, purely classically. Let $h(\Gamma)$ be the ordered integral of the exponentiated gauge potential around a curve Γ far from the monopole at 0, and subtending solid angle Ω at it. Then $h(\Gamma) = \exp(i\, e\, G(P)\, \Omega/4\pi)$, where P is the starting and finishing point of Γ. As Ω increases to 4π, Γ shrinks to a point and so $h(\Gamma) = 1$ [40]. This is the generalized Dirac quantization condition $1 = \exp(i\, e\, G(P))$. As Ω increases from 0 to 4π h provides a closed path in H, thereby defining a homotopy class $\Pi_1(H)$ which is the topological quantum number.

If $G(P) = \sum_{i=1}^{\dim H} g_i T_i$, a general theorem tells us that there exists a gauge rotation sending $g_i \to 0$, $i \notin I$. The remaining coefficients $g_i, i \in I$ are not quite uniquely determined since there is an ambiguity due to the Weyl group, the finite group generated by reflections in the hyperplanes perpendicular to the roots α (for SU(2) this is a sign ambiguity) Modulo this the g_i reflect the long range gauge invariant structure, and because of the Dirac quantization condition satisfy

$$\frac{2e}{4\pi} \sum_{i \in I} g_i T_i = \text{integer}$$

But the T_i can be simultaneously diagonalized and so replaced by all possible weights of H. So $2e\, g_i/4\pi$ must lie on the lattice reciprocal to the lattice $\Lambda(H)$.[41] This new lattice is also the weight lattice of a group, H^ν say, in general different from H [40]. For example if $Z(H) = 1$, $\Lambda(H) = \Lambda_{root}(H)$ so the g_i satisfy $\frac{e}{4\pi} \sum 2 g_i \alpha_i$ = integer. Put $\alpha_i^\nu = \alpha_i/(\alpha_i)^2$ so $\alpha_i = \alpha_i^\nu/(\alpha_i^\nu)^2$. These "dual roots" α_i^ν, are by general theorems also the roots of a Lie algebra, H^ν with the same rank and dimension as H. Now, we see comparing with the formula above for the weight system of a simply connected group, that $g_i = \frac{4\pi}{e} \lambda_i$ where $\lambda_i \in \Lambda_i(H^\nu)$, $\Pi_1(H^\nu) = 1$. In general it always works that $\Pi_1(H^\nu) = Z(H)$, $Z(H^\nu) = \Pi_1(H)$, $(H^\nu)^\nu = H$. e.g.

$SO(3) \leftrightarrow SU(2)$

$SU(N)/Z_N \leftrightarrow SU(N)$

$SU(6)/Z_2 \leftrightarrow SU(6)/Z_3$

Notice that the fundamental representation of H and H^ν cannot both occur for SU(3) say. This is a possible explanation of quark confinement advocated by t'Hooft[42] and Englert and Windey [43].

The conclusion is that the g_i have a very similar structure to the $\langle \Lambda | Q_i | \Lambda \rangle$, but on the basis of quite different arguments, namely a classical field version of the Dirac condition, rather than a canonical quantization

$$\langle \Lambda | Q_i | \Lambda \rangle = \begin{cases} e\hbar\, \Lambda(H) & i \in I \text{ (CSA)} \\ 0 & i \notin I \end{cases} \qquad (10)$$

$$g_i = \begin{cases} \frac{4\pi}{e} \Lambda(H^\nu) & i \in I \\ 0 & i \notin I \end{cases}$$

This formal similarity lead Goddard Nuyts and Olive [40] to conjecture that if these monopoles were solitons, and the quantum field theory of the monopoles were constructed, then i) H^ν would be a symmetry of that theory; ii) the g_i would label the irreducible H^ν representations of the field creating that monopole, iii) H^ν is a gauge symmetry (because of the long range forces).

Then the ambiguity in adding monopoles, $g + \sigma (g')$, $\sigma \in$ Weyl group, would correspond to the Clebsch Gordon series ambiguity in adding irreps., but there is a problem in understanding the missing weight of the irrep. not in the Weyl group orbit of g_i.

The basis for this was analogy with the Sine Gordon-Thirring system mentioned above. t'Hooft has made related arguments, apparently using as basis for his analogy the Kramers Wannier duality of the Ising model on a square lattice [42,23]. This relates two different phases of the same system, and t'Hooft talks of confined and Higgs phases. He uses as order disorder parameters, loop operators creating electric and magnetic flux lines. Mandelstam [44] has tried to construct the latter explicitly, and hence what might be the gauge potential of H^ν.

One would only expect a subset of the possible magnetic charges (10) to correspond to stable monopoles, with the specification depending on the detailed internal structure, supplied by the Higgs fields, but this is not so as Brandt and Neri [45] recently showed. An instability of the Coulomb tail can occur into the non Abelian directions $i \notin I$, unless $g_i = \frac{4\pi}{e} \lambda_i$, $2\lambda\, \alpha^\nu / \alpha^{\nu 2} = 0, \pm 1$. The possibilities (10), $\pm 2, \pm 3 \ldots$ then are all unstable. For example if H^ν = SU(2), only the 2 is stable, if H^ν = SU(3) only the 3 and $\bar{3}$. Then

1) the problem of "missing weights" mentioned above then disappears since the Weyl group acts transitively on the weights of the stable irreps. [46]

2) stable weights, modulo the Weyl group, i.e. H^ν irreps., correspond in a one to one way with the non trivial elements of $\Pi_1(H) \equiv Z(H^\nu)$ which labels the topological quantum number [47].

The second point means that, to any unstable monopole of weight λ_u, there corresponds a stable monopole of weight λ_s such that

$$\lambda_u - \lambda_s = \Sigma\, \eta_i\, \alpha_i \qquad\qquad \eta_i \text{ integer}$$

Since the roots are the weight of magnetic H^ν gluons, this is the group theoretical condition that permits the unstable monopole to decay into the stable one emitting magnetic gluons, and maybe this is the instability mechanism in terms of the GNO conjecture.[40]

I wish to apologise to all those to whose important contributions I have failed to do justice. I should like to close by repeating that the subject poses many interesting unanswered questions covering a very wide scope of mathematics and physics.

REFERENCES

1) E. Schrödinger: Proc. R. Soc. A150, 465 (1935).

2) P. Goddard and D. Olive: Rep. Prog. Phys. 41, 1357 (1978).

3) P.A.M. Dirac: Proc. R. Soc. A133, 60 (1931).

4) T.T. Wu and C.N. Yang: Phys. Rev. D12, 3845 (1975).

5) E.B. Bogomolny: Sov. J. Nucl. Phys. 24, 449 (1976).

6) T.H.R. Skyrme: Proc. R. Soc. A262, 237 (1961)
R. Streater and I.F. Wilde: Nucl. Phys. B24, 561 (1970).

7) S. Coleman: Phys. Rev. D11 2088 (1975).
S. Mandelstam: Phys. Rev. D11 3026 (1975).

8) R. Penrose and M.A.H. MacCallum: Phys. Rep. 6C, 241 (1972).

9) C.N. Yang and R.L. Mills: Phys. Rev. 96, 191 (1954)
R. Shaw: Ph.D Thesis, Cambridge University (1955).

10) P.W. Higgs: Phys. Rev. Lett. 12, 132 (1964); Phys. Rev. Lett. 13, 508, (1964)
Phys. Rev. 145, 1156 (1966).
F. Englert and R. Brout: Phys. Rev. Lett. 13, 321 (1964)
T.W.B. Kibble: Phys. Rev. 155, 1557 (1967).

11) G. 't Hooft: Nucl. Phys. B79, 276 (1974)
A.M. Polyakov: J.E.T.P. Lett. 20, 194 (1974).

12) J. Arafune, P.G.O. Freund and C.J. Goebel. J. Math. Phys. 16, 433 (1975).

13) N.E. Steenrod: The Topology of Fibre Bundles (Princeton, N.J.; Princeton University Press, 1951).

14) M.I.Monastyrskii and A.M.Perelomov: JETP Lett. 21, 43 (1975).
Yu. S. Tyupkin, V.A. Fateev and A.S. Schwarts, JETP Lett. 21, 41 (1975).
S. Coleman: Proc. 1975 Int. School of Physics "Ettore Majorana",
ed. A. Zichichi (New York Plenum, 1975), 297.

15) E. Corrigan, D. Olive, D.B. Fairlie and J. Nuyts: Nucl. Phys. B106, 475 (1976).

16) H. Georgi and S.L. Glashow: Phys. Rev. Lett. 32, 438 (1974).

17) N.P. Chang and J. Perez-Mercader: CCNY-HEP-78-15 (1978)

18) Y. Achiman and B. Stech: Phys. Lett. 77B, 389 (1978).

19) C. Montonen and D. Olive: Phys. Lett. 72B, 117 (1977)
D. Olive: Phys. Rep. 49, 165 (1979)

20) D. Olive: Nucl. Phys. B153, 1 (1979).

21) S. Coleman, S. Parke, A. Neveu and C.M. Sommerfield: Phys. Rev. D15, 554 (1977).

22) M.K. Prasad and C.M. Sommerfield: Phys. Rev. Lett. 35, 760 (1975).

23) B. Julia and A. Zee: Phys. Rev. D11, 2227 (1975).

24) G. 't Hooft: Nucl. Phys. B153, 141 (1979).

25) N. Manton: Nucl. Phys. B126, 525 (1977)
J. Goldberg, J.S. Jang, S.Y. Park and K. Wali: Phys. Rev. 18 542 (1978)
S. Magruder: Phys. Rev. 17, 3257 (1978)
W. Nahm: Phys. Lett. 79B, 426 (1978).

26) E.J. Weinberg: Phys. Rev. 19, 3008 (1979).

27) W. Nahm: Phys. Lett. 85B, 373 (1979).

28) E.J. Weinberg: CU-TP-157 (Columbia preprint).

29) J. Schwinger: Phys. Rev. D12, 3105 (1975).

30) A.M. Polyakov: Phys. Lett. 59B, 82 (1975)
A.A. Belavin, A.M. Polyakov, A. Schwartz and Y. Tyupkin; Phys. Lett. 59B, 85 1975
V.N. Gribov: unpublished
G. 't Hooft: Phys. Rev. Lett. 37, 8 (1976)
C. Callan, R. Dashen and D.J. Gross: Phys. Lett. 63B, 334 (1976)
R. Jackiw and C. Rebbi: Phys. Rev. Lett. 37, 172 (1976).

31) E. Witten: Phys. Lett. 86B, 283 (1979).

32) M. Löhe: Phys. Lett. 70B, 325 (1977).

33) T. Kaluza: Sitzungber Preuss. Akad. Wiss. Berlin, Math. Phys. KA 966 (1921)
O. Klein: Z. Phys. 37, 895 (1926)
W. Thirring: Acta Physica Austriaca Supp IX, 266 (1972).

34) A. D'Adda, R. Horsley and P. DiVecchia: Phys. Lett. 76B 298 (1978)

35) F. Gliozzi, J. Scherk and D. Olive: Nucl. Phys. B122, 253 (1977).
H. Osborn: Phys. Lett. 83B, 321 (1979).

36) B. Zumino: Nucl. Phys. B89, 635 (1975).

37) E.Witten and D. Olive: Phys. Lett. 78B, 97 (1978).

38) A. D'Adda and P. Di Vecchia: Phys. Lett. 73B, 162 (1978)
R. Horsley: Nucl. Phys. B151, 399 (1979).

39) J.F. Schonfeld: University of Minnesota prepring 1979.

40) P. Goddard, J. Nuyts and D. Olive: Nucl. Phys. B125, 1 (1977).

41) F. Englert and P. Windey: Phys. Rev. D14, 2728 (1976).

42) G. 't Hooft: Nucl. Phys. B138, 1 (1978).

43) F. Englert and P. Windey: Phys. Rep. 49, 173 (1979)
Nucl. Phys. B135, 529 (1978).

44) S. Mandelstam: Phys. Rev. D19, 2391 (1979).

45) R. Brandt and F. Neri: NYU preprint.

46) W. Nahm (private communication).

47) S. Coleman (private communication).

INTRODUCTION

TO PAPERS PRESENTED AT THE GAUGE THEORY SESSION

Andrzej Trautman
Institute of Theoretical Physics
Warsaw University
Hoza 69, 00-681 Warszawa, Poland

Most of the papers presented at the gauge theory session complement the invited lectures by D.I. Olive and I.M. Singer. Except for the contribution by B. Schroer, these papers deal with classical aspects of the theory.

W. Thirring describes his views on the gauge approach to gravitation. Following Kibble and Cho, he considers translations to constitute the gauge group of gravitation; the corresponding gauge potentials are identified with the tetrads. An alternative point of view, which was originated by Utiyama [1] is to consider the Lorentz group as the basic group. In the gauge approach to supergravity (P. Van Nieuwenhuizen), translations are also "gauged" to give the tetrads, the remaining generators of the graded Poincaré group yield the linear connection and the gravitino (Rarita-Schwinger) field.

In the fibre bundle approach to gauge fields, which I favour [2,3], gauge configurations are described by connections on principal bundles. This point of view allows a sharp distinction between space-time transformations in the base and "vertical" maps in the bundle, induced by the action of the structure group. The Poincaré group may be taken to be the structure group provided that one considers the bundle of affine frames [4]. In any case, in a theory of gravitation, a gauge situation is described by a connection (linear, affine, conformal...) whereas a choice of tetrads defines a section of the bundle of linear frames; it is thus analogous to fixing the gauge in electrodynamics. Incidentally, an essential -and often overlooked- diffence between gravitation and gauge theories of the Yang-Mills type is due to soldering of the bundle of frames to the base manifold [5,6]. The geometric view of gauge theories is represented in the contributions of M.E. Mayer and L. Vinet. Gary Gibbons describes large classes of gravitational instantons which are important since they dominate the functional integral occurring in the Euclidean approach to quantum gravity. The contribution by M. Mulase extends the description of the Yang-Mills instantons given by Singer.

Some of the simplest gravitational instantons -such as CP_2- do not admit any spin structure. P.G.O. Freund discusses generalized spin structures obtained by extending the Spin(4) group so as to overcome the topological obstruction. Another type of cohomology obstruction, namely the one to linearizability of nonlinear differential equations, is described by D. Sternheimer. J.S.R. Chrisholm shows that electromagnetism with two vector potentials may be used to describe magnetic monopoles of arbitrary strength.

In a beautiful lecture, W. Fritzsch reviews the recent attempts to construct a unified, gauge theory of strong, electromagnetic and weak interactions. SU(5) is among the groups given the most serious attention in this connection.

I gratefully acknowledge financial support from the University of Geneva which enabled me to participate in the IAMP Conference at Lausanne.

REFERENCES

[1] R. Utiyama, Phys. Rev. 101 (1956) 1597.

[2] A. Trautman, Rep. Math. Phys. (Toruń) 1(1970) 29; 10(1976) 297.

[3] A. Trautman, Czech. J. Phys. B29 (1979) 107.

[4] A. Trautman, Symp. Math. 12 (1973) 139.

[5] A. Trautman, Bull. Acad. Polon. Sci., Sér. Sci. Phys. et Astron. 27, No 1 (1979) 7.

[6] A. Trautman, Article in Einstein Commemorative Volume, Ed. by A. Held et al., Plenum Press, New York 1979-80.

THE UNIFICATION OF PHYSICS

Harald Fritzsch

Institute of Theoretical Physics
University of Bern

and

CERN, Geneva

Abstract

The strong, electromagnetic, and weak interactions are described as different manifestations of a grand unified gauge theory. Candidates of unified theories based e.g. on the groups SU(5) and SO(10) are discussed.

At the present time high energy physics explores the structure of the elementary particles at distances of the order of 10^{-15} ... 10^{-16} cm. It has turned out that we can describe the world at those distances or larger rather well by three types of gauge interactions:

I. Gravity (gauge group $\sim$ Poincaré group P)
II. Flavor Interactions (gauge group SU(2)xU(1))
III. Strong Interactions (gauge group $\sim SU(3)^c$).

The three different gauge groups involved are independent of each other, i.e. they commute. The total gauge group G is given by the direct product

$$G = SU(3)^c \times SU(2) \times U(1) \times P.$$

As far as the elementary fermions are concerned, they come in two different varieties:

a) $SU(3)^c$ singlets (leptons)
b) $SU(3)^c$ triplets (quarks).

All lefthanded quarks and leptons are doublets under the weak isospin SU(2), and all righthanded quarks and leptons are singlets.

Quarks and leptons can be grouped together in families according to their mass. The lightest leptons and quarks constitute the following eightfold family:

$$\begin{pmatrix} \nu_e & \vdots & u_r \; u_g \; u_b \\ e^- & \vdots & d_r \; d_g \; d_b \end{pmatrix}$$

(the three colors are denoted by red (r), green (g), and blue (b)). The further, yet heavier fermions constitute the family

$$\begin{pmatrix} \nu_\mu & \vdots & c_r \; c_g \; c_b \\ \mu^- & \vdots & s_r \; s_g \; s_b \end{pmatrix}.$$

Furthermore there exists evidence for a third family:

$$\begin{pmatrix} \nu_\tau & \vdots & t_r \; t_g \; t_b \\ \tau^- & \vdots & b_r \; b_g \; b_b \end{pmatrix}.$$

Thus far one has only evidence for the existence of the τ-lepton, its neutrino, and the b-quark. The t-quark has been searched for at PETRA, thus far without success. The present lower limit on the t-quark mass is of the order of 14 GeV.

There seems to be no difference between the various lepton-quark families, except the masses. Furthermore the dominant weak transitions occur within the same family, i.e. $u \leftrightarrow d$, $c \leftrightarrow s$, $t \leftrightarrow b$, and not $u \leftrightarrow s$ etc. This suggests that there must exist some connection between the fermion masses and the weak interaction.

The world of stable particles which we observe in everyday life consists of the members of the first family. We do not know how many lepton-quark families exist, and nobody knows why nature prefers to have more than one.

If someone is satisfied to describe the world at distances of the order of $\sim 10^{-16}$ cm by the theory described above, he can do so. However he is facing a series of important problems:

a) The number of free parameters is rather large. First of all the three coupling constants of QFD and QCD are essentially free parameters. Furthermore the lepton and quark masses are free parameters, and moreoever all elements of the lepton-quark mass matrix (e.g. the weak mixing angles). In case of the three family scheme denoted above one counts more than twenty parameters.

b) Since the generator of the electric charge contains an Abelean U(1) factor, the electric charges of the quarks and leptons are not quantised, i.e. the electric charge of the d-quark need not be 1/3 of the electron charge.

c) The gauge group $P \times SU(3)^c \times SU(2) \times U(1)$ contains four factors. This is unsatisfactory, and one should aim to reduce the number of factors. The most ideal situation would be to deal only with one gauge group. I am restricting myself to a more modest approach, namely the construction of models which unify the flavor and color interactions. Gravity is left out.

The idea is to interpret the gauge group $SU(3)^c xSU(2)xU(1)$ as a subgroup of a larger gauge group, which is either simple or semisimple such that the various subgroups are related by a discrete symmetry (e.g. parity) [1)]. It is clear that in such a theory one needs to incorporate interactions which are not observed, i.e. the associated gauge bosons must be much heavier than the carriers of the conventional weak interactions, whose masses are believed to be of the order of 10^2 GeV. Thus a unified gauge theory of the color and flavor interactions must exhibit at least two stages of symmetry breaking as shown below:

$$\begin{array}{c} G \\ \downarrow \text{ I.} \\ U(1)xSU(2)xSU(3) \\ \downarrow \text{ II.} \\ U(1)xSU(3). \end{array}$$

At the first stage I. all gauge bosons which are not gluons, W, Z or γ, acquire a large mass, say of the order of M, while at the second stage II. the masses of the W and Z (of the order of m) are generated. One has to have $M >> m$, in order to understand why the subgroup U(1)xSU(2)xSU(3) plays such an important rôle at relatively low energies. At energies much below M the unified interactions can be neglected for many purposes, and we are left with an effective gauge theory, based on the group U(1)xSU(2)xSU(3). Within such an approach the fermions (leptons, quarks) form representations of the large group G, and one may have the hope to learn something about their mass spectrum.

Let us consider the members of the first family: (ν_e, e^-; u, d; e^+, $\bar{u}$, $\bar{d}$). If we construct a unified theory incorporating these fermions one finds two predictions:

$$\sin^2\theta_W = 3/8$$

(θ_W: SU(2)xU(1) mixing angle)

$$\alpha_s = \frac{g_3^2}{4\pi} = \frac{8}{3} \cdot \frac{e^2}{4\pi} \approx 0.02$$

(g_3: $SU(3)^c$ coupling constant).

On the other hand the experimental values of α_s and $\sin^2\theta_W$ measured at energies of a few GeV are

$$\alpha_s \approx 0.3$$

$$\sin^2\theta_W = 0.23 \pm 0.02,$$

i.e. there exists a large gap between the predicted and observed values.

One possible way out is to assume that the unification mass M is a very large mass, and there exist large renormalization effects. Indeed the observed value of α_s is obtained for $M \approx 10^{15}$ GeV. One <u>predicts</u> in this case $\sin^2\theta_W \approx 0.20$, which is not in disagreement with the observed value.

The reader may find it unsatisfactory that in this case the unification of the interactions sets in only at such gigantic energies like 10^{15} GeV. However on the other hand one gains an understanding of the fact that the strong interactions are much stronger than the flavor interactions. At distances corresponding to the unification energy of the order of 10^{15} GeV all interactions are unified, and all coupling constants are of the same order, namely of the order of e. At "normal" energies, say at a few GeV, the strong interactions are much stronger than the flavor interactions, since the gauge group $SU(3)^c$ is larger than the flavor group SU(2)xU(1), hence the renormalization effects for the strong interactions are much larger than those for the flavor interactions. At energies of the order of 1 GeV the strong coupling constant α_s becomes of order one, while the flavor coupling constants are much smaller than one, namely of the order of 1/137.

The group U(1)xSU(2)xSU(3) has rank 4. For this reason the unifying group G must have a rank larger or equal 4. The fermions transform under SU(2)xSU(3) as follows:

$$f = \begin{pmatrix} \nu_e \\ e^- \end{pmatrix}_L , \begin{pmatrix} u \\ d \end{pmatrix}_L , e^+_L, \bar{u}_L, \bar{d}_L + \text{other generations}$$

$$= (2,1) + (2,3) + (1,1) + (1,\bar{3}) + (1,\bar{3}) + \text{other generations.}$$

This representation is complex, and each generation is composed of 15 fermions.

The smallest group which can serve as a unifying group is SU(5) [2], which has rank 4. The (5)-representation of SU(5) decomposes under SU(2)xSU(3) as

$$(5) = (2,1) + (1,3).$$

The 10-representation of SU(5) is obtained as the antisymmetrized product $(5x5)_a$. It is easy to work out its SU(2)xSU(3) content:

$$10 = (5 \times 5)_a = (1,1) + (2,3) + (1,\bar{3}).$$

Thus the fermion representation is obtained as the reducible representation $\bar{5}$ + 10. We make the following comments about this scheme:

a) The basic set of fermions including the electron and its neutrino as well as the light quarks u and d appears in a reducible representation:

$$\bar{5} = \left(\begin{matrix} \nu_e \\ e^- \end{matrix} \;\middle|\; \bar{d} \right)$$

$$10 = \begin{pmatrix} u & \vdots & \bar{u}, e^+ \\ d & \vdots & \end{pmatrix}.$$

b) The other generations of fermions (μ, τ, ...) are simply interpreted as replications of the first one.

c) In the 10-representation of SU(5) appear both quarks and antiquarks. Therefore baryon number could not be a conserved quantity.

d) In SU(5) there is no room for a righthanded counterpart of the lefthanded neutrino.

e) The gauge bosons (24-representation of SU(5)) transform under SU(2)xSU(3) as:

$$24 = (1,8^c) + (3,1^c) + (1,1^c) + (2,3^c) + (2,\bar{3}^c).$$

Besides the gauge bosons belonging to the U(1)xSU(2)sSU(3)^c subtheory (γ, Z, W; gluons) one has 12 additional gauge bosons which are color (anti)triplets as well as weak doublets. We shall denote them by the doublet:

$$\begin{pmatrix} X^c \\ Y^c \end{pmatrix} \quad + \text{ antiparticles} .$$

The adjoint representation of SU(5) can be obtained by multiplying the 5-representation with its complex conjugate: $\bar{5} \times 5 = 24 + 1$. Using the decomposition of the 5 representation in terms of quarks and leptons it is easy to work out the the couplings of X and Y. They couple to a di-antiquark configuration ($\bar{q}\bar{q}$) and to a leptoquark configuration ($q\bar{l}$). The X and Y interactions lead to the decay of the proton into leptons and mesons in the second order of the gauge coupling. As a consequence the masses of the X and Y bosons must be very large (not less than 10^{15} GeV).

f) The decay of the proton is induced by the interactions due to X/Y exchange. The leading decay modes are: $p \to e^+ \pi^0, \rho^0, \omega, \ldots, p \to \bar{\nu}_e \pi^+, \rho^+, \ldots$. The life time of the proton can be calculated in terms of M_X. One obtains:

$$\tau(\text{proton}) \approx 10^3 \ldots 10^4 \cdot M_X^4/M_p^5 .$$

Since the mass of the X-bosons is estimated to be about 10^{15} GeV, the proton life time is expected to be of the order of 10^{30} yrs. However the calculated life time is rather sensitive to M_X, and it is not possible to improve our knowledge about M_X such as to determine it to better than a factor of ten. Thus we can say that within the SU(5) scheme the proton life time is expected to be $\sim 10^{28} \ldots 10^{32}$ yrs.

On the other hand the present experimental limit on the proton life time is $\sim 10^{29}$ yrs, i.e. very close to the theoretical value. For this reason it is important to improve the present experimental limits on the proton stability by 2 - 3 orders of magnitude. If the decay of the proton is found at a level of e.g. $\tau(\text{proton}) \approx 10^{31}$ yrs, it would, of course, not establish that the SU(5) scheme is correct, but it would support the ideas about a unified theory of all interactions.

A more extended scheme involves the gauge group SO(10) [3]. The simplest way to arrive at the group SO(10) is to consider the group $SU(4)xSU(2)_L xSU(2)_R$, where SU(4) is an extension of the color group SU(3) incorporating the quarks and leptons (the lepton number acts as the fourth color) [4]:

$$\underbrace{\begin{pmatrix} u_r & u_g & u_b & \vdots & \nu_e \\ d_r & d_g & d_b & \vdots & e^- \end{pmatrix}}_{SU(4)} \Big\} \quad SU(2)_L xSU(2)_R \, .$$

The group $SU(2)_L xSU(2)_R$ acts on the flavor indices of the lefthanded or righthanded fermions. All generators of $SU(4)xSU(2)_L xSU(2)_R$ which do not belong to the subgroup SU(3)xSU(2)xU(1) are supposed to be broken such that the associated gauge bosons are much heavier than e.g. the masses of the W and Z bosons (of order 10^2 GeV).

The group $SU(4)xSU(2)_L xSU(2)_R$ is isomorphic to SO(6)xSO(4) which can be viewed as a subgroup of SO(10). The fermions transform as a (16)-representation, which is the spinor representation of SO(10). Under $SU(4)xSU(2)_L xSU(2)_R$ one has

$$(16) = (4,\ 2,\ 1) + (\bar{4},\ 1,\ 2).$$

Under SU(5), the (16)plet transforms like $10 + \bar{5} + 1$, i.e. it contains the SU(5) family of lefthanded fields discussed above plus a singlet (a neutral lepton). The latter can be interpreted as the righthanded counterpart of the conventional lefthanded neutrino.

In order to break the SO(10) gauge symmetry down to SU(3)xSU(2)xU(1) one needs several (10) representations of scalars and, eventually, the (126) representation of SO(10) (for details see e.g. ref. (5)). In any case the symmetry breaking in SO(10) is more complicated than in the SU(5) scheme. The interesting new feature of the SO(10) scheme compared to the SU(5) scheme is that it allows the incorporation of the first family of fermions in one irreducible representation. Furthermore one can obtain, by choosing a suitable framework in order to carry out the symmetry breaking, relations between the fermion masses and the weak mixing angles [5], [6] which are in good agreement with the experimental informations available at present.

Besides the schemes mentioned above one has investigated theories based on more extended gauge groups, e.g. $SU(4)^4$ (see ref. (7)) and the exceptional gauge groups E(6), E(7), and E(8) [8]. For details we refer the reader to the original literature.

If one accepts the idea of a unified theory of the strong and flavor interactions, there are still many unsolved problems, for example the problem of incorporating the many flavours observed in nature, or the problem of what determines the huge gap between the grand unified mass scale of the order of 10^{15} GeV and the "normal" mass scale of the order of a few GeV. Finally the unification of the strong and flavor interactions and gravity must be regarded as an unsolved problem.

It is obvious that we are still far away from constructing a unified theory of all interactions, whose completion can be considered as the ultimate goal in physics and would imply the end of the development in fundamental theoretical physics. Nevertheless it is interesting to note that at the present time we have arrived at a stage, at which the construction of an ultimate theory of all interactions has become thinkable.

REFERENCES

1) For reviews see e.g.:
J. Ellis, Proceedings of the Int. Universitätswochen, Schladming (Austria, 1979).
H. Fritzsch, Proceedings of the Int. Gift School, Jaca (Spain, 1979).

2) H. Georgi and S.L. Glashow, Phys.Rev.Lett. 32, 438 (1974).

3) H. Fritzsch and P. Minkowski, Ann.Phys.N.Y. 93, 193 (1975).
H. Georgi, in: Particles and Fields, AIP, N.Y. 1975, p. 575.

4) J.C. Pati and A. Salam, Phys.Rev. D10, 275 (1974).

5) H. Georgi and D.V. Nanopoulos, Harvard preprint HUTP-79/A00).

6) H. Fritzsch, CERN preprint (March 1979), to appear in Nucl.Phys.B.

7) See e.g.: J.C. Pati, Proceedings of the 19th Int. Conference on High Energy Physics, Tokyo 1978, p. 624.

8) See e.g.: P. Ramond, Caltech preprint 1979.

GAUGE THEORIES OF GRAVITATION

W. Thirring
Institut für Theoretische Physik
Universität Wien

When Einstein formulated general relativity he took as Lagrangian the best piece which the differential geometry of that time offered and it worked miraculously well. Ever since physics and mathematics have evolved a great deal and today we think that gauge theories describe the fundamental interactions. Hence one might study the similarities and differences between gauge theories and general relativity to see what the chances of further unifications are. For that I shall emphasize the physical rather than the geometrical aspect.

All classical field theories can be expressed by means of p-forms. Let E_p denote this linear space. One needs the exterior product $\wedge$: $(E_p, E_q) \rightarrow E_{p+q}$, the exterior derivative d: $E_p \rightarrow E_{p+1}$ and in a pseudo-Riemannian manifold the inner product i: $(E_p, E_q) \rightarrow E_{p-q}$ and the duality operation $*$: $E_p \rightarrow E_{m-p}$ (m = 4 is the dimension of space-time M, see [1] for definitions and calculational rules).

The Lagrangian $L \in E_4$ will be constructed from gauge potentials $A_\alpha \in E_1$ and their exterior derivatives dA_α using $\wedge$ and $*$. Then the Euler equations

$$d{*F_\alpha} = -{*J_\alpha} \quad , \qquad {*F_\alpha} := \frac{\partial L}{\partial dA_\alpha} \quad , \qquad {*J_\alpha} := \frac{\partial L}{\partial A_\alpha} \quad , \tag{1}$$

have the form of the inhomogeneous Maxwell equations. They do not only imply charge conservation $d{*J_\alpha} = 0$ but also total charge zero in any compact 3-dimensional submanifold N without boundary

$$\int_N {*J_\alpha} = - \int_{\partial N} {*F_\alpha} = 0 \quad \text{if} \quad \partial N = \emptyset \ . \tag{2}$$

Examples

1. Electrodynamics: There is just one vectorpotential A, $L = \frac{1}{2}\, dA \wedge {*dA} + L_{\text{matter}}$, $F = dA$, and we have the usual Maxwell equations $dF = 0$, $d{*F} = -{*J}$, ${*J} = \partial L_{\text{matter}}/\partial A$.
2. Chromodynamics: Here the gluon potentials A_α are one-forms with values in the Lie-algebra SU(3), $\alpha = 1\ldots 8$ and L is constructed from the covariant exterior derivative $DA_\alpha = dA_\alpha + \frac{1}{2} A_\beta \wedge A_\gamma\, c_{\alpha\beta\gamma}$, $c_{\alpha\beta\gamma}$ being the structure constants in a basis where the Cartan-Killing form is $\delta_{\alpha\beta}$. Then

$$L = \frac{1}{2}\, DA_\alpha \wedge {*DA_\alpha} + L_{\text{matter}} \ , \qquad F_\alpha = dA_\alpha + \frac{1}{2} A_\beta \wedge A_\gamma\, c_{\alpha\beta\gamma} \ ,$$

$$*J_\alpha = c_{\rho\alpha\gamma}\, A_\gamma \wedge *F_\rho + \frac{\partial L_{matter}}{\partial A_\alpha} \ .$$

Thus the conserved colour charge J_α has also a contribution from the gluons.

3. The simplest theory of gravitation: The pseudo metric of M is most conveniently expressed by an orthogonal frame e^α, $i_{e^\alpha} e^\beta = \eta^{\alpha\beta}$, which play the role of the gauge potentials. The currents J_α are then the currents of energy and momentum. For instance,

$$\partial(\tfrac{1}{2}\, dA \wedge *dA)/\partial e^\alpha = dA \wedge i_{e^\alpha} *F - i_{e^\alpha}(\tfrac{1}{2}\, dA \wedge *dA)$$

is in obvious analogy with the Hamiltonian $H = \dot{q}p - L$ of classical mechanics. Its components $T_{\alpha\beta} = i_{e\beta} J_\alpha$ are just the Maxwell energy momentum tensor. Hence we shall consider the e^α as the gauge potentials corresponding to the translations, although the analogy is not perfect; it is further pursued in [2]. Since the translations are an abelian group the covariant and exterior derivative coincide and the simplest Lagrangian seems to be

$$L = \frac{1}{2}\, de_\alpha \wedge *de^\alpha + L_{matter} = L_{grav} + L_{matter} \ ; \quad e_\alpha = \begin{cases} -\,e^\alpha & \text{for } \alpha = 0 \\ +\,e^\alpha & \text{for } \alpha = 1,2,3 \ . \end{cases}$$

Again the field strengths are $F^\alpha = de^\alpha$ and there is a gravitational contribution

$$\frac{\partial L_{grav}}{\partial e^\alpha} = de^\beta \wedge i_{e^\alpha} *F_\beta - i_{e^\alpha} L_{grav}$$

to the energy momentum currents J_α. It has the same structure as in Maxwell's theory. Thus the theory is just 4 × electrodynamics and $e^\alpha(x) = \text{const} + \int D^{ret}(x-x')J^\alpha(x')d^4x'$. Hence, in the linear approximation where $J^\alpha = \partial L_{matter}/\partial e^\alpha$ we get for a point mass at rest only the 0-0-component of the metric differing from the Minkowski metric by $m\kappa/r$. Thus this theory is ruled out empirically since it gives only half the light deflection.

4. Einstein's theory: Here L_g is required to be invariant under local Lorentz transformations

$$e^\alpha \to L^\alpha{}_\beta(x)\, e^\beta \tag{3}$$

which change the Lagrangian 3). One finds that

$$L_g = \frac{1}{2}(de^\alpha \wedge e^\beta) \wedge *(de_\beta \wedge e_\alpha) - \frac{1}{4}(de^\alpha \wedge e_\alpha) \wedge *(de^\beta \wedge e_\beta)$$

changes under (3) only by an exterior derivative since it differs only by

$d(e_\alpha \wedge {}^*de^\alpha)$ from Einstein's Lagrangian $\frac{1}{2} R^*\mathrm{1\!I}$ which is invariant. In this theory the field strength becomes a little more complicated

$$^*F^\alpha = e^\beta \wedge {}^*(de_\beta \wedge e^\alpha) - \frac{1}{2} e^\alpha \wedge {}^*(de^\beta \wedge e_\beta) \quad ,$$

the gravitational energy momentum keeping its form

$$t^\alpha := \frac{\partial L_g}{\partial e^\alpha} = de^\beta \wedge i_{e_\alpha} {}^*F_\beta - i_{e_\alpha} L_g \quad .$$

Thus we see that Einstein's theory has the structure of a gauge theory of translation. Since L_g depends on the e_α explicitly and through the *-operation there is a contribution t^α to the source from the field. t^α is the so-called Landau-Lifschitz pseudo-tensor written in an orthogonal basis. In contradistinction to the Maxwell energy momentum forms they do not transform homogeneously under (3) which leads to all sort of unusual features.

Gauging the translations only there is no reason for requiring invariance under local Lorentz transformations (3). This requirement does not only single out Einstein's Lagrangian but also is essential for a consistent coupling to a Dirac field. Only if the Dirac Lagrangian L_D is made locally Lorentz-invariant the $\delta L_D/\delta e_\alpha$ gives the symmetrized energy momentum tensor and not the canonical one. This suggests that also the Lorentz group should be gauged. Thus we introduce 6 more gauge fields $\omega^\alpha{}_\beta \in E_1$ and the covariant exterior derivatives calculated with the structure constants of the Poincaré group

$$De^\alpha = de^\alpha + \omega^\alpha{}_\beta \wedge e^\beta = \text{torsion}$$

$$D\omega^\alpha{}_\beta = d\omega^\alpha{}_\beta + \omega^\alpha{}_\gamma \wedge \omega^\gamma{}_\beta = \text{curvature} \quad .$$

Now the question is how to construct a Lagrangian with this material. One possibility is to take Einstein's Lagrangian $^*e^\alpha{}_\beta \wedge D\omega_\alpha{}^\beta$ but not to impose $De^\alpha = 0$. This yields the Einstein-Cartan theory [4]. Another Lagrangian, which is more similar to the Yang-Mills-Lagrangian of example 2) is [5] obtained by replacing de^α by De^α in L_g and adding the square of the curvature:

$$L = L_g(e, De) - f^{-1} D\omega^\alpha{}_\beta \wedge {}^*D\omega^\beta{}_\alpha \quad .$$

If one uses in L_g only the first term one obtains in this way a theory proposed by Hehl et al. [6]. Whereas torsion is determined by the spin density locally in the Einstein-Cartan theory it can propagate in these gauge theories and the additional degrees of freedom could give observable effects.

References

[1] W. Thirring, A Course in Mathematical Physics, Vol. II, Classical Field Theory, Springer, New York-Wien 1979

[2] Y. Ne'eman, T. Regge, Gauge Theory of Gravity and Supergravity on a Group Manifold, Rev. Nuovo Cim. 5, 1978

[3] F.G. Basombrio, A Comparative Review of Certain Gauge Theories of the Gravitational Field, preprint, Centro Atómico de Bariloche, 1978

[4] A. Trautman, Symposia Mathematica 12, 139 (1973)

[5] R. Wallner, Notes on Recent U_4 Theories of Gravitation, Vienna preprint, UWThPh-79-12

[6] F. Hehl, J. Nitsch, P.v.d. Heyde, Gravitation and Poincaré Gauge Field Theory with Quadratic Lagrangian, Einstein Commemorative Volume, Plenum Press (1979-80) (A. Hel et al. eds.)

SUPERGRAVITY AS A GAUGE THEORY DERIVED FROM MATTER COUPLING

Peter van Nieuwenhuizen
Institute for Theoretical Physics
State University of New York
Stony Brook, L.I., N. Y. 11794

Supergravity is the gauge theory of supersymmetry (Fermi-Bose symmetry). It is also the unique theory of interacting spin 3/2 fields, or general relativity with a symmetry between fermions and bosons. These definitions are all equivalent since the gauge field of supersymmetry is the real massless spin 3/2 field ψ_μ^a (the gravitino), and because local supersymmetry can only be implemented in Lagrangian field theory if one is in curved spacetime, as we shall see.

Historically, supergravity was derived in refs. (1,2) by first constructing the gauge action, and only afterwards was the problem of matter coupling solved[3]. Also the discovery of auxiliary fields which close the gauge algebra[4] and the establishment of a tensor calculus for supergravity[5] was done without considering matter fields. Today we will rewrite history and start at the matter end. In this way we will rederive all results of simple (N=1) supergravity. Besides being a paedagogical exercise, it is hoped that these new considerations presented here may be useful for extended (N>1) supergravity where only partial results are known.

The simplest matter system in flat spacetime with a global supersymmetry, is the Wess-Zumino model[6]. To begin with we omit the two matter auxiliary fields (F and G) which are needed if one wants a closed global algebra

$$\mathcal{L}^0 = -\frac{1}{2}\left[(\partial_\mu A)^2 + (\partial_\mu B)^2 + \frac{1}{2}\bar{\lambda}\not{\partial}\lambda\right]$$

$$\delta A = \bar{\epsilon}\lambda \;, \delta B = -i\bar{\epsilon}\gamma_5\lambda \;, \delta\lambda = \left[\not{\partial}(A - i\gamma_5 B)\right]\epsilon \tag{1}$$

We use the conventions of ref (3).

For constant ε^a (a=1,4), the Lagrangian varies into a total derivative, and thus the action I^o is invariant and leads to a Noether current. Replacing in (1) ε by local $\varepsilon(x)$, δI^o is equal to $\partial_\mu\bar{\varepsilon}(x)$ times the Noether currents j_N^μ. We cancel δI^o by adding to I^o a coupling I^N between j_N and the gauge field of supersymmetry. Since ε^a is a real anticommuting spinor, the gauge field of supersymmetry is a real massless vector-spinor ψ_μ^a, called gravitino, and we must require that $\delta\psi_\mu{}^a = 2\partial_\mu\varepsilon^a$ + more (the factor 2 is only a convention). The Noether coupling reads

$$I^N = \int d^4x \, \frac{\kappa}{2} \, \bar{\psi}_\mu\left(\not{\partial}(A + i\gamma_5 B)\right)\gamma^\mu\lambda \tag{2}$$

Since the dimension of A is $[A] = 1$, while $[\lambda] = 3/2$, it follows that $[\varepsilon] = -1/2$ so that $[\psi_\mu] = 3/2$. Hence a dimensional coupling appears in (2). We will work from now on order by order in κ.

The only terms left to order κ in $\delta(I^O+I^N)$ come from $\delta(A,B,\lambda)$ in (2). Variation of λ yields the AA and BB variations

$$\delta I^N = \int d^4x \; \kappa \bar{\psi}^\mu \gamma^\nu \left(T_{\mu\nu}(A) + T_{\mu\nu}(B) \right) \epsilon \tag{3}$$

where $T_\mu(A)$ is the energy momentum tensor $\partial_\mu A \partial_\nu A - \frac{1}{2}\delta_{\mu\nu} \; \partial_\lambda A \partial^\lambda A$. Here we discover the need for curved space, since only by coupling minimally to gravity and requiring

$$\delta g_{\mu\nu} = \kappa \bar{\epsilon} \gamma_\mu \psi_\nu + \kappa \bar{\epsilon} \gamma_\nu \psi_\mu \tag{4}$$

do the terms in (3) cancel.

Starting all over again, but now in curved space with $\partial_\mu \lambda \to D^o_\mu \lambda = \partial_\mu \lambda + \frac{1}{2}\omega_\mu{}^{ab}(e) \times \sigma_{ab}\lambda$ with $e^a{}_\mu$ the tetrad field, one finds the same Noether coupling, and again with (4), (3) is cancelled.

The AB terms in δI^N are given by

$$\delta I^N = \int d^4x \left(\bar{\psi}_\mu \gamma_5 \gamma_\tau \epsilon^{\rho\sigma\mu\tau} \right) \left(\gamma_5 \epsilon \right) \left(i\kappa \partial_\rho A \, \partial_\sigma B \right) \tag{5}$$

This term is cancelled by partially integrating $\partial_\rho A$ (and $\partial_\sigma B$, one takes half of both). The terms with $\partial_\rho \varepsilon$ are cancelled by adding a term to the action of the form $\kappa^2 \bar{\psi}\psi A\partial B$ (since $\delta\psi_\mu = \frac{2}{\kappa} \partial_\mu \varepsilon + \cdots$) but the remainder can only be cancelled if one identifies $R^\mu = \varepsilon^{\mu\nu\rho\sigma} \gamma_5 \gamma_\nu D^o_\rho \psi_\sigma$ with the gravitino field equation and adds an appropriate term to $\delta\psi_\mu$. Hence, the gravitino action is deduced to be

$$I^{\frac{3}{2}} = \int d^4x \; -\tfrac{1}{2} \epsilon^{\mu\nu\rho\sigma} \; \bar{\psi}_\mu \gamma_5 \gamma_\nu D^o_\rho \psi_\sigma \tag{6}$$

and, indeed, $\delta\psi_\mu = (\frac{i\kappa}{2} \gamma_5 \varepsilon)(A \overleftrightarrow{\partial}_\mu B)$ cancels (5). It is encouraging that $I^{3/2}$, when linearized and in flat spacetime, is indeed invariant under $\delta\psi_\mu \sim \partial_\mu \varepsilon$ and that $I^{3/2}$ is the unique action with positive energy.

In curved spacetime (6) is not invariant under $\delta\psi_\sigma = \frac{2}{\kappa} D^o{}_\sigma \varepsilon$, but since $[D^o_\rho, D^o_\sigma]\varepsilon$ is proportional to the Riemann curvature and since there are not enough indices in the order $\kappa=0$ terms in $\delta I^{3/2}$ to saturate the Riemann tensor, only its contractions survive.[7] Hence (!) only the Einstein tensor appears in $\delta I^{3/2}$. Thus one can cancel $\delta I^{3/2}$ by adding the Hilbert action $I^2 = -\frac{1}{2} e \kappa^{-2} R$, since its variation is also proportional to the Einstein tensor. One only has to define

$$\delta e^a{}_\mu = \kappa \bar{\epsilon} \gamma^a \psi_\mu \tag{7}$$

Fortunately, this is in agreement with (4). Without proof we state that the complete gauge action is simple $I^2+I^{3/2}$ and is invariant under (7) and $\delta\psi_\mu = \frac{2}{\kappa} D_\mu \varepsilon$, provided one use Palatini formalism: replace every where $\omega_{\mu ab}(e)$ by an independent field $\omega_{\mu ab}$, then solve $\delta(I^2+I^{3/2})/\delta\omega_{\mu ab} = 0$ to obtain $\omega_{\mu ab} = \omega_{\mu ab}(e,\psi)$. One finds in this way torsion induced by gravitinos.

Returning to (2) and varying A,B (and using also (7) in (1)) all variations with $\lambda\lambda$ cancel again if one adds either term to action and transformation laws. (Details can be found in S. Ferrara, F. Gliozzi, J. Scherk and P. van Nieuwenhuizen, Nucl. Phys. B117, 333, 1976). All order κ terms in $\delta(I^o+I^N)$ cancel, if

$$\delta\psi_\mu = \frac{2}{\kappa} D_\mu \epsilon + \frac{i\kappa}{2}\gamma_5\epsilon\left(A\overleftrightarrow{\partial}_\mu B\right) + \frac{\kappa}{4}\left(\bar{\lambda}\gamma_5\gamma^\rho\lambda\right)\left(\sigma_{\mu\rho}\gamma_5\epsilon\right) \tag{8}$$

and if one adds extra order κ^2 terms to the action which happen to be all of the form of an axial current squared. The supercovariant derivative $D_\mu^{cov}A$ has the property that $\delta(D_\mu^{cov}A)$ contains no $\partial_\mu\varepsilon(x)$ terms and is equal to $D_\mu A = \partial_\mu A - \frac{\kappa}{2}\bar{\psi}_\mu\lambda$.

Not all is well. Matter terms in the variation laws of gauge fields preclude the possibility to add invariant actions and still to obtain an invariant action. Repeating the Palatini formalism for $I^2+I^{3/2} + I^o+I^N$ eliminates only the $\lambda\lambda$ terms in (8) but not the AB terms. Thus one is forced to introduce auxiliary fields, i.e., Lagrange multipliers. Since there are two spinor structures in (8), one would need two auxiliary axial-vector fields. There exists such a set of auxiliarly fields,[8] and our procedure might lead to them. We prefer here, however, to restrict our attention to the minimal set of ref (4).

In order to obtain only one spinor structure in (8), we use the freedom there always is in the Noether current, and add an extra term $j_N^\mu(\text{imp})$ so that $\gamma_\mu(j_N^\mu+j_N^\mu(\text{imp})) = 0$ on-shell. Thus we replace (2) by the improved Noether current coupling

$$\mathcal{L}^N+\mathcal{L}^N_{imp} = \frac{e\kappa}{2}\bar{\psi}_\mu\left[\not{\partial}(A+i\gamma_5 B)\right]\gamma^\mu\lambda + \frac{2e\kappa}{3}\bar{\psi}_\mu\sigma^{\mu\nu}D^o_\nu\left[(A+i\gamma_5 B)\lambda\right] \tag{9}$$

where $e = \det e^a_\mu$. Since also $D^o_\mu j^\mu_N(\text{imp}) = 0$ on-shell, we only need to consider to order κ the variations of A,B,λ in this new term in the action. Note now that $\sigma^{\mu\nu} D_\mu\psi_\nu = \frac{1}{4}\gamma_\mu R^\mu$ where R^μ is the gravitino equation defined before. The $\lambda\lambda$ and $A\overleftrightarrow{\delta}B$ terms obtained from (9), are thus removed by adding extra terms to $\delta\psi_\mu$. One finds now only one spinor structure in the gravitino law

$$\delta\psi_\mu = \frac{2}{\kappa}\left(D_\mu + \frac{i\kappa}{2}A_\mu\gamma_5\right)\epsilon - \frac{1}{3}\gamma_\mu i\not{A}\gamma_5\epsilon \tag{10}$$

$$A_\mu = \frac{i\kappa}{8}\bar{\lambda}\gamma_5\gamma_\mu\lambda + \frac{\kappa}{2}A\overleftrightarrow{\partial}_\mu B \tag{11}$$

Hence there is an axial vector auxiliary field A_μ and we consider A_μ in (10) as an independent field, no longer given by (11), but add terms to the action which lead to (11)

$$\mathcal{L}(A_\mu) = \frac{1}{3}A_\mu^2 + \frac{i\kappa}{12}\bar{\lambda}\not{A}\gamma_5\lambda - \frac{\kappa}{3}A^\mu A\overleftrightarrow{\partial}_\mu B \tag{12}$$

Indeed, varying A_μ one finds back (11). The normalization factor 1/3 is arbitrary, and from (10) it follows that if one defines

$$\delta A_\mu = \frac{3i}{2}\bar{\epsilon}\gamma_5\left(R_\mu - \frac{1}{3}\gamma_\mu\gamma\cdot R\right) \tag{13}$$

than all order κ terms with $A\overleftrightarrow{\partial}B$ and $\lambda\lambda$ coming from (9) cancel, while automatically the terms linear in A_μ from $\delta(I^{3/2} + I(A_\mu))$ cancel, too. These are indeed the results of ref. (4).

Note that $D_\mu\bar{\psi}_\nu\sigma^{\mu\nu}$ is proportional to $\bar{R}^\mu$. Thus varying $\delta\lambda = \not{\partial}A\epsilon$ in the last term in (9), one finds terms $\sim \partial_\nu(A)^2\bar{\epsilon}R$. These one cannot eliminate by a suitable $\delta\psi_\mu$, since this would violate the one-spinor-structure of (10). Note, however, that $\delta(eR)$ contains a total derivative $\sim D_\mu\delta\omega_{\nu ab}(e,\psi)$ (which we usually discard) while in $\delta\omega_{\nu ab}(e,\psi)$ one finds terms of the form $\bar{\epsilon}R$. One can indeed cancell the $\partial_\nu(A^2)\bar{\epsilon}R^\mu$ term by adding a term $\frac{\kappa^2}{12}(A^2+B^2)eR$ to the action. This is the onset of nonpolynomiality.

To see how S and P come in, we prefer to first quote the full answer[9], and then to see where one would have made further (re)discoveries. If one couples the Wess-Zumino model with its auxiliary fields F and G to the gauge action with its auxiliary fields S,P,A_μ and requires

(i) that the total action is invariant under the standard transformation rules of ref (4). Then the action is necessarily nonpolynomial (eq.(30) of ref (9))

(ii) that after a Weyl rescaling, the scalar kinemetic terms are as in (1) and the R term is the standard $-\frac{1}{2}eR$(eq(55) of ref(9))

(iii) that afther this Weyl rescaling the action is polynomial (eq(59) of ref (9))

then the action is unique and given by[10]

$$\mathcal{L} = e^{-x}\left[\mathcal{L}^2 + \mathcal{L}^{\frac{3}{2}} + (\mathcal{L}^0 + \mathcal{L}^N)(1-x) + \mathcal{L}^N_{imp} + \mathcal{L}_{ax} + \mathcal{L}(S,P,F,G)\right] \tag{14}$$

where $x = \frac{\kappa^2}{6}(A^2+B^2)$ and $\mathcal{L}^0$ and $\mathcal{L}^N_{imp}$ contains torsion-free derivatives D^0_μ. Furthermore,

$$\mathcal{L}_{ax} = \frac{1}{3}A_\mu^2 - \frac{\kappa}{3}A^\mu\left[A\overleftrightarrow{\partial}_\mu B + \frac{i}{2}\bar{\psi}_\mu\gamma_5(A - iB\gamma_5)\lambda - \frac{i}{4}\bar{\lambda}\not{A}\gamma_5\lambda(1-x)\right]$$
$$+\kappa^2(A\overleftrightarrow{\partial}_\sigma B)\left[-\frac{i}{6}(\bar{\lambda}\gamma_5\gamma^\sigma\lambda)(1-\frac{1}{2}x) - \frac{i}{8}\epsilon^{\mu\nu\rho\sigma}\bar{\psi}_\mu\gamma_\nu\psi_\rho\right] \tag{15}$$

Note that to lowest order in x, $\mathcal{L}_{ax}$ makes all derivatives in $\mathcal{L}^o + \mathcal{L}^N + \mathcal{L}^N_{imp}$ chirally covariant. Also the second and third terms combine to $A\overleftrightarrow{D}_\mu^{cov}B$. Finally

$$\mathcal{L}(S,P,F,G) = \tfrac{1}{2}(F^2+G^2)(1-x) - \tfrac{1}{3}(S^2+P^2) + \tfrac{\kappa}{3}F(AS+BP) + \tfrac{\kappa}{3}G(BS-AP) + \text{terms with } \bar{\psi}\lambda \text{ and } (\bar{\psi}\lambda)^2 \tag{16}$$

We have omitted four-fermion terms in (14). To pick up the discussion where we broke it off, we solve the S and P equation and replace in (16) the S,P terms by $\frac{1}{2}x(F^2+G^2)$.

Suppose we would have started with (1) plus a term $\frac{1}{2}(F^2+G^2)$ in the action. The flat space laws, gravitationally covariantized, read

$$\delta\lambda = [\not{D}^o(A-i\gamma_5 B)]\epsilon + (F+i\gamma_5 G)\epsilon$$
$$\delta F = \bar{\epsilon}\not{D}^o\lambda \quad , \quad \delta G = i\bar{\epsilon}\gamma_5\not{D}^o\lambda \tag{17}$$

Repeating all previous steps, the Noether current and improvement addition remains the same. Invariance of I^o and (2) under (17,7) to order is obtained by replacing D^o_μ by D^{cov}_μ in (17). In particular

$$\delta F = \bar{\epsilon}\not{D}^{cov}\lambda = \bar{\epsilon}\not{D}^o\lambda - \frac{\kappa}{2}\left(\not{\partial}(A-i\gamma_5 B) + (F+i\gamma_5 G)\right)\psi_\mu + O(\kappa^2) \tag{18}$$

Consider now the extra Noether terms in (9). The only new variations to order κ comes from $\delta\lambda = (F+i\gamma_5 G)\epsilon$ and lead to a new term in $\delta\psi_\mu$

$$\delta\psi_\mu = \frac{\kappa}{6}(FA+GB)(\gamma_\mu\epsilon) + \left(\frac{-i\kappa}{6}\right)(BF-AG)(\gamma_\mu\gamma_5\epsilon) \tag{19}$$

Hence, there are a scalar S and a pseudoscalar P auxiliary field and one must add the S and P dependent terms in (16) to the action and finds, in agreement with ref(4)

$$\delta\psi_\mu = \frac{2}{\kappa}\left(D_\mu + \frac{i\kappa}{2}A_\mu\gamma_5\right)\epsilon - \gamma_\mu\eta\epsilon \quad , \quad \eta = -\frac{\kappa}{3}(S-i\gamma_5 P - i\not{A}\gamma_5)$$
$$\delta S = \tfrac{1}{2}\bar{\epsilon}\gamma\cdot R \quad , \quad \delta P = -\tfrac{i}{2}\bar{\epsilon}\gamma_5\gamma\cdot R \tag{20}$$

Complete invariance of the gauge action $\mathcal{L}^2 + \mathcal{L}^{3/2} - \frac{1}{3}(S^2+P^2-A_\mu^2)$ is then obtained simply by replacing R_μ by R_μ^{cov} in (20) and (13) (except that in (13) one must replace A_μ by $A_m = A_\mu e_m{}^\mu$).

These steps are completely analogous to the steps we did already for A_μ. Let us recapitulate. We found A_μ from the requirement that in $\delta\psi_\mu$ there be only one spinor structure. This forced the improved Noether coupling (and brings supergravity near to conformal supergravity). Had we allowed two spinor structures in $\delta\psi_\mu$, we conjectured that another tensor calculus might have emerged. The auxiliary fields F and G in the flat-space globally supersymmetric matter action lead to two new spinor structures in $\delta\psi_\mu$, which led to two new auxiliary fields S and P. There really is no choice, the auxiliary fields follow directly from our analysis.

We will stop here. The reader may enjoy the puzzles of how all further variations of (14) cancell. It is also remarkable that the gauge algebra closes. The cancellation pattern is certainly quite subtle sometimes, but what we have shown is that from the simplest matter model, all features of supergravity can be deduced. The simplest set of auxiliary fields needed the improved Noether coupling, which is not surprising, since the tensor calculus[5] and auxiliary field structure[4] have been reobtained previously from conformed supergravity[11].

REFERENCES

1. D.Z. Freedman, P. van Nieuwenhuizen and S. Ferrara, Phys. Rev. D13, 3214 (1976).
2. S. Deser and B. Zumino, Phys. Lett. 62B, 335 (1976).
3. S. Ferrara, J. Scherk and P. van Nieuwenhuizen, Phys. Rev. Lett. 37, 1035 (1976), idem with P. Breitenlohner, D.Z. Freedman and F. Gliozzi, Phys. Rev. D15, 1013 (1977).
4. S. Ferrara and P. van Nieuwenhuizen, Phys.Lett. 74B, 333 (1978); K.S. Stelle and P.C. West, Phys. Lett. 74B, 330 (1978).
5. S. Ferrara and P. van Nieuwenhuizen, Phys. Lett. 76B, 404 (1978) and Phys. Lett. 78B, 573 (1978); K.S. Stelle and P.C. West, Phys. Lett. 77B, 376 (1978) and Nucl. Phys. B145, 175 (1978).
6. J. Wess and B. Zumino, Phys. Lett. 49B, 52 (1974) and Nucl. Phys. B70, 39 (1974).
7. For spin 5/2 one finds, however, uncontracted Riemann curvatures and one cannot couple consistently spin 5/2 to gravity. See F.A. Berends, B. de Wit, J.W. J.W. van Holten and P. van Nieuwenhuizen.
8. P. Breitenlohner, Nucl. Phys. B124, 500 (1977) and Phys. Lett. 80B, 217 (1979).
9. E. Cremmer, B. Julia, J. Scherk, S. Ferrara, L. Girardello and P. van Nieuwenhuizen, Nucl. Phys. B147, 105 (1979).
10. Put g=0 and $\phi=-3\exp(-x)$ in ref(9). Replace in eq.(30) $\frac{2}{9}A^2_\mu$ by $\frac{1}{18}A^2_\mu$ and multiply in eqs.(63,67) the term $\varepsilon^{abc\mu}\ \bar{\psi}_a\gamma_b\psi_c\ e^{-1}$ by ½. Then one reobtains the results of ref.(3).
11. M. Kaku and P.K. Townsend, Phys. Lett. 76B, 54 (1978); A. Das, M. Kaku and P.K. Townsend, Phys. Rev. Lett. 40, 1215 (1978).

GRAVITATIONAL INSTANTONS: A SURVEY

G. W. Gibbons
D.A.M.T.P.
Silver Street
Cambridge
U.K.

A Review talk given at the "International Conference on Mathematical Physics", Lausanne, August, 1979.

Defn (1) A Gravitational Instanton is a complete, non singular, 4-dimensional Riemannian manifold (signature ++++) which satisfies Einstein's Equations

Gravitational Instantons are believed to dominate the path integral for Euclidean Quantum Gravity [1, 2, 3, 4, 5, 6]. One considers expressions of the form

$$\int_{\mathcal{C}} D[g]\, O[g] \exp - I_{euc} \tag{1}$$

$$I_{euc} = -\frac{1}{16\pi}\int_M R\sqrt{g}\,d^4x - \frac{1}{8\pi}\int_{\partial M} K\sqrt{h}\,d^3x + \frac{1}{8\pi}\int_M \Lambda\sqrt{g}\,d^4x + \frac{1}{8\pi}\int_{\partial M} K_0\sqrt{h}\,d^3x \tag{2}$$

I_{euc} is the Euclidean action of some manifold M, boundary ∂M, metric $g_{\alpha\beta}$ inducing a metric $h_{\alpha\beta}$ on ∂M. $g = \det g_{\alpha\beta}$, $h = \det h_{\alpha\beta}$ and K is the trace of the second fundamental form of ∂M. R is the Ricci scalar of $g_{\alpha\beta}$. $\mathcal{C}$ denotes the boundary conditions that the metric satisfies - corresponding to the quantum mechanical state or density matrix and O (g) is a functional of the metric whose expectation value or matrix element is given by (1). In what follows I shall regard non compact manifolds as the limit of compact manifolds with boundary as the boundary recedes to infinity. (If $R_{\alpha\beta} = \Lambda g_{\alpha\beta}$ and $\Lambda > 0$ there can be no boundary [31]. If $R_{\alpha\beta} = 0$ there can be at most one "end" or "infinity" (N. Hitchin, private communication).) The "integral" in (1) is over all possible manifolds, with all possible topologies subject to $\mathcal{C}$. K_0 is a correction term designed to render zero the action of any flat metric satisfying . Three types of boundary conditions $\mathcal{C}$, are important in physical applications.

I) Asymptotically Euclidean (A.E) together with a weaker, local version - Asymptotically Local Euclidean (ALE). This corresponds to vacuum or zero temperature physics [5, 7, 8].

II Asymptotcally Flat (A.F) together with a weaker, local version - Asymptotically Locally Flat (ALF). This corresponds to finite temperature physics [1, 9].

III Compact without boundary. This is used in the discussion of "spacetime foam"

[3, 4, 10].

Defn (2) $g_{\alpha\beta}$ is ALE if outside a compact set the metric approaches the flat metric on E^4/Γ. E^4 is flat Euclidean space, Γ a discrete subgroup of SO (4) with free action on S^3. $g_{\alpha\beta}$ is A.E if Γ is the identity.

Defn (3) $g_{\alpha\beta}$ is A.F if outside a compact set the metric approaches the standard flat metric on $S^1 \times R^3$.

Defn (4) $g_{\alpha\beta}$ is ALF if outside a compact set the metric approaches

$$dr^2 + \sigma_3^2 + r^2(\sigma_1^2 + \sigma_2^2)$$

where $\{\sigma_i\}$ are left invariant one forms on $S^3/$ and is an isometry of this SU (2) x U (1) invariant metric.

AE Instantons

Thm (1) (Shoen-Yau [11, cf. 8]): There are no AE instantons other than E^4.

The Positive Action Theorem [11] shows that the conformally invariant part of the action (in the sense of [5]) has an absolute minimum at E^4.

ALF Instantons

Thm (2) (Hitchin, private communication) Let (M, $g_{\alpha\beta}$) be ALE and Ricci flat, then $\pi_1(M)$ is finite.

Thm (3) (Hitchin, private communication) Let (M, $g_{\alpha\beta}$) be ALE and half flat ($R_{\alpha\beta\mu\nu} = \pm\frac{1}{2}\epsilon_{\alpha\beta\lambda\rho}R^{\lambda\rho}{}_{\mu\nu} \Rightarrow R_{\alpha\beta} = 0$) then $\pi_1(M) = Z_2$ or $Z_2 \otimes Z_2$ and the universal cover has a discrete subgroup of SU (2) whose action on S^3 is given (for $R_{\alpha\beta\mu\nu} = +\frac{1}{2}\epsilon_{\alpha\beta\lambda\rho}R^{\lambda\rho}{}_{\mu\nu}$) by right actions of SU (2) on itself.

The discrete subgroups of SU (2) are

Z_k	cyclic order k
D^*_k	binary dihedral order 4K
T*	binary tetrahedral
O*	binary octahedral
I*	binary icosohedral

Almost certainly there are half flat metrics for all these groups. Explicit metrics were given in the Z_k case by Gibbons and Hawking [7]. Later Hitchin [13] used "Twistor" methods to give the Z_k case and he now has the D^*_k (private communication).

The Gibbons-Hawking metrics may be written in coordinates (τ, $\underline{x}$) as

$$ds^2 = U^{-1}(d\tau + \underline{\omega}.d\underline{x})^2 + U\,d\underline{x}^2 \tag{3}$$

$$\text{curl}\,\underline{\omega} = \text{grad}\,U \tag{4}$$

$$U = \sum_{i=1}^{i=k} \frac{1}{|\underline{x} - \underline{x}_i|} \tag{5}$$

k = 1 is flat space, E^4 k = 2 is the Eguchi Hanson metric [14, 15, 8] which had been discovered earlier. The Z_k ALE's depend on 3K-6 essential parameters which correspond exactly to the possible infinitesimal variations. They probably exhaust the possible half flat ALE metrics with $\Gamma = Z_R$. He has also constructed scalar Green's function explicitly for these spaces [16]. Atiyah (private communication) subsequently gave a Twistor construction of these Green's functions using sheaf cohomology methods.

Any half flat metric is a local minimum of the action amongst metrics with R = 0 suggesting the following analogue of Theorem I.

Conjecture (1) The only ALE instantons are half flat and their universal cover has $\Gamma \subset SU(2)$.

AF Instantons

The known A.F instantons are the flat metric on $S^1 \times R^3$, the Euclidean Scwarzschild and Euclidean Kerr with imaginary angular momentum [23].

Thm (4) (modified Israel [18]) Let $(M, g_{\alpha\beta})$ be A.F. and Ricci Flat on $R^2 \times S^2$ with a hypersurface orthogonal killing vector, then $g_{\alpha\beta}$ is the Euclidean Schwarzschild.

A. Lapedes (private communication) points out that the analogous modification of Robinson's Theorem [19] for axisymmetric metrics on $R^2 \times S^2$ with another non hypersurface orthogonal killing vector does not go through. Nevertheless on the basis of the Black Hole No Hair Theorems (see [17] for a review) one might

Conjecture (2) The only A.F instantons are flat space, Euclidean Schwarzschild and Euclidean Kerr.

Schwarzschild and Kerr are not local minima of the action amongst A.F metrics with vanishing Ricci scalar [32].

ALF Instantons

If one sets
$$U = 1 + \sum_{i=1}^{i=k} \frac{1}{|\underline{x} - \underline{x}_i|} \tag{6}$$
in equations (3) and (4) one obtains the "multi-Taub-NUT" ALF metrics of Hawking [2]. The boundary at infinity is a cyclic lens space. Presumably these metrics (and those corresponding to D_k^*) can be constructed using Twistor methods. The other groups cannot occur [9]. Page's Green's function construction also works in the multi-Taub-NUT case.

The metric (3) and (4) exhaust the class of metrics with a self-dual killing vector K^α has $K_{\alpha;\beta}$ self-dual.) This is also the class with a killing vector in which the connection forms in the "obvious" basis are self-dual.

The remaining, explicitly known ALF solutions are not half-flat. There are the solutions of Page [21] and its spinning generalization [22]. One might enquire whether more complicated ALF metrics are possible. The answer seems to be no. In the language of [23] such metrics would be multi "bolt" metrics - e.g. multi Schwarzschild. It turns out that these may be ruled out by similar arguments used in black hole physics. Thus it seems that the ALF class of metrics is a little richer than the ALE class but not very much richer.

Compact Instantons

If these are to admit killing vectors then $\Lambda > 0$. The only known examples are

1. S^4
2. CP^2 [24, 25]
3. $S^2 \times S^2$ [25]
4. $CP^2 \# \overline{CP}^2$ [26]

Example 4 is not homogeneous. The manifold is the topological sum of 2 copies of CP^2 with opposite opposite orientations.

If $\Lambda = 0$ the only known examples are half flat

1. $S^1 \times S^1 \times S^1 \times S^1$ this is flat
2. $K3$ with the Yau metric.

The Einstein Kahler metric on $K3$ is known only implicitly from a theorem of Yau [27] or approximately by gluing together 16 the Eguchi Hanson solutions [25, 28]. $K3$ exhausts the possible compact half-flat metrics. One is tempted to

Conjecture (3) The only compact, non flat Ricci flat metric is the Yau metric on $K3$.

If $\Lambda < 0$ there are many implicit examples given by Yau's theorem [27]. None of these is known explicitly. The only semi explicitly known instantons with $\Lambda < 0$ are

1. Spaces of constant curvature ("anti de Sitter space/discrete group)
2. Spaces of constant holanosphere sectional curvature (i.e. set Λ negative in the CP^2 metric and factor by a discrete group)
3. Metric products of 2-2 spaces of constant negative curvature.

An interesting class of Yau examples are complex hypersurfaces of degree k in CP^3, with $k \geq 4$ [30]. $k = 4$ corresponds to $K3$.

Taking more complicated algebraic subvarieties of CP provides an enormous class of examples. There exist important inequalities limiting their behaviour however. Define for each compact Einstein space the number f by

$$\Lambda \left(\int_M \sqrt{g}\, d^4x \right)^{\frac{1}{2}} = -f \tag{7}$$

The classical action I_{euc} is

$$I_{euc} = -f^2/8\pi\Lambda \tag{8}$$

The fact that the Weyl tensor satisfies $C_{\alpha\beta\gamma\delta} C^{\alpha\beta\gamma\delta} \geq | C^{\alpha\beta\gamma\delta} {}^* C_{\alpha\beta\gamma\delta} |$

yields [29]

$$2\chi - 3|\tau| \geq f^2/16\pi^2 \tag{9}$$

χ is the Euler number and τ the signature with equality if the Weyl tensor is self-dual. This implies

$$\chi \geq f^2/12\pi^2 \tag{10}$$

with equality for constant curvature. If $g_{\alpha\beta}$ is Kahler we have

$$3\tau = C_1^2 - 2C_2 = C_1^2 - 2\chi \tag{11}$$

where C_1^2 and C_2 are the first and second Chern numbers. So if $g_{\alpha\beta}$ is Einstein Kahler

$$f = 2\pi^2 C_1^2 \tag{12}$$

and

$$\chi \geq f^2/16\pi^2 \tag{13}$$

with equality if Weyl tensor is self-dual (i.e. constant holomorphic sectional curvature, e.g. CP^2).

For algebraic subvarieties of CP^n we have (N. Hitchin private communication)

$$2\pi^2\chi \leq f^2 \leq 4\pi^2\chi$$

equality on the left is for hypersurfaces ($\chi = R(R^2 - 8R + 22$ $\tau = \frac{1}{3}R(R+2)(R-3)$) and on the right for products of 2-spaces of constant negative curvature genus g_1 and g_2 ($\chi = 4(1-g_1)(1-g_2)$, $\tau = 0$).

These results and some qualitative arguments have encouraged Hawking [4, 10] to

<u>Conjecture</u> As χ increases most compact instantons have $\Lambda < 0$ and $f \propto \sqrt{\chi}$

Friedan (private communication) has shown that amongst metrics with constant Ricci scalar all known compact instantons except $S^2 \times S^2$ are local minima of the action. For $S^2 \times S^2$ there is just one direction in which the action decreases.

REFERENCES

1. S. W. Hawking, G. W. Gibbons, Phys.Rev. 13 2752 (1977).
2. S. W. Hawking, Phys.Letts. 60A 81 (1977).
3. S. W. Hawking, article in "General Relativity" edited by S. W. Hawking and W. Israel, C.U.P. 1979.
4. S. W. Hawking "Euclidean Quantum Gravity" lecture held at the NATO Summer School at Cargese 1978, to be published by Plenum Press, S. Deser editor.
5. G. W. Gibbons, M. J. Perry and S. W. Hawking, Nucl.Phys. B 138 141 (1978).
6. G. W. Gibbons and M. J. Perry, Nucl.Phys. B 146 90 (1978).
7. G. W. Gibbons and S. W. Hawking, Phys.Letts. 78B 430 (1978).
8. G. W. Gibbons and C. N. Pope, Comm.MAth.Phys. 66 267 (1979).
9. G. W. Gibbons, H. Romer and C. N. Pope, Nucl.Phys. in Press.
10. S. W. Hawking, Nucl.Phys. B 144 349 (1978).
11. R. Shoen and T. S. Yau, Phys.Rev.Letts. 42 547 (1979).
12. G. W. Gibbons and S. W. Hawking, unpublished.
13. N. Hitchin, Math.Proc.Camb.Phil.Soc.
14. T. Eguchi and A. Hanson, Phys.Letts. 74B 249 (1978).
15. V. Belinsky, G. W. Gibbons, D. N. Page and C. N. Pope, Phys.Letts. 76B 433 (1978)
16. D. N. Page, Phys.Letts. in Press.
17. B. Carter, article in same volume as ref. 3 above.
18. W. Israel, Phys.Rev. 164 1776 (1967).
19. D. C. Robinson, Gen.Rel.Grav. 8, 695 (1977).
19. D. C. Robinson, Phys.Rev.Letts.
20. P. Tod and R. Ward, to appear.
21. D. N. Page, Phys.Letts. 78B 249 (1978).
22. G. W. Gibbons, to be published.
23. G. W. Gibbons and S. W. Hawking, Commun.Math.Phys.
24. P. G. O. Freund and T. Eguchi, Phys.Rev.Letts. 37 1251 (1976).
25. G. W. Gibbons and C. N. Pope, Commun.MAth.Phys. 61 239 (1978).
26. D. N. Page, Phys.Letts. 80B 55 (1978).
27. S. T. Yau, Proc.Natt.Acad.Sci. 74 1798 (1971). See also Asterique. 58 (1978).
28. D. N. Page, Phys.Letts. 79B 235 (1978).
29. N. Hitchin, J.Diff.Geom. 9 435 (1974).
30. A. Back, M. Forger and P. Freund, Phys.Letts. 77B 181 (1978).
31. J. Milnor, "Morse Theory" Princeton University Press.
32. D. N. Page, Phys.Rev. D18 2733 (1978).

INSTANTONS AND LINE GEOMETRY

M. MULASE

Research Institute for Mathematical Sciences
Kyoto University
KYOTO 606, JAPAN

Let M be a conformally flat 4-dimensional oriented Riemannian manifold, and let Λ^2 be the bundle of 2-forms on M. Since the Hodge star operator $*$ acts on Λ^2 as an involution, Λ^2 splits into a direct sum of the ± 1 eigenspaces of $*$;

$$\Lambda^2 = \Lambda^2_+ \oplus \Lambda^2_- .$$

It is known that the sphere bundle $S(\Lambda^2_-)$ of Λ^2_- has a complex structure defined by the Riemannian metric of M ([1]). Put $Y = S(\Lambda^2_-)$, and let $\pi : Y \longrightarrow M$ denote the natural projection. This sphere bundle Y is a 3-dimensional complex manifold of Kodaira dimention $-\infty$. Each fibre of π is a complex submanifold of Y and is isomorphic to $\mathbb{P}^1$, the 1-dimensional complex projective space, which we call a line. The normal bundle of a fibre in Y is isomorphic to

$$\mathcal{O}_{\mathbb{P}^1}(1) \oplus \mathcal{O}_{\mathbb{P}^1}(1),$$

the direct sum of two hyperplane section bundles on $\mathbb{P}^1$. Since we have

$$H^1(\mathbb{P}^1, \mathcal{O}_{\mathbb{P}^1}(1) \oplus \mathcal{O}_{\mathbb{P}^1}(1)) = 0 \quad \text{and} \quad \dim_{\mathbb{C}} H^0(\mathbb{P}^1, \mathcal{O}_{\mathbb{P}^1}(1) \oplus \mathcal{O}_{\mathbb{P}^1}(1)) = 4,$$

Y contains a 4-dimensional complex family of lines by Kodaira's deformation theory.

So let X denote the connected component of the Douady space of Y which contains the points corresponding to the fibres of π. Roughly speaking, X is a parametrizing space of all lines in Y. It is a 4-dimensional complex analytic space. We define an injection $\iota : M \longrightarrow X$ by sending a point m in M to the corresponding point in X of the fibre $\pi^{-1}(m)$, and this map gives a complexification of M.

A generic point in X corresponds to a line in Y. Let $\Gamma \subset X \times Y$ denote the graph of this correspondence, and let $\alpha : \Gamma \longrightarrow X$ and $\beta : \Gamma \longrightarrow Y$ be the natural projections. The correspondence is given by $\tau = \alpha \circ \beta^{-1}$. This map α is proper and a projection map of a $\mathbb{P}^1$-bundle on some Zariski open subset of X. Now we have the whole stage of instantons:

Diagram 1.

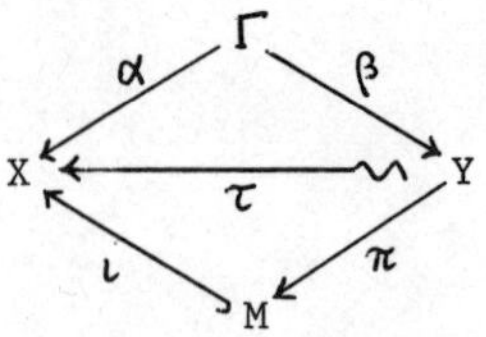

This diagram is commutative in the sense that for every point m in M we have

$$\beta \circ \alpha^{-1} \circ \iota(m) = \pi^{-1}(m).$$

Next let $P \longrightarrow M$ be an $SU(2)$-principal fibre bundle defined on M, and let A

denote a real analytic connection on P. A connection on P is given by a 1-form on M with values in the associated $\mathfrak{su}(2)$-bundle $\mathfrak{g}$ of P which is defined by the adjoint representation of SU(2). The 2nd Chern class $c_2(P) \in H^4(M,\mathbb{Z}) \cong \mathbb{Z}$ can be regarded as an integer. We deal only with anti-self-dual connections, which are connections satisfying the equation

$$*(dA + 1/2[A,A]) = -(dA + 1/2[A,A]),$$

defined on a bundle P with the positive 2nd Chern class. We call them instantons.

An anti-self-dual connection A corresponds injectively to a rank-2 complex algebraic vector bundle E(A) on Y. Since α is a proper map,

$$\widetilde{E}_A = \alpha_* \beta^* E(A)$$

is a coherent analytic sheaf on X. The restriction of E(A) to a fibre of π is isomorphic to the trivial bundle, hence the restriction of E(A) to a general line in Y must be trivial. Because of this fact $\widetilde{E}_A$ has rank 2 and is actually a vector bundle on some neighborhood of M in X. From the commutativity of Diagram 1, the restricted bundle $\widetilde{E}_A|_M$ is isomorphic to the rank-2 vector bundle associated with P given by the usual 2-dimensional representation of SU(2). Then the connection A can be exteded as a complex analytic connection $\widetilde{A}$ of $\widetilde{E}_A$ defined on a neiborhood U of M in X.

The anti-self-dual equation is interpreted as a partial integrability condition on X. To explain this, we need certain 2-dimensional subspace in X. For a point y in Y, let $\tau(y)$ denote the parametrizing space of all lines passing through the point y. Set theoretically this is given by $\tau(y) = \alpha \circ \beta^{-1}(y)$ and in our case it is a 2-dimensional subspace of X. Then we have

Theorem.
A connection A is anti-self-dual if and only if the restriction $\widetilde{A}|_{\tau(y)\cap U}$ is integrable for every y in Y.

We can reconstruct E(A) by using $\widetilde{E}_A$. The pull-back bundle $\alpha^*(\widetilde{E}_A|_U)$ defined on $\alpha^{-1}(U)$ has a connection $\alpha^*\widetilde{A}$. Since this connection $\alpha^*\widetilde{A}$ is integrable along each fibre $\beta^{-1}(y)\cap\alpha^{-1}(U)$ of $\beta : \alpha^{-1}(U) \to Y$, we can define a bundle $E_{flat}(A)$ on Y whose fibre at $y \in Y$ is a vector space of all flat sections of the bundle

$$\alpha^*(\widetilde{E}_A|_U)|_{\beta^{-1}(y)\cap\alpha^{-1}(U)} \text{ w.r.t. the connection } \alpha^*\widetilde{A}|_{\beta^{-1}(y)\cap\alpha^{-1}(U)}.$$

This bundle $E_{flat}(A)$ is isomorphic to E(A).

Now let us consider the case (1) M is S^4, the real 4-sphere with the canonical metric, and (2) M is T^4, the real 4-torus with the flat metric. If $M = S^4$, then $Y = \mathbb{P}^3$, X = Gr(1,3) (the complex Grassmann manifold which parametrizes lines in $\mathbb{P}^3$) and Γ = Fl(0,1,3) (the flag manifold). If $M = T^4$, then Y is the total space of certain complex analytic deformation family of complex 2-tori whose parametrizing space is $\mathbb{P}^1$, $X = (\mathbb{C}^*)^4$ and $\Gamma = (\mathbb{C}^*)^4 \times \mathbb{P}^1$. In this case, Y is not a Kähler manifold because the Douady space $X = (\mathbb{C}^*)^4$ is not compact. In the both cases, $\Gamma \to X$ is a $\mathbb{P}^1$-bundle on X.

For a general instanton A on $M = S^4$ or T^4, the coherent sheaf $\widetilde{E}_A = \alpha_*\beta^*E(A)$ is locally free everywhere on X, and A is continued analytically onto X as a meromorphic connection $\widetilde{A}$ of this bundle $\widetilde{E}_A$. The singularities of $\widetilde{A}$ form an analytic subset J(A), which is a non-singular divisor of X, and $\widetilde{A}$ has simple poles along J(A).

Diagram 2.

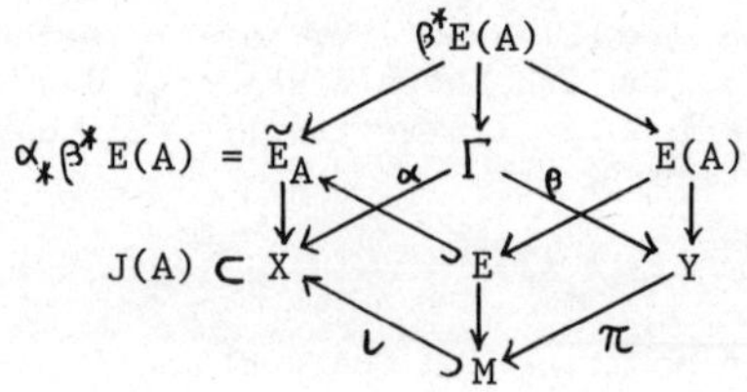

E: the rank-2 vector bundle associated with P

This divisor J(A) has a geometrical meaning. Take a point $x \in X$. The corresponding line $\tau^{-1}(x)$ in Y of x is called a jumping line of the bundle E(A) iff the restricted bundle of E(A) to $\tau^{-1}(x)$ is not trivial. The divisor J(A) of poles of an instanton A coincides with the set of all jumping lines of E(A). It does not meet M in X.

In the case of (1), we have $\deg J(A) = c_2(E(A)) = c_2(P)$ = the instanton number (note that $\deg J(A) \in H^2(Gr(1,3),\mathbf{Z}) = \mathbf{Z}$), and $c_1(\tilde{E}A) = -c_2(E(A))$. Moreover E(A) is uniquely determined by the divisor J(A). It means that an instanton solution of the (anti-) self-dual Yang-Mills equation is uniquely determined by its poles on the complexified domain $X = Gr(1,3)$.

A self-dual connection B defined on S^4 corresponds to a vector bundle over the dual projective space $\mathbb{P}^{3*}$. The complexification X of S^4 is the dual Grassmann manifold $Gr^*(1,3)$ which parametrizes lines in $\mathbb{P}^{3*}$. There is an isomorphism from Gr(1,3) to $Gr^*(1,3)$ which sends a line in $\mathbb{P}^3$ to the dual line in $\mathbb{P}^{3*}$. Then a self-dual instanton B is extended onto Gr(1,3) as a meromorphic connection $\tilde{B}$. This connection $\tilde{B}$ is integrable along each 2-dimensional subspace of Gr(1,3) which parametrizes all lines in a hyperplane of $\mathbb{P}^3$.

Finally let us consider the case that M is $\mathbb{P}^2$ with the canonical Kähler metric. As a Riemannian manifold, $\mathbb{P}^2$ is not conformally flat, however, we have Y, X and Γ ([1]). They are given by $Y = Fl(0,1,2)$ (the 3-dimensiona flag manifold contained in $\mathbb{P}^{2*}\times\mathbb{P}^2$) and $X = \mathbb{P}^{2*}\times\mathbb{P}^2$. The injection $\iota: \mathbb{P}^2 \longrightarrow \mathbb{P}^{2*}\times\mathbb{P}^2$ is defined by $\iota(z) = (\bar{z},z) \in \mathbb{P}^{2*}\times\mathbb{P}^2$ for $z \in \mathbb{P}^2$. A generic fibre of the projection map α is $\mathbb{P}^1$, but the fibre on a point contained in the degree (1,1) divisor of $\mathbb{P}^{2*}\times\mathbb{P}^2$ which coincides with Fl(0,1,2) w.r.t. the natural inclusion is isomorphic to $\mathbb{P}^1 \vee \mathbb{P}^1$, the one-point union of two lines.

Let A be a real analytic connection defined on an SU(2)-principal bundle over $M = \mathbb{P}^2$ with positive 2nd Chern class c_2, and $\tilde{A}$ the analytic extension of A. If $\tilde{A}$ satisfies the partial integrability condition

$$(*) \qquad \tilde{A}\big|_{\tau(\zeta)} \text{is integrable for all } \zeta \in Fl(0,1,2),$$

then A is anti-self-dual. The converse is not true in general. So let us suppose that A satisfies the condition (*). Then we have a vector bundle $\tilde{E}_A = \alpha_*\beta^*E(A)$ on $X = \mathbb{P}^{2*}\times\mathbb{P}^2$ as well as the former cases, and $\tilde{A}$ is a meromorphic connection on this bundle with simple poles along some divisor J(A) of degree (c_2,c_2). This divisor J(A) coincides with the set of all jumping lines of E(A).

Reference .

[1] M.F. Atiyah, N.J. Hitchin and I.M. Singer: Self-duality in four-dimensional Riemannian geometry, Proc. R. Lond. A. 362, 425-461 (1978).

GEOMETRIC ASPECTS OF SYMMETRY BREAKING IN GAUGE THEORIES

(Higgs Models Without Higgs Bosons)

Meinhard E. Mayer
University of California, Irvine, CA 92717 USA

1. INTRODUCTION. The Brout-Englert-Higgs-Kibble (BEHK) mechanism for spontaneous symmetry breaking in gauge theories has been instrumental in the success of the electroweak unification and the various models of "grand unification" which are so popular today. The introduction of the Higgs bosons (sometimes in quite large numbers, and with predicted masses rapidly approaching macroscopic values, predictions which are hardly refutable by experiments) is neither unique nor elegant, and makes many of us uncomfortable. It is therefore not surprising that alternatives to the BEHK model are actively pursued, either as "dynamical" mechanisms, exploiting the nonlinearities of the problem, or considering the Higgs bosons as excitations of more fundamental fermions (much as their prototype, the Ginzburg-Landau field, was related to electron pairs in BCS theory).

If we take the fiber-bundle approach to gauge theories seriously, symmetry breaking should be interpreted as the reduction of the structure group G of the principal bundle P (the hidden symmetry group) to a subgroup H (the manifest, surviving, symmetry subgroup, or the holonomy group of the vacuum), and at the same time, a reduction of the connection and curvature corresponding to the gauge field to one invariant under H. In the reduction process the action of the Yang-Mills field (norm of the curvature form) breaks up into a part invariant under H which also maintains its conformal invariance (masslessness) and a part corresponding to the complement of the Lie algebra h of H in g (the Lie algebra of G) which is no longer conformally invariant (a classical manifestation of the "masses" of the appropriate intermediate bosons, such as the $W^{\pm}$ and the Z^0 in the Weinberg-Salam model). If one wishes to preserve the other advantages of the BEHK model (renormalizability) one should look for a reduction mechanism which in a certain gauge and approximation reproduces the quartic Ginzburg-Landau Lagrangian of that model, the role of the Higgs bosons being played by certain extra components of the connection.

With this in mind, I first analyze the conditions under which a

G-principal bundle P reduces to an H-principal bundle Q over the same (4-dimensional) base space B. It is well known that a necessary and sufficient condition for this is the existence of a cross section σ of the associated bundle E = P/H with fiber G/H. These facts and the notation are summarized briefly in Section 2.

In Sec. 3 I propose a model in which the section σ is interpreted as a nonlinear field, derived from an action principle whose Lagrangian has been investigated in connection with harmonic maps.

In Sec.4 I discuss a new variant of geometric symmetry breaking, in which the original bundle (with group G) is reinterpreted as a principal bundle with structure group H over the base space E = P/H ("fiber-flipping"), then extending this bundle to a G-bundle (which is always possible), and finally achieving the symmetry reduction by an integration over the symmetric space G/H. The latter step, together with a particular choice of gauge, reduces the Yang-Mills action to an expression equivalent to the Higgs Lagrangian, the role of the Higgs particles being played by the $\dim\mathfrak{g} - \dim\mathfrak{h}$ components of the connection in the extended bundle over E. It is gratifying to see that this reproduces the BEHK model in all its features, the only drawback being that the "Higgs" particles are in the adjoint, rather than fundamental, representation of G.

2. REDUCTION OF BUNDLES AND CONNECTIONS. Lack of space forces me to summarize briefly the facts about reduction of structure groups of principal bundles and connections. The reader can find all the details in ref. 2, the notation of which I will follow as far as possible.

Given two principal bundles P(B, G) and Q(B, H) with the same base space B and H a closed subgroup of G, and a morphism $f: Q \to P$ such that $f(qh) = f(q)h$ ($h \in H \subset G$, $q \in Q$, $f(q) \in P$) we say that Q (or f) is a reduction of P and P is an extension of Q.

Fact 1. Extension of the structure group is always possible.

Fact 2. The bundle P has a reduction to Q iff the associated bundle E(B,G/H,G,P) = P/H (with fiber G/H) has a cross section $\sigma: B \to E$.

In the context of gauge theory B is either Minkowski space, or $\mathbb{R}^4$ (or its compactification S^4) and G is a compact Lie group. The Yang-Mills potential is the pullback Y of the $\mathfrak{g}$-valued connection 1-form ω on P to the base space B, and the field strength M is the pullback to B of the curvature 2-form Ω, both being pulled back by the locally trivializing section s: $Y = s^*\omega$, $M = s^*\Omega$. Under a change of trivialization (gauge transformation) $s' = gs$ we have $Y' = \mathrm{Ad}(g^{-1})Y + g^{-1}dg$, $M' = \mathrm{Ad}(g^{-1})M$, where g is a G-valued function on B. The definition $M = DY = dY + \frac{1}{2}[Y,Y]$ implies the Bianchi identity $DM = dM + [Y,M] = 0$, and the Yang-Mills equation $D{*}M = {*}J$ follows from the action principle. Here $*J$ is the current 3-form (dual of the usual current vector) of the particle fields - not a closed form by itself. The particle fields are sections of associated vector bundles or, equivalently, G-equivariant functions on P with values in the vector space V of the representation defining the associated vector bundle; cf.,e.g.,[1,3].

The subgroup H describing the residual symmetry (the isotropy subgroup of the vacuum) may be considered as the (restricted) holonomy group of the connection. The reduced bundle is then the holonomy bundle (the set of points in P which can be reached from a given point by horizontal curves.

Considering the associated bundle $E = P/H$ with fiber G/H and a section $\sigma : B \to E$ which gives rise to a reduction to a subbundle Q with group H, a connection ω in P is reducible to Q (i. e., will split according to $\mathfrak{g} = \mathfrak{h} + \mathfrak{m}$, H-equivariantly for the $\mathfrak{h}$-component) iff σ is parallel with respect to the transport induced by the connection in E.

One of the models discussed below simplifies considerably if the curvature splits in such a way that the bracket of two $\mathfrak{m}$-components is in $\mathfrak{h}$, i. e., if G/H is a symmetric space (this is the case for the models based on SU(2)/U(1), SU(N)/SU(n)U(N-n), but is not true in some popular grand unifications).

3. SYMMETRY BREAKING BY HARMONIC SECTIONS. In the BEHK mechanism[4] of spontaneous symmetry breaking of a gauge theory with the action

$$S = \tfrac{1}{4}\int_B M \wedge {*M} + \int_B \bar{\psi}\not{D}\psi\, d(\mathrm{vol}), \tag{1}$$

where ψ denotes the Dirac field(s) and $\not{D}$ the covariant Dirac-Weyl operator, the Dirac field(s) being sections of a bundle associated to P by a representation, one adds to the action the Ginzburg-Landau term

$$S_{GL} = \int D\varphi \wedge {*D\varphi} + V[\varphi], \quad V[\varphi] = -\mu\|\varphi\|^2 + \lambda\|\varphi\|^4, \tag{2}$$

where the Higgs field φ is a section of a bundle associated to P by a representation (usually the fundamental representation). The "potential" V has a nontrivial extremum φ_0 if $\mu>0$, assumed invariant under the subgroup H of G. By appropriate gauge-fixing the total action is then rewritten in a form where the components of Y belonging to $\mathfrak{m} = \mathfrak{g} - \mathfrak{h}$ "acquire mass", as do the surviving Higgs particles (cf. ref.4).

Instead of introducing the scalar fields φ "by hand",one obvious possibility is to consider the sections $\sigma : B \to E$, required for the reduction of P to Q as new "nonlinear fields". They may be considered also as G/H-valued functions on P, equivariant under G. In line with the general principles one must introduce an action term for σ, postulating that the extrema of the action determine the particular σ. There is essentially only one invariant one can form out of the exterior covariant differential of σ (this is required in order to couple it invariantly to the connection Y), namely the "harmonic map" functional considered in another context by Eells and Sampson [5]:

$$S_{HM} = \int_B D\sigma \wedge {*D\sigma}. \tag{3}$$

Specializing to a basis shows that this action is identical to the action of a 4-dimensional sigma model. Such a model has been proposed independently by Misner [6] and myself [7] and is related to earlier work.[8] Fixing the section σ produces the required "mass-terms" in S, but it is not clear in what sense such a theory is renormalizable - the main attraction of the BEHK model.

4. SYMMETRY BREAKING BY FIBER-FLIPPING. A more attractive method of symmetry breaking, which I call the "fiber-flipping method" for reasons which become clear from the figure below, makes use of the details of the reduction process described in Sec. 2. In the figure below are represented the various bundles involved in the process, where for graphical simplicity all manifolds and groups have been represented by line segments or rectangles (one may also consider the figure more literally, replacing B by a trivializing open subset U, P and Q respectively by $U \times G$, $U \times H$, and E by $U \times G/H$; alternatively, you may view the lines as representing the Lie algebras g, h, and the vector space m, corresponding to G/H, when considering the connection and its curvature). In particular, over U, we may consider $U \times G = P|U$ as $(U \times G/H) \times H$, i. e., by "flipping" part of the fiber it is locally the principal bundle over the base space $B \times G/H$ with structure group H. The second drawing shows clearly how a choice of section σ in the "new" base space picks out in the bundle space P a manifold which becomes the reduced bundle Q in the third frame of the figure. The fourth picture represents an extension of the bundle $P \to E$ (second picture)

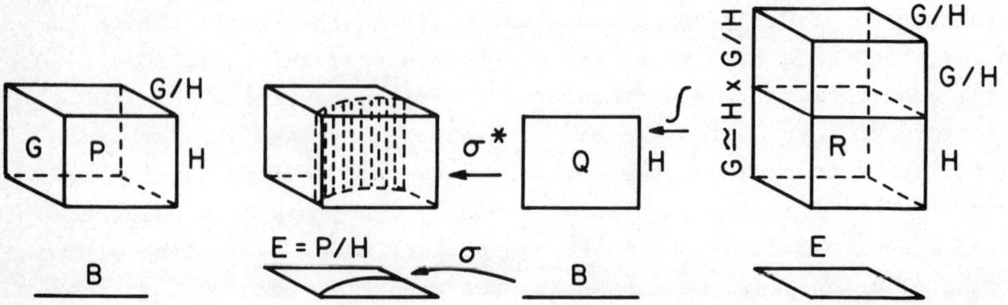

to a principal bundle with structure group G over E (achieved by another "flipping" of the portion G/H into the vertical direction). We denote this new bundle by R and extend the original connection in P to a connection ω with curvature Ω on R. The pullbacks to E will be denoted by (the script letters) $\mathcal{Y}$, $\mathcal{M}$. The latter are forms on the 4 + dim m -dimensional base space E, and the appropriate action is

$$S_{\mathcal{YM}} = \tfrac{1}{4}\int_E \mathcal{M} \wedge {}^*\mathcal{M} \wedge \zeta, \qquad (4)$$

where ζ is the volume m-form of G/H (the measure of G/H). If G/H is a symmetric space the connection and curvature forms $\mathcal{Y}$, $\mathcal{M}$ can be split along the Lie algebra h and the vector space m , with the brackets of the m-components ending up in h. Then by an appropriate gauge-fixing one can reduce the action (4) in such a way that the m-part corresponds to a reduced Yang-Mills action (invariant under H) and a part which involves the covariant differentials of the m-components of $\mathcal{Y}$. The final reduction is achieved by carrying out an averaging over the homogeneous space G/H, made possible by the existence of a finite invariant measure on this space. One ends up with an action which, if one identifies the G/H components of $\mathcal{Y}$ in m with fields which are "scalars" in B, looks exactly like the Higgs action (1) + (2), with the quartic functional coming from the corresponding terms in (4). The details are too lengthy and will be published elsewhere; one should note only that the "Higgs" components end up belonging to the adjoint representation of G, which is a disadvantage for many physical models.

5. DISCUSSION. Although I have not succeeded in producing a model which is both geometrical and satisfying the requirements of phenomenology, the model described in Sec. 4 (which is related to earlier work by Witten, Forgacs, and Manton [9], and to the "dimensional reduction in supergauge theories discussed in the talks of Scherk and Julia [10]) is attractive, since it does not require too many ad hoc assumptions. One might be able to relax the requirement that G/H be symmetric, replacing it with the milder reductivity condition. It is not clear how to take the new "Higgs" states out of the adjoint representation.

BRIEF BIBLIOGRAPHY (for further references cf. 1,4)

1. M. E. Mayer, Lect. Notes in Physics, Vol. 67, Springer, 1977.
2. S.Kobayasi and K. Nomizu, Found. of Differ. Geometry, vol.I,1963.
3. A. Trautman, Bull. Acad. Pol. Sci. 27,7 (1979) and earlier work.
4. cf.,e. g., L. O'Raifeartaigh, Rep. Prog. Phys. 42, 159-223 (1979).
5. J. Eells Jr. and J.H.Sampson, Amer. J. Math. 86,109 (1964).
6. C. W. Misner, Phys. Rev. D18, 4510 (1978).
7. M. E. Mayer, Proc. VIII Intern. Symp. Group Theor. Meth. to appear.
8. J. Madore, CMP 56,115 (1977). J. Harnad, S. Shnider, L. Vinet, this volume, p. 295
9. E. Witten, PRL 38,121 (1977). P. Forgacs and N. S. Manton, preprints.
10. J. Scherk and B. Julia, this volume, pp. 342, 367.

INVARIANCE CONDITIONS FOR GAUGE FIELDS

by

John HARNAD*, Steven SHNIDER** and Luc VINET*

* C.R.M.A. Université de Montréal, Montréal, Québec, Canada.

** Dept. of Mathematics, McGill University, Montreal, Quebec, Canada

Invariance considerations have proved very useful in the context of gauge theories especially in the search for classical solutions to field equations. (See for example [1] and [2].) The problem arises in this context of how to give a general definition for invariance of gauge fields under space-time transformation groups and a way to construct the most general symmetric field configurations.

Let H be the gauge group with Lie algebra h, M a differentiable manifold (e.g. space-time) and U an open set in M on which the gauge potential is represented by an h valued 1- form ω. Let G be a transformation group acting differentiably on M, $f_g: M \to M$, $g \in G$; and suppose for the moment that U is G invariant. <u>The local condition of invariance under G is</u>:

$$f_g^*(\omega)_x = \mathrm{Ad}\rho^{-1}(g,x)(\omega)_x + \rho^{-1}(g,x)d_x\rho(g,x) \tag{1}$$

where ρ: $G\times U \to H$ defines a gauge transformation and satisfies the composition rule:

$$\rho(gg',x) = \rho(g',x)\rho(g,g'x) \tag{2}$$

(If U is not G-invariant, the domain of definition of ρ must be restricted to a neighborhood of the identity in G and the composition rule restricted accordingly).

A change of gauge replaces ρ and ω by equivalent ρ' and ω'. We are interested in determining a characterization of the invariance condition which is independent of this choice of gauge. To do so, the problem must be formulated invariantly in terms of fibre bundles. In this language, the transformation ρ defines (locally) a lift of the G-action on U to that on a principal H-bundle $E|_U$ over U by:

$$\tilde{f}_g\sigma(x) = \sigma(f_gx)\rho^{-1}(g,x) \tag{3}$$

where $\sigma: U \to E|_U$ is the local section defining the gauge. The transformation properties of ω and ρ on a covering of M by overlapping open sets U guarantee the global action on a bundle E over all of M. The gauge field is given by a connection form $\tilde{\omega}$ on E (whose local pull-back by σ is ω) and the invariance conditions (1) become simply

$$\tilde{f}_g^{*}\tilde{\omega} = \tilde{\omega} \tag{4}$$

The problem of finding all classes of inequivalent ρ is equivalent to that of classifying all principal H-bundles with G-actions as automorphisms projecting to the given G-action on M. Solving the invariance conditions (1) amounts to determining all invariant connections on the given bundles. Formulated thus, these problems turn out to have known solutions for the case of homogeneous G-spaces and the results may be extended straightforwardly to more general G-spaces provided the orbit structure is uniform and the isotropy and gauge groups are compact. The main theorems together with some simple corollaries clarifying the results are presented below.

Let $x \in M$ and G_x be the isotropy group at x, the structure of E over G/G_x is determined by:

Proposition 1. [5] There is a one-to-one correspondance between

a) equivalence classes of principal H-bundles E over G/G_x admitting a G action projecting to left multiplication on G/G_x (up to G-equivalent isomorphism) and

b) conjugacy classes of homomorphisms $\lambda: G_x \to H$.

Three corollaries follow. The first states when the bundle is trivial over the entire orbit and hence allows a global definition of ρ.

Corollary 1 The bundle E is trivial over G/G_x iff the homomorphism $\lambda: G_x \to H$ extends to a smooth function $\Lambda: G \to H$ such that

$$\Lambda(gg_1) = \Lambda(g)\lambda(g_1) \quad \text{for} \quad g \in G,\ g_1 \in G_x \tag{5}$$

The two other corollaries analyze the possible reductions of ρ in terms of intrinsic properties of the bundles. Corollary 2 establish when the transformation function ρ can be chosen independent of the point in the orbit so that the inhomogeneous term in condition (1) disappears. Corollary 3 gives the criterion for ρ to be the identity, obviously the simplest transformation function possible.

Corollary 2 The following two conditions are equivalent:

a) The bundle E is trivial with gauge function $\rho(g,kG_x)$ independent of the point kG_x,

b) The homomorphism $\lambda: G_x \to H$ extends smoothly to a homomorphism $\Lambda: G \to H$.

Corollary 3 The transformation function $\rho(g,kG_x)$ reduces to the trivial function $\equiv e$ iff it is trivial when restricted to the isotropy group G_x.

When corollary 3 applies, the invariance condition (1) simplifies to the ordinary invariance condition $f_g^*\omega = \omega$ for 1-forms on M. One case in which this always occurs is when the G-action on M is free, i.e. $G_x = e$. We would like to point out that the constructive nature of the proofs of prop. 1 and its corollaries permits the explicit construction of expressions for the transformation functions ρ. (See [3] for details.)

We now come to the generalization of these results when the base space is not a homogeneous space. Suppose that for all $x \in M$, there is a smooth embedding of an open set $S \subset \mathbb{R}^k$ $(k = \dim M - \dim G/G_0)$ into M, $\varphi: S \to M$ with $\varphi(0) = x$ and $\varphi(S)$ intersecting each orbit in a unique point and further that the isotropy group of all the points $\varphi(S)$ is the same. We call such a situation a simple G-action and such an imbedding a special cross-section. We then have:

Theorem 1 Let M be a manifold with simple G action and compact isotropy groups. Let E be a principal H bundle with G action projecting on the G-action on M. Assume H is compact. Let $\varphi: S \to M$ be a special cross-section through $x \in M$ and $U = G.\varphi(S) \subset M$. Then there is an isomorphism

$$E|_U \simeq E_\lambda \times S \quad \text{for some} \quad \lambda: G_x \to H.$$

E_λ is the bundle determined by the homomorphism λ.

This theorem together with prop.1 and its corollaries completely analyzes the structure of bundles with G-action over a neighborhood of an orbit in any space with a simple G action. On these bundles the G-invariant connections are characterized by a generalization of Wang's theorem (see [4]):

Theorem 2 Let $\mathfrak{g}$ be the Lie algebra of G, $\mathfrak{g}_0$ the Lie algebra of $G_0 \subset G$ and h the Lie algebra of H. The G invariant connections on $E_\lambda \times S$ are determined by

(i) A family of linear maps $A_s: \mathfrak{g} \to h$ depending smoothly on s (the cross-section variable) and satisfying

$$A_s(\xi) = \lambda_*(\xi) \quad \text{for} \quad \xi \in \mathfrak{g}_0 \quad \text{and} \quad \lambda_*$$

the homomorphism $\lambda_*: \mathfrak{g}_0 \to h$ determined by the differential of λ and $A_s(\mathrm{Ad}g^{-1}\xi) = \mathrm{Ad}\,\lambda(g)^{-1}(A_s(\xi))$ for $\xi \in \mathfrak{g}$ and $g \in G_0$.

(ii) A one form μ on S with values in the subalgebra of h of elements invariant under the adjoint action of $\lambda(G_0)$. Again, the theorem is constructive in nature and its application allows the explicit determination of all invariant gauge fields for group actions satisfying the requirements of theorem 1. Cases with a more complicated group orbit structure may also be treated, by using continuity arguments, but no general theorem is known. Detailed examples illustrating these results and their applicability in solving field equations are given in [2,3].

References

[1] R. Jackiw, C. Nohl and C. Rebbi, "Classical and Semi-Classical Solutions of Yang-Mills Theory" in D. Boal and A. Kamal Particles and Fields, Plenum Press 1978.

[2] J. Harnad, S. Shnider and L. Vinet, J. Math. Phys. 20, 931 (1979).

[3] J. Harnad, S. Shnider and L. Vinet "Group Actions on Principal Bundles and Invariance Conditions for Gauge Fields" McGill preprint (1979).

[4] S. Kobayashi and K. Nomizu, Foundations of Differential Geometry Wiley Interscience, 1969.

[5] G.E. Bredon, Introduction to Compact Transformation Groups, Academic Press, 1972.

NONLINEAR GROUP REPRESENTATIONS AND THE LINEARIZABILITY OF NONLINEAR EQUATIONS

Daniel STERNHEIMER

C.N.R.S. Paris and Physique Mathématique, Université de Dijon, FRANCE

Basic definitions and some results in the recently developed theory of nonlinear Lie group representations in Banach and Fréchet spaces are presented. Using cohomological methods, this framework permits a study of the linearizability of covariant nonlinear evolution equations. Formal linearizability is proved under some conditions on the linear part of the representation, for massive and for massless Poincaré covariant equations. In particular, pure Yang-Mills equations supplemented with a relativistic gauge condition are formally linearizable.

The relation between linear differential equations, covariant under the action of a Lie group G, and the (linear) representations of this group is rather well-known. In the nonlinear case such a connection has not been developed — as a matter of fact, even a systematic theory of nonlinear group representations in linear spaces was lacking. The basic ingredients for such a theory have been developed recently by Flato, Pinczon and Simon [1], relying in part on earlier works on extensions of linear group representations and 1-cohomology [2], and using the notions of analytic functions and formal series on Banach or Fréchet spaces (see e.g. ref. 3). This permits to exploit symmetries of nonlinear equations and transform them into a simpler form (e.g. linear) instead of a direct study for each equation.

A similar procedure is used in classical mechanics : the Hamilton-Jacobi method. (For generalization to field theory, it seems that a reduction to a completely integrable system such as the harmonic oscillator would be preferred). If we remember that quantum mechanics can be treated in an autonomous manner [4] using a deformed product of functions over classical phase space and that quantum fields can be viewed as functionals over classical fields (initial data), the advantages of performing well-chosen transformations on the initial data space, which will induce transformations of the quantum fields (in the same way as general coordinate transformations induce linear transformations on function spaces), becomes evident. Due to lack of space, we shall only indicate some essential definitions and results, referring to the quoted literature for details.

1 - Nonlinear Lie group and algebra representations in Banach and Fréchet spaces.

If E is a Fréchet space, $L_n(E)$ the space of symmetric multilinear maps from E^n to E (also considered as linear maps $\hat{\otimes}_\pi^n E \to E$, where π denotes the projective tensor product topology), then to each $f^n \in L_n(E)$ one can associate a monomial $\hat{f}^n$ on E by $\hat{f}^n(\phi) = f^n(\phi,\ldots,\phi), \phi \in E$. Thus one defines the space F(E) of formal series (with fixed point) $f = \sum_{n\geq 1} \hat{f}^n$, with the usual composition law, and formal representations of a group G as homomorphisms $S : G \to F(E)$ such that the maps $g \to S_g^n(\phi)$ are measurable. A one-to-one map Λ between F(E) and the linear operators $L(\tilde{E})$ on the "Fock" space $\tilde{E} = \bigcup_{n\geq 1} (\bigoplus_{p=1}^n \hat{\otimes}_\pi^p E)$ can be defined, the composition of formal series being mapped into the product of operators. Its differential $d\Lambda$ defines a Lie bracket on F(E), which corresponds to the commutator in $L(\tilde{E})$, whence the definition of a formal representation of a Lie algebra $\underset{\sim}{g}$ as a homomorphism $\underset{\sim}{g} \to F(E)$. In a natural way one then defines analytic functions and representations. Two (formal or analytic) representations (S,E) and (S',E) of G are said equivalent if there exists a (formal or analytic) series A such that $S'_g = A\, S_g\, A^{-1}$ for all $g \in G$.

A useful notion is that of a smooth analytic representation, for which the terms of degree >1 do not bring more differentiability requirements than the linear term, that is the function

$$g \to S^1_{g^{-1}}\, S_g$$

from G to a space of analytic functions on the Banach space E is differentiable in some neighbourhood of the identity. Since the successive "layers" in S are built by successive extensions of the linear representation S^1, and extension 1-cocycles can be taken differentiable up to a 1-coboundary, one can show that there always exist an equivalent smooth representation (S',E) to an analytic one.

Then a kind of "Lie theory" is developed : passage from (nonlinear) smooth Lie group to Lie algebra representations, and vice-versa when the linear part is the differential of a (linear) group representation. This theory provides an easy proof of the linearizability [5] of the action of semi-simple Lie group around a fixed point in a finite-dimensional manifold. Several other cases have been studied, including inhomogeneous classical groups in finite-dimensional space [6] where an interesting phenomenon of quantization of coupling constant (between two nonlinear terms) appears.

2 - Applications to covariant nonlinear evolution equations

A nonlinear evolution equation $\frac{d}{dt}\phi_t = P_o(\phi_t)$ is said covariant under a Lie algebra $\underset{\sim}{g}$ if there exists a representation (T, H_∞) of $\underset{\sim}{g}$ such that $T^1 = dU$, U being a linear representation of the corresponding Lie group G on the Banach space H and H_∞ its Fréchet space of differentiable vectors, and if $P_o = T(X_o)$ for some $X_o \in \underset{\sim}{g}$ (the time-translations generator in the case of the Poincaré group).

If (S,E) is a formal representation of a Lie group G in a Fréchet space E of differentiable vectors, we let G act on $\hat{\otimes}^n_\pi E$ via $\hat{\otimes}^n S^1$ and define in this way the cohomology spaces

$$H^{1,n}_\infty = H^1_\infty(G, L(\hat{\otimes}^n_\pi E, E))$$

If $H^{1,n}_\infty = 0$ for all $n \geq 2$ there exists $A \in F(E)$ such that S is formally linearizable, i.e. $S_g = A\, S^1_g\, A^{-1}$; the equation $(d\,\phi_t/dt) = P_0(\phi_t)$ is then formally linearizable, with solution $\phi_t = S_{\exp tX_0}\phi$.

The vanishing of these cohomology spaces is easier to prove using infinitesimal methods, for which the "Lie theory" mentioned above is essential. When G is the Poincaré group [7], $H^{1,n}_\infty = 0$ if S^1 is unitary irreducible with mass $m^2 > 0$. Hence equations of the type $(\partial\phi/\partial t) = (-\Delta + m^2)^{1/2}\phi + f(\phi)$ are formally linearizable. When both energy signs are present, linearizability can be shown [8] for initial data with either positive or negative energy ; this is the case of $(\Box + m^2)\phi = f(\phi)$ for instance. The problem of convergence of the linearization operator A, which is obtained algorithmically in the form

$$A = \prod_{n=2}^{\infty} (1 - \hat{B}^n) \ , \ B^n \in L_n(E)$$

is more delicate since it requires in particular an implicit function theorem in Fréchet spaces.

In the massless case, more can be said on less (i.e. on the time-translations only) : $H^1(\mathbb{R} P_0, L_n(E)) = 0$ when $\mathbb{R} P_0$ acts on E by $t \to U^L(\exp tX_0)$, U^L being a unitary induced representations of the Poincaré group on $E = \mathcal{S}(M,F)$ or $C^\infty(M,F)$, where M is the vertexless future cone C_+ or past cone C_-, or $C_+ \cup C_-$, and F is a finite-dimensional space (namely, any finite number of massless particles is allowed, with both energy signs). Therefore $(\partial/\partial t)\ \phi_t = P_0(\phi_t)$ is formally linearizable when P^1_0 generates $U^L(\exp tX_0)$. This is in particular the case of Yang-Mills equations in Minkowski space supplemented with a relativistic gauge condition such as $\partial_\mu A^\mu = 0$ or $\partial_\mu \Box A^\mu = 0$ (conformally covariant), etc.. This is also the case of Einstein equations (without matter) and others.

REFERENCES

[1] M. FLATO, G. PINCZON and J. SIMON : Ann. Scient. Ec. Norm. Sup. 10, 405 (1977).
[2] G. PINCZON and J. SIMON : Reports On Math. Phys. (1979).
[3] L. NACHBIN : Topology on Spaces of Holomorphic Mappings. Springer Verlag (1969).
[4] F. BAYEN, M. FLATO, C. FRONSDAL, A. LICHNEROWICZ and D. STERNHEIMER : Ann. Phys. (NY) 111, 61, 111 (1978).
[5] V. GUILLEMIN and S. STERNBERG : Trans. Amer. Math. Soc. 130, 110 (1968).
[6] J. SIMON, G. PINCZON : Lett. Math. Phys. 2, 499 (1978).
[7] M. FLATO, J. SIMON : Lett. Math. Phys. 2, 155 (1977).
[8] M. FLATO, J. SIMON : J. Math. Phys. 21, (1980).
[9] M. FLATO, J. SIMON : Lett. Math. Phys. 3, 279 (1979).

Determinants, Green Functions and Induced Action

B.Schroer

Institut für Theoretische Physik, Freie Universität Berlin

In quantum field theories with Lagrangians which are bilinear in the matter fields, relevant physical properties often only become exposed after integration over the matter fields. The rules for obtaining a functional representation for correlation functions in terms of integration over gluon fields only are well known[1], however there are some subtle modifications[2] if topology and zero eigenvalues for the matter field equations came into play e.i. in gauge theories. The functional integral representation for quarks field correlation functions is of the form

$$\langle \psi(x_1) \ldots\ldots \rangle = \left(A_\mu \text{independent factor}\right) \cdot \int [dA_\mu]\, e^{-S_{ind}(x_1,\ldots)}$$

where the induced interaction consists of three parts.

$$S_{ind}(x_1,\ldots) = S_0 + \Gamma + \Gamma_{source}(x_1,\ldots)$$

The first two parts are the Yang Mills or Maxwell action S_o as well as the modified (by removal of zero modes) determinant of the quark field; they are usually referred to as the effective action. Apart from the zero mode modification of QCD model :

$$\Gamma = \text{"} tr \ln \frac{i\not{D}}{i\not{\partial}} \text{"}$$

this part is universal, i.e. does not depend on the correlation function. The Γ_{source} originates from the (exponentially written) external field dependence of the quark Green Functions and therefore depends on the specific correlation function under consideration. A powerful mathematical formalism (the De Witt Seeley heat equation - or ζ-function formalism) was used to explicitly compute S_{ind} in abelian and nonabelian two-dimensional gauge theories. For (generalized) abelian theories with vanishing quark mass the complicated nonquadratic A_μ terms which appear as an effect of nontrivial topology in Γ cancel against similar terms in Γ_{ind} and the remaining quadratic functional can be integrated reproducing the θ-vacuum-expectation values of the (generalized) Schwinger model. In QCD_2 the S_{ind} remains complicated (nonpolynomial) and was only studied by quasiclassical approximations. These saddle point method applied to correlation functions of gauge invariant composite fields becomes exact in the limit of large euclidean seperations; they asymptotically approach the correlation functions of the corresponding abelians "torus" gauge theory. A detailed account of the methods and results has been given elsewhere[3]. The reader who is interested in details and furthergoing investigation is referred to the literature[4)5)6)7)].

References

1) P.T.Mathews and A.Salam, Phys.Rev.90,690 (1953).
2) N.K.Nielsen and B.Schroer, Nucl.Phys. B120,62, (1977) and Phys.Lett.66B,373(1977).
3) "Green Functions, Determinants and Induced Actions in Gauge Theories", K.D.Rothe and B.Schroer, FU preprint, Lectures given at the International Summer Institute in Kaiserslautern August 13-24,1979, to be published in the proceedings of this Summer Instituts.
4) M.Hortaçsu, K.D.Rothe and B.Schroer, FUB/HEP March 1979, to be published in Phys.Rev.D (Nov.79).
5) N.K.Nielsen, K.D.Rothe and B.Schroer, FUB/HEP May 1979, to be published in Nucl.Phys.B.
6) "Do Quark Correlation Functions exist in Confining Gauge Theories" K.D.Rothe and B.Schroer FUB/Oct.1979.
7) "Zero Energy Eigenstates for the Dirac Boundary Problem" M.Hortaçsu, K.D.Rothe and B.Schroer, FUB/HEP Oct.1979.

SPIN GAUGE THEORY OF ELECTRIC AND MAGNETIC SPINORS

by

J.S.R. Chisholm and Ruth S. Farwell
University of Kent, Canterbury, Kent, England.

The work is in two sections:

(a) A theory of four-component "generalised" electrons, with the Dirac electron as a special case;
(b) An eight-component dual theory of modified generalised electrons and monopoles; it was not possible to include monopoles in a four-component theory.

Our aim has been to formulate a dual gauge theory of electric and magnetic sources, without unphysical singularities such as strings and patching. The gauge theory has some definite differences from standard gauge theories, and we call it a "spin gauge theory". The basic physical problem is that magnetic sources have div $\underline{H} \neq 0$, so that the relation $\underline{H}$ = curl $\underline{A}$, where $\underline{A}$ is the magnetic vector potential, must fail somewhere. To meet this difficulty with purely classical sources, Cabibbo and Ferrari[1] introduced two potentials, the usual A_μ and the dual "magnetoelectric potential" B_μ; the corresponding fields are

$$F_{\mu\nu} = \partial_\mu A_\nu - \partial_\nu A_\mu, \quad G_{\mu\nu} = \partial_\mu B_\nu - \partial_\nu B_\mu.$$

Their work was further elucidated by Han and Biedenharn.[2] The total electromagnetic field

$$K_{\mu\nu} = F_{\mu\nu} - \tilde{G}_{\mu\nu} \equiv F_{\mu\nu} + \tfrac{1}{2}\eta_{\mu\nu\rho\sigma}G^{\rho\sigma}$$

is invariant under the 8-parameter set of transformations

$$A_\mu \to A_\mu + a\partial_\mu \Theta + c\partial^\alpha \Lambda_{\alpha\mu} \tag{1a}$$

$$B_\mu \to B_\mu + b\partial_\mu \Phi + c\partial^\alpha \tilde{\Lambda}_{\alpha\mu}, \tag{1b}$$

the Λ terms defining "mixing transformations"; $\Lambda_{\mu\nu}$ is an anti-symmetric tensor with $\Box\Lambda_{\mu\nu} = 0$.

We use transformations of type (1) as local spin space invariance transformations of a generalised electron Lagrangian. We define covariant derivatives as spin space operators by

$$\gamma^\mu D_\mu\psi \equiv \gamma^\mu(\partial_\mu\psi) - (\alpha-i\beta)(iA_\mu-\gamma_5 B_\mu)\gamma^\mu\psi \tag{2a}$$

and its conjugate

$$\bar\psi \mathbb{D}_\mu\gamma^\mu \equiv (\partial_\mu\bar\psi)\gamma^\mu + \bar\psi(\alpha+i\beta)(iA_\mu+\gamma_5 B_\mu), \tag{2b}$$

where α,β are constants and $\bar\psi = \psi^\dagger\Gamma_4$ with Γ_4 a constant matrix. Then

$$L_1^e = \tfrac{1}{2}i[\bar\psi(\gamma^\mu D_\mu\psi) - (\bar\psi \mathbb{D}_\mu\gamma^\mu)\psi] - M\bar\psi\psi \equiv L_o^e + \alpha\bar\psi A\!\!\!/\psi + \beta\bar\psi\gamma_5 B\!\!\!/\psi,$$

where L_o^e is the symmetrised free fermion Lagrangian, is invariant under the infinitesimal local spin space transformations

$$\psi \to S\psi,\ \bar\psi \to \bar\psi S^{-1}, \tag{3}$$

where

$$S = I - i\Lambda_{\mu\nu}(x)\sigma^{\mu\nu} - i\Theta(x)I - \gamma_5\Phi(x), \tag{4}$$

together with

$$\gamma^\mu \to S\gamma^\mu S^{-1} \tag{5}$$

and (with fixed γ^μ)

$$A_\mu\gamma^\mu \to A_\mu\gamma^\mu + (\beta^2+\alpha^2)^{-1}\{\alpha\partial^\alpha\tilde\Lambda_{\alpha\mu}\gamma_5\gamma^\mu + \beta\partial^\alpha\Lambda_{\alpha\mu}\gamma^\mu - \alpha\partial_\mu\Theta\gamma^\mu + \beta\partial_\mu\Phi\gamma_5\gamma^\mu\} \tag{6a}$$

$$B_\mu\gamma^\mu \to B_\mu\gamma^\mu + (\beta^2+\alpha^2)^{-1}\{-\alpha\partial^\alpha\Lambda_{\alpha\mu}\gamma_5\gamma^\mu + \beta\partial^\alpha\tilde\Lambda_{\alpha\mu}\gamma^\mu - \beta\partial_\mu\Theta\gamma_5\gamma^\mu - \alpha\partial_\mu\Phi\gamma^\mu\}. \tag{6b}$$

The transformations (6) have several important properties: (i) they satisfy duality, (ii) they preserve parity, unlike the well-known transformations of $\underline{E} + i\underline{H}$, (iii) they preserve the conjugacy of D_μ and $\mathbb{D}_\mu$, (iv) since they are of form (1), $K_{\mu\nu}$ remains invariant and a spin scalar.

We have ensured invariance under an eight-parameter set of transformations by introducing only two gauge fields, with eight components. The reason for this is that the spin differential operator $\partial\!\!\!/$ converts (S,T,P) type terms in (4) to (V,A) type terms. This economy in numbers of gauge fields can only happen for Clifford algebras in 2^n dimensions. Equations (2) and the transformations (6) deal essentially with spin space operators $\partial\!\!\!/$, $A\!\!\!/$ and $B\!\!\!/$, and we must regard these operators as being defined within x-space and spin space simultaneously. It seems to us that we should define "spin dynamics", in which $\partial\!\!\!/$ is the single dynamical group generator, within a Clifford algebra formalism. In this spirit, we can combine (5) and (6) to give

$$A\!\!\!/ \to SA\!\!\!/S^{-1} + \text{derivative terms}$$

$$\not{B} \rightarrow S\not{B}S^{-1} + \text{derivative terms}$$

so that $\not{A}$, $\not{B}$ transform as A_μ, B_μ do in standard gauge theories. The transformations are very simple in terms of Hertz tensors.[2,3]

The invariant Lagrangian

$$L^e = L_o^e + \alpha\bar{\psi}\not{A}\psi + \beta\bar{\psi}\gamma_5\not{B}\psi + \tfrac{1}{4}K_{\mu\nu}K^{\mu\nu} - \tfrac{1}{2}K_{\mu\nu}(\partial^\mu A^\nu - \partial^\nu A^\mu) + \tfrac{1}{2}\tilde{K}_{\mu\nu}(\partial^\mu B^\nu - \partial^\nu B^\mu) \tag{7}$$

gives the generalised electron equation

$$(i\not{\partial} - M)\psi + \alpha\not{A}\psi + \beta\gamma_5\not{B}\psi = 0 \tag{8}$$

and its conjugate, the definition of $K_{\mu\nu}$, and the generalised Maxwell equations: for quantisation, (7) needs minor modification. If $\alpha = e$, $\beta = 0$, (8) is the Dirac equation; so changing the local spin frame introduces a B-field into the Dirac equation; and equations (3) - (6) define an eight-parameter invariance algebra of the equation.

The fact that γ^μ is transformed by (5) requires modification of Noether's theorem to the form

$$\partial_\mu \left[\frac{\partial L}{\partial \psi,_\mu} \delta\psi + \delta\bar{\psi} \frac{\partial L}{\partial \bar{\psi},_\mu} + \frac{\partial L}{\partial A_{\nu,\mu}} \delta A_\nu + \frac{\partial L}{\partial B_{\nu,\mu}} \delta B_\nu \right] = -\frac{\partial L}{\partial \gamma_\mu} \delta\gamma_\mu ,$$

the last term implying replacement of γ_μ by $\delta\gamma_\mu$. Since $\delta\gamma_\mu = 0$ only for the Θ-variation, we have only the one usual conservation law. The other seven variations give "partial conservation laws".

It proved impossible to define monopoles, with $\not{A}$ and $\not{B}$ interchanged, in the four-component theory. We define

$$\underline{\Psi} = \begin{pmatrix} \psi \\ \tilde{\psi} \end{pmatrix},$$

where $\tilde{\psi} = CT_4\psi^*$ is the charge conjugate of a generalised electron ψ. Introducing a set $\{\rho_r\}$ of Pauli matrices acting on $\underline{\Psi}$ allows us to define a square root of the charge conjugation operator; this acts on $\underline{\Psi}$ to give the eight-component monopole spinor χ (four independent components), for which B_μ is the Coulomb-Ampere potential. The Lagrangian exhibits dual symmetry, but this symmetry is broken by the gauge transformations of $\not{A}$, $\not{B}$. Also, the theory becomes singular when $\alpha = \beta$, imposing a second asymmetry.

References

1. N. Cabibbo and E. Ferrari, Nuovo Cim. 23, 1147 (1962)
2. M.Y. Han and L.C. Biedenharn, Nuovo Cim. 2A, 544 (1971)
3. J.S.R. Chisholm and Ruth S. Farwell, Kent Preprint, April 1979

SPIN STRUCTURES AND GAUGE THEORY

Peter G. O. Freund
University of Chicago, Chicago, Illinois 60637

The electroweak gauge theory and QCD, or the more ambitious "grand unified" gauge theories[1] do not provide theoretical criteria for the selection of the gauge group itself or for the symmetry multiplet assignments of spinor matter fields. Here I would like to review briefly a body of recent work[2-8] that sheds some light on these problems from a quite unexpected source: the global consistency requirements that arise when coupling spinor fields to gravity.

Given a vierbein we can always locally define spinors on a Riemann 4-manifold. A global definition on the whole manifold is possible only if its second Stiefel-Whitney class w_2 vanishes. This is connected with the properties of closed curves in the orthonormal frame bundle that lie entirely in one fiber and correspond to vierbein rotations of 2π so that spinors change sign. Such curves cannot be contracted to a point while staying in the fiber. But by deforming them through distant regions of the bundle their contraction to a point can still be achieved if $w_2 \neq 0$. The instructions for spinor signs are then self-contradictory, and no global definition of spinors is possible. The simplest procedure for avoiding this problem is to construct "generalized spin structures" on the manifold.[4,5,10] Essentially, one assigns spinors to multiplets of some gauged internal symmetry and absorbs the undetermined sign in the gauge indeterminacy.[11]

The problem is thus to devise a procedure that will upgrade the orthonormal frame bundle (an SO(4) bundle) to a bundle that allows the global definition of spinors. If $w_2 \neq 0$ then the SO(4) bundle cannot just be upgraded to a Spin(4) $\equiv$ $\equiv$ SU(2)XSU(2) bundle and the question is what group to use instead of Spin(4). The simplest possibility[5] turns out to be $\overline{\mathrm{Spin}}(4) = \mathrm{Spin}(4)\mathrm{X}_{Z_2}\mathrm{SU}(2)$. Indeed, there exists a homomorphism. λ: Spin(4) $\to$ Spin(4)XSU(2) defined by the identity map to Spin(4) and projection to one of its SU(2) factors as the map to SU(2). We now specify $\overline{\mathrm{Spin}}(4)$ by identifying $(g,h)\sim(-g,-h)$ for $g\in$Spin(4), $h\in$SU(2). The groups SO(4) and $\overline{\mathrm{Spin}}(4)$ differ from Spin(4) and Spin(4)XSU(2) each by a Z_2 factor. The just described λ-homomorphism then induces an SO(4) $\to$ $\overline{\mathrm{Spin}}(4)$ homomorphism that can be used to upgrade the SO(4)-bundle into a $\overline{\mathrm{Spin}}(4)$ bundle and thus to consistently define spinors on any Riemann 4-manifold once one has gauged SU(2). But for this to work there is one more requirement. Any representation Λ of $\overline{\mathrm{Spin}}(4)$ on a vector space can be pulled back to a representation of Spin(4)XSU(2) $\equiv$ $(\mathrm{SU}(2))^3$. But the element $(-1, -1, -1)$ of $(\mathrm{SU}(2))^3$ must map to the identity in Λ. Label Λ by three spins (j_1, j_2, j_3). This then requires $e^{i\pi(2j_1+2j_2+2j_3)}=1$, or $j_1+j_2+j_3$ = integer. But the ordinary spin-statistics connection tells us that j_1+j_2 = half odd integer (integer) for fermions (bosons). Thus we find that the "internal spin" j_3 must be half-odd integer (integer) for fermions (bosons) and thus establish an "internal spin" - statistics connection.

This construction can be readily generalized to other nonabelian Lie groups[5-8] that contain a Z_2 in their center. For any such group G replace $\overline{\mathrm{Spin}}(4)$ in the above argument by $\mathrm{Spin}_G(4) = \mathrm{Spin}(4) \times_{Z_2} G$. (the construction above had G = SU(2)). G = SU(2)XSU(2) gives an amusing phenomenology[5] if identified as the $SU(2)_L \times SU(2)_R$ of electroweak gauge theory. We shall rather go on directly to "grand" unification theories. Such theories as a rule postulate some "very" large simple group that encompasses $SU(3)_{color} \times (SU(2) \times U(1))_{electroweak}$. Typically, SU(N) and SO(N) groups have been considered. Since the "grand" unification group is the maximal gauged internal symmetry group, it is the only candidate for G in $\mathrm{Spin}_G(4)$. Now for SU(N) groups only those with even N have a Z_2 in their center (those with N = odd have Z_N but no Z_2 in their center). This rules out the famous SU(5) of Georgi and Glashow.[12] (For SO(N) groups rather than a selection in N, we are instructed to take the twofold coverings Spin(N) which is possible for all N > 3). We thus see the theoretical group selection criterion at work. Weak as it is, it manages to rule out the much discussed SU(5)-theory. Recently, based on maximally extended supergravity,[13] we have studied[14] the phenomenology of a grand SU(8) unification group. Remarkably, the favored alternative has the spin 1/2 fermions (leptons and quarks) in <u>56</u> representations of SU(8) which satisfy the "internal spin" - statistics connection as do all the other Bose and Fermi multiplets in that case. The alternative (less favored) assignments of spin 1/2 fermions to 56, $\overline{28}$, $\overline{8}$ multiplets is ruled out by the "internal-spin" - statistics connection. We thus see that this general criterion again has some bite.

It is remarkable that considerations of world topology thus lead us[11] to statements about the nature and multiplet structure of gauged internal symmetries.

This work was supported in part by the United States National Science Foundation.

REFERENCES

1. See, e.g., A. Salam, Proc. 19th Int. Conf. High Energy Physics, Tokyo, 1978, S. Homma, M. Kawaguchi and M. Miyazawa eds. Phys. Soc. of Japan, p. 933.
2. T. Eguchi and P.G.O. Freund, Phys. Rev. Lett. 37, 1251 (1976).
3. S. Hawking, Phys. Lett. 60A, 84 (1977).
4. S. Hawking and C.N. Pope, Phys. Lett. 73B, 42 (1978).
5. A. Back, P.G.O. Freund and M. Forger, Phys. Lett. 77B, 181 (1978).
6. M. Forger and M. Hess, Comm. Math. Phys. 64, 269 (1979).
7. S.P. Avis and C.J. Isham, preprint ICTP/78-79/21.
8. S. Hawking and M. Roček, private communication.
9. R. Geroch, J. Math. Phys. 9, 1739 (1968).

10. For <u>compact</u> Riemann 4-manifolds $Spin^c(4) = Spin(4) X_{Z_2} U(1)$ structures can always be defined, F. Hirzebruch and H. Hopf, Math. Ann. 136, 156 (1958); for the case of $P_2(C)$ the relevant U(1) gauge field is that considered by A. Trautman, Int. J. Theor. Phys. 16, 561 (1977).

11. The obvious alternative is to limit the functional integration that defines the quantum theory to spin-manifolds. Here we explore mainly the consequences of not making this very restrictive assumption.

12. H. Georgi and S.L. Glashow, Phys. Rev. Lett. 32, 438 (1974).

13. E. Cremmer, B. Julia and J. Scherk, Phys. Lett. 76B, 409 (1978); E. Cremmer and B. Julia, Phys. Lett. 80B, 48 (1978) and preprint LPTENS 79/6.

14. T. Curtright and P.G.O. Freund, preprint EFI 79/25 and unpublished.

RECENT RESULTS ON DIFFERENTIABLE DYNAMICAL SYSTEMS

David RUELLE

IHES. 91440 Bures-sur-Yvette. France

Differentiable dynamical systems.

We may think of dynamical systems as describing time evolution. In the simplest case the time will be discrete (a natural integer n) and the evolution given by

$$x_{n+1} = f(x_n)$$

where x_n is a m-dimensional vector and f a differentiable function on $\mathbb{R}^m$.

For instance if $m = 1$ we may take

$$x_{n+1} = ax_n(1-x_n)$$

this quadratic transformation of the line will be discussed by J.-P. Eckmann. If $m = 2$ and x_n has components x'_n, x''_n we may take

$$x'_{n+1} = x''_n + 1 - ax'^2_n$$

$$x''_{n+1} = bx'_n$$

This is the Hénon model, which has been much studied numerically, in particular for $a = 1.4$ and $b = .3$. J. Curry will speak about it.

A continuous time dynamical system is described by a differential equation

$$\frac{d}{dt} x_t = X(x_t)$$

Instead of $\mathbb{R}^m$ we may work on a general differentiable manifold (sphere, torus, etc.), or perhaps a Hilbert space, but the main features of the problems in which we are interested already occur for $\mathbb{R}^m$.

Ergodic theory.

Let ρ be a probability measure with compact support in $\mathbb{R}^m$ (for simplicity) and invariant under the map f . If ρ is ergodic, the ergodic theorem asserts that for ρ-almost all x , and every continuous function φ ,

$$\lim_{n\to\infty} \frac{1}{n} \sum_{k=0}^{n-1} \varphi(f^k x) = \int \rho(dy)\varphi(y)$$

There exists a multiplicative ergodic theorem corresponding to the situation where φ (or rather $\exp\varphi$) is replaced by a matrix-valued function T . Write

$$T_x^n = T(f^{n-1}x). \ldots . T(fx).T(x) .$$

Then for ρ-almost all x , and all $u \in \mathbb{R}^m$

$$\lim_{n\to\infty} \frac{1}{n} \log \|T_x^n u\| = \chi(x,u)$$

exists. For fixed x , $\chi(x,u)$ may take at most m different values called characteristic exponents. For ergodic ρ , the characteristic exponents are constant ρ-almost everywhere. The upper characteristic exponent is

$$\chi^+ = \lim_{n\to\infty} \frac{1}{n} \log \|T_x^n\|$$

We shall now apply these notions to the case where $T(x)$ is the matrix of partial derivatives $\frac{\partial f}{\partial x}$. Then T_x^n is the matrix of partial derivatives $\partial f^n/\partial x$.

If $\chi^+ > 0$, a small error in x_o will yield an error on x_n growing exponentially with n :

$$\delta x_n = \frac{\partial f^n}{\partial x} \delta x_o \sim \exp n\chi^+$$

This we call sensitive dependence on initial condition.

It is important not to confuse χ^+ with the entropy $h(\rho)$, measuring the exponential rate of mixing determined by f with respect to the measure ρ . For a definition of $h(\rho)$ see for instance Billingsley [1] . One has

$$h(\rho) \leq \Sigma \quad \text{positive char. exponents w.r.t.}\rho$$

(the characteristic exponents have to be counted with multiplicity). This was noted by Margulis (unpublished) when ρ is smooth, then proved in general by Ruelle [6] *). In particular, if $h(\rho) > 0$, there is always one strictly positive characteristic exponent : $\chi^+ \geq h(\rho)/(m-1)$.

If f has a differentiable inverse the entropy with respect to f^{-1} is the same as with respect to f , and the characteristic exponents have the opposite

*) Also S. Katok (unpublished) when f is invertible.

sign. Thus if $h(\rho) > 0$ there is a strictly negative characteristic exponent $\chi^- \leq -h(\rho)/(m-1)$.

Suppose that the derivatives of f are Hölder continuous, and that ρ is ergodic.

- If all characteristic exponents are < 0 , then ρ is carried by an attracting periodic orbit (Ruelle [5]).
- If all characteristic exponents are $\neq 0$, the periodic points are dense in the support of ρ (Katok [2]).

Stable manifolds.

Consider the vectors u such that $\chi(x,u) < 0$, they form a linear space V_x . One may consider this fact as the infinitesimal version of a stable manifold theorem which is most easily formulated if we assume that $\mathbb{R}^m$ is replaced by a compact manifold M , that $f : M \mapsto M$ has Hölder continuous derivatives, and a differentiable inverse f^{-1} (i.e. f is a diffeomorphism). Under these conditions there is a set $\Lambda \subset M$, with $f\Lambda = \Lambda, \rho(\Lambda) = 1$ for every f-invariant probability measure, and if $x \in \Lambda$ the set

$$\mathcal{V}_x = \{y \in M : \limsup_{n\to\infty} \frac{1}{n} \log d(f^n x, f^n y) < 0\}$$

is a differentiable manifold contained in Λ *). (Here $d(x,y)$ is the distance of x and y for some Riemann metric, $\mathcal{V}_x$ is tangent to V_x at x , but may turn around so as to be dense in parts of M). If there is at least one strictly negative characteristic exponent, the manifold $\mathcal{V}_x$ is not reduced to a point. $\mathcal{V}_x$ is called a stable manifold. The stable manifolds for f^{-1} are called unstable manifolds.

Pesin theory.

The greatest progress in the study of differentiable dynamical systems in recent times was made by Ia. B. Pesin [4]. Pesin considers a compact manifold M with a diffeomorphism f which is twice differentiable, and preserves a smooth measure ρ . He first proves the existence of stable manifolds almost everywhere with respect to ρ (actually, his results can be extended to non smooth ρ as we indicated above). The stable manifolds do not form a continuous, but only a measurable family. However, what is measurable becomes continuous when restricted to a suitable subset of measure arbitrarily close to 1. Intersecting the bunch of stable manifolds by two transversal manifolds, one gets a natural map from a subset of one transversal

---

*) For a proof see Ruelle [5].

to a subset of the other transversal. This map is absolutely continuous i.e. it sends sets of measure 0 to sets of measure 0 . From this Pesin derives the following striking fact. The set M^* of points at which all characteristic exponents are different from zero is (up to a set of ρ-measure 0) a countable union of ergodic components. In other words the decomposition of ρ into ergodic components is discrete rather than continuous where the characteristic exponents do not vanish. This applies for instance in the situation of Moser's twist theorem. If one could prove that the "hyperbolic points" (i.e. those with nonzero characteristic exponents) form a set of positive measure *) one would have a decomposition of this set into countably many ergodic components. For a number of further results in the line of Pesin theory (but not assuming a smooth invariant measure) see Katok [2] .

Asymptotic measures.

In Hamiltonian mechanics there is a natural measure invariant under time evolution, and it is smooth so that Pesin theory applies. For dissipative systems (for instance in hydrodynamics) one does not know a priori what invariant measure to use (there are in general uncountably many different ergodic measures). One finds however often that one invariant measure is stable under small random perturbations, and that the other measures are simply not seen. The random perturbations could for instance be the roundoff errors in the case of a dynamical system generated by a computer. In some cases the measure which is stable under small random fluctuations can be mathematically determined (for Axiom A diffeomorphisms, see Kifer [3]). One finds that this asymptotic measure ρ is selected by the condition that its conditional measures on unstable manifolds are absolutely continuous with respect to the measure defined by the Riemann metric**). This implies (Walters and Katok, unpublished), that

$$h(\rho) = \Sigma \text{ positive char. exponents}$$

The difference between the entropy and the sum of the positive characteristic exponents reaches therefore here its maximum value, which is zero.

Hydrodynamic turbulence.

My personal interest in differentiable dynamical systems with sensitive dependence on initial condition comes from the belief that they are needed in the description of hydrodynamic turbulence. This requires however going over from finite dimensional manifolds to infite dimensional functional spaces. I have recently been able to extend the multiplicative ergodic theorem and the stable (unstable) manifold theorem to Hilbert spaces (R. Mañe - private communication - appears to have similar

*) Unfortunately no such proof exists !

**) This condition is unfortunately not universal, there are counterexamples.

extensions to Banach spaces). It seems therefore that one is slowly getting closer to the measures which describe turbulence in the realistic setting of partial differential evolution equations of the Navier-Stokes type.

References

1 P. Billingsley. Ergodic theory and information, John Wiley, New York, 1965.

2 A. Katok. Lyapunov exponents, entropy and periodic orbits for diffeomorphisms. Preprint.

3 Iu. I. Kifer. On the limiting behavior of invariant measures of small random perturbations of some smooth dynamical systems. Dokl. Akad. Nauk SSSR 216 N°5, 979-981 (1974). English translation. Soviet Math. Dokl. 15, 918-921 (1974). On small random perturbations of some smooth dynamical systems. Izv. Akad. Nauk SSR. Ser. Mat. 38 N°5, 1091-1115 (1974). English translation. Math. USSR Izvestija 8, 1083-1107 (1974).

4 Ia. B. Pesin. Lyapunov characteristic exponents and ergodic properties of smooth dynamical systems with an invariant measure. Dokl. Akad. Nauk SSSR 226 N°4, 774-777 (1976). English translation Soviet Math. Dokl. 17 N°1, 196-199 (1976). Invariant manifold families which correspond to nonvanishing characteristic exponents. Izv. Akad. Nauk SSSR, Ser. Mat. 40 N°6, 1332-1379 (1976). English translation Math. USSR Izvestija 10 N°6, 1261-1305 (1976). Lyapunov characteristic exponents and smooth ergodic theory. Uspehi Mat. Nauk 32 N°4 (196), 55-112 (1977). English translation Russian Math. Surveys 32 N°4, 55-114 (1977).

5 D. Ruelle. Ergodic theory of differentiable dynamical systems. Publ. Math. IHES. 50.

6 D. Ruelle. An inequality for the entropy of differentiable maps. Bol. Soc. Bras. Mat. 9, 83-87 (1978).

A number of further references can be found in my review at the Rome conference in 1977 (Springer Lecture Notes in Physics N° 80). See also Bifurcation Theory and Applications in Scientific Disciplines, Ann. N.Y. Acad. Sci. 316 (1979).

ON SOME SYSTEMS MOTIVATED BY THE LORENZ EQUATIONS: NUMERICAL RESULTS

J. H. Curry
National Center for Atmospheric Research
Boulder, CO 80307/USA
and
Department of Meteorology
Massachusetts Institute of Technology

1. Introduction

All of the work reported in this article is motivated by a system of three ordinary differential equations which was introduced into the literature by Lorenz [1] in his study of atmospheric prediction and convection. The three-variable model is of interest not only because it provides a simple example of a strange attractor but also because the careful analysis of Lorenz suggested several areas of further study. Among other things, Lorenz reduced (heuristically) the study of a complicated attracting set to the simpler problem of studying a transformation which maps the unit interval into itself; we refer the reader to the articles of Eckmann and Lanford in this volume for more details concerning maps of the unit interval.

Motivated by work on the Lorenz equations, Hénon [2] discovered a polynomial transformation which maps the plane into itself and has some of the properties of the Poincaré map of Lorenz's system. In Section 2 a description of Hénon's transformation is given along with a discussion of its dynamical behavior for several parameter values. A principle tool which we shall use is the frequency spectrum. In Section 3 we describe a 14-variable generalization to Lorenz's equation which has a completely different sequence of bifurcations leading to chaotic behavior. In the final section we mention some recent experimental work and its connection to the 14-variable model of Section 3.

2. Hénon's transformation

Given a system of differential equations in three variables, it is possible to study its dynamics by constructing a surface of section transversal to the flow and considering the resulting transformation which carries the section into itself; such a map is called a Poincaré map. (Lanford [3] has carried out this analysis for Lorenz's system.)

The Hénon transformation is an invertible polynomial mapping of the plane into itself which shares some of the properties of the Poincaré map of Lorenz's system and is defined as follows:

$$T(x,y) = (1 + y - ax^2, bx) \quad . \qquad (1)$$

T has two fixed points whose coordinates are given by

$$x = \frac{(b - 1) \pm\sqrt{(b - 1)^2 + 4a}}{2a} \quad , \quad y = bx \quad . \qquad (2)$$

These stationary solutions are real provided $a > (1-b)^2/4$, in which case the stationary solution in the left half plane is always unstable while that in the right is unstable when $a > 3(1-b)^2/4$.

In [2] a and b are 1.40 and 0.3, respectively. What Hénon found is apparently a strange two-dimensional attractor. Fig. 1a is a picture of the Hénon attractor. In Figs. 1b-1d we see successive enlargements of the region indicated. The structure which is observed on magnification is a family of lines which suggest that locally the attractor is the product of a Cantor set and interval. It is possible to prove using the computer, modulo certain technical facts, that there is indeed a Cantor set in the neighborhood of the Hénon attractor, but such Cantor sets are not attracting.

Hénon was also able to show that for the parameter values he considered there is a compact region M which is mapped into itself under the action of T, M contains the object pictured in Fig. 1a. In [5] Feit proves that for a large range of parameter values such an M exists.

Numerical experiments indicate that if several thousand points are placed on the eigenspace associated with the expanding direction of the fixed point and then iterated by T, one recovers the object pictured in Fig. 1a. This suggests that the Hénon attractor is the closure of the unstable manifold of the fixed point in M. If the Hénon attractor is the closure of its unstable manifold, then, since T contracts volumes (determinant = -b), Figs. 1a and 1b suggest that the unstable manifold may have arbitrarily sharp bends and therefore may not satisfy Smale's Axiom A [6].

Non Axiom A systems are not yet well-understood, however, there are some results concerning such systems. Newhouse, for example, has proven results which suggest that Hénon may have discovered an extremely long, stable periodic orbit [7] and because of errors inherent in computing, the process never exactly repeats. Hence the graphics in Figs. 1b-1d.

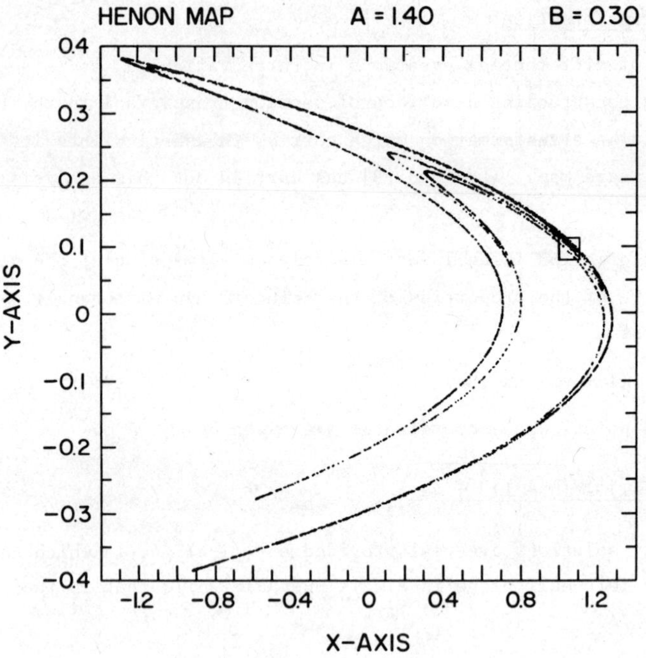

Figure 1a. Hénon's attractor a = 1.40, b = 0.3

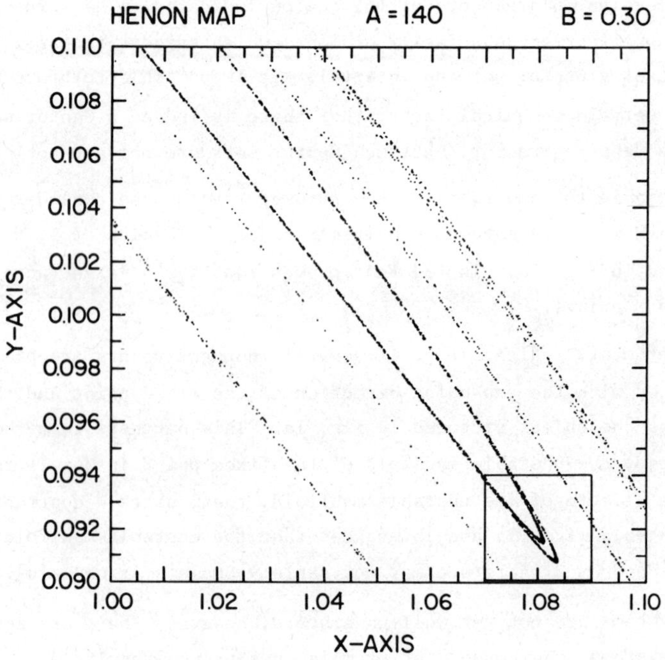

Figure 1b. An enlargement of the boxed section from 1a

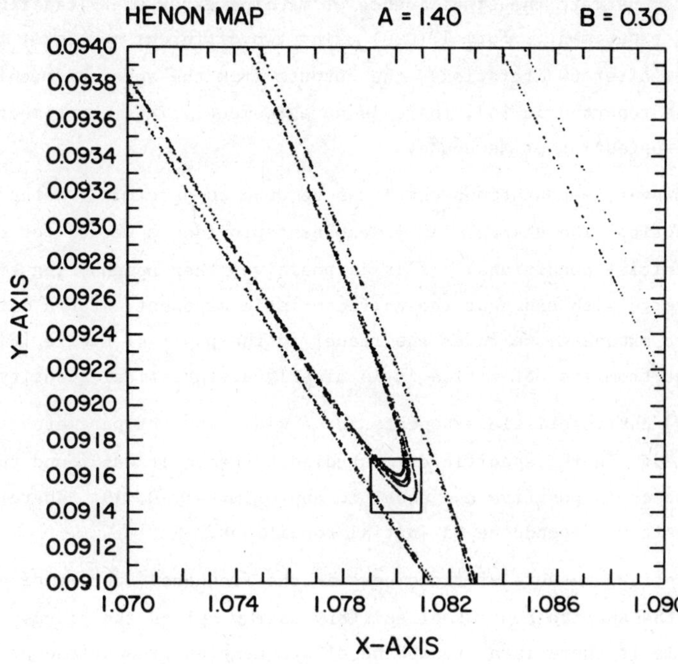

Figure 1c. An enlargement of the boxed section from 1b

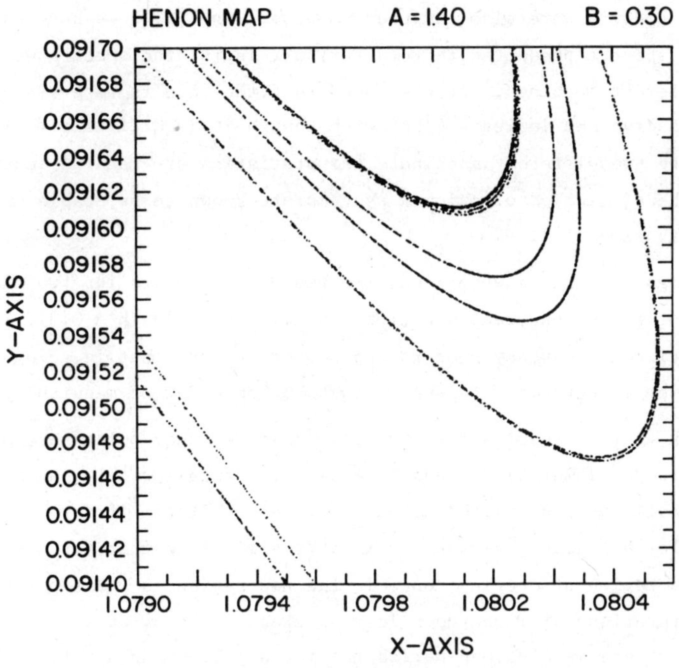

Figure 1d. An enlargement of the boxed section from 1c

In order to demonstrate the significance of machine error when iterating T, consider the following experiment. Form $T^n(0,0)$ using two different machines; there is a priori no reason why, after 60 iterations, the outputs from the separate machines should agree. Indeed, as was reported in [4], there is no agreement. This experiment provides support for the suggestion of Newhouse.

There are, however, computations which can be done to determine whether a process is random or chaotic. The characteristic exponent provides a method for determining sensitivity to initial conditions. If it is positive, then neighboring trajectories exponentially diverge with exponent the characteristic exponent. For a definition of the characteristic exponent, we refer the reader to the paper of Ruelle [8] in this volume; we remark that from its definition it is clearly a statistical quantity.

In [5] and [4] characteristic exponents for a wide range of parameter values of T have been calculated. In the specific case studied by Hénon it was found that the characteristic exponent is positive and equal to approximately 0.419. Therefore what Hénon found has sensitive dependence on initial conditions.

Another statistical measure of randomness is the frequency spectra of a process. Recall that if the spectra consist of solitary narrow spikes the process is (multiply) periodic, while if there is a broad band of frequencies present the process has continuous spectra and is not periodic.

Fig. 2 is the frequency spectra of the Hénon transformation ($a = 1.40$, $b = 0.3$). In [4] frequency spectra were also computed; however, in Fig. 2 we have twice the resolution of the spectra presented there. Further, in [4] the spectra were also computed using two different machines. It was found that the results were virtually identical. This latter observation indicates that even though it is difficult to follow individual trajectories on two different machines, the statistics are machine independent. We remark that the statistics of Axiom A systems are known to be stable under small stochastic perturbations [9].

Finally, in Figs. 3a and 4a we present the phase space plots for two additional parameter values for the Hénon transformation $a = 1.15$ and 1.16 ($b = 0.3$), in Figs. 3b and 4b their associated frequency spectra can be seen. Note that in going from Fig. 3 to Fig. 4 the power in certain frequencies jumps by an order of magnitude.

A careful look at the dynamics for $a = 1.15$ indicates that points are mapped from the upper branch to the lower branch and vice versa. Hence for this parameter value the process is noisily periodic with period two. We see little if any evidence of this in Fig. 3b. For $a = 1.16$ the dynamics is very similar to that for $a = 1.15$. However, there is now a region of overlap; once in the new region, a phase point remains there for several iterations of T and upon leaving executes behavior similar to that for $a = 1.15$. Why there is a strong period two for $a = 1.16$ and not for $a = 1.15$ is not clear. We remark that there is an a_c ($1.15 < a_c < 1.16$) for which the stable and unstable manifolds of the fixed point cross; C. Simo [23] has calculated a_c for the fixed point and other periodic points as well. When this happens we expect complicated new orbits to come into existence [6] or [7]. This may explain the sudden increase in power for certain frequencies.

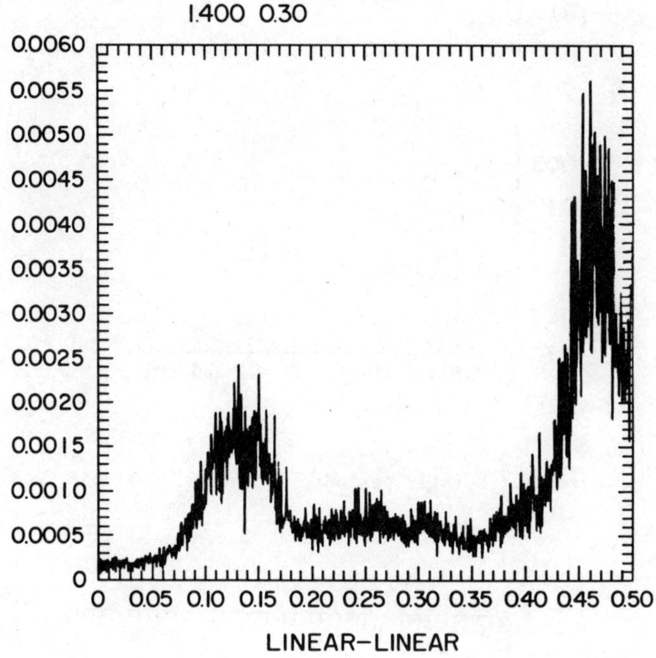

Figure 2. Frequency spectra of Hénon's map $a = 1.40$, $b = 0.3$ linear-linear

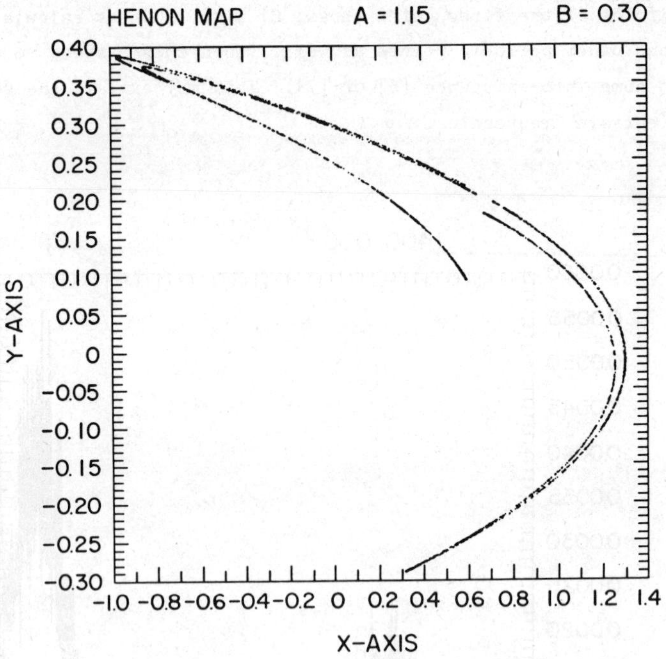

Figure 3a. Noisily periodic attractor a = 1.15, b = 0.3

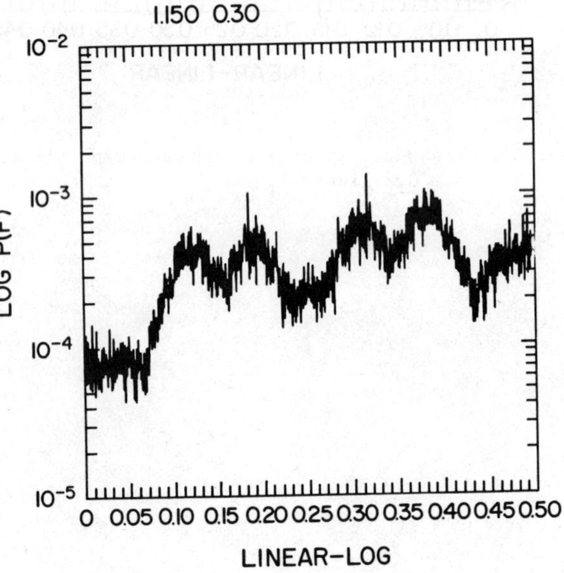

Figure 3b. Frequency spectra of Fig. 3a (linear-log)

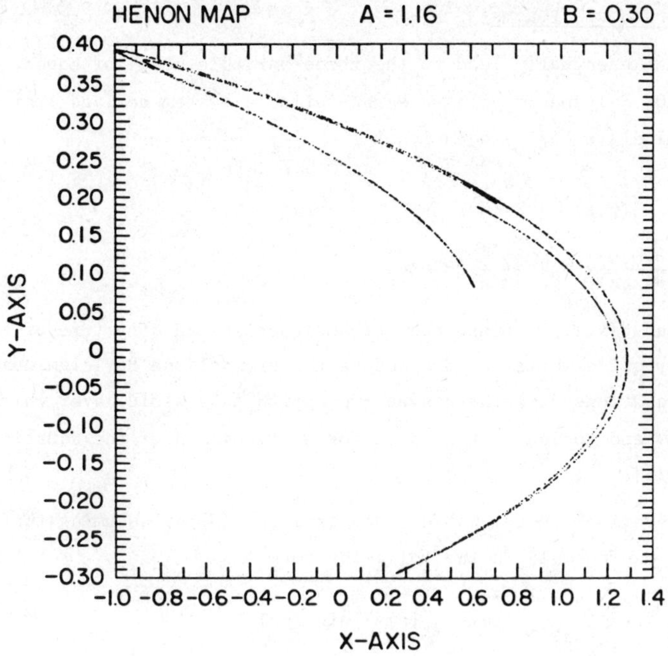

Figure 4a. Attractor after homoclinic intersection
$a = 1.16$, $b = 0.3$

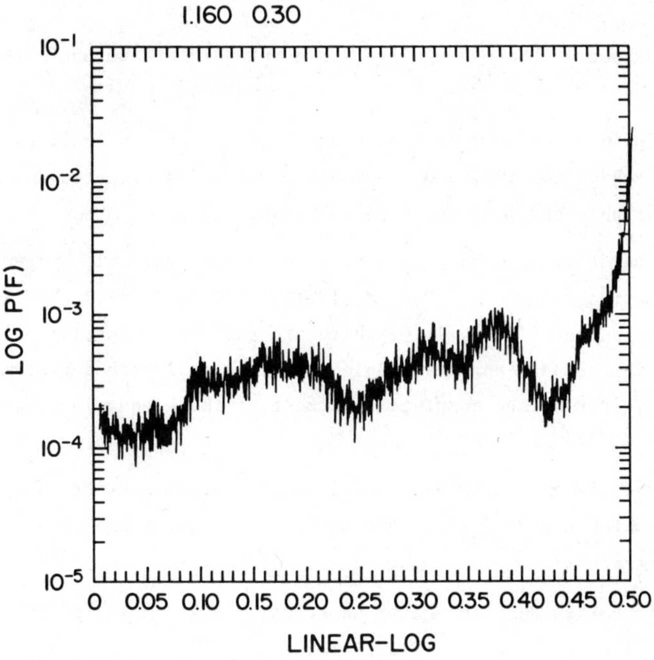

Figure 4b. Frequency spectra of Figures 3 and 4a (linear-log)

3. Higher dimensional systems

The equations that eventually lead to the three-variable model of Lorenz are due to Saltzman. In [10] Saltzman studies a seven-variable system derived from the following system of partial differential equations:

$$
\begin{aligned}
(\Delta\psi)_t &= -\frac{\partial(\psi,\Delta\psi)}{\partial(x,z)} + \sigma\Delta^2\psi + \sigma\frac{\partial\theta}{\partial x} \\
(\theta)_t &= -\frac{\partial(\psi,\theta)}{\partial(x,z)} + R\frac{\partial\psi}{\partial x} + \Delta\theta
\end{aligned}
\tag{3}
$$

where $\psi(x,z,t)$ and $\theta(x,z,t)$ denote the stream function and departure of temperature from its linear profile while σ and R denote the Prandtl and Rayleigh numbers respectively. These equations describe convective motion in a fluid layer which is uniformly heated from below and cooled at the top. For a discussion of the equations in the above form see [11].

Saltzman's seven-variable model was derived from Eq. (3) by assuming that $\psi(x,z,t)$ and $\theta(x,z,t)$ are expressible in the following form

$$
\begin{aligned}
\psi(x,z,t) &= \sum_{\substack{(m,n)\in\Lambda \\ m\neq 0}} \psi_{mn}(t)\,\sin(amx)\,\sin(nz) \\
\theta(x,z,t) &= \sum_{(m,n)\in\Lambda} \theta_{mn}(t)\,\cos(amx)\,\sin(nz) \quad .
\end{aligned}
\tag{4}
$$

Λ denotes a subset of the nonnegative plane integer lattice and a is a geometric constant.

Saltzman noted that for some initial conditions all but three of the seven variables in his model tended to zero, and the nonzero variables executed aperiodic behavior. This three-variable model is the Lorenz system.

In our previous studies we have concentrated on a 14-variable system of ordinary differential equations with $\Lambda = \{(0,2),(1,1),(2,2),(1,3),(3,1),(3,3),(2,4),(0,4)\}$. Since Λ contains the subset $\{(0,2),(1,1)\}$ which are the same modes (up to geometry) which are included in Lorenz's three-variable model, we call such a system a generalized Lorenz system. For a more complete treatment of these equations, we refer the reader to [12].

In treating the 14-variable model we are particularly interested in the sequence of bifurcations which ultimately lead to an attractor which has some characteristics of the Lorenz attractor.

It is convenient when describing the numerical result to introduce the parameter

$$
r = \frac{R}{6.75}
\tag{5}
$$

where the value in the denominator is the critical Rayleigh number below which convection is impossible, r is the bifurcation parameter and σ is fixed and equal to ten.

Numerical Results

In reporting the results of our numerical experiments it is convenient to divide the parameter range into several intervals. In each succeeding interval the dynamics is more complex than the preceding one. The intervals are:

Interval I	$1 \leqq r \leqq 44.3$
Inverval II	$44.3 < r < 45.19$
Interval III	$r > 45.19$.

Interval I. As r exceeds 1, the origin in the 14-dimensional phase space becomes unstable and a symmetric pair of stable fixed points bifurcate off, we label them C_1 and C_2. As r crosses the sequence of values defined by

$$r = \frac{2(0.5\ m^2 + n^2)^3}{(6.75)(m)^2} \tag{6}$$

for $(m,n) \varepsilon \Lambda$ new symmetric pairs of unstable stationary solution bifurcate from the origin.

For $r \varepsilon (1,43.48)$ the model exhibits very simple dynamics. There are at least seven stationary solutions, two of which are stable, C_1 and C_2. Further, given any initial condition in which the Lorenz components are nonzero, the solution curve will converge to either C_1 or C_2.

When r passes 43.48 the C_i become unstable by having a pair of complex conjugate eigenvalues cross into the open right half plane and the stability of each C_i is transferred to an attracting closed cycle. For r = 43.48 the imaginary part of the eigenvalues of interest is ±35.672. From the Hopf Bifurcation Theorem [13] we expect the period to be $2\pi/|35.672| \approx 0.17613$. The computed period is 0.174, which is less than 2% from the predicted value. As r is increased to 44.3, the Hopf orbit undergoes a subharmonic bifurcation which causes its period to double. We conclude this subsection by remarking that for $r \approx 44.2$ we find the unexpected appearance of a stable-unstable closed orbit pair. This pair of hyperbolic orbits suddenly appears in phase space and is not associated with the bifurcation of any of the seven unstable critical points. The orbits persist until around r = 45.10, when they coalesce.

Interval II. We denote by γ a closed orbit, by Σ a 13-dimensional local section transverse to γ, $x \varepsilon \gamma$, and by $P_r(x)$ the associated Poincaré map.

It is not difficult to define a Poincaré map using the computer. Σ was chosen to be the section defined by θ_{04} = constant, the solution was required to intersect Σ to an accuracy of 10^{-10}. We shall concentrate on the behavior of the flow in a neighborhood of C_1. Recall that the stability of γ can be determined by studying the spectrum of DP the derivative of the Poincaré map. If the spectrum is contained inside the unit circle γ is stable, while if a portion lies outside γ is unstable.

In [14] we have described a method for finding numerically a point on a closed orbit. The method has been applied to the 14-variable model. For the subharmonic orbit which bifurcates from the Hopf orbit, we find that when $r = 44.87$ a complex conjugate pair of eigenvalues cross out of the unit circle for the spectrum of the associated Poincaré map.

Fig. 5 is the projection of Σ onto the (ψ_{11},θ_{11}) plane for $r = 44.96$. This figure provides evidence that the γ mentioned above has undergone a bifurcation and transferred its stability to an invariant two-dimensional torus. The Naimark-Sacker, Ruelle, Takens bifurcation theorem [13] does not provide any information concerning the nature of the flow on the torus in this situation.

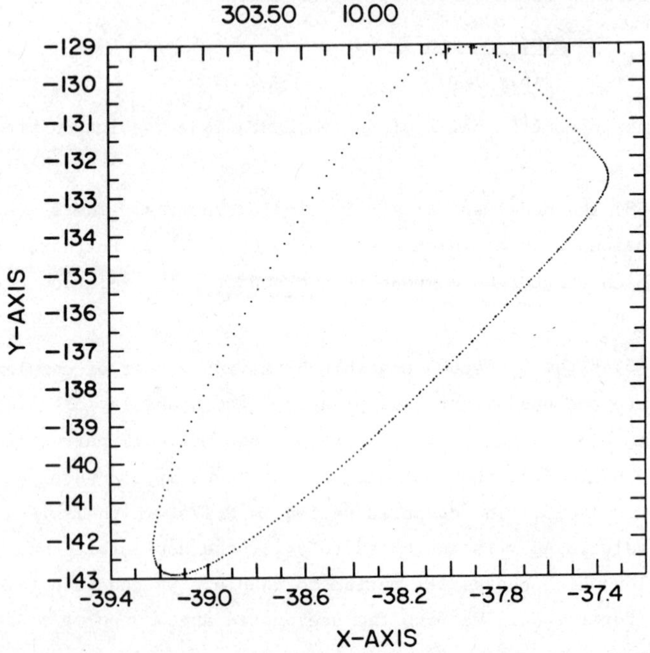

Figure 5. Projection of Σ onto the (ψ_{11},θ_{11}) plane ($r = 44.96$) the motion is on a two torus

In Fig. 6 ($r = 45.148$) we have a projection of Σ onto the (ψ_{11},θ_{11}) plane. The additional loop in the torus should be noted. Recall that a two-torus in dimension greater than three does not have an inside and outside [7]. As r is increased to 45.19, more and more loops appear inside the projected torus.

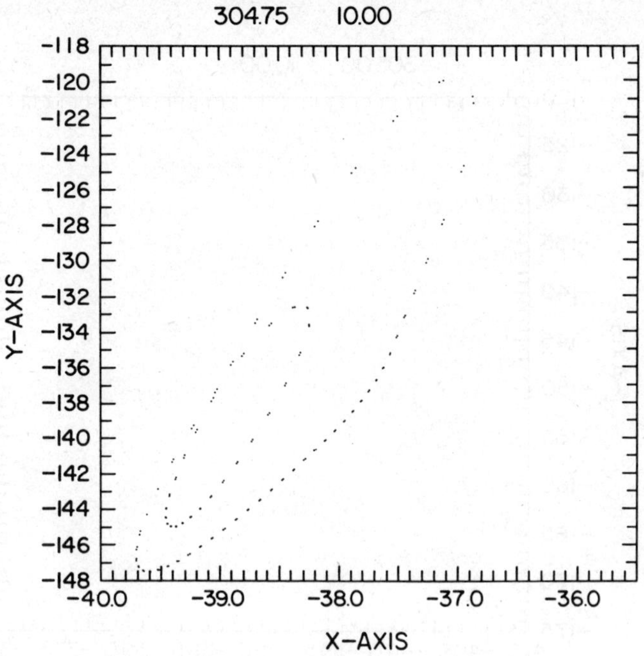

Figure 6. Projection of Σ onto the (ψ_{11},θ_{11}) plane (r = 45.15) the torus has developed loops

<u>Interval III</u>. As r exceeds 45.19 the flow on the torus becomes unstable and we find behavior similar to that which was observed by Lorenz in his three-variable model. We remark, however, that the dynamics in this regime are extremely complicated and not at all well-understood. We refer the reader to [12] for more detail. Fig. 7 provides evidence that the flow on the torus in unstable, r = 45.19.

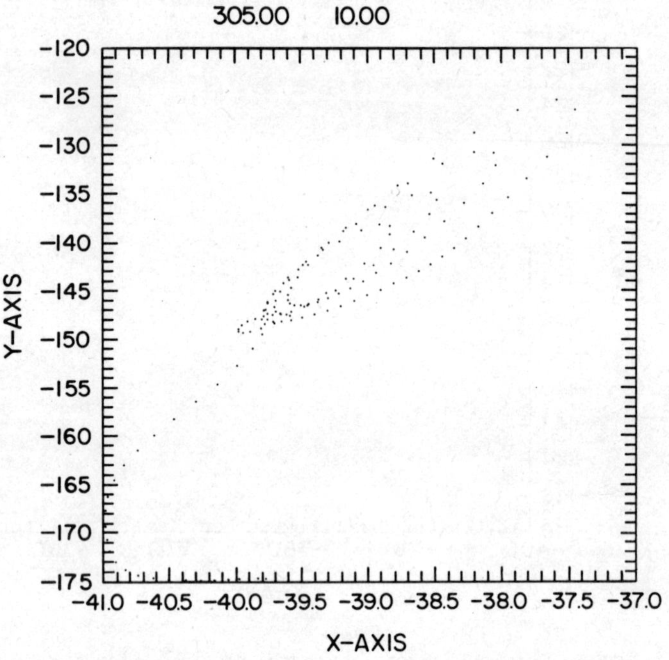

Figure 7. Projection of Σ onto (ψ_{11}, θ_{11}) plane (r = 45.19) flow on the torus is unstable

4. Periodic forcing

In the previous section we noted that Eq. (3) provides a description of convective motion in a fluid layer which is uniformly heated from below and cooled at the top. The first quantitative experiments on this problem were done by Bénard in 1900 [17]. For a review of the state-of-the-art prior to 1960 we refer to [18], for a recent review see [19].

Recently Gollub and Benson [20] have redone the transition to turbulence experiments of Bénard. By computing the power spectra of the velocity field of the fluid, their experiments indicate that the transition to turbulence occurs in the following way. There is first a laminar state which, as the Rayleigh number is increased, loses its stability. The spectra in [20] then indicate sharp peaks corresponding to a single frequency and its harmonics.

As R was increased further, they saw a subharmonic bifurcation which caused a halving of the original frequency. As R crossed another critical value, they found a range for which the motion was quasiperiodic with two frequencies. Beyond the quasiperiodic regime they found complicated turbulent behavior. This sequence of bifurcations is qualitatively similar to those which we noted in the previous section for our 14-variable model.

Gollub and Benson then introduced a small amplitude periodic modulation in the temperature at the lower boundary, i.e., they modulated the Rayleigh number. It is reported that when the flow is in the quasiperiodic regime a small perturbation of the type just described does induce turbulence well below the threshold for the unforced problem provided the forcing amplitude is sufficiently large. Further, the effective Rayleigh number never exceeds the value which is necessary for the unperturbed flow to transition to turbulence.

Motivated by their work, I have introduced a small amplitude periodic modulation into the 14-variable model [21]. There it is also found that the flow for the nonautonomous system became unstable well below the threshold which occurred in the autonomous case.

The modulating frequency in [20] was chosen from a small neighborhood of the difference of the two naturally occurring frequencies; hence, the behavior may be a simple resonance. But our numerical calculations indicate that there is a large frequency range both resonance and nonresonance for which the flow for the nonautonomous system will become chaotic. Just as in [20] the effective Rayleigh number never exceeds the value which is necessary for the unperturbed flow to become chaotic.

The recent work of Newhouse, Ruelle and Takens [22] may provide an explanation for the behavior observed in [20] and [21]. In [22] it is proved that in the neighborhood of any quasiperiodic motion with three frequencies there is a motion which will be chaotic. By adding a perturbing frequency in the parameter range where there is a torus we are certainly in a situation for which the results of [22] is suggestive at the very least.

Acknowledgments

This work was done while the author was a visitor at the National Center for Atmospheric Research, which we thank for their hospitality and for the use of their computation center. (The National Center for Atmospheric Research is sponsored by the National Science Foundation.) J. H. Curry is on leave from the Department of Mathematics, University of Colorado, Boulder, CO 80307/USA. Thanks to H. Howard for typing several drafts of this manuscript. This work was supported by the Air Force Geophysics Laboratory under contract #AF F19628-78-C-0032.

REFERENCES

1. Lorenz, E. N., 1963: Deterministic nonperiodic flow. J. Atmos. Sci., 20, 130.
2. Hénon, M., 1976: A two-dimensional mapping with a strange attractor. Comm. Math. Phys., 50, 69.
3. Lanford, O. E., 1979: Statistical and qualitative theory of dissipative systems. CIME Lecture, 1977.
4. Curry, J. H., 1979: On Henon's transformation. Comm. Math. Phys., (to appear).
5. Feit, S. D., 1978: Characteristic exponents and strange attractors. Comm. Math. Phys., 61, 249.
6. Smale, S., 1967: Differentiable dynamical systems. BAMS, 73, 747.
7. Newhouse, S., 1979: Dynamical systems. CIME Lecture, 1978.
8. Ruelle, D., 1979: Lecture in this volume.
9. Ruelle, D., 1977: Lecture presented to New York Academy of Sciences.
10. Saltzman, B., 1963: Finite amplitude free convection as an initial value problem-I. J. Atmos. Sci., 19, 329.
11. Curry, J. H., 1978: Finite dimensional approximation to the Boussinesq equations. SIAM J. Math. Analysis, 10, 71.
12. Curry, J. H., 1978: A generalized Lorenz system. Comm. Math. Phys., 60, 193.
13. Marsden, J., and M. McCracken, 1976: The Hopf bifurcation and its application. In: Lecture notes in applied mathematical sciences, Vol. 19, Berlin-Heidelberg-New York, Springer.
14. Curry, J. H., 1979: An algorithm for finding closed orbits. In Proc. Int'l. Conf. Global Theory of Dyn. Systems.
15. Ruelle, D., and F. Takens, 1971: On the nature of turbulence. Comm. Math. Phys., 20, 167.
16. Lanford, O. E., 1973: In Springer lecture notes in Math., vol. 322.
17. Bénard, H., 1900: Les Tourbillions dans une nappe liquide. Revue géneral des Sciences pures et appliquées, 11, 1261 and 1309.
18. Saltzman, B., 19??: Thermal Convection, Dover Publishing.
19. Busse, F. H., 1978: Nonlinear properties of thermal convection. Reports Prog. Phys., 41, 1930.
20. Gollub, J., and S. Benson, 1979: Phys. Rev. Lett., 41, 948.
21. Curry, J. H., 1979: Chaotic response to periodic modulation in a model of a convecting fluid. (submitted).
22. Newhouse, S., D. Ruelle, and F. Takens, 1978: Comm. Math. Phys., 64, 35.
23. Simó, C., 1979: On the Henon-Pomeau attractor. To appear in J. Stat. Physics.

PROPERTIES OF CONTINUOUS MAPS OF THE INTERVAL TO ITSELF *

P. Collet, Harvard University
J.-P. Eckmann, University of Geneva

Continuous maps of an interval into itself are among the simplest dynamical systems one can study. We review here some of the properties which seem most relevant in the context of natural sciences. Given a function f which maps $[-1,1]$ into itself, and an initial point $x_o \in [-1,1]$, we can define $x_n = f(x_{n-1})$ for $n = 1,2,\ldots$. We are interested in the set of points thus obtained. In order to visualize these iterates on can recur to the graphical method shown in Fig. 1.

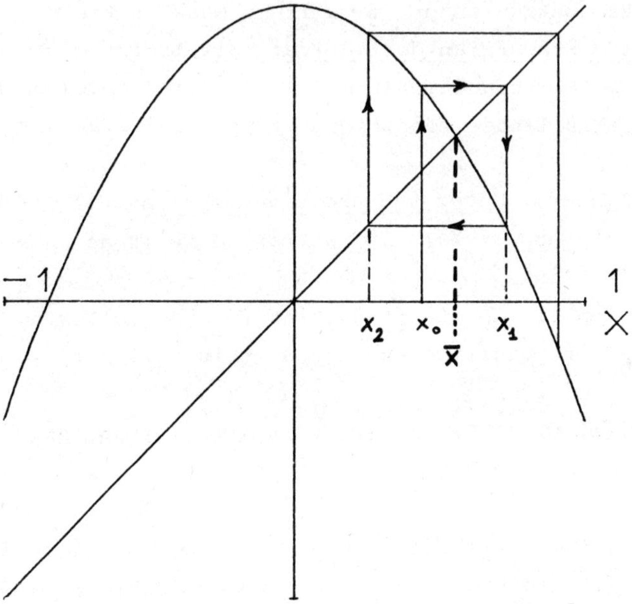

Fig. 1. The function $f(x) = 1 - 1.4\,x^2$.

It is clear that the point marked $\bar{x}$ is a fixed point of the map $x \mapsto f(x)$. As is well known, in order to examine the local stability of the fixed point, one analyzes $f'(\bar{x})$. If $|f'(\bar{x})| < 1$ then the

*Supported by NSF Grant PHY-77-18762 and by the Fonds National Suisse.

fixed point is stable; i.e. if x is near $\bar{x}$, then $f^n(x) \to \bar{x}$ as $n \to \infty$. If $|f'(\bar{x})| > 1$, the fixed point is called unstable. A fixed point $\bar{x}_p$ of f^p is called a periodic point of period p, and $|f^{p'}(\bar{x}_p)| \lessgtr 1$ is used to decide on its stability. An easy, but important, consequence of the chain rule of differentiation is

$$f^{p'}(x_o) = \prod_{j=o}^{p-1} f'(f^j(x)) = \prod_{j=o}^{p-1} f'(x_j) .$$

In particular, if one of the points x_j equals 0 and if x_o is periodic of period p, then it is stable.

In order to simplify the discussion, we shall henceforth assume that $f(0) = 1$, and that $\operatorname{sign} f'(x) = -\operatorname{sign} x$, and $f'(0) = 0$. This special role played by 0 should be borne in mind below. An important simplifying assumption about f will be made in the sequel. We assume f has <u>negative Schwarzian derivative</u>; this is for $f \in C^3[-1,1]$ equivalent to the weaker condition that $|f'|^{-\frac{1}{2}}$ is convex on $[-1,0)$ and on $(0,1]$. (This latter condition is sufficient for our purpose).

<u>THEOREM 1</u>. [Singer]. <u>Under the above assumptions, if f has a stable periodic orbit, then $f^n(0)$ is attracted to it as</u> $n \to \infty$.

This observation is the starting point for the following strategy, which will be our guideline throughout this review :

THE STUDY OF THE ORBIT OF 0 SHOULD ALREADY REVEAL MANY PROPERTIES OF f.

If we do this, e.g. for $f(x) = 1 - x^2$ (Fig. 2a) and for $f(x) = 1 - 1.5436\, x^2$ (Fig. 2b), $x_o = 0$, we get the following histograms for 10000 points (with 200 intervals on $[-1,1]$).

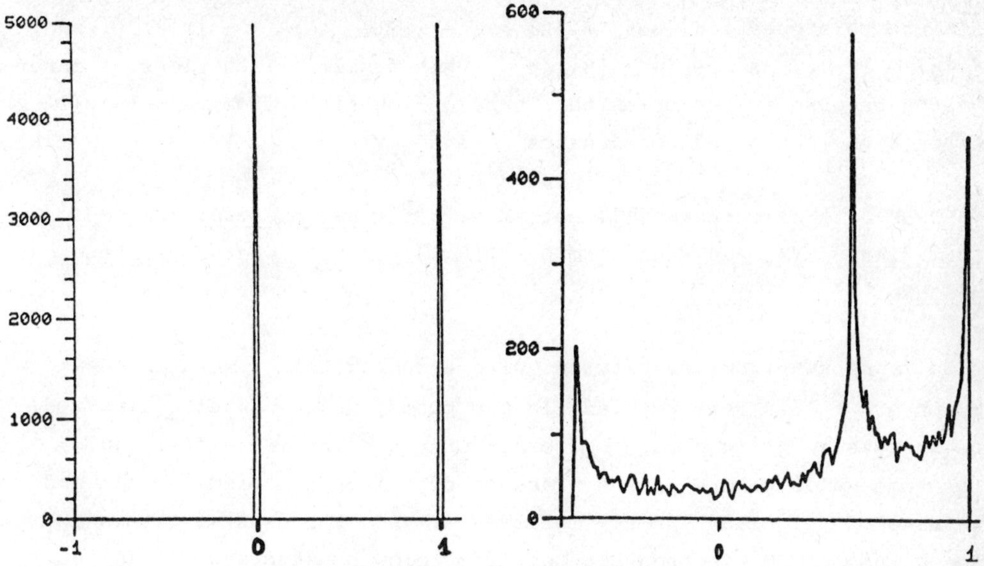

Fig. 2a,b. Two typical histograms.

So apparently, while in Fig. 2a we are faced with a stable periodic orbit of period 2, something very drastic and irregular happens in Fig. 2b. The following theorem covers in fact the case of Fig. 2b as we shall see below.

THEOREM 2. [Misiurewicz]. Assume f has no stable periodic orbit.

I. If $\{f^n(0)\}_{n=1,2,\ldots}$ avoids a neighborhood of zero, then f has an invariant measure which is absolutely continuous with respect to the Lebesgue measure.*

II. If $\{f^n(0)\}_{n=1,2,\ldots}$ consists of finitely many different points, then the density of the above measure is piecewise convex.

Now if $f(x) = 1 - 1.544\ldots x^2$, then $f^3(0) = f^4(0)$, and $|f'(f^3(0))| > 1$, so that a combination of Theorem 1 and 2.II explains Fig. 2b (cf [Ruelle]).

It seems that the condition of Theorem 2.I is relatively hard to be satisfied. Therefore it is a legitimate and important question to know whether the conclusion of Theorem 2.I holds "often". We now

---

* A measure ν is called invariant, if $\nu = \nu \circ f^{-1}$.

analyze this question. Let $\hat{f}_\mu(x) = 1 - 2|x|$, for $1 \geqslant |x| > \mu$ and $\hat{f}_\mu(x) = 1 - \mu - x^2/\mu$ for $|x| \leqslant \mu$. This family of functions is interesting because it resembles the functions obtained from $(4-\mu^2)y(1-y)$ through the change of coordinates $y = \sin^2(\pi x/2)$.

THEOREM 3. [Collet-Eckmann]. Let $M = \{\mu | \hat{f}_\mu$ has no stable periodic orbit$\}$. Then $L(M) > 0$ and $\lim_{\mu \to 0} L(M \cap [0,\mu])/\mu = 1$, where L is Lebesgue measure.

This means that aperiodicity is quite a common thing, when the graph of $\hat{f}_\mu$ is very pointed. In the proof, which is quite involved, we analyze in great detail the orbit of the point 0. If $\hat{f}^n_{\mu'}(0) = 0$ for some μ', we are in the presence of a stable periodic orbit and we discard μ' (and a small interval around it). Thus we obtain $M' \subset M$. The volume of M' is then bounded from below by studying $d\hat{f}^n_\mu(0)/d\mu$ to measure how fast the critical values μ' of μ are traversed.

A very important notion in the study of dynamical systems is sensitive dependence on initial conditions. This notion says that the orbits of two arbitrarily close points will eventually separate by some amount which is bounded below. An intuitively appealing formulation, due to [Guckenheimer] is : f has sensitive dependence on initial conditions if there is an $\varepsilon > 0$ and an interval $I \subset [0,1]$ such that for every $x \in I$ and every neighborhood U of x the following is true : There is a $y \in U$ and an n such that $|f^n(x) - f^n(y)| > \varepsilon$. Again, sensitive dependence on initial conditions can be decided on the basis of the orbit of 0. We give a simplified version of the result.

THEOREM 4. [Guckenheimer]. If there is no neighborhood V of 0 contained in $[-\bar{x},\bar{x}]$ (where $f(\bar{x}) = \bar{x}$) such that 0 returns with a period to V (i.e. $\forall\, p\, \exists m$ such that $f^{mp}(0) \notin V$), then f has sensitivity with respect to initial conditions. If $V = [-\bar{x},\bar{x}]$, where $f(\bar{x}) = \bar{x}$, then f is topologically conjugate to a piecewise linear map $L_\rho : x \mapsto 1 - \rho|x|$, with $2^{\frac{1}{2}} < \rho \leqslant 2$.

Remark : The set M' in Theorem 3 is constructed in such a way that for $\mu \in M'$ the conclusions of Theorem 4 hold for $\hat{f}_\mu$.

We next turn our attention to some <u>discrete aspects</u> of maps on the interval, which reflect the ordering of points on the line. A very useful concept is the socalled <u>kneading sequence</u>, introduced in [Milnor-Thurston]. Let $x_n = f^n(0)$ be the successive images of the critical point. Associate to it the sequence of symbols $I_n \in \{L,C,R\}$ according to whether x_n is <0, $= 0$, >0, respectively (=left, center, right). Surprisingly, this is almost a topological invariant. We give the easiest version of a much more complete result.

<u>THEOREM 5</u>. [Guckenheimer]. *If f and g have the same kneading sequence (i.e. $I_n(f) = I_n(g)\ \forall n$), and if f has no stable periodic orbit, then f is topologically conjugate to g (i.e. $\exists$ homeomorphism h such that $h^{-1}\circ f\circ h = g$).*

We now turn again to parametrized families of maps, and assume that $\mu \to f_\mu$ is continuous from $\mathbb{R}$ to C^1. Consider now the set of all periodic kneading sequences, which may occur as kneading sequences of a map and which are of the form

$$S = \{I_1,\ldots,I_{p-1}, C, I_1,\ldots,I_{p-1}, C,\ldots\}$$

with $I_j \in \{L,R\}$, $j = 1,\ldots,p-1$. There is a linear ordering of such sequences, defined as follows :

$$\begin{array}{l} \text{If } I_j(S_1) = I_j(S_2) \text{ for } j = 1,\ldots,n, \text{ and} \\ \text{if } I_{n+1}(S_1) < I_{n+1}(S_2) \quad (\text{where } L < C < R) \end{array}$$

then we say $S_1 < S_2$ if the set $\{I_j(S_1)\}_{j=1}^{n}$ contains an even number of R's, and $S_2 < S_1$ otherwise.

<u>THEOREM 6</u>. [Metropolis-Stein-Stein, Guckenheimer, Lanford]. *Suppose $\mu_1 < \mu_2$ and suppose the kneading sequence of f_{μ_i} is S_i, $i = 1,2$ with $S_1 < S_2$. If $S_1 < S_3 < S_2$ then there is a μ_3 satisfying $\mu_1 < \mu_3 < \mu_2$ such that the kneading sequence of f_{μ_3} is S_3.*

In the above ordering, the sequences

$$\begin{array}{l} S = (I_1,\ldots,I_{k-1}, C,\ldots) \text{ and} \\ S_* = (I_1,\ldots,I_{k-1}, X, I_1,\ldots,I_{k-1}, C,\ldots), \end{array}$$

where $X = R$ if $\{I_j\}_{j=1}^{k-1}$ has an even no. of R's, $X = L$ otherwise, are contiguous; that is, as we vary μ there is no other period between a period of length k and length $2k$, and hence by induction, $k, 2k, 4k, 8k, \ldots$ are contiguous. Let $\mu_1, \mu_2, \ldots, \mu_j, \ldots$ be the values of μ for which a transition from $k \to 2k$, $2k \to 4k, \ldots$ $2^{j-1}k \to 2^j k$ takes place. Then we have the following.

FACT 7. [Feigenbaum]. The $\lim_{n\to\infty} \mu_n = \mu_\infty$ exists and $|\mu_n - \mu_\infty|$ is asymptotically proportional to $(4.66920\ldots)^{-n}$ (independently of k and of the family f_μ), as $n \to \infty$.

This striking observation relates thus the discrete aspects of maps on the interval to analytic aspects. (See at the end about the consequences of this regularity).

In the paper [Collet-Eckmann-Lanford] we have explained on a mathematically sound basis why the above result is true. Let us sketch those aspects of the proof which are related to the renormalization group analysis [Collet-Eckmann], for the case $k = 1$. Note that a stable period of (shortest) length 2^n for g is a stable fixed point for g^{2^n}. We thus wish to study the map $\mathcal{T}_0 : \psi \to \psi \circ \psi$. Since we want to preserve the condition $\psi(0) = 1$, it is more appropriate to study $\mathcal{T}\psi(x) = -a^{-1}\psi \circ \psi(-ax)$, $a = -\psi(1)$, instead (this only changes a scale and does not affect stable periods). This is a complicated non-linear map on a function space, and it is simplified by first looking for a fixed point and then linearizing around the fixed point. We shall look for fixed points of the form $\mathcal{T}\psi = \psi$, with $\psi(x) = f(|x|^{1+\varepsilon})$, and small $\varepsilon > 0$ (the case of interest in applications is $\varepsilon = 1$, see Lanford's contribution in these proceedings). Our varying ε is motivated by the ε-expansion of statistical mechanics. We have then

THEOREM 8. [Collet-Eckmann-Lanford]. For sufficiently small $\varepsilon > 0$, $\mathcal{T}$ has a fixed point ψ_ε, $\psi_\varepsilon(x) = f_\varepsilon(|x|^{1+\varepsilon})$, with

$$f_\varepsilon(t) = 1 - (1 + \lambda_\varepsilon)t + O(\varepsilon^2 \log\varepsilon),$$
$$\lambda_\varepsilon = -\varepsilon\log\varepsilon + O(\varepsilon^2).$$

The derivative $D\mathcal{T}$ at ψ_ε (on a Banach space of functions analytic in $|x|^{1+\varepsilon} = t$ for $|t| < 2$), has one simple eigenvalue $\delta(\varepsilon) > 2$, and the remainder of the spectrum in a small disk around 0.

Note : $\delta(\varepsilon = 1) = 4.66920\ldots$.

THEOREM 9. [C-E-L]. The Fig. 3 below holds for any fixed small $\varepsilon > 0$ and allows a proof of Feigenbaum's observations (with $\delta(\varepsilon = 1)$ replaced by $\delta(\varepsilon)$) for a family of functions $\psi_\mu(x) = f_\mu(|x|^{1+\varepsilon})$, $f_\mu(t)$ analytic in $|t| < 2$.

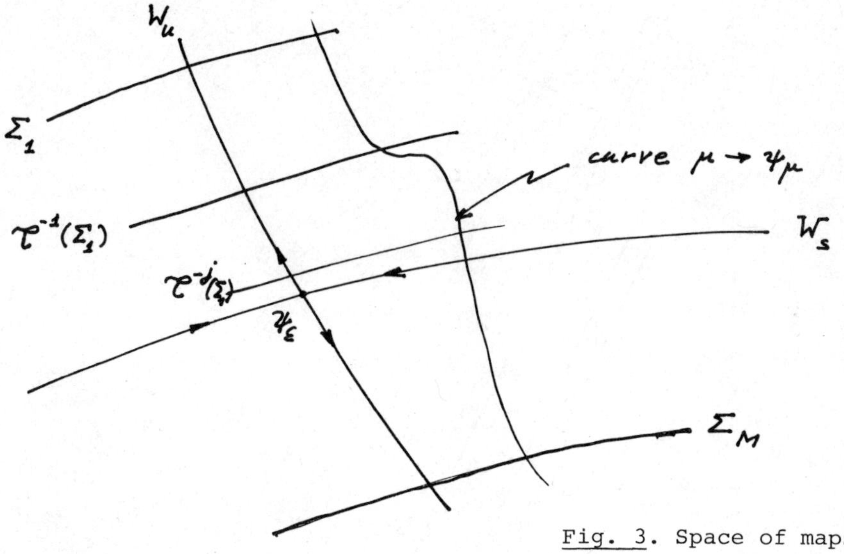

Fig. 3. Space of maps

Explanation : W_u is the unstable manifold, W_s the stable manifold for $\mathcal{T}$. Σ_1 is the manifold of ψ's with $\psi(1) = 0$; since $\psi(0) = 1$ these are the maps for which $x = 0$ has period 2. Now it is easy to see that $\mathcal{T}^{-1}(\Sigma_1)$ is a set of ψ for which 0 has period 4. Similarly, the manifold $\mathcal{T}^{-n}(\Sigma_1)$ is a set of ψ for which 0 has period 2^{n+1}. The theorem follows then by intersecting the curve $\mu \to \psi_\mu$ with these manifolds, and due to the fact that $\mathcal{T}$ expands by $\delta(\varepsilon)$ on W_u, and that $\mathcal{T}$ can be linearized [C-E-L]. We can similarly define Σ_M, the set of ψ for which $\psi^3(0) = \psi^4(0)$, cf. Theorem 2.II. Then for $\psi_\mu \in \mathcal{T}^{-j}(\Sigma_M)$, $j = 0,1,2,\ldots$; ψ_μ will have an absolutely continuous invariant measure, (cf [C-E-L] for more details).

We end this talk by the observation that the phenomenon of Feigenbaum is not restricted to maps on [0,1], it occurs for maps on $\mathbb{R}^n$ and for flows. A proof of the case for $\mathbb{R}^n$ is in preparation [Collet-Eckmann-Koch], cf Fig. 4.

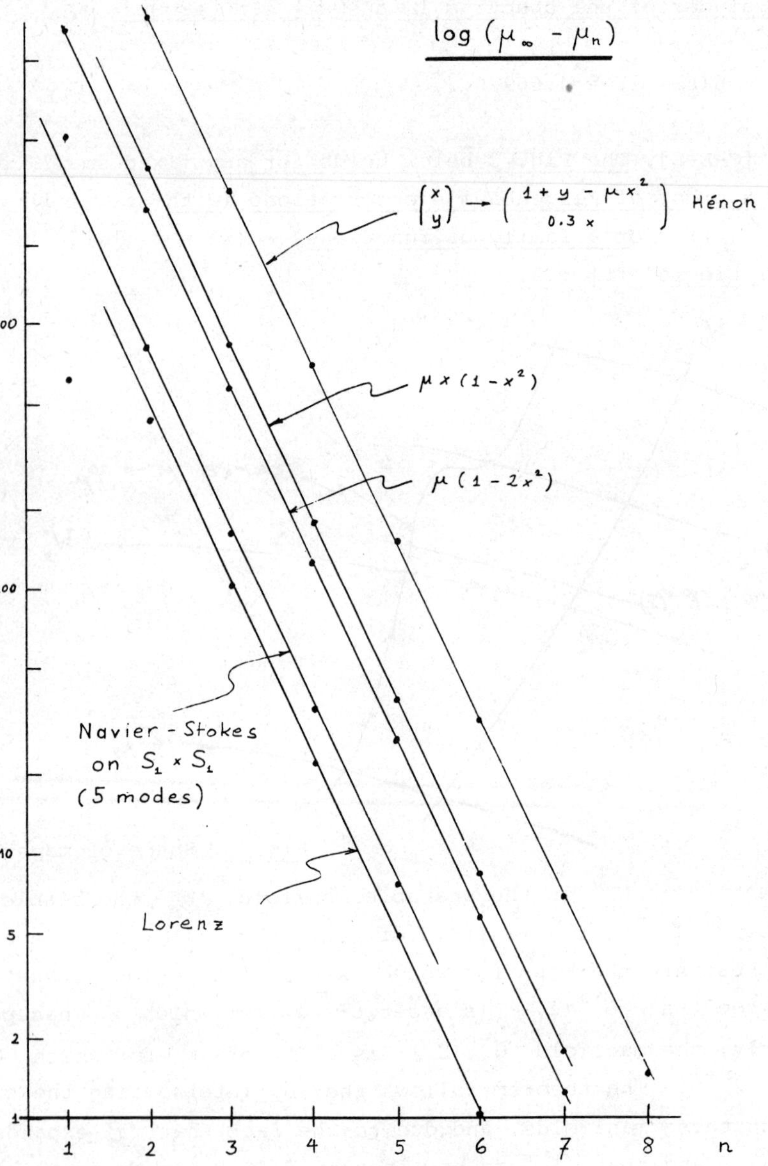

Fig. 4. Experiments

(Calculations for Navier-Stokes and Lorenz done by [Franceschini] et al.) Note also the following predictive character of the geometrical picture Fig. 3. If some system (physical or numerical) shows two successive period doubling bifurcations for values μ_o and μ_1 of a parameter, then one may expect other bifurcations near $\mu_j = \mu_o(\delta^{1-j}-1)/(\delta - 1) + \mu_1(\delta - \delta^{1-j})/(\delta - 1)$, and "chaotic behaviour" near $\mu_j \sim (\delta\mu_1 - \mu_o)/(\delta - 1) - \delta^{1-j}\cdot(\mu_o - \mu_1)\cdot 0{,}33241/(\delta - 1)$, and period $3\cdot 2^j$ near $\mu_j' \sim (\delta\mu_1 - \mu_o)/(\delta - 1) - \delta^{1-j}(\mu_o - \mu_1)\cdot$ $\cdot 0{,}803/(\delta - 1)$ (this is not the sequence $3, 2\cdot 3, 4\cdot 3,\ldots$ described above), etc.

REFERENCES

D. Singer. Stable orbits and bifurcations of maps on the interval. SIAM J. Appl. Math. 35, 260 (1978).

M. Misiurewicz. Absolutely continuous measures for certain maps of an interval. Preprint IHES, Bures-sur-Yvette (1979).

D. Ruelle. Applications conservant une mesure absolument continue par rapport à dx sur [0,1]. Commun. Math. Phys. 55, 47 (1977).

P. Collet, J.-P. Eckmann. Abundance of chaotic behaviour for maps on the interval. Preprint. University of Geneva (1979).

J. Milnor, W. Thurston. On iterated maps of the interval I, II. Preprint, Princeton (1977).

J. Guckenheimer. Sensitive dependence to initial conditions for one dimensional maps. Preprint IHES, Bures-sur-Yvette (1979).

N. Metropolis, M.L. Stein, P.R. Stein. On finite limit sets for transformations on the unit interval. J. Comb. Theory (A) 15, 25 (1973).

J. Guckenheimer. On the bifurcation of maps of the interval. Inventiones Math. 39, 165 (1977).

O.E. Lanford III. Private communication.

M. Feigenbaum. Quantitative universality for a class of non-linear transformations. J. Stat. Phys. 19, 25 (1978).

P. Collet, J.-P. Eckmann, O.E. Lanford. Universal properties of maps on an interval, to appear.

P. Collet, J.-P. Eckmann. A renormalization group analysis of the hierarchical model in statistical mechanics. Lecture Notes in Phys. Vol. 74.

V. Franceschini, C. Tebaldi. Sequences of infinite bifurcations and turbulence in a 5-modes truncation of the Navier-Stokes equations. J. Stat. Phys., to appear.

V. Franceschini. A Feigenbaum sequence of bifurcations in the Lorenz model. Preprint. University of Modena (1979).

Remarks on the accumulation of period-doubling bifurcations

Oscar E. Lanford III

Department of Mathematics
University of California
Berkeley, California 94720

This note may be regarded as an appendix to J.-P. Eckmann's contribution, giving a brief status report on the proof of the Feigenbaum conjectures for $\varepsilon = 1$. For appropriate mappings ψ of [-1,1] into itself, we define

$$\mathfrak{T}\psi(x) = -\frac{1}{a}\,\psi\circ\psi(-ax)\ , \quad a = -\psi(1)$$

The problem is to show that $\mathfrak{T}$ has a fixed point ϕ,

$$\phi \sim 1 - 1.4\,x^2$$

and that the derivative of $\mathfrak{T}$ at ϕ has spectrum inside the unit circle except for a single positive eigenvalue larger than one.

We will look for a fixed point ϕ which is even and analytic so we write

$$\phi(x) = f(x^2)\ .$$

We then want to solve $\hat{\mathfrak{T}} f = f$, where

$$(\hat{\mathfrak{T}}f)(t) = -\frac{1}{a}\, f([f(a^2t)]^2)\ ;\ a = -f(1)$$

We will apply Newton's method in the Banach space of functions bounded and analytic in some domain Ω. The domain is chosen on the basis of convenience; a disk of radius 3 centered at 1 seems to be satisfactory.

The argument goes as follows:

1. Find (by numerical computation) an explicit polynomial f_o such that

$$\varepsilon = \|\hat{\mathfrak{J}}f_o - f_o\|$$

is very small. The f_o I have been using is of degree 20 . Strictly speaking, the construction of f_o is not part of the proof and does not need to be justified; it is simply an educated guess of a good approximation to a fixed point.

2. Show that $D\hat{\mathfrak{J}}(f_o) - \mathbb{1}$ is invertible; denote $\|(D\hat{\mathfrak{J}}(f_o) - \mathbb{1})^{-1}\|$ by ℓ .

3. Let $r_o = 2\ell\varepsilon$ and let

$$m = \sup \{\|D^2\hat{\mathfrak{J}}(f)\| : \|f-f_o\| \leq r_o \}$$

An elementary computation shows that if $m < 1/2\ell^2\varepsilon$ then Newton's method with f_o as a first approximation converges to a solution f of $\hat{\mathfrak{J}}f = f$, and $\|f - f_o\| \leq r_o$.

Thus, if we can find f_o such that ε is very small ($< 10^{-10}$) and if ℓ is not too big (< 20) , the existence of a fixed point will follow from very crude estimates on $D^2\hat{\mathfrak{J}}(f)$ for f near f_o . The latter estimates are straightforward to obtain using the explicit formula for $D^2\hat{\mathfrak{J}}(f)$. As a by-product, this argument shows that f_o is an excellent approximation to a true fixed point.

How do we bound ε and ℓ ? The idea is to construct strict bounds which can be evaluated by finitely many arithmetic operations and to do the arithmetic on a computer. The construction of the bound for ℓ is more interesting than that for ε , but we will discuss only the latter since it illustrates cleanly the issues involved.

Since f_o is a polynomial of degree 20, $\hat{\mathfrak{J}}(f_o) - f_o$ is a polynomial of degree 800 . We need to estimate its supremum over a disk. It can easily be shown that the supremum of a polynomial of degree n over a disk is majorized by its supremum over any number $m > 2n$ points distributed uniformly around a circle concentric with the disk and of a slightly larger radius. Thus to estimate ε we have only to evaluate $|\hat{\mathfrak{J}}f_o - f_o|$ at 1601 properly chosen points in the complex plane (and this number can be reduced to 801 by exploiting the reality of f_o).

A problem which arises immediately in making such estimates airtight is that floating point arithmetic on a computer is inherently imprecise because of roundoff error. It is possible, however, to write programs which automatically generate strict error bounds on all arithmetic operations as they are done (so-called interval arithmetic). This in fact defines the status of my proof of the existence of a fixed point. All the necessary bounds have been checked ignoring roundoff error, and the process of writing programs which generate error bounds is, I estimate, half finished. Verification of the spectral properties of $D\hat{\mathfrak{J}}(f)$ should be feasible by extensions of these methods.

AN OVERVIEW OF SUPERSYMMETRY AND SUPERGRAVITY

J. Scherk

Laboratoire de Physique Théorique, Ecole Normale Supérieure
24 rue Lhomond 75231 Parix Cedex 05 FRANCE

ABSTRACT: Supersymmetry is defined in terms of transformations which leave an action invariant and transform fermions into bosons and vice-versa. Graded Lie algebras (GLA's) are the basic mathematical structures used in supersymmetry / supergravity (SG). Several GLA's used in SS/SG models as well as their representations, are discussed: the (extended) Poincaré GLA, the (extended) De Sitter GLA, the (extended) conformal GLA. The N = 8 model and its relevance (or irrelevance) to physics are discussed in great detail.

1. SUPERWORD DICTIONARY

Let us consider a small and incomplete dictionary of the word "super" as many people ask: "Why do you call your subject "supersymmetry"?". If you read the Dictionary,[1] you will find a long list of words starting with super-, many of which are scientific, some others being of other origins. A list of some of these words, together with their definitions is given in Table I.*

---

* Les Editions Larousse decline all responsibilities for these definitions, except the last one, the others being purely a figment of the author's imagination.

TABLE I

Super : Word of American origin. Opposite of "regular". In Switzerland "super" is a small additional investment compared to "regular"($\sim$1.08 SF/ liter versus $\sim$ 1.04 SF/liter)

Scientific words :

-nova, -star, -helix (A.D.N.), -conductivity, -fluidity, -aerodynamics, -phosphate, -polyamide, -sonic, -structure.

Political :

-powers (in 1979 : Monaco, Liechtenstein), -phenix (mythological bird unknown to the Antiquity, of Gallic origin)

Touristic :

-Tignes (where is the CERN staff today?)

Amer. Colloq. :

"Gee, it's super !", "Super - Duper"

Pop Music :

Supertramp

Daily life :

-markets (Migros, Coop), -man (comic strip by Friedrich Nietzsche)

Poetic touch :

The French poet Jules SUPERvielle (Montevidéo 1884 - Paris 1960) wrote in 1925 a collection of poems called "Gravitations"

Supersymmetry, supergravity and their offsprings are just a few new words to be added to this list. Supersymmetry (flat space) is described by the words contained in the next table :

TABLE II

Supersymmetric transformation laws (SSTL) : a set of continuous transformations which transform classical, commuting, Bose real or complex integer spin fields into classical anticommuting Grassmann Fermi variables half integer spin fields and vice versa.

Supersymmetric model (SM) : a classical Lagrangian field theory, in flat space time, whose action is invariant under N = 1 global SSTL

Extended supersymmetric model (ESM) : same as before with $N > 1$.

Supersymmetry (SS) : the field of $M \cap \Phi$ which studies SSTL, SM's and ESM's

Superspace : a set of coordinates z^A generalizing ordinary space-time

$z^A = (x^\mu, \theta^i_\alpha)$ x^μ : D Bose commuting coordinates ;

θ^i_α : $(i = 1, \ldots N)$ Fermi, anticommuting coordinates

Superfield : a function $\Phi(z^A)$ with or without indices.

Supermultiplet : an irreducible representation of S.S. (simple N = 1 or extended $N > 1$) in terms of fields.

Matter supermultiplet : a supermultiplet with J_{MAX} =1, or 1/2.

(J_{MAX} = 1 : vector supermultiplets ; J_{MAX} = 1/2 : scalar supermultiplets)

Supersymmetric Yang-Mills theories (SSYM) : the self interacting ($g \neq 0$) field theory based on a vector supermultiplet.

Goldstino : The $m = 0$, $J = 1/2$ fermions associated with the spontaneous breaking of S.S.

Gluinos : $SU(3)_c$ octets with J = 1/2

Superalgebras : customarily called by $M \cap \Phi$ addicts graded Lie algebras (GLA's)

A simple example of a supersymmetric field theory is given in the next table[(2)]:

TABLE III

N = 1, S.S.Y.M.

a. Spectrum : A^i_μ J = 1, real,
χ^i J = 1/2, Majorana

b. Coupling constant : g

c. Infinitesimal S.S.T.L. parameter : ϵ J = 1/2, constant, Majorana

d. S.S.T.L. : $\delta A^j_\mu = i \bar{\epsilon} \gamma_\mu \chi^j \; ; \; \delta \chi^j = \sigma^{\mu\nu} F^j_{\mu\nu} \epsilon$

e. Lagrangian : $\mathcal{L} = -\frac{1}{4} F^i_{\mu\nu} F^{\mu\nu i} + \frac{i}{2} \bar{\chi}^j \gamma^\mu \mathcal{D}_{\mu jk} \chi^k$

f. Definitions : $F^i_{\mu\nu} = \partial_\mu A^i_\nu - \partial_\nu A^i_\mu + g f^i_{jk} A^j_\mu A^k_\nu \; ; \; \mathcal{D}_{\mu ij} = \partial_\mu \delta_{ij} + g f^k_{ij} A^k_\mu$

If we now go to curved space-time, we have the words referring to supergravity, which are defined in Table IV. Simple supergravity, its Lagrangian and transformations laws, are described in Table V :[3]

TABLE IV

Supergravity models (S.S.M.) : S.S.M's in curved space-time, invariant under N = 1 local S.S.T.L.

Extended supergravity models (E.S.G.M.) : same as before with $N > 1$

Supergravity : the field of $M \cap \Phi$ which studies local, S.S.T.L.'s, S.G.M.'s and E.S.G.M.'s.

Pure supergravity : S.G.M. or E.S.G.M. uncoupled to any matter super-multiplets, with a self coupling κ, of dimension M^{-1}.

Gravitino : the gauge particles of spin 3/2, associated with local, simple (or extended) S.S.T.L.

Super Higgs effect : the analog of the Higgs effect for supersymmetry whereby when S.S. is spontaneously broken, a gravitino (or several) "eats up" a goldstino (or several) and becomes massive.

TABLE V

N = 1 Supergravity

a. Spectrum : V_μ^a $J = 2$ real field

ψ_μ $J = 3/2$ Majorana field

b. Coupling constant : κ

c. Infinitesimal S.S.T.L. parameter :

ϵ ; $J = 1/2$; x -dependent ; Majorana

d. S.S.T.L.:

$$\delta V_\mu^a = -i\kappa\, \bar{\epsilon}\, \gamma^a \psi_\mu$$

$$\delta \psi_\mu = \kappa^{-1} D_\mu \epsilon$$

e. Lagrangian :

$$\mathcal{L} = -\frac{1}{4\kappa^2} V V^{\mu a} V^{\nu b} R_{\mu\nu ab} - \frac{1}{2} \epsilon^{\mu\nu\rho\sigma} \bar{\psi}_\mu \gamma_5 \gamma_\nu \mathcal{D}_\rho \psi_\sigma$$

f. Definitions :

$$R_{\mu\nu ab} = \partial_\mu \omega_{\nu ab} + \omega_{\mu a}{}^{c} \omega_{\nu cb} - (\mu \leftrightarrow \nu)$$

$$\mathcal{D}_\nu = \partial_\nu + \frac{1}{2} \omega_{\nu ab} \sigma^{ab}$$

Now, we go into the technical facts about SS/SG, and more words crop up. The basic facts of life about Dirac, Weyl, and Majorana spinors, are summarized in Table VI :(For the theorems of this section see ref. (4).

TABLE VI

Clifford algebra : $\{\gamma^\mu, \gamma^\nu\} = 2\eta^{\mu\nu}$

$\eta^{\mu\nu} = \mathrm{diag}\,(+\cdots+\ (t\ \text{times});\ -\cdots-\ (s\ \text{times}))$

$D = s + t$

Irreducible representation of the γ matrices : $2^{[D/2]}$ dimensional

Majorana representation (M.R.): a representation in which the γ matrices are i X real matrices.

Theorem : a M.R. exists in even D if and only if $s - t = 0,2 \mod 8$

Charge configuration matrix : C : $C\gamma^\mu C^{-1} = -\gamma^{\mu T}$

Majorana spinor :

A spinor such that $\chi = C\gamma^0\chi^*$ if $t = 1$. Majorana spinors exist only if a M.R. exists. In the M.R. repr. $C = \gamma^0$ and $\chi = \chi^*$

Weyl spinors : Set for even D: $\gamma^{D+1} = \eta\,\gamma^0\gamma^1\cdots\gamma^{D-1}$ η: $(\gamma^{D+1})^2 = +1$

a Weyl spinor is such that $\gamma^{D+1}\chi = \pm\chi$

Majorana-Weyl spinors : χ is both Majorana and Weyl. M+W spinors exist in D even if and only if $s - t = 0 \mod 8$

2. SUPERALGEBRAS USED BY SUPERSYMMETRISTS

The superalgebras are the skeleton of supersymmetry. Those used in supersymmetry are of 3 types : the superconformal algebra (S.C.A.: $15 + N^2$ Bose generators, 8N Fermi generators) ; the super-De Sitter algebra(S.D.S.A.: $10 + \frac{N(N-1)}{2}$ Bose generators ; 4N Fermi generators) ; the super Poincaré algebra (S.P.A. : same number of generators as the S.D.S.A. except in the presence of central charges).

The field theories which are the offsprings of these superalgebras are given in Table VII :

TABLE VII : SUPERALGEBRAS AND THEIR OFFSPRINGS

S.C.A. :

Flat space : S.S. $\lambda\Phi^4$ theories
S.S.Y.M. theories (N = 1,2,4)

Curved space : S.C. supergravity

S.D.S.A. :

Curved space : extended (N > 1) or simple N = 1 De Sitter supergravity, with SO(N) gauge group

S.P.A. :

Flat space : all supersymmetric models in flat space (N = 1) in D = 6, 10 : S.S.Y.M. theory

Curved space : D = 4 : S.G. and E.S.G. theories without SO(N) gauging. D = 11 , N = 1 supergravity

The most interesting models are the extreme cases, such as the N = 8 supergravity theory which will be discussed later or the N = 4 supersymmetric Yang Mills theory (G.S.O. model)(4). This last model is known to have remarkable convergence properties. It is renormalizable, which is unsurprising, but further,in the Fermi-Feynman gauge it is fully 1-loop finite(5) (no renormalization is needed). The one-loop corrections to the propagator vanish identically (finite and infinite parts), and finally the Gell-Mann-Low $\beta[g]$function vanishes identically for 1 and 2 loops(6). This makes the G.S.O. model a very attractive one for any $M \cap \Phi$ physicist , while Nature may be aware of if when coupled to N = 4 supergravity : indeed the G.S.O. model exhibits no spontaneous breaking of supersymmetry, while coupled to supergravity it does(7). An advantage of the G.S.O; model compared to the SO(8) theory is that the gauge group is arbitrary and can be fitted for Nature. Another application of the S.C.A. algebra this time in curved space is the superconformal supergravity(8), whose Bose sector includes the Weyl theory of gravity $R_{\mu\nu}^2 - \frac{1}{3} R^2$ and has 4 derivatives in the action, which leads to dipole ghosts and is usually rejected on these grounds, as well as on the classical ground of one word.

The S.C.A. has no dimensional constant entering it and thus the field theories based on it have dimensionless coupling constants (S.C. gravity, S.S.Y.M.). The S.D.S.A. contains one dimensional coupling constant m of the dimensions of a mass. Translations do not commute and give a rotation times m^2. The Universe described by the S.D.S.A. is thus not a flat universe, but a De Sitter (SO(3,2) rather than SO(4,1)) universe. The radius of the De Sitter universe is $R_0 \sim m^{-1}$ (in $\hbar = c = 1$ units). The constant m plays the role of a "mass" term $m \bar{\psi}_\mu \sigma^{\mu\nu} \psi_\nu$ for the gravitino field, and a cosmological constant $\frac{m^2}{K^2}$ occurs in the Lagrangian, where K is Newton's constant. In the theories with $N \geq 2$, the vector fields such as A_μ couple with the dimensionless minimal coupling constant g and $Km \sim g$ (we will see more of such relations later)[9]. The SO(3,2) universe is the simple, maximally symmetric solution of the field equations, and the actual size of the universe puts a stringent bound on g: $g \leq 10^{-120}$!

Actually, another point of view is that $g \sim 1$ and $\frac{g^2}{K^4} \sim (10^{19}\mathrm{GeV})^4$ is acceptable : it is the space-time foam picture of Wheeler, Hawking and Townsend[10]. As they point out, the De Sitter theory has solutions where space time is flat or nearly flat at large distances (1cm) but very strongly curved at distances of the order of Planck's length 10^{-33}cm. The physical space-time may well be a statistical ensemble of fluctuating space time foams of arbitrary sizes and topologies.

Finally, we have the S.P.A., with or without central charges. These have the dimension of a mass, and representations of the S.P.A. with central charges occur for massive supermultiplets only (for instance the N = 2, $J_{MAX} = 1/2$ supermultiplet) or for classical solutions of the N = 2,4, supersymmetric Yang-Mills theories.

In these theories, we have respectively 2 and 6 central charges. In the N = 2 model, the 2 charges are the electric and magnetic charges and the relation[11]

$$M^2 - m^2 (Q^2 + G^2) = 0$$

holds. As Olive pointed out[12], this is suggestive of identifying mQ and mG with P_5, P_6 in a Minkowskian ($s = 5$, $t = 1$) 6 dimensional space time where $M_6^2 = 0$. Similarly, in the N = 4 model we have 6 central charges Q_i, and these can be identified with 6 extra-momenta through the relation $P_i = m Q_i$, and the mass relation $M_{10}^2 = 0$ holds true[13].

These relations are no surprise as the N =2,4 ; D = 4 theories can be obtained by dimensional reduction from the N = 1, D = 6,10

S.S.Y.M. field theories.

3. REPRESENTATIONS OF THE SUPERPOINCARE ALGEBRA :

We shall discuss only the representations of the S.P.A. in D = 4 (s = 3; t = 1), in the massless case, in terms of fields. One can show that only one of the Q_α^i is relevant (say Q_1^i) and that it decreases the helicity by 1/2. The particle contents of a supermultiplet is thus easily found.

On Table VIII, we see the particle contents of N = 1, ..., 10 supersymmetric free field theories with $J_{MAX} \leqslant 5/2$. Scalar multiplets exist up to N = 2, vector multiplets up to N = 4, supergravity multiplets up to N = 8 and hypergravity J_{MAX} = 5/2) multiplets up to N = 10.

TABLE VIII

J \ N	1				2				3			4			5	6	7	8	9	10
5/2																			1	1
2				1				1			1			1	1	1	1	1	10	10
3/2			1	1			1	2		1	3		1	4	5	6	8	8	45	45
1		1	1			1	2	1	1	3	3	1	4	6	10	16	28	28	120	120
1/2	1	1			2	2	1		4	3	1	4	7	4	11	26	56	56	210	210
0	2				4	2			6	2		6	8	2	10	10	70	70	256	256
G	SO(1)				SO(2)				SO(3)			SO(4)			SO(5)	SO(6)	SO(7)	SO(8)	SO(9)	SO(10)

Representations contents of N = 1, ..., 10 supersymmetry with 0 mass. Ogievetsky multiplets (J_{MAX} = 3/2) are shown up to N = 4, but exist up to N = 6 ; hypergravity multiplets (J_{MAX} = 5/2) exist from N = 1 to N = 10 but are shown only for N = 9, 10.

Explicit constructions are needed to show that interacting field theories based on these multiplets exist. This is true up to J_{MAX}= 2, but no interacting field theory (hypergravity) based on J_{MAX}= 5/2 exists [15]. The very meaning of the word "hyper"* suggests that these

* Hyperbole : Greek demagog of Samos, threatened the comic poets Alcibiade and Nicias of ostracism, was turned into ridicule by them and was himself finally ostracized in 417 B.C.

theories may not exist.

4. WHAT IS SUPER ABOUT SUPERSYMMETRY/GRAVITY ?

The super prefix associated with these theories is due first to their unification power and to their renormalizability properties.

a) Unification:

- Fields of different spins are unified in the same representation, overcoming the previous NO-GO theorems.
- Internal and space-time symmetries are unified.
- Fermions and bosons play symmetrical roles
- The dichotomy between fields and sources is also abolished.

b) Renormalizability :

- If the Bose sector of a S.S. field theory is renormalizable, so is the whole theory
- Further in that case the S.S. field theory is more convergent than its Bose sector[16]. Ex. : the N = 4 S.S.Y.M. theory
- Non supersymmetric theories of gravity in interaction with matter ($\mathcal{J}$ = 0, 1/2; 1) fields are one loop non renormalizable, while pure extended S.G. theories based on the S.P.A. (N = 1, ..., 8) are 1 and 2 loop finite[17]. S.C. gravity is also renormalizable, but so is conformal gravity[18].

5. EXTRA SPATIAL DIMENSIONS AND SS/SG

The S.P.A. exists for any s, t ($D = s+t$). If we keep $D = 4$ and increase N we meet the limits $N = 2,4,8$ for the existence of multiplets with $J_{MAX} = 1/2, 1, 2$. Similarly, if we keep $N = 1$ and increase D we meet the limits $D = 6,10,11$ for the existence of the same multiplets[19][20].

An example of a S.S. field theory in $D = 10$ ($s = 9$, $t = 1$) is easy to provide[7]. Let us take the example of the $N = 1$ S.S.Y.M. theory given in Table III. If we keep $N = 1$ and put $D = 10$ in the notations, the theory is still S.S. invariant, provided that χ^i are Majorana-Weyl spinors, which is possible since $s-t = 8$. The vector A_μ^i in $D = 10$ gives in $D = 4$ a vector A_μ^i, and six scalars ϕ_K^i ($K = 1,\ldots 6$) and the MW spinor χ^i reduces to 4 Majorana spinors χ_j^i ($j = 1,\ldots 4$), which is the field contents of the $N = 4$, $D = 4$ S.S.Y.M. theory[4].

Dimensional reduction consists in starting with a S.S. field theory in $D \geq 5$. One lets the size of the internal dimensions shrink to zero, in which case only the excitations of the fields which are constant in the extra directions, or have a simple y dependence (of the type $\psi(x,y) \sim \exp imy\, \psi(x)$) survive[21]. The resulting theory in $D = 4$ is still supersymmetric invariant, but S.S. may be spontaneously broken. The mass m appears as the momentum in the 5th direction.

Well-known examples are the $J_{MAX} = 1$, $N = 1$, $D = 10$ theory reduced to the $J_{MAX} = 1$, $N = 4$, $D = 4$ theory[4], and the $J_{MAX} = 2$, $N = 1$, $D = 11$ theory[22] reduced[23] to the $J_{MAX} = 1$, $N = 8$, $D = 4$ theory.

In generalized dimensional reduction[7], a phase $\exp imy$ dependence is introduced, as in the Kaluza-Klein theory. The N = 8 theory can then be generalized starting from the $D = 5$, $N = 8$, $J_{MAX} = 2$ theory which has a rank-4 group of invariance, namely $Sp(8)$[24]. Going from 5 to 4 dimensions, 4 mass parameters m_i ($i = 1,\ldots,4$) can be introduced and a 4 parameter family of $N = 8$ theories is obtained which contain the C.J. theory[23] as a special case ($m_i = 0$). In these theories S.S. is spontaneously broken, the Fermi-Bose mass degeneracy is lifted but the breaking is soft as the following mass relations still hold :

$$\sum_{J=0}^{J=2} (-1)^{2J} (2J+1)(M_J^2)^K = 0$$

for $K = 0, 1, 2, 3$.

These mass relations imply that the one loop correction to the cosmological constant[25] is still fully finite (but apparently non-zero). The spectrum of the $N = 8$ theory is described in the next section.

6. PHYSICAL CONTENTS OF THE N = 8 MODEL

In the spontaneously broken $N = 8$ theory, all massive states are complex, and the theory has therefore a U(1) symmetry. This symmetry is actually a local one. Its gauge field is the vector field A_μ^λ obtained by reducing the metric tensor from 5 to 4 dimensions : $2\kappa A_\mu^\ell = V_\mu^5$. In the next section, we shall see that the coupling of this vector field leads to antigravity, and we shall denote the particle associated with the quantized A_μ^ℓ field by the Egyptian letter ℓ (read "shen"). *

If all m_i are unequal, the only symmetry of the theory is U(1) (apart from supersymmetry, coordinate invariance, etc..). If we set all the $m_i = m$, the spectrum is SU(4) X U(1) degenerate and at zero mass, there are 15 + 1 gauge bosons. As a gauge group SU(4) is too small to include $SU(3)_c$ X SU(2) x U(1), but if we disregard the weak interactions we obtain a model which includes $SU(3)_c$ (strong interactions) X U(1) (electromagnetism) X U(1) (antigravity). Actually, the breaking of SU(4) into $SU(3)_c$ X U(1) is automatic if we set $m_i = m$ (i = 1,2,3) ; $m_4 = M$. The electric charge is taken to be -1/3 for the triplet of J = 3/2 "graviquarks" of mass m , 0 for the J =3/2 singlet gravitino of mass M . Once this is fixed, all masses and charges are derived by taking the product of representations. ++

The model contains only one electron of mass $3m$, but no μ, τ and ν 's. This is not too surprising and one could tentatively attribute the family structure to composite bound states as well as the SU(2) of weak interactions.

A more humble attitude is to take the model just as a model not as a theory. As it stands, it is not too bad : it is 1 and 2 loop finite (unlike Einstein theory coupled to leptons, quarks, and $SU(3)_c$ X SU(2) X U(1) bosons) ; it includes a massless graviton, 8 massless gluons, a photon (γ) ; an electron (e^-) ; a d quark (Q = - 1/3) ; a u quark and a c quark (Q = 2/3).

The exotic particles are : the sexy quarks ($\underset{\sim}{6}$; Q = 1/3) ; the gluinos of Fayet ($\underset{\sim}{8}$ (Q = 0) ; the antigraviton ℓ ; massive vector triplets, a triplet of d -gravi quarks ($\underset{\sim}{3}$, Q = -1/3), a set of 2 neutral gravitinos ($\underset{\sim}{1}$;(0)), massless scalar gluons (2octets), massless

* The symbol ℓ suggested itself to the author as an inverted γ , to distinguish it from the photon but to suggest its similarity with it. W. Nahm pointed out to the author that this symbol exists at least in two languages. In the Cypriot syllabary, it reads "ro". The Egyptian hieroglyphs include the signs : γ , phonetic value š s (shess), ℓ , phonetic value š n (shen). They are inversions of each other and depict parts of a rope. (W. Nahm, private communication)

++ These symmetries were discovered by the authors of ref.24 but were not published.

singlet scalar particles (2 of them), massive sextets and singlet scalars, massive scalar quarks ("sarks").

7. ANTIGRAVITY AND SUPERGRAVITY : A CRAZY IDEA ?

Let us consider the scattering of two particles of mass m, m' having also a coupling to a massless vector field A_μ^{ℓ} (distinct from the electromagnetic potential) with charges g, g'. In the static limit the total potential is given by :

$$V(r) = 4K^2 mm' r^{-1} \left[mm' - K^{-2} gg' \right]$$

Antigravity is defined by the cancellation which would occur if one had systematically the relation between g and m : $g = K\epsilon m$ where ϵ = +1 for a particle, - 1 for an antiparticle, 0 for a particle which is its own antiparticle, like the Z^0.

In 1977, it was guessed (26) that since in N = 2 S.G. there is a vector A_μ, it had to couple minimally to fields with a gauge coupling constant $g \sim Km$. The coupling was actually written in lowest order in K in ref(27). Then in 1978, K. Zachos (28) coupled N = 2 S.G. to the multiplet (1/2(2), 0(4)) with a mass m and found $|g|$ = Km as well as antigravity. In 1979, the spontaneously broken N = 8 theory was found (7) (24) and it was discovered that a vector A_μ coupled to all the fields of the model with strength $|g| = 2Km$, a relation true for the 256 states of the model. *

If there is such an antigravitational force in Nature (29), and this is an inescapable consequence of supersymmetry if $N>1$, why don't we see it ? If we look at the spontaneously broken N = 8 model, we may find a beginning of an answer. Suppose we consider the static force between 2 protons (uud bound states). As the graviton couples not to the mass, but to $T_{\mu\nu}$ it sees the total energy of the quarks and thus the mass of the proton, which is mostly kinetic energy of the quarks and gluons. So its contribution to the force is given by $K^2 r^{-2} M_p^2$. The antigraviton ℓ is coupled to $\bar{\chi}\gamma^\mu\chi$ x Km where m is the mechanical mass of the quark in question, and $\bar{\chi}\gamma^\mu\chi$ is the conserved e.m.current. Hence, the ℓ sees directly the quark mass and not the proton mass and its contribution is $\pm K^2 r^{-2} (2m_u + m_d)^2$. Already, we see that the $\frac{\ell}{\text{graviton}}$ contribution is of the order $\left(\frac{m_u}{M_p}\right)^2$ which is small ($\sim 10^{-4}$) for the u and d quarks.

Finally, if we compute the relative difference between the acceleration of a proton and a neutron, we find that it is given by

$3 m_u (m_d - m_u)/M_p^2$ which is even smaller, depending upon the u, d mass difference. In the limit of exact SU(2) symmetry for strong

*In spite of the factor 2, antigravity still occurs as there is a also a Brans-Dicke scalar.

interactions, this difference vanishes so that antigravity is in good shape.

If one looks closer however, one finds that the ℓ exchange leads to serious problems with the equivalence principle. Two atoms of atomic numbers A, A', having Z, Z' protons and of charge zero fall with different accelerations towards the earth. The force between 2 such atoms is given by :

$$F = 8\pi G r^{-2} \left[M(Z,A)\, M(Z',A') - M^\circ(Z,A)\, M^\circ(Z',A') \right]$$

The negative term is due to the ℓ exchange ; one has :

$$M(Z,A) = Z(M_p + m_e) + (A-Z) M_n$$

$$M^\circ(Z,A) = Z\,(2m_u + m_d + m_e) + (A-Z)(m_u + 2m_d)$$

So :

$$\gamma(Z,A) = 8\pi G r^{-2} M(Z,A) \left[1 - \frac{M^\circ(Z,A)\, M^\circ(Z',A')}{M(Z,A)\, M(Z',A')} \right]$$

If A', Z' represent the earth, we can safely replace $\frac{M^\circ}{M}(Z',A')$ by $\frac{3m_u}{M_p}$ so that the acceleration of (Z,A) towards the earth is given by :

$$\gamma(Z,A) = \frac{8\pi G}{r^2} M_E \left(1 - \frac{M^\circ(Z,A)}{M(Z,A)} \frac{3m_u}{M_p} \right)$$

The relative difference in acceleration of two atoms is given by

$$\frac{\gamma(Z,A) - \gamma(Z',A')}{\gamma(Z,A)}$$

$$= (Z'A - ZA') \frac{3m_u}{M_p} \left[m_e (M_n - m_u - 2m_d) + \tfrac{3}{2} (m_u + m_d)(M_n - M_p) + \tfrac{1}{2} (m_u - m_d)(M_n + M_p) \right] M^{-1}(Z,A) M^{-1}(Z',A')$$

In the last bracket the dominant contribution is given by $m_e M_n$. so that the $\frac{\delta\gamma}{\gamma}$ is given by

$$\frac{\delta\gamma}{\gamma} \sim (Z'A - ZA') \frac{3m_u}{M_p} \frac{m_e M_n}{M M'}$$

For H, He_4 we find a $\frac{\delta\gamma}{\gamma} \sim 10^{-6}$ which is unacceptable.

To save antigravity, one must assume that the ℓ acquires a mass, presumably through the Higgs mechanism. This is rather likely since it is universally coupled to scalar fields through the action :

$$\mathcal{L} = -\frac{1}{4} F^{\ell}_{\mu\nu} F^{\mu\nu\,\ell} - \frac{1}{2} \left| (\partial_\mu - i\kappa m A^\ell_\mu)\, \phi \right|^2 + m^2 \phi^* \phi - V(\phi)$$

At the classical level $\langle\phi\rangle \neq 0$ and $m_\ell = \kappa\, m_\phi \langle\phi\rangle$.
If $\langle\phi\rangle \neq 0$ is due to SU(2) X U(1) breaking, one has typically $\langle\phi\rangle \sim 1 - 100$GeV and with $\langle\phi\rangle \sim 1$ GeV one finds $m_\ell \sim 10^{-19}$GeV. This gives to the ℓ a

Compton wavelength of the order of 1 km which is reasonable. In the case where $m_\ell \neq 0$ the potential between an atom (Z,A) and the earth is given by : *

$$V(Z,A) = \frac{8\pi G}{R_E} M_E \left(1 - \frac{M^\circ}{M}(Z,A)\, \frac{3m_\nu}{M_p} \exp - m_\ell R_E\right)$$

Usually, masses are thought to be fixed parameters. However, one knows that they depend upon external conditions such as the temperature T. If one could "heat the vacuum" enough, the phase where $\langle\phi\rangle = 0$, $m_\ell = 0$ might be restored.

An antigravity device of this type however still belongs to the field of UFOlogy[30] or Science Fiction[31], and apparently not to the field of mathematical physics.

SUMMARY

1. SUPERWORD DICTIONARY
2. SUPERALGEBRAS USED BY SUPERSYMMETRISTS
3. REPRESENTATIONS OF THE SUPERPOINCARE ALGEBRA
4. WHAT IS SUPER ABOUT SUPERGRAVITY ?
5. EXTRA SPATIAL DIMENSIONS AND SUPERSYMMETRY/GRAVITY
6. PHYSICAL CONTENTS OF THE N = 8 MODEL
7. ANTIGRAVITY AND SUPERGRAVITY : A CRAZY IDEA ?

* This formula would be correct if the Earth was a point like object. Taking account of its actual size leads (for an homogeneous sphere) to multiplying the last term in this expression by $f(m_\ell R_E)$ where $f(x) = 3x^{-3}[x\,\mathrm{ch}\,x - \mathrm{sh}\,x]$. The altitude from the surface now appears rather than the distance from the center. This leads to an upper bound on m_ℓ^{-1} : $m_\ell^{-1} \leqslant 2.5$ meters.

REFERENCES

(1) Le Larousse en 3 volumes
(2) A. Salam and J. Strathdee, Phys. Lett. 51B (1974) 353, S. Ferrara and B. Zumino, Nucl. Phys.B79 (1974) 413
(3) D.Z. Freedman, S. Ferrara, P. Van Nieuwenhuizen, Phys. Rev.D13., 3214 (1976) and D14 (1976) 912
S. Deser and B. Zumino, Phys. Lett. 62B (1976) 335
(4) F. Gliozzi, J. Scherk and D. Olive, Phys. Lett. 65B (1976) 282 ; Nucl. Phys. B122 (1977) 253
(5) M.T. Grisaru, W. Siegel and M. Rocek, June 1979 preprint
(6) E.C. Poggio, H.N. Pendleton, Phys. Lett. 72B (1977) 200
D.R.T. Jones, Phys. Lett. 72B (1977) 199
(7) J. Scherk and J.H. Schwarz, Phys. Lett. 82B, 1 (1979) 60; Nucl. Phys. B153 (1979) 61
(8) S. Ferrara, M. Kaku, P.K. Townsend and P. Van Nieuwenhuizen, Nucl. Phys. B129 (1977) 125
(9) A. Das and D.Z. Freedman, Nucl. Phys. B120 (1977) 221
(10) J.A. Wheeler, Relativity group and topology, ed. S.S. and C.M. De Witt (Gordon and Breach, New York 1964)
S.W. Hawking, Nucl. Phys. B 144 (1978) 349
S.W. Hawking, D.N. Page and C.N. Pope, Phys. Lett. 86B (1979) 175
P.K. Townsend, Phys. Rev.D 15 (1977) 2795
(11) E. Witten, D. Olive, Phys. Lett. 78V (1978) 97
(12) D. Olive, Nucl. Phys. B153 (1979) 1
D. Olive, invited talk at the international conference on Mathematical physics, E.P.F.L. Lausanne, (1979)
(13) H. Osborn, Phys. Lett. 83B (1979) 321
(14) V.I. Ogievetsky and E. Sokatchev, JETP Lett. 23 (1976) 58
(15) Free spin 5/2 field theories without ghost exist, see for instance F.A. Behrends, J.W. Van Holten, P. Van Nieuwenhuizen, and B. De Witt Phys. Lett. 83B (1979) 188 ; C. Aragone, S. Deser, Ph. Lett. 868 (1979) 161
(16) J. Iliopoulos and B. Zumino, Nucl. Phys.B76 (1974) 310
(17) For a review on the finiteness issue in S.G. see : P. Van Nieuwenhuizen and M.T. Grisaru, Proc. of Orbis Scientiae 1977 in High energy physics, Kursunoglu, Perlmutter and Scott eds. Plenum Press, N.Y., London (1977)
(18) K. Stelle, Phys. Rev.D16 (1977) 953
(19) L. Brink, J. Scherk, J.H. Schwarz, Nucl. Phys. B121 (1977) 77
(20) W. Nahm, Nucl. Phys. B135 (1978) 149
(21) E. Cremmer, J. Scherk, Nucl. Phys.B103 (1976) 399
(22) E. Cremmer, B. Julia, J. Scherk, Phys. Lett. 76B (1978) 409
(23) E. Cremmer, B. Julia, Phys. Lett. 80B (1978) 48 and LPTENS 79/6 to be published in Nucl. Phys. B
(24) E. Cremmer, J. Scherk, and J.H. Schwarz, Phys. Lett. 84B (1979) 83
(25) B. Zumino, Nucl. Phys. B89 (1975) 535
(26) J. Scherk, La Recherche (1977)October issue, 878
(27) S. Ferrara, J. Scherk, B. Zumino, Nucl. Phys. B121 (1977) 393

(28) K. Zachos, Phys. Lett. 76B (1978) 329
PhD Thesis, Caltech, submitted April 17 (1979)
(29) J. Scherk, LPTENS 79/17 preprint,
(30) UFOlogy, James McCampbell, LC.C.C.N. 73-93488, Joymac-Hollmann (1973)82
(31) A.C. Clarke, Profiles of the future, Paris books (1962) 64

SUPERSPACE ASPECTS OF SUPERSYMMETRY AND SUPERGRAVITY

S. Ferrara
Laboratori Nazionali di Frascati, INFN, Italy,
and
International Centre for Theoretical Physics, Trieste, Italy.

Abstract

Superspace, as an extension of Minkowski space-time with spinning degrees of freedom is reviewed. Superspace techniques useful for the construction of globally and locally supersymmetric field theories are discussed.

I. INTRODUCTION

Supersymmetric field theories [1] involve multiplets of fields with different Lorentz transformation laws. These fields transform into each other under infinitesimal transformations involving anticommuting Grassmann parameters ϵ_α which belong to the $(\frac{1}{2},0) \oplus (0,\frac{1}{2})$ self-conjugate representation of the spinor group SL(2,C). The simplest collection of fields having the above properties is the so-called chiral (scalar) multiplet [2] (A,B,χ,F,G). (A,B) are scalars, (B,G) pseudoscalars and χ a spin-$\frac{1}{2}$ Majorana field. The infinitesimal transformation law of these fields is most conveniently written using complex fields $z = \frac{A - iB}{2}$, $H = \frac{F + iG}{2}$ and a (two-component) Weyl spinor χ_α $(\alpha = 1,2)$:

$$\delta z = \epsilon^\alpha \chi_\alpha \quad ,$$

$$\delta\chi_\alpha = 2i(\sigma^m)_{\alpha\dot\beta}\, \bar\epsilon^{\dot\beta} \frac{\partial}{\partial x^m} z + 2\epsilon_\alpha H \quad , \tag{1}$$

$$\delta H = - i \frac{\partial}{\partial x^m} \chi^\alpha \sigma_{\alpha\dot\beta}\, \bar\epsilon^{\dot\beta} \ .$$

We use Weyl spinors in Van der Waerden notations [3]. Undotted and dotted indices refer to $(\frac{1}{2},0)$ and $(0,\frac{1}{2})$ representations of SL(2,C). The spinor metric $\varepsilon^{\alpha\beta}$ satisfies $\varepsilon^{\alpha\beta} = -\varepsilon^{\beta\alpha} = -\varepsilon_{\alpha\beta}$, $\varepsilon^{12} = 1$ (same for dotted indices). $(\sigma_\mu)_{\alpha\dot\beta} = (1,\sigma_i)$ and $\chi^* = \chi_{\dot\alpha}$, $\chi^*_{\dot\alpha} = \chi_\alpha$. Weyl spinors are related to four-component spinors through the chiral projections $\frac{1}{2}(1 \pm i\gamma_5)$.

The transformation laws in (1) realize the simplest representation of the supersymmetry algebra. The latter is the 14-dimensional graded Poincaré algebra with bosonic generators P_m , M_{mn} and spinorial generators Q_α , $\overline{Q}_{\dot\alpha}$, the grading representation of the Poincare algebra

$$\{Q_\alpha,\bar{Q}_{\dot\alpha}\} = 2\sigma^m_{\alpha\dot\alpha} P_m \qquad [Q_\alpha,P_m] = [\bar{Q}_{\dot\alpha},P_m] = 0 \ ,$$
$$\{Q_\alpha,Q_\beta\} = \{\bar{Q}_{\dot\alpha},\bar{Q}_{\dot\beta}\} = 0 \qquad [Q_\alpha,M_{mn}] = (\Sigma_{mn})_{\alpha\beta} Q^\beta \ . \tag{2}$$

The multiplet in (1) describes scalar and spinor fields. More complicated multiplets exist as well. They are used to describe supersymmetric interactions with a gauge invariance. This gauge invariance can be associated with an internal symmetry (Yang-Mills interactions) or with a space-time symmetry (supergravity) and the corresponding (gauge) field multiplets describe massless particle doublets of helicity content $(\frac{1}{2},1)$ and $(\frac{3}{2},2)$, respectively, for Yang-Mills and supergravity theories.

The most natural mathematical framework for the construction and investigation of supersymmetric Lagrangian field theories has been settled by Salam and Strathdee [4] with the introduction of the concept of superspace and superfields. Superspace is an extension of ordinary space-time (Minkowski space) to spinning space-time. The base manifold of superspace has points parametrized by co-ordinates

$$Z^M = (x^m,\theta^\mu,\bar\theta^{\dot\mu}) \ , \qquad m = 1,\ldots,4 \ , \quad \mu,\dot\mu = 1,2 \ . \tag{3}$$

Latin indices denote vectors, Greek indices spinor indices. x^m are commuting space-time co-ordinates and θ^μ $(\bar\theta^{\dot\mu})$ anticommuting Grassmann variables:

$$[x^m,x^n] = [x^m,\theta^\mu] = \{\theta^\mu,\theta^\nu\} = \{\theta^\mu,\bar\theta^{\dot\nu}\} = 0 \tag{4}$$

or more symbolically $[Z^M,Z^{M'}\} = 0$.

From a group-theoretical point of view, superspace may be regarded as the quotient space G/H in which G is the graded Poincaré group and H is the homogeneous Lorentz group. In extended supersymmetry with N spinorial charges Q^i_α (Q^i_α) i = 1,...,N, the Grassmann variables are supplemented with internal symmetry indices as well θ^i_α $(\bar\theta^i_{\dot\alpha})$ and superspace is a 4 + 4N graded manifold. However, we shall not consider extended superspace in this review.

Supersymmetry transformations are realized as motions in superspace [4]

$$\delta x^m = -i\epsilon\sigma\bar\theta + i\theta\sigma\bar\epsilon \ ,$$
$$\delta\theta^\mu = \epsilon^\mu \ , \ \delta\bar\theta^{\dot\mu} = \bar\epsilon^{\dot\mu} \ . \tag{5}$$

We shall refer to (5) as a supertranslation of the point $Z^M = (x^m,\theta^\mu,\theta^\mu)$ - On the other hand, an ordinary translation with parameter a^m shifts the superco-ordinate as follows:

$$\delta x^m = a^m \ , \ \delta\theta^\mu = 0^\mu \ , \ \delta\bar\theta^{\dot\mu} = \bar{0}^{\dot\mu} \ . \tag{6}$$

The composition rule of supertranslations is obtained by performing the commutator of two infinitesimal charges of parameters ϵ^μ_1 $(\bar\epsilon^{\dot\mu}_1)$, ϵ^μ_2 $(\bar\epsilon^{\dot\mu}_2)$:

$$[\delta_2,\delta_1]Z^M = ([\delta_2,\delta_1]x^m \ , \ [\delta_2,\delta_1]\,\epsilon^\mu \ , \ [\delta_2,\delta_1]\,\bar\epsilon^{\dot\mu}) =$$
$$= (-2i\,\epsilon_1\sigma\,\bar\epsilon_2 + 2i\,\epsilon_2\sigma\,\bar\epsilon_1 \ , \ 0 \ , \ 0) \ . \tag{7}$$

The composition law (7) reflects the basic anticommutators given by (2). It is the manifestation of the fact that an infinitesimal space-time displacement can be obtained in superspace by performing two infinitesimal supertranslations. If we regard the supersymmetry algebra as the "square root" of the Poincaré algebra, we could equally say that superspace is the "square root" of Minkowski space-time.

II. SUPERFIELDS

We can elucidate the action of supersymmetry transformations in superspace giving more precise concepts. The superspace previously introduced furnishes a representation of the algebra given in (2) in terms of differential operators. The set of motions defined by (5) and (6) is obtained by the left-action on an element of the graded Lie group generated by Q_α, $\bar{Q}_{\dot\alpha}$, P_m,

$$L(x,\theta,\theta) = \exp i(\theta^\alpha Q_\alpha + \bar\theta_{\dot\alpha}\bar{Q}^{\dot\alpha} - x^m P_m) \quad , \tag{8}$$

with another element of the group $G(a^m,\epsilon,\bar\epsilon)$

$$G(a^m,\epsilon,\bar\epsilon)\, L(x^m,\theta,\theta) = L(x + a - i\epsilon\sigma\bar\theta + i\theta\sigma\bar\epsilon, \theta + \epsilon, \bar\theta + \bar\epsilon) \quad , \tag{9}$$

where we have used the Hausdorff formula.

The infinitesimal generators of this motion are

$$P_m = i\frac{\partial}{\partial x^m} \; , \; Q_\alpha = \frac{\partial}{\partial\theta^\alpha} - i\sigma^m_{\alpha\dot\beta}\,\bar\theta^{\dot\beta}\frac{\partial}{\partial x^m} \; , \; \bar{Q}_{\dot\alpha} = -\frac{\partial}{\partial\theta^{\dot\alpha}} + i\theta^\beta\,\sigma^m_{\beta\dot\alpha}\frac{\partial}{\partial x^m} \quad . \tag{10}$$

A scalar superfield is a scalar function in superspace

$$\Phi'(x',\theta') = \Phi(x,\theta) \quad ,$$

where in the infinitesimal

$$\delta\phi = \left[\epsilon\frac{\partial}{\partial\theta} + \bar\epsilon\frac{\partial}{\partial\bar\theta} + i(\theta\sigma^m\bar\epsilon - \epsilon\sigma^m\bar\theta)\frac{\partial}{\partial x^m}\right]\phi \quad . \tag{11}$$

Because $\theta^\alpha\theta^\beta + \theta^\beta\theta^\alpha = 0$ it is equivalent to a finite collection of ordinary fields defined over Minkowski space. A scalar field unifies into a single object 16 Bose and 16 Fermi fields. Covariant derivative can also be introduced [3)]

$$\partial_a = \frac{\partial}{\partial x^a} \; , \; D_\alpha = \frac{\partial}{\partial\theta^\alpha} + i\sigma^m_{\alpha\dot\beta}\,\bar\theta^{\dot\beta}\frac{\partial}{\partial x^m} \; , \; \bar{D}_{\dot\alpha} = -\frac{\partial}{\partial\bar\theta^{\dot\alpha}} - i\theta^\beta\,\sigma^m_{\beta\dot\alpha}\frac{\partial}{\partial x^m} \quad . \tag{12}$$

They have the property that they commute with all the generators P_m, Q_α, $\bar{Q}_{\dot\alpha}$. Moreover, they satisfy the algebra

$$\{D_\alpha,\bar{D}_{\dot\alpha}\} = -2i\sigma^a_{\alpha\dot\alpha}\partial_a \; , \; \{D_\alpha,D_\beta\} = 0 \quad , \quad [D_\alpha,\partial_a] = 0 \quad . \tag{13}$$

The commutation relations given by (13) show that "flat" superspace has non-vanishing (super-)torsion although its (super-)curvature vanishes. Note that the "flat" (super-) vierbein $\tilde{V}_A^M$ (we refer early letters to tangent space and late letters to the basic manifold) defined through the relation

$$D_A = \tilde{V}_A^M \frac{\partial}{\partial Z^M} \quad , \tag{14}$$

where

$$D_A = \partial_a \; , \; D_\alpha \; , \; \bar{D}_{\dot{\alpha}} \; , \tag{15}$$

$$\frac{\partial}{\partial Z^M} = \frac{\partial}{\partial x^m} \; , \; \frac{\partial}{\partial \theta^\alpha} \; , \; - \frac{\partial}{\partial \theta^\alpha} \tag{16}$$

is also non-trivial due to relation (13).

We can introduce different parametrizations of superspace using group elements [3]

$$\begin{aligned} L_1(x,\theta,\theta) &= \exp(i\theta Q - ixP) \; \exp(i\bar{\theta}\bar{Q}) \quad , \\ L_2(x,\theta,\theta) &= \exp(i\bar{\theta}\bar{Q} - ixP) \; \exp(i\theta Q) \quad . \end{aligned} \tag{17}$$

$L_{1(2)}$ and L are connected by the relation

$$L(x,\theta,\theta) = L_1(x + i\theta\sigma\bar{\theta},\theta,\bar{\theta}) = L_2(x - i\theta\sigma\bar{\theta},\theta,\bar{\theta}) \; . \tag{18}$$

The motion induced by the left-action of a group element $G(a,\epsilon,\bar{\epsilon})$ on $L_{1(2)}$ is given by

$$x'_m = x_m + a_m + 2i\theta\sigma_m\bar{\epsilon} + i\epsilon\sigma_m\bar{\epsilon} \quad , \quad \theta' = \theta + \epsilon \quad , \quad \bar{\theta}' = \bar{\theta} + \bar{\epsilon} \; , \tag{19}$$

$$x'_m = x_m + a_m - 2i\epsilon\sigma_m\bar{\theta} - i\epsilon\sigma_m\bar{\epsilon} \quad , \quad \theta' = \theta + \epsilon \quad , \quad \bar{\theta}' = \bar{\theta} + \bar{\epsilon} \; . \tag{20}$$

Correspondingly, we can introduce superfields $\Phi_{1(2)}$ transforming according to (19), (20). From (19),(20), it follows that type I (or type II) superspace has complex bosonic co-ordinates. The curved space analogue of (19), (20), has been used recently by Ogievetsky and Sokatchev [5] in order to have a geometrical description of supergravity in terms of an axial vector superfield. One of the advantages of type I or II superfields is that the expression of covariant derivatives become extremely simple in these bases. For instance

$$\bar{D}_{\dot{\alpha}} = - \frac{\partial}{\partial \bar{\theta}^{\dot{\alpha}}} \quad \text{on} \quad \Phi_1 \quad \text{and} \quad D_\alpha = \frac{\partial}{\partial \theta^\alpha} \quad \text{on} \quad \Phi_2 \quad .$$

This shows that a solution of the covariant constraint

$$\bar{D}_{\dot{\alpha}}\Phi = 0 \quad \text{(chirality condition)} \tag{21}$$

is simply given by $\Phi_1(x,\theta,\theta) = \Phi_1(x,\theta)$, so we get

$$\Phi(x,\theta,\theta) = \Phi_1(x + i\theta\sigma\bar{\theta},\theta) \quad . \tag{22}$$

If we expand Φ_1 in power series of θ we get

$$\Phi_1(x,\theta) = z(x) + \theta^\alpha \chi_\alpha(x) + \theta^\alpha \theta_\alpha H(x) \quad , \tag{23}$$

and we obtain the scalar multiplet with transformation laws as given by (1).

A superfield Φ can be taken to be real $\Phi = \Phi^*$; in that case we can no longer impose a condition like (21) and the superfield is essentially irreducible. In the literature it has been called a vector superfield (because it contains a vector field) and its general expansion is

$$\begin{aligned}\phi(x,\theta,\theta) = {} & C + i\theta\chi - i\bar{\theta}\bar{\chi} + \frac{i}{2}\theta\bar{\theta}(M + iN) - \frac{i}{2}\theta\bar{\theta}(M - iN) \\ & - \theta\sigma^m\bar{\theta}v_m + i\theta\bar{\theta}\bar{\theta}\bar{\lambda} - \frac{1}{2}\theta\theta\partial_m\chi\sigma^m\bar{\theta} - i\bar{\theta}\bar{\theta}\theta\lambda \\ & + \frac{1}{2}\bar{\theta}\bar{\theta}\theta\sigma^m\partial_m\bar{\chi} + \frac{1}{2}\theta\theta\bar{\theta}\bar{\theta}(D + \frac{1}{2}\Box C) \ .\end{aligned} \tag{24}$$

We have so far confined our considerations to superfields with no (external) Lorentz indices. Such an index is irrelevant for the previous considerations and can be added without modifications of the previous results. We can extend our analysis to more general superfields as follows. A general (quantum) superfield can be written as

$$\Phi(x,\theta) = L(x,\theta)\Phi \ L^{-1}(x,\theta) \quad , \tag{25}$$

by definition $\Phi = \Phi(0,0)$ is the superfield at the (super) origin. The latter is an object which transforms according to a (representation) of the stability (super) algebra H of the point $(x,\theta) = (0,0)$. In the case of the algebra given in (2)

$$H = M_{mn} \quad \text{is six-dimensional (Lorentz algebra)} \ . \tag{26}$$

For the superconformal algebra [2]

$$H = M_{mn}, K_m, S_\alpha, D, \Pi \quad \text{is sixteen-dimensional.} \tag{27}$$

If Φ is a representation of H we can induce on $\Phi(x,\theta)$ a representation of the whole algebra. If X is a generator of H we can compute the action of X on $\Phi(x,\theta)$ as follows:

$$[X,e^{-T}\Phi e^T] = e^{-T}[Y,\phi]e^T + e^{-T}[X,\phi]e^T \quad , \tag{28}$$

$$Y = \int_0^1 d\lambda \ e^{\lambda T}[T,X]e^{-\lambda T} \quad , \quad T = i(\theta^\alpha Q_\alpha + \bar{\theta}_\alpha\bar{Q}^\alpha - x\cdot P) \ , \tag{29}$$

where the infinite chain of commutators which defines Y stops after a finite number of steps due to the O'Raifeartaigh theorem. The superfield formalism is particularly convenient for working out tensor products of supersymmetry representations. In fact multiplication of representations is merely reduced to superfield multiplications which is an almost trivial operation. Such a multiplication in conjunction with the concept of Berezin integration [6] over anticommuting variables is one of the essential ingredients for the construction of supersymmetric field theories. If we label a chiral left-handed (right-handed) superfield and vector superfield with subscripts L(R), V , one can check the following properties of superfield multiplications:

$$\begin{aligned}
&\phi_L\psi_L = (\phi\psi)_L \;, \quad \phi_R\psi_R = (\phi\psi)_R \;, \quad \phi_L^* = \phi_R \;, \\
&\phi_L\psi_R + \phi_R\psi_L = (\phi\psi)_{V(SYMM)} \;, \quad i(\phi_L\psi_R - \phi_R\psi_L) = (\phi\psi)_{V(ANTISYMM)} \;, \\
&\phi_V\psi_V = (\phi\psi)_V \;, \quad DD\phi = (DD\phi)_R \;, \\
&DD\phi_L = (DD\phi_R^*)_R \;.
\end{aligned} \tag{30}$$

Integration of superfields is defined through the Berezin integration recipe:

$$\int d\theta_i \theta_j = \delta_{ij} \;. \tag{31}$$

This implies that

$$\int d^4\theta \;\; \Phi(x,\theta,\theta) = \Phi_{LAST}(x) \;, \tag{32}$$

where Φ_{LAST} means the coefficient of the $\theta\theta\bar{\theta}\bar{\theta}$ monomial in the θ expansion of Φ. It follows that an invariant action constructed out of a superfield Φ is given by [7)]

$$\int d^4x \, d^4\theta \; \Phi = \int dZ \; \Phi = \frac{1}{4}\int d^4x \; \bar{D}\,\bar{D}\; DD\; \Phi \;. \tag{33}$$

Eq.(25) is the starting point for the construction of a Lagrangian in the superfield formulation of supersymmetric Lagrangian field theories.

III. GAUGE THEORIES

The most striking application of superspace is in the context of gauge theories. In fact when a gauge invariance principle is imposed in supersymmetry, it turns out that the resulting geometrical structure of superspace manifests some unconventional features with respect to ordinary gauge theories over Minkowski space. The geometry is a constrained geometry. This is the case both for the gauging of an internal symmetry (Yang-Mills) and for a space-time symmetry (supergravity). Let us consider first the gauging of an internal symmetry group G . Then it is known that the appropriate superfield [8),9)] for the description of the vector potential is a Lie algebra valued vector superfield V and the gauge parameter is a (Lie algebra valued) chiral superfield $\Lambda(\bar{D}_{\dot{\alpha}}\Lambda = 0)$. The (finite) Yang-Mills gauge transformation on V is given as follows:

$$e^V \rightarrow e^{-i\Lambda^+} e^V e^{i\Lambda} \;, \tag{34}$$

and the superfield strength is given by the following spinorial chiral multiplet [8)]:

$$W_\alpha = \bar{D}_{\dot\alpha}\bar{D}^{\dot\alpha}(e^{-V} D_\alpha e^V) \qquad (\bar{D}_{\dot\alpha} W_\alpha = 0) \quad . \tag{35}$$

It is simple to show that under Yang-Mills transformation W transforms as follows:

$$W_\alpha \to e^{-i\Lambda} W_\alpha e^{i\Lambda} \quad . \tag{36}$$

The Yang-Mills superspace Lagrangian is given by

$$\mathrm{Tr}(W_1^\alpha W_{1\alpha}) \tag{37}$$

and the invariant action is obtained by integration over $d^4x\, d^2\theta$.

It is important that the vector potential V , the field strength W_α and the chiral parameter Λ can be understood in terms of a genuine Yang-Mills theory in superspace with a constrained geometry. If we define (Lie algebra valued) vector potentials V_A and field strengths F_{AB} in superspace

$$F_{AB} = D_A V_B - D_B V_A + [V_A, V_B\} + \tilde{T}_{AB}^{\ \ C} V_C \tag{38}$$

the gauge parameter Λ is an unconstrained "vector" superfield Λ . It turns out that in order to reduce the number of field components of V_A in such a way that the theory is equivalent to the previous formulation (in terms of V) the following constraints on the field strength (38) must be imposed [10),11)]:

$$F_{\alpha\beta} = F_{\dot\alpha\dot\beta} = F_{\alpha\dot\beta} = 0 \ . \tag{39}$$

These constraints imply that the remaining components of the curvature $F_{\alpha m}$, $F_{\dot\alpha m}$, F_{mn} are only functions of a chiral superfield W_α and that the vector potential V_A is only a function of a real vector superfield V . The remaining gauge transformation on V restricts the superfield Λ to be a chiral superfield $\bar{D}_{\dot\alpha}\Lambda = 0$. Note that Eqs.(39) imply $V_\alpha = e^{-V} D_\alpha e^V$ (and $W_\alpha = \bar{D}\bar{D}V_\alpha$ as given by (27)), which expresses the fact that V_A is a pure gauge along the spinorial (horizontal) directions. We learn from the geometrical description of the supersymmetric Yang-Mills theory that the kinematical constraints on the curvatures given by (39) are crucial in order to lower the spin content of the vector potential from highest spin-2 to highest spin-1 as required for a consistent description of the theory.

IV. SUPERGRAVITY

The theory of supergravity in the superspace approach can be formulated in close analogy with the case of the Yang-Mills supersymmetric gauge theory. One can start with a general affine superspace whose points are parametrized by co-ordinates $Z^M = (x^m, \theta^\mu)$. (Here $\mu = 1,\ldots,4$ and θ^μ stands for $\theta^\mu, \bar\theta^{\dot\mu}$, $\mu = 1,2$.)

Under general co-ordinate transformations we have

$$Z^M \to Z^M + f^M(Z) \quad , \tag{40}$$

$f^M(Z)$ is an arbitrary function on superspace and M refers to a (super) world index.

At each point in superspace one erects local tangent frames and defines supertetrads (super-Vierbein)

$$V_M^A(Z) \quad , \quad A = (a,\alpha) \quad , \quad M(m,\mu) \quad , \tag{41}$$

and their inverse V_A^M . Flat indices are raised and lowered with the flat metric of the group of tangent space. The appropriate group for the superspace formulation of supergravity turns out to be the Lorentz group.[12] In the tangent space there are two invariant tensors

$$\eta_{AB} = \begin{pmatrix} \eta_{ab} & 0 \\ 0 & 0 \end{pmatrix} \quad , \quad \eta'_{AB} = \begin{pmatrix} 0 & 0 \\ 0 & \gamma^0_{\alpha\beta} \end{pmatrix} \quad , \tag{42}$$

which have no inverse. Then it follows that in this approach an invertible metric cannot be defined. This is in contrast with the metric approach considered by Nath and Arnowitt [13] in which the tangent space group is taken to be the orthosymplectic group Osp(4/4) with invariant (invertible) metric

$$\eta_{AB} = \begin{pmatrix} \eta_{ab} & \\ & \kappa\gamma^0_{\alpha\beta} \end{pmatrix}$$

(κ being a dimensional constant).

One can define on an affine space covariant derivatives

$$\mathcal{D}_M = \partial_M - \Phi_M \tag{43}$$

in terms of a superconnection $\Phi_M = \Phi_M^{\ ab} X_{ab}$, which is Lie algebra valued over the Lorentz Lie algebra.

Covariant derivatives with tangent-space indices $\mathcal{D}_A = V_A^{\ M} \mathcal{D}_M$ obey

$$[\mathcal{D}_A, \mathcal{D}_B\} = -R_{AB}^{\ \ ab} X_{ab} - T_{AB}^{\ \ C} \mathcal{D}_C \tag{44}$$

$T_{AB}^{\ \ C}$ and $R_{AB}^{\ \ ab}$ are respectively, the supertorsion and the supercurvature. Torsion and curvatures satisfy two sets of Bianchi identities which follow from the Jacobi identities for the $\mathcal{D}_A$'s

$$[[\mathcal{D}_A, \mathcal{D}_B\}, \mathcal{D}_C\} + \text{cyclic} = 0 \quad . \tag{45}$$

Up to now the geometry is not sufficiently specified in order to correctly reproduce the supergravity theory. As first stressed by Wess and Zumino, additional kinematical constraints [12] must be introduced in order that the super-Vierbein $V_A^{\ M}$ be suitable for the description of the component fields of supergravity. The correct constraints to be imposed are the following restrictions on the supertorsion [14]:

$$T_{\alpha\beta}^{\ \ c} = 2i\ \gamma^c_{\alpha\beta} \ , \quad T_{\alpha\beta}^{\ \ \gamma} = T_{ab}^{\ \ c} = T_{\alpha b}^{\ \ c} = 0 \tag{46}$$

$(\alpha,\beta,\gamma = 1.,,,4, \quad a,b,c = 1,\ldots,4)$.

As a consequence of the Bianchi identities (44) these constraints imply [15] that all components of the supercurvature $R_{AB}^{\ \ ab}$ and the remaining components of the supertorsion $T_{\alpha b}^{\ \ \gamma}$, $T_{ab}^{\ \ \gamma}$ can be expressed in terms of three superfields

$$G_{\alpha\dot{\beta}} \text{ (hermitian) }, \quad W_{\alpha\beta\gamma} \text{ (symmetric) and } R \ , \tag{47}$$

which satisfy the following differential identities:

$$\bar{\mathcal{D}}_{\dot{\delta}} W_{\alpha\beta\gamma} = 0 \ , \ \bar{\mathcal{D}}_{\dot{\alpha}} R = 0 \ , \tag{48}$$

$$\mathcal{D}^{\alpha} W_{\alpha\beta\gamma} = \bar{\mathcal{D}}^{\dot{\delta}}_{\beta} G_{\gamma\dot{\delta}} + \mathcal{D}^{\dot{\delta}}_{\gamma} G_{\beta\dot{\delta}} \ , \quad \mathcal{D}^{\alpha} G_{\alpha\dot{\beta}} = \bar{\mathcal{D}}_{\dot{\beta}} R^* \ . \tag{49}$$

The analogy with the Yang-Mills case is now clear. The constraints (46) are the analogue of the constraints (39) and the three supermultiplets in (47) are the analogue of the multiplet in (35). Eqs.(48) and (49) are the analogue of $\bar{D}_{\dot{\alpha}} W_{\alpha} = 0$. We observe that the chirality condition $\bar{D}_{\dot{\alpha}}\phi = 0$ is not in general a covariant statement in curved superspace because left (right)-handed spinorial covariant derivatives no longer anticommute (see (44)). However, the kinematical restrictions on the (super-) torsion given in (46) imply $T^{c}_{\alpha\beta} = T_{\alpha\beta}^{\ \ \gamma} = T_{\alpha\beta}^{\ \ \dot{\gamma}} = 0$ and also $R_{\alpha\beta}^{\ \ \dot{\gamma}\dot{\delta}} = 0$ $(\alpha,\beta,\gamma,\delta = 1,2)$ (through Bianchi identities). Then the chirality constraint is always invariant if the superfield Φ carries only undotted indices. This is in contrast with flat superspace where no restriction on the Lorentz structure of a chiral superfield exists. This result has been also derived in the component formalism in Ref.16.

The three multiplets given by (47) are the supersymmetric completion of the multiplets containing the Einstein tensors, the Weyl tensor and the curvature scalar. It is important to point out that the torsion constraints also imply constraints on the super-Vierbein $V_A^{\ M}$ itself. In a suitable supersymmetric gauge a solution of these constraints can be given in terms of a (real) axial superfield $V_{\alpha\dot{\alpha}}$ (endowed with a vector index). This superfield, proposed [17] as the natural object which is coupled to the supercurrent superfield of globally supersymmetric theories [18], also emerges in a natural way in the unconstrained geometrical approach pursued by Ogievetsky and Sokatchev [5] and in another similar (although more general) approach considered by Siegel and Gates [19), 20)]. The axial superfield $V_{\alpha\dot{\alpha}}$ in supergravity is the analogue of the superfield V (see Eq.(34)) of supersymmetric Yang-Mills theories.

Other approaches which use a set of kinematical constraints different from those of Wess and Zumino (Eq.(46)) have been considered by Brink, Gell-Mann, Ramond and Schwarz [21] and J.G. Taylor [22]. All these approaches turn out to be equivalent as far as the description of pure supergravity is concerned. However, differences may emerge when coupling to matter multiplets or non-minimal interactions are also considered.

Finally it is important to point out that the minimal auxiliary fields which emerge in the component formulation of supergravity [23),24)] are nothing but the (Lorentz) irreducible parts of $T_{\alpha b}^{\ \ \gamma}(x,\theta = 0)$, in the formulation of Wess-Zumino [25]. This has recently been used, by means of a technique developed in Ref.21, to establish the connection between the gauging of the graded Poincaré group in ordinary space and (super) co-ordinate transformations in superspace [26].

REFERENCES

1) For complementary reviews, see for instance:
P. Fayet and S. Ferrara, Phys. Rep. 32C, 249 (1977);
A. Salam and J. Strathdee, Fortschr. Phys. 26, 57 (1978).
2) J. Wess and B. Zumino, Nucl. Phys. B70, 39 (1974).
3) For conventions and notations see:
S. Ferrara, J. Wess and B. Zumino, Phys. Letters 51B, 239 (1974).
4) A. Salam and J. Strathdee, Nucl. Phys. B76, 477 (1974).
5) V. Ogievetsky and E. Sokatchev, Phys. Letters 79B, 222 (1978), and Dubna preprints (1978, 1979).
6) F.A. Berezin, The Method of Second Quantization (Academic Press, New York 1966).
7) A. Salam and J. Strathdee, Phys. Rev. D11, 521 (1975).
8) S. Ferrara and B. Zumino, Nucl. Phys. B79, 413 (1974).
9) A. Salam and J. Strathdee, Phys. Letters 51B, 353 (1974).
10) J. Wess, Topics in Quantum Field Theory and Gauge Theories, Salamanca, 1977 (Springer-Verlag, Berlin 1978), Vol.81.
11) M.F. Sohnius, Nucl. Phys. B136, 461 (1978).
12) J. Wess and B. Zumino, Phys. Letters 66B, 361 (1977).
13) R. Arnowitt and P. Nath, Phys. Letters 56B, 117 (1975).
14) J. Wess and B. Zumino, Phys. Letters 74B, 51 (1978).
15) R. Grisaru, J. Wess and B. Zumino, Nucl. Phys. B152, 255 (1979).
16) M. Fischler, Stony Brook preprint (1979).
17) V.I. Ogievetsky and E. Sokatchev, Nucl. Phys. B124, 309 (1978);
S. Ferrara and B. Zumino, Nucl. Phys. B134, 301 (1978).
18) S. Ferrara and B. Zumino, Nucl. Phys. B87, 207 (1975).
19) W. Siegel, Nucl. Phys. B142, 301 (1978);
W. Siegel and S.J. Gates, Nucl. Phys. B147, 77 (1978).
20) S.J. Gates and W. Siegel, Harvard preprint HUTP-79/A034.
21) L. Brink, M. Gell-Mann, P. Ramond and S.H. Schwarz, Phys. Letters 74B, 336 (1978); Phys. Letters 76B, 417 (1978).
22) J.G. Taylor, Phys. Letters 78B, 577 (1978); 79B, 399 (1978); 80B, 52 (1978);
J. Bedding, C. Pickup, J.G. Taylor and S. Downes-Martin, Phys. Letters 83B, 59 (1979).
23) S. Ferrara and P. van Nieuwenhuizen, Phys. Letters 74B, 333 (1978).
24) K. Stelle and P. West, Phys. Letters 74B, 330 (1978).
25) J. Wess and B. Zumino, Phys. Letters 79B, 394 (1978).
26) S. Ferrara and P. van Nieuwenhuizen, Stony Brook preprint (September 1979).

SYMMETRIES OF SUPERGRAVITY MODELS

Bernard JULIA

Laboratoire de Physique Théorique de l'Ecole Normale Supérieure
24, rue Lhomond, 75231 PARIS CEDEX 05, FRANCE

INTRODUCTION :

On the one hand, mathematical physicists seem to be using functional and complex analysis more frequently than group theory and algebra ; on the other hand, supersymmetry has been studied mostly from the second point of view ; yet one can predict that analysis will become more and more important there too. For instance, one must investigate systematically possible cancellations of quantum corrections in supersymmetric models, a task that requires powerful analytical tools ; furthermore, a large class of these models possesses a Kaehlerian structure[1] to be described with complex function theory etc... The subject of my talk is to review supergravity theories and especially their symmetries. After mentioning the non-linear σ-model form of pure Einstein gravity[2], I shall discuss the " κ series" construction of locally supersymmetric actions and illustrate it by giving the 11 dimensional supergravity Lagrangian discovered in(3) (κ is the gravitational coupling constant). By standard "dimensional reduction"[2] one obtains a (CPT invariant) local field theory in four dimensions, where the supersymmetry charges act very simply on zero-mass states. We shall elaborate only on the so-called "spontaneous breaking" of supersymmetry[4] and refer to (2) for a discussion of the "standard" (straightforward) dimensional reduction which does not introduce masses nor symmetry breakdown. One finds a new type of algebra with a non-central charge acting on the fields and the example of N = 8 supergravity obtained from d = 11 supergravity illustrates this. We also review the generalized duality transformations that preserve the classical equations of motion and the mysterious exceptional symmetry groups of these equations. We conclude with three conjectures.

I. N = 0 SUPERGRAVITY : EINSTEIN'S GRAVITATION AS A NON LINEAR σ-MODEL

N designates the number of spinor charges, it happened that studying $N \geq 1$ supergravity theories shed some new light on the N = 0 case (which is the only established theory of the family). It is well-known that general relativity can be described by two (locally) equivalent formalisms : in the metric formalism (in first order form) the independent variables are the metric $g_{\mu\nu}$ and a symmetric connection $\Gamma^{\rho}_{\mu\nu}$ and the action reads :

$$\mathcal{L}_0 = \frac{1}{4\kappa^2} \sqrt{|g_{\mu\nu}|}\, g^{\rho\sigma} g_{\upsilon}{}^{\tau} \left(\partial_\rho \Gamma^{\upsilon}_{\sigma\tau} + \Gamma^{\kappa}_{\sigma\tau} \Gamma^{\upsilon}_{\kappa\rho} - (\rho \leftrightarrow \tau) \right) .$$

In the moving frame formalism (Weyl) the variables are a moving frame $e_\mu{}^\alpha$ and a Lorentz connection $\omega_{\mu\alpha}{}^\beta$ and the action is :

$$\mathcal{L}_0 = - \frac{1}{4\kappa^2} \det(e_\mu{}^\alpha)\; R_{\mu\nu\alpha}{}^\beta(\omega)\, e^{\mu\alpha} e^{\nu}{}_\beta .$$

The second formalism allows the introduction of fermions transforming only under the Lorentz group, and its global properties (topology for example) are different from those of the first one. In both cases, $\omega_{\mu\alpha}{}^\beta$ and $\Gamma^\rho_{\mu\nu}$ can be eliminated algebraically : they are auxiliary fields; but how comes it that $e_\mu{}^\alpha$ (16 numbers) describes as many degrees of freedom (2) as $g_{\mu\nu}$ (10 numbers) ? The answer lies in the additional SO(1,3) Lorentz invariance which identifies all ($e_\mu{}^\alpha \Lambda_\alpha{}^\beta$)'s ($\Lambda_\alpha{}^\beta$ = any Lorentz transformation), i.e. the propagating fields are extracted from a field which takes values in the coset space $GL(4,\mathbb{R})/SO(3,1)$ (2); the symmetric parametrization of this coset space is just the metric field of signature (+ − − −). One calls such a theory a generalized σ-model by opposition to usual σ-(or chiral) models for which the field is a group element.

II. FROM GLOBAL TO LOCAL SUPERSYMMETRY :

1) Global algebras :

We shall restrict ourselves to a discussion of the supersymmetry algebras ; they can be exponentiated by using anticommuting parameters to form Lie groups but only infinitesimal transformations will be necessary for what follows. The so-called SO(N) supersymmetry theories have invariant actions under the Poincaré group times the internal group SO(N) times a N-vector of spinor charges whose anticommutation rules are :

$$[Q_\varepsilon{}^a , \bar{Q}_\eta{}^b]_+ = \delta^{ab} (\gamma^\alpha)_{\varepsilon\eta} P_\alpha$$

ε, η = Majorana spinor indices
$a, b = 1, \ldots, N$; $\alpha = 0, \ldots, 3$ Lorentz index.

The "little algebra" for fixed momentum $P_\alpha = p_\alpha$ ($p^2 = 0$ here) can be discussed in more detail (5), it fixes the spectrum of states and suggests the field contents of the Lagrangian. Choosing the conventions of (5) for a representation of the γ matrices with $\gamma^5 = \begin{bmatrix} -1 & 0 \\ 0 & 1 \end{bmatrix}$, a Majorana spinor reads $\begin{bmatrix} Q_1 \\ Q_2 \\ Q_2^* \\ -Q_1^* \end{bmatrix}$ and for $p^\mu = (\omega, 0, 0, \omega)$ the only non-vanishing anticommutators are $[Q_2{}^a , Q_2{}^{b*}]_+ = 2\omega\,\delta^{ab}$.
The N Q_2^{a*}'s lower helicity by one half unit, so, assuming the spin two state to be a SO(N) singlet, one finds the global supersymmetry multiplets : $|2\rangle$, $|3/2\rangle^a$, $|1\rangle^{[ab]}$, $|\tfrac{1}{2}\rangle^{[abc]}$, and so on, with anti-symmetric sets of indices because of the anticommutativity of the Q_2^{a*}'s among themselves. Thus requiring at most one spin-two field (the would-be graviton) forbids N to be larger than 8. Actually for N = 8 we can check that CPT is built into the supersymmetry algebra : the $|-2\rangle^{abcdefgh}$ state is nothing but $|-2\rangle\,\varepsilon^{abcdefgh}$, the CPT conjugate state to $|+2\rangle$; more generally $\frac{1}{8!}\varepsilon^{abcdefgh}(Q_2^a + Q_2^{a*}) \cdots (Q_2^h + Q_2^{h*})$, exchanges states of helicity $+s$ and $-s$. This is a necessary condition if we aim at a local covariant field theory with usual connection between spin and statistics.

We must now assign fields to these states : this is straighforward for spin 0 and 1/2 but for higher spins we must introduce gauge invariances if we insist on having conventional tensors, i.e. vectors $A_\mu^{[ab]}$ and not bispinors $\psi_{(\varepsilon\eta)}^{[ab]}$ .... The common procedure is to introduce at first abelian vector fields gauging $\frac{N(N-1)}{2}$ U(1) groups, N Rarita-Schwinger fields $\psi_{\mu\varepsilon}^a$ and the linearized Einstein action for the moving frame field $\kappa h_{\mu\alpha} = e_{\mu\alpha} - \delta_{\mu\alpha}$.

2) Gauging supersymmetry :

Already in the global supersymmetry case one must introduce auxiliary (non-propagating) fields in order to realize the supersymmetry algebra linearly,in the simplest example of a Majorana spinor plus a complex scalar field one needs a complex auxiliary scalar field. This also implies that the algebra closes on the classical fields without using the equations of motion and that it does not depend on the form of the action. In the local supersymmetry (supergravity) case, one does not yet know the auxiliary fields for N $\geqslant$ 3, one must consequently be satisfied with an algebra that depends on the action and that closes only on shell.

Clearly, the spin 3/2 part of the N $\psi_{\mu\varepsilon}^{a}$ fields can serve as gauge fields for the N supersymmetry charges, the idea is to couple them to the supersymmetry currents with a coupling constant κ but thenglobal supersymmetry implies a coupling of the energy-momentum tensor to the metric field $h_{\mu\alpha}$. These changes in the action partially destroy the supersymmetry invariance which must then be restored order by order in κ by modifying both the action and the transformation laws.

The result is a theory of gravitation having also a local supersymmetry invariance. Let us detail the algebra for N = 1 supergravity : the commutators of two supersymmetry transformations with parameters $\varepsilon_1(x)$ and $\varepsilon_2(x)$ are sums of a general coordinate transformation with displacement $\xi^{\mu}(x) = i e^{\mu}{}_{\alpha}\, \bar{\varepsilon}_1(x)\, \gamma^{\alpha}\, \varepsilon_2(x)$, a local Lorentz transformation with <u>field dependent</u> parameter $\lambda_{\alpha\beta} = \xi^{\nu}\omega_{\nu\alpha\beta}$ and a <u>field dependent</u> supersymmetry transformation with parameter $\varepsilon_3 = -\kappa\, \xi^{\mu}\psi_{\mu}$. Adding the auxiliary fields S, P, A_{μ}, one gets an algebra that closes without using the equations of motion and the only modification is a new value for the parameter of the local Lorentz rotation :

$$\lambda'_{\alpha\beta} = \lambda_{\alpha\beta} + \frac{\kappa}{3}\left[-i\,\varepsilon_{\alpha\beta\gamma\delta}\,\xi^{\gamma}A^{\delta} + 4\,\bar{\varepsilon}_2\,\sigma_{\alpha\beta}(S - i\gamma_5 P)\,\varepsilon_1\right]$$

(we used the conventions of (7) in this last formula).

Dimensional analysis reveals that bosonic fields : $\kappa\varphi$, κA_{μ} , $e_{\mu}{}^{\alpha} = \delta_{\mu}{}^{\alpha} + \kappa h_{\mu}{}^{\alpha}$ are dimensionless as fermionic combinations $\kappa\psi$, $\kappa\psi_{\mu}$ have dimension 1/2 and can appear at most to the fourth power. Thus,resumming the spin 2 terms to have only $e_{\mu}{}^{\alpha}$ appearing in Einstein's action (times κ^2) and deducing from gauge invariance that only $\kappa F_{\mu\nu}$ (dimension 1) appears, the only potentially troublesome terms that could appear are <u>non polynomial</u> functions of the scalar fields[6]. By going to 11 dimensions the authors of (3) avoided the problem of scalar fields which can appear non polynomially (in any space-time dimension).

3) N = 1 supergravity in 11 dimensions :

Inspired by the fermionic dual model spectrum of states in 10 dimensions and its conjectured supersymmetry, one is led to a multiplet of three gauge fields in 11 dimensions A_{MNP} , $\psi_{M\zeta}$ (32 component Majorana vector-spinor) and a moving frame $e_{M}{}^{A}$. The dimensional reduction[2] of N = 1 supergravity from 11 to 10 dimensions should lead to the small slope limit of the dual model[3]; its reduction to 4 dimensions leads to N = 8 supergravity[8], and to 5 dimensions it gives another interesting theory[4]. The iterative construction stops after a finite number of steps (in 11 dimensions) and one obtains the following Lagrangian :

$$\mathcal{L} = -\frac{e}{4\kappa^2} R(\omega) - \frac{ie}{2} \bar{\psi}_M \Gamma^{MNP} D_N\left(\frac{\omega+\hat{\omega}}{2}\right) \psi_P - \frac{e}{48} F_{MNPQ} F^{MNPQ}$$
$$+ \frac{\kappa e}{192} \left(\bar{\psi}_M \Gamma^{MNABCD} \psi_N + 12 \bar{\psi}^A \Gamma^{CD} \psi^B\right)\left(F_{ABCD} + \hat{F}_{ABCD}\right)$$
$$+ \frac{2\kappa}{(12)^4} \varepsilon^{ABCDEFGHIJK} F_{ABCD} F_{EFGH} A_{IJK} .$$

The last term does not violate the (generalized) gauge invariance of the action and is topological in character. The generalized gauge structure is abelian ; $F_{MNPQ} = 4\, \partial_{[M} A_{NPQ]}$ is invariant under 55 U(1)'s : $A_{NPQ} \to A_{NPQ} + \partial_{[N} \Lambda_{PQ]}$. It does not commute with space-time transformations. The symmetries of this theory which have been discovered so far are : 1 supersymmetry spinor of 32 Majorana charges, general coordinate transformations containing $GL(11, \mathbb{R})$ as a subgroup, and local Lorentz invariance SO(1,10). We shall now discuss the fate of this action under dimensional reductions.

III. SYMMETRIES AS SIGNATURES FOR EXTRA-DIMENSIONS :

1) Simple dimensional reduction :

We refer to (2) for details, but the principle of standard dimensional reduction is quite simple. One starts with tensor fields $T^{MN\cdots}$ in 4 + p dimensions depending a priori on the p internal coordinates as well as on x^μ $\mu = 0, \ldots, 3$, the space-time coordinates. If the original Lagrangian is hermitian, it is a consistent unitary truncation to restrict oneself to fields that are independent of the internal coordinates. One can assume that the extra-dimensions curl up to form an internal space of very small radius compared to usual Compton wavelengths ; if it is an hypertorus, Fourier analysis in the internal dimensions leads to a tower of particles of increasing (four dimensional) masses for each (4+p) dimensional field :

$$m_4^2 = \sum_{i=1}^{p} \frac{n_i^2 4\pi^2}{L_i^2} + m^2_{4+p} \qquad \text{(for a scalar field).}$$

The low energy truncation amounts to ignore all non zero n_i's and implies a global SO(p) symmetry (SO(7) resp. SO(6) in (8) resp. (4)). So, part of the symmetry group of the four dimensional theory could be blamed on a higher dimensional ancestor, yet strong evidence for extra-dimensions can only come from the observation of the very massive recurrences (which do not share this symmetry).

Let us illustrate this type of dimensional reduction by the SO(8) supergravity example[8]. First of all, the field contents in four dimensions is familiar :

Number of degrees of freedom	11 dimensions	4 dimensions		
Field	$e_M{}^A$	$e_\mu{}^\alpha$	$7 \cdot B_\mu{}^{a'}$'s	$28\ e_m{}^{a'}$'s
off shell	$66-11=55$	$10-4=6$	$7 \times (4-1=3)$	28
on shell	44	$= 2$	$+ 14$	$+ 28$
Field	A_{MNP}		$21\ A_{\mu ij}$'s	$35 A_{ijk}$'s $+ 7 \varphi^{i}$'s ($\sim A_{\mu\nu i}$) $A_{\mu\nu\rho}$ auxiliary
off shell	$165-55+11-1=120$		$21 \times (4-1=3)$	42
on shell	84	$=$	42	$+ 42$
Field	ψ_M	$8\ \psi_\mu{}^{a'}$'s		$56\ \chi_a{}^{a'}$'s
off shell	$176-16=160$	$8 \times (8-2=6)$		2×56
on shell	128	$= 16$		$+ 112$

(A,M) ; (α, μ) ; (a,i) are indices in 11 dimensions, 4 dimensions or internal (flat resp. curved) and a' takes 8 values and comes from the splitting of a 32 component spinor into 8 4-component ones. We find the expected 8 spin 3/2 fields, 28 vectors, 56 spin 1/2's and 70 scalars of N = 8 supergravity.

Secondly, we a priori have a larger symmetry than the above mentioned global SO(7) : we have local SO(7) times $GL(7,\mathbb{R})$ as a residue of the original invariances. The problem in[8] was to restore the expected global SO(8) in four dimensions : it required some hard work strongly motivated by the presence of 8 supersymmetry charges in four dimensions. The outcome was the discovery of a much larger symmetry group of the equations of motion, namely local SU(8) times the global (non compact) $E_7(+7)$.

2) Dualities and first order formalism :

It is important to note that in our version of supergravity, the vector fields are all abelian ; furthermore, they have no minimal coupling to matter fields, but only Pauli-type couplings (via their field strengths), consequently the equations of motion of the vector fields look very much like Bianchi identities, they are respectively :

$\partial_\mu \tilde{G}^{\mu\nu} = 0$ and $\partial_\mu \tilde{F}^{\mu\nu} = 0$ where $\tilde{G}^{\mu\nu} = F^{\mu\nu}$ + non-linear coupling terms $\simeq \frac{\delta \mathcal{L}_{INT}}{\delta F_{\mu\nu}}$.

The manifest SO(N) invariance of the action of extended supersymmetric models may sometimes be extended to an SU(N) or U(N) invariance of the equations of motion. For real fields a complex invariance is possible only if one can find a real representation of the imaginary unit i namely $\begin{bmatrix} 0 & -1 \\ 1 & 0 \end{bmatrix}$ for scalars and pseudoscalars, $i\gamma_5$ (real) for Majorana spinors, and $\frac{1}{2}\varepsilon_{\mu\nu}{}^{\rho\sigma}$ for vector field strengths (if one neglects the couplings) more precisely it is the rotation between each $F_{\mu\nu}$ and its $G_{\mu\nu}$. The canonical Lagrangian is manifestly dissymetric as it involves only the A_μ potentials for $F_{\mu\nu} = 2\,\partial_{[\mu} A_{\nu]}$. In (8) we proposed a first order formalism with constraints where the variables are more numerous. One starts off with potentials A'_μ's and dual potentials B'_μ's as well as tensors $F''_{\mu\nu}$ and $G''_{\mu\nu}$, related by a constraint which can be solved to give one as a function of the other. Both the constraint and the first order action are symmetric under SU(8) local and E_7 global transformations whereas the second order (usual) action is only invariant under SO(8) local and $SL(8,\mathbb{R})$ global. The dissymmetry at the Lagrangian level stems from the elimination of only one half of the potentials ; the usual Lagrangian comes by eliminating all pseudo-vector potentials , B'_μ , but some other choices of 28 vector fields among the 56 A'_μ's and B'_μ's would lead to new actions. One example of this arbitrariness has been studied in detail : it is the case of N = 4 supergravity one form of which has an invariant Lagrangian under SU(4) (and not only the equations of motion which do not change) (9). So much for the symmetric theory of (8). Let us now study possible "broken" supersymmetries also obtained from higher dimensions.

3) "Spontaneous breakdown" of extended supersymmetries :

The whole idea of spontaneous symmetry breaking is to preserve nice features like renormalizability or predictive power, and yet account for states which do not transform linearly under the symmetry. Supersymmetry must be broken. One does not observe Fermi-Bose mass degeneracies in practice, but a spontaneous breakdown is hard to achieve, especially in the case of extended supersymmetry. A general technique for breaking usual supersymmetry has been put forward by J. Scherk and J. Schwarz in (10) using a modified dimensional reduction. It preserves a modified algebra.

Let us first discuss their example of N = 1 supergravity reduced from 4 to 3 dimensions. If the higher dimensional theory has some global invariance, chiral invariance in our example, one can replace the former requirement of independence on the fourth coordinate by a more general ansatz with periodicity of the Fermi fields only up to chiral transformations along the closed fourth dimension. One still truncates the theory by introducing only one period (the lowest mass scale) :

$$\psi_\mu (x_0, x_1, x_2, x_3) = e^{im\gamma^5 x^3} \psi'_\mu (x_0, x_1, x_2)$$

Because the moving frame is supposed to be independent of x^3, usual supersymmetry is broken in 3 dimensions.

Here, we seem to have added an extra parameter to the theory, violating the spirit of spontaneous symmetry breaking ; indeed it is not the traditional form of breaking : the "broken" Lagrangian still has an invariance but it is not the same invariance as in the symmetric case ; it is this new invariance which is spontaneously broken and which is not an invariance of the vacuum. The most obvious way to see that the algebra has changed is to note that the "electric" charge operator acts on the 2 supersymmetry charges and is a non-central charge. One can recover a global flat space superalgebra only in the limit m = 0. Before going to the example of N = 8 supergravity, we must realize that by this trick predictive power is reduced a little by extra arbitrary parameters and that quantum finiteness must be checked again. In (4) E. Cremmer et al. managed to introduce four mass parameters in a modified "spontaneously broken" N = 8 supergravity. They started from the dimensional reduction of 11 dimensional supergravity to 5 dimensions which exhibits also a large invariance group : $E_6{}^{(+6)}$ global times USp(8) local. It is an observation of E. Cremmer that, more generally, reducing d = 11 supergravity to d' dimensions (2 $\underset{(<)}{\leq}$ d' $\leq$ 8) leads to exceptional global groups of invariances $E_{11-d'}(11 - d')$! Then generalizing the method of (10) they obtained a number of mass parameters equal to the rank of the maximal compact group of invariances of the action, in this case USp(8). Finally it is encouraging that even after symmetry breaking the first order quantum correction to the cosmological term is finite(4).

IV PHENOMENOLOGY AND 3 CONJECTURES FOR N = 8 SUPERGRAVITY :

All 8 supersymmetries could be broken by the above method, and lead through the super-Higgs mechanism to 8 massive spin 3/2 fields. Another crucial step towards phenomenology would be to transform the 28 U(1)'s into a non abelian group SO(8) for example ; there is a difficulty there because this would seem to bring in a large cosmological term and negative scalar self-interactions (11). At any rate, generalizing the above dualities in the case of non zero minimal coupling of the vectors might be interesting : one expects then duality to be non-local.

Much remains to be done and most urgently more group theory, more differential and algebraic topology. We shall conclude by proposing three open questions :

1) Can one exhibit a 12 dimensional structure and use it to simplify dimensional reduction ? Could it explain the exceptional symmetry group and the nature of the topological term in the 11 dimensional supergravity action ?

2) Are there any more fermionic symmetries ? We mentioned the generalized non-linear σ-model structure of Einstein's action in section I and it is a fact that the 70 scalars parametrize the coset space $E_7/SU(8)$; is the full theory a supersymmetric super-coset model ?

3) Especially important for phenomenology is the problem of appearance of a dynamical kinetic term for the 63 auxiliary fields gauging SU(8) local. Can quantum corrections in four dimensions introduce these new degrees of freedom ?

REFERENCES :

(1) B. Zumino - CERN preprint TH 2733
See also subharmonic functions appearing in matter couplings to N = 1 supergravity : E. Cremmer et al., Nucl. Phys. B147 (1979) 105

(2) B. Julia - Extra-dimensions : recent progress using old ideas, LPTENS preprint 79/15 and references therein .

(3) E. Cremmer, B. Julia and J. Scherk, Phys. Lett. 76B (1978) 409

(4) E. Cremmer, J. Scherk and J.H. Schwarz , Phys. Lett. 84B (1979) 83 and references therein .

(5) D. Freedman, Lecture at the 1978 Cargèse Institute on Gravitation and references therein .

(6) S. Ferrara et al., Nucl. Phys. B117 (1976) 333

(7) P. Van Nienwenhuizen, Lectures at the 1978 Cargèse Institute on Gravitation .

(8) E. Cremmer and B. Julia, Phys. Lett. 80B (1978) 48 and Nucl. Phys. B (1979) to appear .

(9) E. Cremmer et al., Phys. Lett. 74B (1978) 61

(10) J. Scherk and J.H. Schwarz, Phys. Lett. 82B (1979) 60

(11) J. Scherk, Lectures at the 1978 Cargèse Institute on Gravitation

A REVIEW OF DERIVATIONS

Derek W. Robinson

Department of Pure Mathematics
University of New South Wales

P.O. Box 1, Kensington, Australia

Abstract; We give a general review of progress in the study of (unbounded) derivations with particular emphasis on criteria for the derivations to be generators of continuous groups of *-automorphisms

0. Introduction

A symmetric derivation δ of a C*-algebra $\mathfrak{A}$ is a linear operator from a *-subalgebra $D(\delta)$, the domain of δ, into $\mathfrak{A}$ which satisfies the derivation property

$$\delta(AB) = \delta(A)B + A\delta(B) \ , \quad A,B \in D(\delta)$$

and the symmetry property

$$\delta(A^*) = \delta(A)^* \ , \quad A \in D(\delta) \ .$$

Since all the derivations we consider are symmetric we consistently omit the qualifying adjective. Note that the range $R(\delta)$ of a derivation δ is a subspace of $\mathfrak{A}$ but it is not generally a subalgebra.

Derivations are of interest because they arise as generators of continuous groups of *-automorphisms. (Recall that such groups are automatically isometric) Let τ; $t \in \mathbb{R} \longrightarrow \tau_t \in \mathrm{Aut}(\mathfrak{A})$ denote a one-parameter group of *-automorphisms of $\mathfrak{A}$ and assume that τ is continuous in some topology. The generator δ of τ is defined as the linear operator

$$\delta; \ A \longmapsto \delta(A) = \lim_{t \to 0} (\tau_t(A) - A)/t$$

where the limit is taken in the topology in which τ is continuous and the domain $D(\delta)$ of δ is defined to be the set of $A \in \mathfrak{A}$ for which the limit exists. Conversely τ can be constructed from δ and formally one has $\tau_t = \exp(t\delta)$ but some care has to be exercised in the definition of the exponential.

There are various cases of interest. One can consider C*-algebras or W*-algebras and there are a variety of possible types of continuity for τ. For simplicity we restrict our attention to three cases.

Norm Continuity

The group τ is norm continuous if

$$\|\tau_t - \iota\| = \sup_{A \in \mathfrak{A}} \|\tau_t(A) - A\| / \|A\| \underset{t=0}{\longrightarrow} 0$$

Although norm continuous automorphism groups do possess much interesting algebraic structure they are rather simplistic in the sense that their generators are automatically bounded (see Section 1). In this respect the next form of continuity is more interesting.

Strong Continuity

The group τ is strongly continuous if

$$\|\tau_t(A) - A\| \underset{t=0}{\longrightarrow} 0$$

for all $A \in \mathfrak{A}$. Such groups are called C_0-groups.

In this context there are two relevant and not at all evident remarks.

Remark 1. Kallman has proved that each strongly continuous one-parameter group of *-automorphisms of a W*-algebra $\mathfrak{A}$ is automatically norm continuous (In fact Kallman needed a countability hypothesis on $\mathfrak{A}$ but this was eliminated by Elliott). Thus strongly continuous groups are only of interest in the C*-algebra context.

Remark 2. Let T be a one-parameter group of mappings on a Banach space X. There is a classic theorem of Yosida which states that strong continuity of T is equivalent to weak, or $\sigma(X,X^*)$-, continuity, i.e., is equivalent to the conditions

$$f(T_t A) \underset{t=0}{\longrightarrow} f(A)$$

for all $A \in X$ and all $f \in X^*$.

The third kind of continuity, σ-weak, or $\sigma(\mathfrak{A}, \mathfrak{A}_*)$-continuity, is only of interest for W*-algebras $\mathfrak{A}$ because it depends upon the existence of a pre-dual $\mathfrak{A}_*$. (Recall that a Banach space X is said to have a predual X_* if X_* is a Banach space satisfying $X = (X_*)^*$. Moreover Sakai has proved that a C*-algebra has a predual if and only if it is a W*-algebra).

σ-weak Continuity

The group τ is a σ-weakly ($\sigma(\mathfrak{A}, \mathfrak{A}_*)$-) continuous group of automorphism of the W*-algebra $\mathfrak{A}$ if

$$f(\tau_t(A)) \underset{t=0}{\longrightarrow} f(A)$$

for all $A \in \mathfrak{A}$ and $f \in \mathfrak{A}_*$. These groups are called C_0^*-groups.

Remark 3. It is possible to study properties intermediate to strong, or $\sigma(\mathfrak{A}, \mathfrak{A}^*)$-, continuity and $\sigma(\mathfrak{A}, \mathfrak{A}_*)$-continuity. For example if F is any subspace of the dual $\mathfrak{A}^*$ then one can examine $\sigma(\mathfrak{A},F)$-continuous groups, i.e., the τ for which

$$f(\tau_t(A)) \underset{t=0}{\longrightarrow} f(A)$$

for all $A \in \mathfrak{A}$ and $f \in F$. To obtain an interesting theory it is however necessary to restrict the possible F. This general kind of continuity has been discussed by Bratteli and Robinson.

Thus the relevant cases are summarized in the following chart

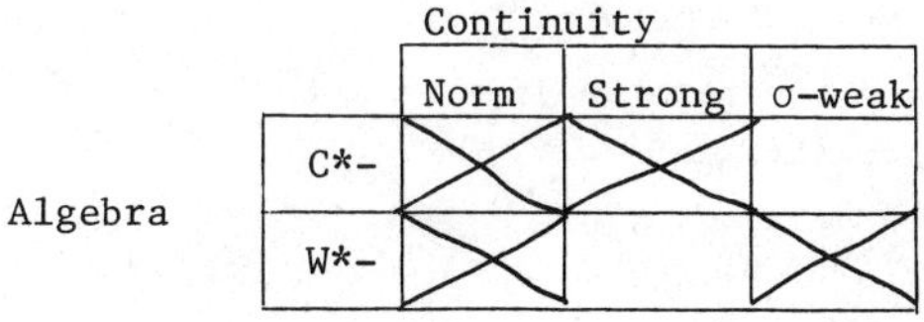

References; The text of this review is largely based upon Chapter 3 of
Operator Algebras and Quantum Statistical Mechanics
by O. Bratteli and D.W. Robinson,
Springer-Verlag Berlin-Heidelberg-New York (1979).

The specific results we have quoted can be found in the following references

Kallman, R.R.; Amer. J. Math. 91 (1969) 785-806.

Kallman, R.R.; Proc. Amer. Math. Soc. 24 (1970) 336-340.

Elliott, G.A.,; J. Func. Anal. 11 (1972) 204-206.

Yosida, K; Functional Analysis, Springer-Verlag, Berlin-Heidelberg-New York (1968), Chapter IV.

Sakai, S: C*-algebras and W*-algebras, Springer-Verlag, Berlin-Heidelberg- New York (1971), Theorem 1.16.7.

1. Norm Continuity

The discussion of norm continuity is greatly simplified by the following classic Banach space result.

Theorem 1 Let $T = \{T_t\}_{t\in\mathbb{R}}$ be a one-parameter group of bounded linear operators on the Banach space X.

The following conditions are equivalent

1. T is norm continuous
2. There is a bounded operator S such that
$$\lim_{t\to 0} \| (T_t - I)/t - S \| = 0$$
3. There is a bounded operator S such that
$$T_t = \sum_{n\geqslant 0} \frac{t^n}{n!} S^n$$

This result was proved by Nagumo in 1936. The proof is relatively straightforward and it has one easy consequence which is of interest.

Corollary 2 If $T = \{T_t\}_{t\in\mathbb{R}}$ is a C_0(or C_0^*-) group of bounded linear operators on the Banach space X then T is norm continuous if, and only if, there exist $\varepsilon,\delta > 0$ such that

$$\|T_t - I\| \leqslant 1 - \varepsilon \quad \text{for all} \quad 0 \leqslant t < \delta.$$

If in Theorem 1 X is replaced by a C*-algebra $\mathfrak{A}$ and T is replaced by a group τ of *-automorphisms then the bounded generator δ if τ is automatically a derivation. This follows by differentiating the relations

$$(*) \qquad \tau_t(AB) = \tau_t(A)\tau_t(B) \; , \quad \tau_t(A^*) = \tau_t(A)^* .$$

Conversely if δ is a bounded derivation of $\mathfrak{A}$ then the one-parameter group τ that it generates is automatically is group of *-automorphisms, e.g.

$$\begin{aligned}\frac{d}{dt}\tau_{-t}(\tau_t(A)\tau_t(B)) &= -\tau_{-t}(\delta(\tau_t(A)\tau_t(B))) \\ &= -\tau_{-t}(\delta(\tau_t(A)\tau_t(B))) + \tau_{-t}(\delta(\tau_t(A))\tau_t(B) + \tau_t(A)\delta(\tau_t(B)) \\ &= 0\end{aligned}$$

and

$$\begin{aligned}\frac{d}{dt}(\tau_{-t}(\tau_t(A^*)^*)) &= -\tau_{-t}(\delta(\tau_t(A^*)^*) + \tau_{-t}((\delta\tau_t(A^*))^*) \\ &= 0\end{aligned}$$

hence the automorphism properties are valid.

Thus the simplest algebraic version of Theorem 1 is the following

Corollary 3 Let $\mathfrak{A}$ be a C*-algebra. The following are equivalent

1. δ is the generator of a norm-continuous one-parameter group of *-automorphisms of $\mathfrak{A}$.
2. δ is a bounded derivation of $\mathfrak{A}$.

There is, however, a much richer algebraic structure which is less evident. The key point is that each bounded derivation of a W*-algebra is both spatial and inner.

Theorem 4 If δ is a bounded derivation of a W*-algebra M then there exists a self-adjoint $H \in M$ with $\|H\| \leqslant \|\delta\|/2$ such that

$$\delta(A) = i[H,A] , \qquad A \in M,$$

and hence the group τ generated by δ has the covariant form

$$\tau_t(A) = e^{iH}Ae^{-itH}, \qquad A \in M.$$

This theorem has various implications for C*-algebras once one establishes that each bounded derivation of a C*-algebra $\mathfrak{A}$ of bounded operators can be extended to the weak closure of $\mathfrak{A}$. For example

Theorem 5 If δ is a bounded derivation of a C*-algebra $\mathfrak{A}$ and $\pi(\mathfrak{A})$ is a representation of $\mathfrak{A}$ then there exists a self-adjoint $H_\pi \in \pi(\mathfrak{A})^-$ such that the norm-continuous one-parameter group τ generated by δ is given by

$$\pi(\tau_t(A)) = e^{itH_\pi}\pi(A)e^{-itH_\pi} , \qquad A \in \mathfrak{A} , \; t \in \mathbb{R} .$$

In particular one sees that the norm-continuous one-parameter groups of *-automorphisms are always unitarily implemented by unitary groups $U_\pi(t) = \exp\{itH_\pi\}$ which are weakly inner, i.e. which lie in the weak closure of $\pi(\mathfrak{A})$. In particular there are no non-trivial norm continuous one-parameter groups of *-automorphisms of abelian C*-algebras.

These results are relatively old. More recently it has been possible to classify the C*-algebras for which bounded derivations generate inner automorphism groups if inner is suitably defined. First recall that a multiplier of $\mathfrak{A}$ is a $B \in \mathfrak{A}^{**}$ such that $B\mathfrak{A} \subseteq \mathfrak{A}$ and $\mathfrak{A}B \subseteq \mathfrak{A}$.

Theorem 6 Let $\mathfrak{A}$ be a separable C*-algebra. The following are equivalent

1. every bounded derivation δ generates a norm continuous one-parameter group of *-automorphisms of the form.

$$A \in \mathfrak{A} \mapsto \tau_t(A) = e^{itH}Ae^{-itH}$$

where $H = H^*$ is a multiplier of $\mathfrak{A}$ with $\|H\| \leqslant \|\delta\|/2$.

2. $\mathfrak{A}$ has the form $\mathfrak{A} = \mathfrak{A}_1 \oplus \mathfrak{A}_2$ where $\mathfrak{A}_1$ is a C*-algebra with continuous trace and $\mathfrak{A}_2$ is the direct sum of simple C*-algebras.

References;

Nagumo's result, Theorem 1, can be found in the book by Hille and Phillips together with the basic Banach space theory of one-parameter groups.

Hille, E. and R.S. Phillips; Functional Analysis and Semigroups, Amer. Math. Soc., Providence, R.I. (1957).

Theorem 4 and Corollary 5 were proved by Kadison and Sakai. This and much related material is described in Sakai's book.

Sakai, S; C*-algebras and W*-algebras, Springer Verlag, Berlin-Heidelberg-New York.

Theorem 6 is basically due to Elliott with a subsequent refinement by Akemann and Pedersen.

Elliott, G.A.; Some C*-algebras with outer derivations III, Ann. Math. 106(1977) 121-143.

Akemann, C.A. and G.K. Pedersen; Central Sequences and Inner Derivations of Separable C*-algebras (To appear in the Amer. Jour. Math.)

2. Strong Continuity

The characterizations of generators of C_0-groups of *-automorphism are more complicated. We first need two definitions.

Definition 7 An operator S on a Banach space X is dissipative if for each $A \in D(S)$ there exists a non-zero $\eta \in X^*$ satisfying

1. $\eta(A) = \|\eta\|\|A\|$
2. $\mathrm{Re}\ \eta(SA) \leqslant 0$.

The existence of an $\eta \in X^*$ satisfying property 1 follows from the Hahn-Banach theorem.

Property 2 can be understood by assuming that S generates a C_0-group T of isometries. Then $|\eta(T_tA)| \leqslant \|\eta\|\|A\| = \eta(A)$.
Consequently

$$\mathrm{Re}\ \eta((T_t - I)A)/t \leqslant 0\ , \qquad t > 0$$

and $\mathrm{Re}\ \eta(SA) \leqslant 0$.

Definition 8 Let S be an operator on the Banach space X. Then $A \in X$ is analytic (entire analytic) for S if $A \in D(S^n)$, $n = 1,2,\ldots$ and

$$z \in \mathbb{C} \mapsto \sum_{n\geq 0} \frac{z^n}{n!} \|S^n A\|$$

is an analytic (entire analytic) function

It is easy to construct analytic elements for generators of C_0-, or C_0^*-, groups by averaging with suitable analytic functions, i.e. by replacing $A \in X$ with

$$A_f = \int dt f(t) T_t A$$

There are twelve characterization of generators which exploit several different ideas.

Theorem 9 Let $\mathfrak{A}$ be a C^*-algebra with identity $\mathbb{1}$ and δ a norm-densely defined, norm-closed, operator on $\mathfrak{A}$. It follows that δ is a generator of a C_0-group of *-automorphisms of $\mathfrak{A}$ if, and only if, it satisfies one of the twelve combinations (A_i, B_j, C_k) of the following conditions

A_1; δ is a derivation

A_2; $\mathbb{1} \in D(\delta)$ and $\delta(\mathbb{1}) = 0$

B_1; $(I + \alpha\delta)(D(\delta)) = \mathfrak{A}$, $\alpha \in \mathbb{R}\setminus\{0\}$

B_2; The self-adjoint analytic elements for δ are dense in the self-adjoint elements of $\mathfrak{A}$.

C_1; $\|(I + \alpha\delta)(A)\| \geq \|A\|$, for all $\alpha \in \mathbb{R}$, $A \in D(\delta)$

C_2; $(I + \alpha\delta)(A) \geq 0$ implies $A \geq 0$, for all $\alpha \in \mathbb{R}$ and $A \in D(\delta)$

C_3; δ and $-\delta$ are dissipative

This theorem synthesizes a variety of different ideas. We will describe some of the features involved in constructing the group.

Firstly if A_1 is valid then it suffices to prove that δ generates a C_0-group τ of mappings of $\mathfrak{A}$ viewed as a Banach space since these mappings will automatically be *-automorphisms by the argument used in the discussion of Corollary 3. Hence we are reduced to a Banach space problem. Now let us consider the six possibilities with A_1.
(A_1, B_1, C_1); This is the Hille-Yosida theorem. The conditions state that the resolvents $(I + \alpha\delta)^{-1}$, $\alpha \in \mathbb{R}$, are bounded operators with norm one and this suffices to construct the group by strong limits

$$\tau_t(A) = \lim_{n\to\infty} (I - \tfrac{t}{n}\delta)^{-n}(A)$$

(A_1, B_1, C_2); These conditions state that the $(I + \alpha\delta)^{-1}$, $\alpha \in \mathbb{R}$, are everywhere defined, hence bounded, operators which map positive elements of $\mathfrak{A}$ into positive elements of $\mathfrak{A}$. Hence they are contractions and condition C_1 is satisfied (To draw this last conclusion one must know that $\mathbb{1} \in D(\delta)$ and $\delta(\mathbb{1}) = 0$. But the first condition follows because δ is a norm-closed derivation (see Section 4) and the second follows because

$\delta(\mathbb{1}) = \delta(\mathbb{1}^2) = 2\delta(\mathbb{1})$.)

(A_1,B_1,C_3); This is the Lumer-Phillips theorem for groups of isometries. Condition C_1 can be deduced from C_3,

(A_1,B_2,C_1); This type of theorem was first proved by Lumer and Phillips and independently by Nelson. In fact B_2 can be replaced by the simpler assumption.

B_2'; δ has a dense set of analytic elements.

Basically one defines τ by setting

$$\tau_t(A) = \sum_{n\geq 0} \frac{t^n}{n!} \delta^n(A)$$

for each analytic A and small t. Next one proves that τ is isometric and hence τ can be extended to large t by iteration. The isometric property follows because

$$\tau_t(A) = \lim_{n\to\infty} (I + \frac{t}{n}\delta)^n (A)$$

and hence

$$\|\tau_t(A)\| \geq \liminf_{n\to\infty} \|(I + \frac{t}{n}\delta)^n(A)\| \geq \|A\|$$

by Condition C_1. Consequently

$$\|A\| = \|\tau_{-t}(\tau_t(A))\| \geq \|\tau_t(A)\| \geq \|A\|$$

and $\|\tau_t(A)\| = \|A\|$.

(A_1,B_2,C_3); Again C_1 follows from C_3. (Hence one could replace B_2 by B_2'.)

(A_1,B_2,C_2); The argument is similar to the above but it relies upon the fact that an invertible positivity preserving map which leaves the identity invariant is isometric. (It is not known whether B_2 can be replaced by B_2').

Now consider the six cases for which A_2 is valid. In each such case one can construct a C_0-group τ of isometries of $\mathfrak{A}$ with generator δ and condition A_2 implies that

$$\tau_t(\mathbb{1}) = \mathbb{1}, \qquad t \in \mathbb{R}.$$

It remains to conclude that the τ_t are *-automorphisms. This is a consequence of the theory of positive maps and Jordan morphisms. Firstly <u>every invertible isometric map ϕ of a C*-algebra which leaves the identity fixed is automatically a Jordan isomorphism</u>, i.e. both ϕ and ϕ^{-1} satisfy

$$\phi(A^*) = \phi(A)^*, \quad \phi(AB + BA) = \phi(A)\phi(B) + \phi(B)\phi(A).$$

Secondly <u>every C_0-group τ_t of Jordan automorphisms is a group of *-automorphisms</u>.

<u>References;</u>

The standard theory of C_0-groups on Banach space is described in

Hille, E. and R.S. Phillips, Functional Analysis and semi-groups, Amer. Math. Soc. Colloq. Publ. <u>31</u> Providence (1957).

Reed, M. and B. Simon; Methods of Mathematical Physics II, Academic Press, New York

San Francisco-London (1975).

Dissipative operators were introduced in
Lumer, G. and R.S. Phillips; Pac. J. Math. 11 679-689 (1961)
This paper also contains a version of an analytic vector theorem. More general results are contained in
Nelson, E: Ann. Math. 70 572-615 (1960).

Parts of Theorem 9 were given by various authors
Bratteli, O. and D.W. Robinson; Commun. Math. Phys. 42 253-268 (1975), 46 11-30 (1976).
Kishimoto, A.; Commun. Math. Phys. 47 25-32 (1976).
Powers, R. and S. Sakai; J. Func. Anal. 19 81-95 (1975), Commun. Math. Phys. 39 273-288(1975).

3. σ-weak Continuity

There is an analogue of Theorem 9 for C_0^*-groups

Theorem 10 Let M be a W*-algebra and assume that M is abelian or that M is a factor. Let δ be a $\sigma(M,M_*)$-densely defined, $\sigma(M,M_*)$-$\sigma(M,M_*)$-closed, operator on M with domain $D(\delta)$ containing $\mathbb{1}$. It follows that δ is the generator of a C_0^*-group of *-automorphisms of M if, and only if, it satisfies one of the eight combinations (A_i, B_j, C_k) of the following conditions.

A_1; δ is a derivation

A_2; $\delta(\mathbb{1}) = 0$

B_1; $(I+\alpha\delta)(D(\delta)) = M, \quad \alpha \in \mathbb{R}\backslash\{0\}$

B_2; the self-adjoint part of the unit sphere of analytic elements is σ-weakly dense in the self-adjoint part of the unit sphere of M

C_1; $\|(I+\alpha\delta)(A)\| \geqslant \|A\|, \quad \alpha \in \mathbb{R}, A \in D(\sigma)$

C_2; $(I+\alpha\delta)(A) \geqslant 0$ implies $A \geqslant 0$ for all $\alpha \in \mathbb{R}$, $A \in D(\sigma)$.

For general W*-algebras the theorem remains true if condition A_2 is removed and B_2 is replaced by the weaker condition

B_2'; the set of analytic elements for δ is σ-weakly dense in M.

The construction of the automorphism group in Theorem 9 relied upon Banach space arguments to obtain the group and supplementary algebraic arguments to obtain the automorphism properties. The same holds true in Theorem 10 but the Banach space arguments are carried out on the pre-dual M_* of M. We will not elaborate further.

Note that M is restricted to be abelian, or to be a factor, in the first part of the theorem because the previous key result for groups of Jordan automorphisms is no longer valid if strong continuity is replaced by σ-weak continuity. There are C_0^*-groups τ_t of Jordan automorphisms which are not groups of *-automorphisms. These take the form $\tau = E\tau^{(1)} + (1-E)\tau^{(2)}$ where $\tau^{(1)}$ is a group of *-morphisms, $\tau^{(2)}$ is a group of *-anti-morphisms and E is a projection in the center of M.

References;

Bratteli, O. and D.W.Robinson, Ann. Inst. H. Poincaré 25A 139-164 (1976).

4. General Properties of Derivations

There are several interesting and somewhat surprising properties of closed derivations δ which basically arise from a functional analysis on the domain $D(\delta)$. There are two approaches, Fourier analysis, or complex analysis, but the following result is useful in both cases.

Lemma 11 Let δ be a norm-closed derivation of a C*-algebra $\mathfrak{A}$ with identity $\mathbb{1}$. If $A = A^* \in D(\delta)$ and λ is not in the spectrum of A then $A(\lambda\mathbb{1} - A)^{-1} \in D(\delta)$ and

$$\delta(A(\lambda\mathbb{1} - A)^{-1}) = \lambda(\lambda\mathbb{1} - A)^{-1}\delta(A)(\lambda\mathbb{1} - A)^{-1}$$

If, moreover, $\mathbb{1} \in D(\delta)$ then $(\lambda\mathbb{1} - A)^{-1} \in D(\delta)$ and

$$\delta((\lambda\mathbb{1} - A)^{-1}) = (\lambda\mathbb{1} - A)^{-1}\delta(A)(\lambda\mathbb{1} - A)^{-1}$$

This is straightforward to deduce. For example if $|\lambda|$ is larger than $\|A\|$ then $A(\lambda\mathbb{1} - A)^{-1}$ may be approximated in norm by polynomials in λ^{-1}. The action of δ on these polynomials is readily calculated and then the conclusion $A(\lambda\mathbb{1} - A)^{-1} \in D(\delta)$ is obtained by taking limits and using the assumption that δ is norm-closed. This also yields the action of δ. The result for small λ is obtained by an analytic continuation argument.

This result has an immediate corollary which we already used in discussing Theorem 9.

Corollary 12 Let δ be a norm-closed derivation of a C*-algebra $\mathfrak{A}$ with identity $\mathbb{1}$. The following are equivalent

1. $\mathbb{1} \in D(\delta)$
2. there is a positive invertible $A \in D(\delta)$.

In particular if δ is norm-densely defined then $\mathbb{1} \in D(\delta)$

$1 \Rightarrow 2$ by taking $A = \mathbb{1}$

$2 \Rightarrow 1$ by remarking that $A(\varepsilon\mathbb{1} + A)^{-1} \in D(\delta)$

for all $\varepsilon > 0$, $\|A(\varepsilon\mathbb{1} + A)^{-1} - \mathbb{1}\| \longrightarrow 0$ as $\varepsilon \longrightarrow 0$, and

$$\limsup_{\varepsilon \to 0} \|\delta(A(\varepsilon\mathbb{1} + A)^{-1})\| \leqslant \limsup_{\varepsilon \to 0} \varepsilon\|A^{-1}\|\|\delta(A)\| = 0$$

Thus $\mathbb{1} \in D(\delta)$ because δ is norm closed and $\delta(\mathbb{1}) = 0$.

If $A = A^* \in D(\delta)$ and f is a function analytic in an open simply connected set containing the spectrum of A then it has a Cauchy representation

$$f(A) = (2\pi i)^{-1} \int_C d\lambda f(\lambda)(\lambda\mathbb{1} - A)^{-1}$$

for a suitable contour C. It is then simple to deduce from Lemma 11 that $f \in D(\delta)$, if δ is norm closed. In particular this type of argument shows that if $A \in D(\delta)$ is positive and invertible then $A^{\frac{1}{2}} \in D(\delta)$. This is not generally true if A is not invertible. In fact stability of the domain of a derivation under the square root operation has striking consequences.

Proposition 13 If δ is a derivation of a C*-algebra $\mathfrak{A}$ with identity $\mathbb{1} \in D(\delta)$ and if $A \in D(\delta)$, $A \geqslant 0$, imply $A^{\frac{1}{2}} \in D(\delta)$ then δ and $-\delta$ are dissipative. In particular δ is norm-closable and satisfies

$$\|(I + \alpha\delta)(A)\| \geqslant \|A\|, \qquad \alpha \in \mathbb{R}$$

An even more striking result is true if δ is closed.

Proposition 14 If δ is a norm-closed norm densely defined derivation of a C*-algebra $\mathfrak{A}$ and $A \in D(\delta)$, $A \geqslant 0$, imply $A^{\frac{1}{2}} \in D(\delta)$ then δ is bounded

There are, however, many derivations of UHF algebras with domains invariant under the square root operation. Assume that the UHF algebra $\mathfrak{A}$ is the norm closure of an increasing family $\{\mathfrak{A}_\alpha\}$ of full matrix algebra $\mathfrak{A}_\alpha$ and δ is a derivation with

$$D(\delta) = \bigcup_\alpha \mathfrak{A}_\alpha$$

then δ satisfies the hypothesis of Proposition 13. Thus referring to Theorem 9 one sees that the closure $\bar{\delta}$ of δ generates an automorphism group if, and only if

$$\overline{R(I + \alpha\delta)} = \mathfrak{A}$$

for all $\alpha \in \mathbb{R}$.

The second way to analyze the domains $D(\delta)$ is by Fourier analysis and for this the following result is basic

Lemma 15 Let δ be a norm-closed derivation of a C*-algebra $\mathfrak{A}$ with identity $\mathbb{1}$ and assume $\mathbb{1} \in D(\delta)$. If $A = A^* \in D(\delta)$ then $\exp\{tA\} \in D(\delta)$ and

$$\delta(e^{tA}) = t\int_0^1 ds\, e^{stA}\delta(A)e^{(1-s)tA}$$

This allows one to construct functions f such that $f(A) \in D(\delta)$ by Fourier representation. We will not attempt to prove the most general result but end this section by mentioning a rather strange discrepancy between the behaviour for abelian and non-abelian $\mathfrak{A}$.

If δ is a norm-closed derivation of an abelian C*-algebra, $A \in D(\delta)$, and f is a once continuously differentiable function then it follows by polynomial approximation that $f(A) \in D(\delta)$ (and in fact $\delta(f(A)) = \delta(A)f'(A)$). This is not generally true. One can find counterexamples if $\mathfrak{A}$ is non-abelian.

References;

Functional Analysis of the domain of closed derivation was begun in

Bratteli,O., and D.W. Robinson, Commun. Math. Phys. 42 253-268 (1975).

Powers, R.; J. Fun. Anal. 18 85-95 (1975).
Sakai, S.; Amer. J. Math. 98 427-440 (1976).
Proposition 14 is taken from
Ôta, S.: J. Func. Anal. 30 238-244 (1978).
The counterexample mentioned at the end can be found in
McIntosh, A.; J. Func. Anal. 30 264-275 (1978).

5. Spatial derivations

Each bounded derivation of a von Neumann algebra M has the form

$$\delta(A) = i[H,A] , \qquad A \in M$$

for some bounded self-adjoint H (and H can even be chosen in M). Certain unbounded derivations have an analogous form.

Definition 16 A derivation δ of a C*-algebra of bounded operators on a Hilbert space H is spatial if there exists a symmetric operator H, on H, with domain D(H) such that $D(\delta)D(H) \subseteq D(H)$ and

$$\delta(A) = i[H,A] , \qquad A \in D(\delta)$$

As an example suppose that $\omega(A) = (\Omega, A\Omega)$ is a state over $\mathfrak{A}$. One can show that the conditions

$$|\omega(\delta(A))|^2 \leqslant \text{const.}\{\omega(AA^*) + \omega(A^*A)\}, \qquad A \in D(\delta),$$

are equivalent to δ being spatial and $\Omega \in D(H)$. In particular this criterion applies if ω is invariant under δ, i.e., if

$$\omega(\delta(A)) = 0 , \qquad A \in D(\delta).$$

There are various simplifications that occur for spatial derivations especially in association with invariant states, e.g.,

Proposition 17 Let δ be derivation and ω a δ-invariant state. Moreover assume that the cyclic representation $(H_\omega, \pi_\omega, \Omega_\omega)$ associated with ω is faithful. It follows that δ is norm-closable and its closure $\bar{\delta}$ generates a C_0-group of *-automorphisms of $\mathfrak{A}$ if, one of the following conditions is satisfies

1. $R(I \pm \bar{\delta}) = \mathfrak{A}$
2. δ possesses a dense set of analytic elements.

Since ω is invariant δ defines a spatial derivation of $\pi_\omega(\mathfrak{A})$ and the first, and easy, part of the proof of the proposition is the deduction that δ can be implemented by a self-adjoint H. Subsequently one has to prove that

$$(*) \qquad e^{itH}\pi_\omega(\mathfrak{A})e^{-itH} = \pi_\omega(\mathfrak{A})$$

But this follows by applying Theorem 10 to $M = L(H_\omega)$ and then using Theorem 9. Thus one is led to study derivations implemented by self-adjoint operators and it is natural to ask to what extent (*) is a natural consequence of algebraic structure. A principal result in this direction is the following

Theorem 18 Let M be a W*-algebra with a cyclic and separating vector Ω and let H be a self-adjoint operator on H such that $H\Omega = 0$. Define

$$D(\delta) = \{A \in M: \ i[H,A] \in M\}$$

The following conditions are equivalent

1. $e^{itH}Me^{-itH} = M, \quad t \in \mathbb{R}$

2. $D(\delta)\Omega$ is a core for H and H commutes strongly with the modular operator Δ associated with the pair (M,Ω)

3. H is essentially self-adjoint on $D(\delta)\Omega$ as an operator on the graph Hilbert space $H_{\#}$ associated with $\Delta^{\frac{1}{2}}$.

(The space $H_{\#}$ mentioned in condition 3 is the space $D(\Delta^{\frac{1}{2}})$ equipped with the scalar product

$$(\psi,\phi)_{\#} = (\Delta^{\frac{1}{2}}\psi, \Delta^{\frac{1}{2}}\phi) + (\psi,\phi).$$

A reasonably straightforward calculation with the modular operator and the modular conjugation show that H restricted to $D(\delta)\Omega$ defines a symmetric operator on $H_{\#}$).

This result is not easy to prove. Basically one begins with the case that $\omega(A) = (\Omega,A\Omega)$ is a trace state and then uses crossed product techniques to obtain the general result. We will not elaborate.

Finally we remark that the automorphism property (*) can be derived from a positivity preserving property.

Theorem 19 Let δ be a spatial derivation of a W*-algebra M implemented by a self-adjoint H. Assume M has a cyclic vector Ω such that $\Omega \in D(H)$ and $H\Omega = 0$. Define $D(\bar{\delta})$ by

$$D(\bar{\delta}) = \{A \in M; \ i[H,A] \in M\}$$

and assume $D(\bar{\delta})$ is a core for H. If

either $\quad H \geqslant 0$

or $\quad \Omega$ is separating for M

then the following are equivalent

1. $e^{itH}Me^{-itH} = M, \quad t \in \mathbb{R}$
2. $e^{itH}M_+\Omega \subset \overline{M_+\Omega}, \quad t \in \mathbb{R}$

where M_+ denotes the positive elements in M and the bar denotes weak closure.

The semibounded case $H \geqslant 0$ is of interest in ground state problems of mathematical physics and a classic result of Borchers which states that under the assumptions of the theorem

$$e^{itH}Me^{-itH} = M, \quad t \in \mathbb{R}$$

is equivalent to

$$e^{itH} \in M , \qquad t \in \mathbb{R}$$

i.e. the automorphisms are inner

References;

Bratteli, O. and D.W. Robinson; Ann. Inst. H. Poincaré 25A 139-164 (1976), Commun. Math. Phys. 46 31-35 (1976).
Bratteli, O. and U. Haagerup; Commun. Math. Phys. 59 79-95 (1978).

6. Stability properties

To conclude we comment on various aspects of stability of automorphism groups and their generators. Many results can be obtained by adaptation of known Banach space results but there are two contexts in which the algebraic structure leads to stronger results.

a. Perturbation Theory

The theory of perturbations of C_0-semigroups of contractions can be applied to the C_0-groups of *-automorphisms of C*-algebras. There appears to be no such theory for C_0^*-groups although some results can be taken over from the C_0-theory. For example if S is the generator of a C_0-group, or a C_0^*-group, and P is bounded then S + P is a generator of the same type. In the C_0-case the result remains valid if P is a dissipative operator with S-bound less than one i.e. if $D(P) \supseteq D(S)$ and

$$\|PA\| \leqslant a\|A\| + b\|SA\|$$

for all $A \in D(S)$, some $a \geqslant 0$, and $b < 1$. No analogue appears to hold in the C_0^*-case.

One basic problem that arises if one applies the theory of relatively bounded perturbations to derivations is the problem of existence. There appear to be no non-trivial examples of derivations δ_1, and δ_2, satisfying the relative bound relations $D(\delta_1) \supseteq D(\delta_2)$ and

$$\|\delta_1(A)\| \leqslant a\|A\| + b\|\delta_2(A)\| .$$

Apart from this difficulty the following result is striking.

Theorem 20 <u>Let δ be the generator of a C_0-group of *-automorphisms of the C*-algebra $\mathfrak{A}$ and δ' a derivation with the same domain. Then δ' is δ-bounded and δ' is dissipative. Hence $\delta + \lambda\delta'$ generates a C_0-group of *-automorphisms for all sufficiently small λ.</u>

b. Approximation Theory

Approximation theory is the name we give to the recently developed theory of comparison of pairs of C_0-, or C_0^*-, groups. If T and T' are two such groups with generators S and S' one attempts to characterize relationships between S and S' by properties of T and T', e.g. one can show that for C_0-groups <u>S - S' is bounded if, and only if, $D(S) \cap D(S')$ is a core for S and</u>

$$\|T'_t - T_t\| = 0(t) \quad \text{as} \quad t \to 0$$

and S - S' is S-bounded if, and only if, D(S) ∩ D(S') is a core for S and

$$\|(T'_t - T_t)A\| = 0(t) \quad \text{as} \quad t \to 0$$

for all $A \in D(S)$. (In fact these examples do not completely characterize the properties of S - S' in terms of T - T' because the core condition is also necessary.) Instead of attempting to describe the complete development of this field we give two results which depend on the algebraic structure

Theorem 21 Let τ, τ' be two C_0^*-groups of *-automorphisms of the W*-algebra M with generators δ, δ'. The following conditions are equivalent

1. $\|\tau_t - \tau'_t\| = 0(t)$ as $t \to 0$
2. $D(\delta) = D(\delta')$ and the σ-weak closure of $\delta - \delta'$ is bounded.

If M_* is separable the following are equivalent

1. there exist $0 < \varepsilon_1 < \frac{1}{4}$ and $\delta_1 > 0$ such that

$$\|\tau_t - \tau'_t\| < \varepsilon_1 \quad \text{for} \quad |t| < \delta_1$$

2. there exist $0 < \varepsilon_2 < \frac{1}{4}$, $\delta_2 > 0$, an inner automorphism γ of M and a bounded derivation δ_p of M such that

$$\delta = \gamma(\delta' + \delta_p)\gamma^{-1}$$

and

$$\|\tau_t \circ \gamma \circ \tau_{-t} - \gamma\| < \varepsilon_2 \quad \text{for} \quad |t| < \delta_2$$

References;

For Theorem 20 see
Longo, R.: Automatic relative boundedness of derivations in C*-algebras (Rome preprint)
Batty, C.J.K.; Small perturbations of C*-dynamical systems (Edinburgh preprint).

Approximation theory was begun in
Buchholz, D. and J.E. Roberts; Commun. Math. Phys. 161-177 (1976).
Robinson, D.W.; J. Func. Anal. 24 280-290 (1977).

The characterizations of bounded S - S' were partly given in this last reference and completely given in
Batty, C.J.K.: A characterization of relatively bounded perturbations (Edinburgh preprint).

Theorem 21 is taken from
Bratteli, O., Herman, R., and D.W. Robinson; Commun. Math. Phys. 59 167-196 (1978).

THERMAL EQUILIBRIUM AND CORRELATION INEQUALITIES

M. Fannes

Instituut voor Theoretische Fysica, Universiteit Leuven

B-3030 Leuven, Belgium

This paper consists in three sections, the first [1] and the last [2] resulting from a collaboration with A. Verbeure and the second with H. Spohn and A. Verbeure [3].

We show in the first section how one can derive correlation inequalities from the variational principle of thermodynamics for quantum spin lattice systems. These inequalities express therefore a condition of global thermodynamical stability (GTS). As it can be shown that a state which satisfies these inequalities is KMS [4] one can prove that GTS implies KMS. A similar approach has been used to prove that local thermodynamical stability (LTS) implies KMS [5].

The remaining sections are devoted to applications of the correlation inequalities to more concrete situations. In the second we characterize the structure of the limiting Gibbs states for discrete models of the mean-field type. Our results are sharp enough to yield in specific cases, such as the BCS or Dicke maser models, the existence (uniqueness) of the limiting Gibbs state.

Finally we study in the third section the condensed phase of the imperfect Bose gas. The occurence of condensation in and only in the ground state is proven for sufficiently low (high) temperatures (densities). We also show the existence (uniqueness) of the limiting Gibbs state in the case of condensation.

I. Derivation of Correlation Inequalities from Global Thermodynamical Stability [1].

Let $\mathcal{A}$ be the C^*-albebra of quasi-local observables for a quantum spin system on a lattice $\mathbb{Z}^\nu$ [6]. Lattice translations are described in a natural way by a group $\{\tau_k \mid k \in \mathbb{Z}^\nu\}$ of automorphisms of $\mathcal{A}$. To fix the dynamics we specify the local hamiltonians:

$$H_\Lambda = \sum_{\{X \mid X \subset \Lambda\}} \phi(X) \qquad \Lambda \subset \mathbb{Z}^\nu$$

where the potential $\{\phi(X) \mid X \subset \mathbb{Z}^\nu\}$ satisfies the following conditions:

i) $\phi(X) = \phi(X)^* \in \mathcal{A}_X$, the subalgebra of $\mathcal{A}$ of the elements living on the set X

ii) $\tau_k(\phi(X)) = \phi(X+k)$ translation invariance

iii) $\|\phi\| = \sum_{\{X \mid 0 \in X\}} \frac{\|\phi(X)\|}{\#(X)} < \infty$

Definition I.1. A state ω of $\mathfrak{A}$ is GTS at inverse temperature β, if

i) ω is translation invariant

ii) ω minimizes the expression

$$f(\omega) = \omega(A\phi) - \frac{1}{\beta} s(\omega)$$

where $A\phi = \sum_{\{X \mid 0 \in X\}} \frac{\phi(X)}{\#(X)}$ and $s(\omega)$ is the entropy density. ∎

In order to derive the correlation inequalities we will 'slightly' perturb the GTS state and express that this does not decrease the value of $f(\omega)$. However the perturbations we need cannot be local as they should preserve the translation invariance of the state. We therefore construct a class of translation invariant semigroups of completely positive unity preserving maps of $\mathfrak{A}$ with generators of the Lindblad type [7].

Theorem I.2. Let $v \in \mathfrak{A}_{loc}$ be a local element, then

$$\gamma : \mathfrak{A}_{loc} \to \mathfrak{A}_{loc} : \quad x \to \sum_{k \in \mathbb{Z}^\nu} L_{\tau_k(v)}(x)$$

where $\quad L_y(x) = y^* x y - \frac{1}{2}(y^* y x + x y^* y)$

is well defined and generates a translation invariant strongly continuous semigroup $\{\exp \gamma t \mid t \in \mathbb{R}^+\}$ of completely positive unity preserving maps of $\mathfrak{A}$. ∎

Let ω now be a GTS state at inverse temperature β ; this implies that

$$f(\omega) \leq f(\omega \circ \exp \gamma t)$$

By computing $\lim_{t \to 0^+} \frac{1}{t}\{f(\omega \circ \exp\gamma t) - f(\omega)\}$ we get our main result of this section:

Theorem I.3. Let ω be GTS at inverse temperature β then for any $v \in \mathfrak{A}_{loc}$

$$\lim_{\Lambda \to \mathbb{Z}^\nu} \beta\, \omega(v^* [H_\Lambda, v]) \geq \omega(v^* v) \ln \frac{\omega(v^* v)}{\omega(v v^*)}$$

∎

II. Equilibrium States for Mean-Field Models [3].

Discrete mean-field models are essentially characterized by the invariance of the dynamics under permutations of the 'spins'. So we take as algebra of observables the quasi-local albebra $\mathfrak{A}$ built on $\mathbb{N}_0$ (with N degrees of freedom at each site). The group $\mathcal{P}$ of permutations of finite subsets of $\mathbb{N}_0$ is described on the level of the observables by a group $\{\pi_p \mid p \in \mathcal{P}\}$ of automorphisms of $\mathfrak{A}$.

We specify our local hamiltonians by

$$H_\Lambda = \sum_{i \in \Lambda} A_i + \frac{1}{2\#(\Lambda)} \sum_{\substack{i,j \in \Lambda \\ i \neq j}} B_{ij}$$

where A_i is a copy at the ith site of a selfadjoint $A \in \mathcal{B}(\mathbb{C}^N)$ and B_{ij} is a copy at the ith and jth site of a selfadjoint symmetric $B \in \mathcal{B}(\mathbb{C}^N) \otimes \mathcal{B}(\mathbb{C}^N)$.

Let $\omega_{\beta,n}$ be the Gibbs state defined by $H_{\{1,\dots,n\}}$ on $\mathfrak{A}_{\{1,\dots,n\}}$ and ω_β a w^*-limit point of the $\omega_{\beta,n}$ (which exists by w^*-compactness of the state space of $\mathfrak{A}$).

Clearly ω_β is permutation invariant and can therefore be written as [8,9]:

$$\omega_\beta = \int d\mu(\rho)\ \omega_\rho \tag{1}$$

where ρ is a density matrix on $\mathbb{C}^N$ and ω_ρ is the product state of $\mathcal{A}$ induced by ρ.

The state $\omega_{\beta,m}$ can also be characterized by the correlation inequalities [4]:

$$\beta\ \omega_{\beta,m}(x^* [H_{\{1,\dots,m\}}, x]) \geq \omega_{\beta,m}(x^* x) \ln \frac{\omega_{\beta,m}(x^* x)}{\omega_{\beta,m}(x x^*)}$$

$$x \in \mathcal{A}_{\{1,\dots,m\}}$$

This leads together with (1) to the following correlation inequalities for ω_β:

Lemma II.1. Let $x \in \mathcal{A}_\Lambda$, then:

$$\beta \int d\mu(\rho)\ \omega_\rho(x^* [H_\rho^\Lambda, x]) \geq \omega_\beta(x^* x) \ln \frac{\omega_\beta(x^* x)}{\omega_\beta(x x^*)} \tag{2}$$

where $H_\rho^\Lambda = \sum_{i\in\Lambda} H_{\rho,i}$ and $H_{\rho,i}$ is a copy at the ith site of $H_\rho = A + B_\rho \in \mathcal{B}(\mathbb{C}^N)$ where $B_\rho = \mathrm{Tr}_2(\mathbb{1}\otimes\rho)\, B$ and Tr_2 is the partial trace over the second Hilbert space in $\mathbb{C}^N \otimes \mathbb{C}^N$.

■

In formula (1) the state ω_β is written as an integral over disjoint states, this allows us to 'concentrate' the correlation inequality (2) on the product states that enter in the integral decomposition (1) leading to our main result in this section:

Theorem II.2. For μ-almost all ρ

$$\beta\ \omega_\rho(x^* [H_\rho^\Lambda, x]) \geq \omega_\rho(x^* x) \ln \frac{\omega_\rho(x^* x)}{\omega_\rho(x x^*)} \qquad x \in \mathcal{A}_\Lambda$$

and so $\rho = \dfrac{e^{-\beta H_\rho}}{\mathrm{Tr}\, e^{-\beta H_\rho}}$ $\quad \mu$-a.e. (3)

■

Theorem II.2. shows that finding the equilibrium states of mean-field models amounts to solve the non-linear one-site equation (3).

Some further information about the limiting Gibbs states can be obtained:

i) for high temperatures ($\beta < 1/2\|B\|$) equation (3) admits a unique solution. Therefore all limiting Gibbs states coincide and one obtains the existence of the limit Gibbs state.

ii) using the variational principle of thermodynamics one can still further reduce the support of the measure $\mu(\rho)$ in (1):
if $f_0 = \inf_\omega f(\omega) = \inf_\omega e(\omega) - 1/\beta\, s(\omega)$ where the infimum is taken over the permutation invariant states of $\mathcal{A}$ and $e(\omega)$ and $s(\omega)$ are the energy and entropy density of ω then
$f(\omega_\rho) = f_0$ $\quad \mu$-a.e.

iii) in specific models such as the BCS model and the Dicke maser model (which can be treated in an analogous way by replacing at each site $\mathcal{B}(\mathbb{C}^N)$ by $\mathcal{B}(\mathcal{H})$ where $\mathcal{H}$ is a separable infinite dimensional Hilbert space) one finds for $\beta > \beta_{\text{critical}}$ a solution ρ_{norm} of (3) corresponding to a normal phase and also solutions $\rho(\alpha)$ $\alpha \in [0,2\pi[$ corresponding to a superconducting or superradiant phase. The solutions $\rho(\alpha)$ transform under the action of a group $\{\gamma_\alpha \mid \alpha \in [0,2\pi[\}$ of automorphisms of $\mathcal{A}$ (gauge group) as

$$\omega_{\rho}(\alpha) \circ \gamma_{\alpha'} = \omega_{\rho}(\alpha+\alpha' \bmod 2\pi)$$

In this case however the limiting Gibbs state is invariant under the gauge group and one has an example of a broken symmetry. The normal solution can be ruled out by ii) and one finds

$$\omega_\beta = \frac{1}{2\pi}\int_0^{2\pi} d\alpha \;\; \omega_{\rho(\alpha)}$$

which leads again to the uniqueness of the limiting Gibbs state below the critical temperature.

III. The Condensed Phase of the Imperfect Bose Gas [2].

We consider in this section only ν-dimensional systems with $\nu \geq 3$ as condensation can be ruled out for $\nu = 1,2$ [10]. Let Λ be the centered cubic box in $\mathbb{R}^\nu$ with side L, $\mathcal{H}_F(\Lambda)$ the boson Fock space constructed on Λ and $a^*(\varphi)$, $a(\varphi)$, $\varphi \in \mathcal{L}^2(\Lambda, dx)$ the canonical creation and annihilation fields:

$$[\, a(\varphi), a(\psi)] = 0 \qquad\qquad [\, a(\psi), a^*(\varphi)] = \langle \psi | \varphi \rangle$$

The local hamiltonians of the imperfect Bose gas are then given by:

$$H_L(\mu_L) = T_L - \mu_L N_L + \frac{\lambda}{2} \frac{N_L^2}{L^\nu} \tag{1}$$

where T_L is the usual kinetic energy with periodic boundary conditions and N_L is the particle number operator. The coupling constatnt λ is strictly positive and the chemical potential μ_L is adjusted in such a way that

$$\omega_{\beta,L}(N_L) = \rho L^\nu \tag{2}$$

where $\omega_{\beta,L}$ is the Gibbs state at inverse temperature β determined by (1) and $\rho > 0$ is the mean particle density.

In order to state our first main result we need the following definitions:

- $\tilde{\Lambda} = \{ k \mid k \in (\frac{2\pi}{L}\mathbb{Z})^\nu \}$ (the dual of the torus Λ)
- $\{\varphi_k^L \mid k \in \tilde{\Lambda}\}$ is the orthonormal basis of $\mathcal{L}^2(\Lambda, dx)$ determined by the functions

$$\varphi_k^L(x) = \frac{1}{L^{\nu/2}} e^{ikx} \qquad x \in \Lambda$$
$$= 0 \qquad x \notin \Lambda$$

- $N_{L,k} = a^*(\varphi_k^L)\, a(\varphi_k^L)$

Theorem III.1.

(i) $$\varliminf_{L\to\infty} \frac{\omega_{\beta,L}(N_{L,0})}{L^\nu} \geq \rho - \frac{1}{(2\pi)^\nu}\int_{\mathbb{R}^\nu} dk \frac{1}{e^{\frac{\beta k^2}{2}}-1}$$

(ii) $$\lim_{\delta\to 0} \varlimsup_{L\to\infty} \frac{1}{L^\nu} \sum_{0<|k|<\delta} \omega_{\beta,L}(N_{L,k}) = 0 \qquad \blacksquare$$

Theorem III.1. (i) implies that there is a macroscopic occupation of the ground state for sufficiently low (high) temperatures (densities) and (ii) shows that the ground state alone has such an occupation. The essential ingredients for the proof of this theorem come from repeated applications of the correlation inequalities.

Take now as algebra of observables the inductive limit of the $\mathcal{B}(\mathcal{H}_F(\Lambda))$ and suppose that ω_β is a w^*-limit point as $L\to\infty$ of the $\omega_{\beta,L}$. As one can prove that ω_ρ is locally normal it is sufficient to compute ω_β on some smaller algebra such as the C^*-algebra generated by $\{W(\varphi) \equiv \exp i(a^*(\varphi)+a(\varphi)) \mid \varphi \in \mathcal{D}\}$ where $\mathcal{D}$ is the space of infinitely differentiable complex valued functions on $\mathbb{R}^\nu$ with compact support.

It is now possible, as in section II, to derive correlation inequalities for ω_β and to solve them explicitly. This leads then to:

Theorem III.2. Let $\omega_\beta = w^*\text{-}\lim_\alpha \omega_{\beta,L_\alpha}$

(i) $\lim_\alpha \omega_{\beta,L_\alpha}\left(\frac{N_{L_\alpha,0}}{L_\alpha^\nu}\right) = \rho_0$ exists

(ii) if $\rho_0 > 0$ then for $\varphi \in \mathcal{D}$:

$$\omega_\beta(W(\varphi)) = \exp\{-\tfrac{1}{2}(\varphi, K\varphi)\} \frac{1}{2\pi}\int_0^{2\pi} d\alpha \exp 2i\rho_0^{1/2}|\hat{\varphi}(0)|\cos\alpha$$

where $\hat{\varphi}$ denotes the Fourier transform of φ and

$$\widehat{K\varphi}(k) = \frac{1}{e^{\frac{\beta k^2}{2}}-1}\hat{\varphi}(k) \qquad \blacksquare$$

It is perhaps worthwhile to remark that our proof of Theorem III.2. implies also that the Bogoliubov approximation is valid for the imperfect Bose gas.

[1] M. Fannes, A. Verbeure; J. Math. Phys., 19, 558 (1978)

[2] M. Fannes, A. Verbeure; The Condensed Phase of the Imperfect Bose Gas, Leuven preprint

[3] M. Fannes, H. Spohn, A. Verbeure; Equilibrium States for Mean-Field Models, Leuven preprint KUL-TF-79/004

[4] M. Fannes, A. Verbeure; Commun. Math. Phys. 55, 125 (1977) and Commun. Math. Phys. 57, 165 (1977)

[5] G.L. Sewell; Commun. Math. Phys. 55, 50 (1977)

[6] D. Ruelle; 'Statistical Mechanics', Benjamin New York (1969

[7] G. Lindblad; Commun. Math. Phys. 48, 119 (1976)

[8] E. Størmer; J. Funct. Analysis 3, 48 (1969)

[9] R.L. Hudson, G.R. Moody; Z. Wahrscheinlichkeitstheorie verw. Gebiete, 33, 343 (1976)

[10] P.C. Hohenberg; Phys. Rev. 158, 383 (1967)

ON THE POSSIBLE TEMPERATURES OF A C*-DYNAMICAL SYSTEM

Ola Bratteli,
School of Mathematics, University of New South Wales,
P.O. Box 1, Kensington, N.S.W., 2033, Australia.
(On leave from University of Oslo)
(This is a report on joint work with G.A. Elliott and R.H. Herman, [1])

There are several good reasons for using the KMS condition as a characterization of equilibrium for an infinite quantum system, [2]. The main purpose of this note is to show that the KMS condition alone, without any further specification of the system, may allow almost any structure on the set of equilibrium states at varying temperatures.

In this setting the system is represented by a C*-dynamical system $(\mathfrak{A}, \gamma)$, where $\mathfrak{A}$ is a C*-algebra with identity, and $t \in \mathbb{R} \to \gamma_t$ is a one-parameter group of *-automorphisms which for simplicity is assumed to be strongly continuous. If ω is a state on and β is a real number, ω is said to be a (γ, β)-KMS state if

$$\omega(A\gamma_{i\beta}(B)) = \omega(BA)$$

for all $A,B \in \mathfrak{A}$ such that B is entire analytic for γ. We use the terminology that ω is a $(\gamma, +\infty)$-KMS state if ω is a ground state, i.e.

$$-i\omega(A^*\delta(A)) \geqslant 0$$

for all A is the domain of the generator δ of γ (see [2] for a justification of this). The notion of $(\gamma, -\infty)$-KMS state is defined by the converse inequality.

The set K_β of (γ, β)-KMS states is a compact, convex subset of the state-space $E_{\mathfrak{A}}$ of $\mathfrak{A}$ for any $\beta \in \mathbb{R} \cup \{\pm \infty\}$. K_β is a simplex if $|\beta| < +\infty$ and a face in $E_{\mathfrak{A}}$ for $\beta = \pm\infty$. $K_{+\infty}$ is in general not a simplex, a certain condition of asymptotic abelianess is necessary and sufficient for this, [2].

The structure of the map

$$\beta \in [0, \infty] \to K_\beta$$

has been analyzed in detail in several models, most notably in quantum lattice spin systems. The C*-algebra $\mathfrak{A}$ has then a unique trace state τ, and there exists a sequence $\{H_n\}_{n\geq 1}$ of local Hamiltonians, $H_n^* = H_n \in \mathfrak{A}$, such that

$$\gamma_t(A) = \lim_{n\to\infty} e^{itH_n}Ae^{-itH_n}$$

for all $A \in \mathfrak{A}$. If ω_β is a weak*-limit of the sequence $\omega_{\beta,n}$ defined by

$$\omega_{\beta,n}(A) = \tau(e^{-\beta H_n}A)/\tau(e^{-\beta H_n})$$

as $n \to \infty$, then ω_β is a (γ, β)-KMS state, and in particular $K_\beta \neq \phi$ for all β. Under mild restrictions on the sequence $\{H_n\}_{n\geq 1}$ one knows that K_β consists of only one point for small β, i.e. for $\beta \in [0, \frac{1}{T_c}\rangle$ where T_c is a critical temperature. For one-dimensional models with short range interactions one has $T_c = 0$. For the two-dimensional Ising model T_c has a finite value, and K_β has affine dimension one for $\beta \in \langle \frac{1}{T_c}, +\infty\rangle$, see Aizenmann's lecture at this conference. For the three-dimensional Ising model, K_β is known to be infinite dimensional for large β. K_∞ is infinite dimensional for the Ising model in all dimensions, [2].

The main purpose of the present note is to show that the general features of the map $\beta \to K_\beta$ which are valid for lattice systems does not hold for general C*-dynamical systems, and presumably almost any "field" $\beta \to K_\beta$ can occur. Unfortunately we do not have a precise notion of "field" for the moment, and will give three partial results from [1]. If $\mathfrak{A}$ is an arbitrary C*-algebra this result would not be too surprising, since one could superpose simpler systems. However, the result is true even for simple C*-algebras.

THEOREM A Let F be a closed subset of $\mathbb{R} \cup \{\pm\infty\}$. There exists a C*-dynamical system $(\mathfrak{A}, \gamma)$ such that $\mathfrak{A}$ is simple, and there is a (γ, β)-KMS state if and only if $\beta \in F$, and this state is unique.

THEOREM B Let $F_1 \supseteq F_2 \supseteq \ldots \supseteq F_n$ be a finite sequence of closed subsets of $\mathbb{R} \cup \{\pm\infty\}$ such that $\pm\infty \notin F_2$. There exists a C*-dynamical system $(\mathfrak{A}, \gamma)$ such that $\mathfrak{A}$ is simple, and K_β is a k-1-dimensional simplex if $\beta \in F_k \backslash F_{k+1}$, $k = 0,1,\ldots,n$. (use the conventions $F_0 = \mathbb{R} \cup \{\pm\infty\}$, $F_{n+1} = \phi$, the -1-dimensional simplex $= \phi$).

THEOREM C Let $K_{\beta_1},\ldots,K_{\beta_n}$ be n compact, metrizable simplexes indexed by $\beta_1,\ldots,\beta_n \in \mathbb{R}$. There exists a C*-dynamical system $(\mathfrak{A}, \gamma)$ such that $\mathfrak{A}$ is simple, and there is a (γ, β)-KMS state if and only if $\beta \in \{\beta_1,\ldots,\beta_n\}$, and the set of (γ, β_k)-KMS-states is affinely isomorphic to K_{β_k} for $k = 1,\ldots,n$.

Furthermore, in all of the theorems A,B,C, the system $(\mathfrak{A}, \gamma)$ can be constructed such that $\mathfrak{A}$ is separable nuclear with unit, and γ is periodic with period 2Π.

The proof of Theorem A goes in two steps:

Step 1. One constructs an approximately finite dimensional C*-algebra $\mathfrak{B}$ and an automorphism α of $\mathfrak{B}$ and a projection E in $\mathfrak{B}$ such that

1. If $\beta \in F$ there exists a unique (infinite, lower semicontinuous) trace τ_β on $\mathfrak{B}$ such that $\tau_\beta(E) = 1$ and $\tau_\beta \circ \alpha = e^{-\beta}\tau_\beta$. If $\beta \notin F$, no such trace exists.
2. There are no globally α-invariant ideals in $\mathfrak{B}$ (In the case $\pm\infty \notin F$, $\mathfrak{B}$ can be taken to be simple, otherwise not).
3. α transforms each nonzero projection in $\mathfrak{B}$ into a non-equivalent projection.

Step 2. 2 and 3 implies that the C*-crossed product $C^*(\mathfrak{B}, \alpha)$ of $\mathfrak{B}$ by α is simple. $C^*(\mathfrak{B}, \alpha)$ is the closure of the linear span of elements of the form $A\,U^n$, where $A \in \mathfrak{B}$ and U is a unitary operator such that $\alpha(A) = U\,A\,U^*$ for all $A \in \mathfrak{B}$.

Define
$$\gamma_t(AU^n) = e^{int}AU^n\ , \quad A \in \mathfrak{B}$$
and
$$\mathfrak{A} = E\,C^*(\mathfrak{B}, \alpha)E$$
Then $(\mathfrak{A}, \gamma)$ satisfies the conclusion of Theorem A. In particular if $\Phi : \mathfrak{A} \to E\,\mathfrak{B}\,E$ is the projection defined by
$$\Phi(A) = \frac{1}{2\pi}\int_0^{2\pi} dt\ \gamma_t(A), \quad A \in \mathfrak{A}\ ,$$
then the unique (γ, β)-KMS state ω_β for $\beta \in F$ is given by
$$\omega_\beta(A) = \tau_\beta(\Phi(A)), \quad A \in \mathfrak{A}\ .$$

To get Theorem B, C one modifies step 1,1. in the obvious way.

To construct $\mathfrak{B}$, α, E one uses the following characterisation of dimension groups

[4], of AF-algebras

THEOREM, [3]. An ordered abelian group G is the dimension group (= K_0-group) of an AF-algebra if and only if the following two properties are valid.

1. (G is unperforated) If $g \in G$ and $ng \geqslant 0$ for some $n \in \mathbb{N}$, then $g \geqslant 0$.
2. (G has the Riesz interpolation property) If $g_1, g_2, g_3, g_4 \in G$ and

$$g_1, g_2 \leqslant g_3, g_4$$

then there exists a $g_5 \in G$ such that

$$g_1, g_2 \leqslant g_5 \leqslant g_3, g_4$$

By a theorem in [4], the dimension group together with a hereditary subset of its positive cone is a complete invariant for AF-algebras. Therefore, to construct $\mathfrak{B}$, α, E it is enough to construct a dimension group G, an order-automorphism α of G and an element $g_0 \in G$ corresponding to E with the correct properties. The construction involves many arbitrary choices, but one can for example let $G = \mathbb{Z}[x, x^{-1}, (1-x)^{-1}]$, i.e. G is the additive group of all polynomials in x, x^{-1} and $(1-x)^{-1}$ with integer coefficients. If F^1 is the set of $t \in [0, 1]$ such that $t(1-t)^{-1} = e^{-\beta}$ for some $\beta \in F$ one defines an order on G by saying that $p \in G$ is (strictly) positive if and only if $p(t) > 0$ for all $0 < t < 1$ in a neighbourhood of F^1. It is then not hard to verify that G has the Riesz interpolation property. The order automorphism α is defined as multiplication by $x(1-x)^{-1}$ and the element g_0 as the constant function 1.

To cope with the cases B and C one consider additive groups of functions from $\mathbb{R}$ which in the point $\beta \in \mathbb{R}$ assumes values in the real affine functions over K_β.

[1] Bratteli, O., G.A. Elliott and R.H. Herman, On the possible temperatures of a dynamical system, Penn. State Univ. preprint (1979).

[2] Bratteli, O. and D.W. Robinson, Operator algebras and quantum statistical mechanics, Vol. II, Springer Verlag, Berlin-Heidelberg-New York, to appear.

[3] Effros, E., D. Handelman and C.L. Shen, Dimension groups and their affine representations, Univ. of Penn. preprint (1979).

[4] Elliott, G.A., On the classification of inductive limits of sequences of semisimple finite-dimensional algebras, J. Algebra 38 (1976), 29-44.

A REVIEW ON SEMIGROUPS OF COMPLETELY POSITIVE MAPS

David E. Evans*

School of Mathematics, The University, Newcastle-upon-Tyne, NE1 7RU, England.

We review recent progress on one-parameter semigroups of completely positive maps on operator algebras, and its application to irreversible Markovian dynamics in quantum systems.

We recall that the positive cone of a C*-algebra A is $A_+ = \{a^*a: a \in A\}$, and a linear map T between C*-algebra A and B is positive if $T(A_+) \subseteq B_+$. If one considers C*-algebras and positive contractions then the isomorphisms for this category are precisely the Jordan isomorphisms, namely the *-linear bijections which preserve the Jordan product $a \circ b = (ab+ba)/2$. Thus in order to study the C*-structure of an operator algebra, it is not enough to look at the positive cone A_+, and positive maps. Instead, one can consider the whole sequence of matrix algebras $M_N(A) \simeq A \otimes M_N$, together with their positive cones $M_N(A)_+$. A map T between C*-algebras A and B is said to be N-positive if $T \otimes 1: A \otimes M_N \to B \otimes M_N$ is positive, and completely positive if it is N-positive for all N. Then the isomorphisms in the category of C*-algebras and completely positive contractions preserve all the C*-structure, i.e. they are the linear *-bijections which preserve the usual product ab. (In fact, one only needs 2-positivity here). Completely positive contractions arise naturally in quantum theory as those operations which are probability reducing (in the Schrodinger picture) and remain so in interaction with N-body systems [1], see also [2]. Moreover if one takes the viewpoint that reversible dynamics is given by a group $\{\alpha_t: t \in \mathbb{R}\}$ of morphisms (i.e. *-automorphisms) of a C*-algebra B, and if one restricts to a subsystem represented by a C*-subalgebra A, via a conditional expectation N:B→A, then one obtains $T_t = N\alpha_t|_A$ (*), which is necessarily completely positive by Tomiyama's theorem [3] on projections of norm one; [4]. In general, there will be memory effects and the family $\{T_t: t \geq 0\}$ will not be Markovian. However, rigorous derivation of the semigroup law $T_{t+s} = T_t T_s$; $t, s \geq 0$, has been obtained in a variety of models when taking weak or singular coupling limits (see [5] for a review of this). Here we consider only Markovian dynamics and define a dynamical semigroup [1] $\{T_t: t \geq 0\}$ on an operator algebra A to be a semigroup of completely positive contractions on A (normally with some continuity assumptions). We review some dilation problems, ergodic and spectral properties of the semigroup, but we are mainly concerned here with the infinitesimal generator $L = \lim_{t \downarrow 0} (T_t - 1)/t$. (For reviews biased in other directions see [4, 5, 6]). First we emphasise that complete positivity is a much stronger property than positivity, However, if T is a positive map between C*-algebras A and B, with either A or B commuta-

* Supported by the Science Research Council

tive, then T is automatically completely positive [7]. Thus for classical systems, the distinction does not arise. Conversely, if every positive map from a C*-algebra A into another, B, is completely positive, then either A or B is commutative [8].

For reversible systems, where e^{tL} extends to a group of automorphisms, L is a derivation, as in the talk of D.W. Robinson. In this case if A is represented on a Hilbert space H, one is interested in obtaining a hamiltonian h (a self adjoint operator on H) such that

$$e^{tL} = \mathrm{Ad}(e^{ith}), \quad t \in \mathbb{R}, \quad \text{or} \quad L = i\,\mathrm{ad}(h) \qquad (1)$$

(with suitable domain interpretation). This can certainly be done if L is bounded (with bounded h); and moreover if the algebra A is a von Neumann algebra or say a simple unital C*-algebra, then one can even choose h in A itself [10]. The situation when L is unbounded is more complicated (see [11]), however, if $\{e^{tL} : t \in \mathbb{R}\}$ is a weakly continuous group of *-automorphisms of B(H), then there exists a self adjoint operator h on H such that $e^{tL} = \mathrm{Ad}(e^{ith})$. We seek analgous spatial descriptions for dynamical semigroups. First, what are the obvious norm continuous semigroups of completely positive maps on a C*-algebra $A \subseteq B(H)$? Let $K \in B(H)$ be such that the Lyapunov transform $L_o(x) = Kx + xK^*$, $x \in A$ leaves A globally invariant. Then $e^{tL_o} = e^{tK}(\cdot)e^{tK^*}$, $t \in \mathbb{R}$ defines a group of completely positive maps on A. Moreover, if Ψ is a completely positive map on A, then $e^{t\Psi} = 1 + t\Psi + t^2\Psi^2/2 + \dots$ is certainly completely positive for $t \geq 0$. Thus by the Trotter product formula, $L = L_o + \Psi$ generates a semigroup of completely positive maps. In fact,

$$e^{tL} = e^{tL_o} + \int_o^t e^{sL_o}\, \Psi\, e^{(t-s)L}\, ds\,, \quad t \geq 0,$$

so that $e^{tL} \geq e^{tL_o} = e^{tK}(\cdot)e^{tK^*}$, $t \geq 0$; where if φ, φ^1 are linear maps between C*-algebras, we write $\varphi \geq \varphi^1$ if $\varphi - \varphi^1$ is completely positive. Thus we see that if $\{e^{tL} : t \geq 0\}$ is a norm continuous semigroup of completely positive maps on a C*-algebra $A \subseteq B(H)$, and $K \in B(H)$, then $e^{tL} \geq e^{tK}(\cdot)e^{tK^*}$, $t \geq 0$ if and only if $L \geq K(\cdot) + (\cdot)K^*$, as maps into B(H). (The implication $\Rightarrow$ is easily seen by differentiation). Now for an arbitrary dynamical semigroup T_t on an operator algebra $A \subseteq B(H)$ we seek a strongly continuous contraction semigroup G_t on H such that

$$T_t \geq G_t(\cdot)G_t^*\,, \quad t \geq 0. \qquad (2)$$

[Note that we do not insist on $G_t A G_t^* \subseteq A$]. We regard this as a suitable analogue for irreversible systems of (1). Note that if T is an automorphism of an irreducible C*-algebra $A \subseteq B(H)$ and G is a non zero contraction on H such that $T \geq G(\cdot)G^*$, then $[T(a_i)^*T(a_j)] \geq [(a_iG^*)^*(a_jG^*)]$ $\forall\, a_1 \dots a_n \in A$, and so there exists a well defined contraction C on H such that $C\,T(a) = aG^*$, $a \in A$. Thus $C = G^*$, andso $G^*T(a) = aG$. It is easily checked that $G^*G, GG^* \in A'$, so that by irreducibility, G can be taken unitary and $T = Ad(G)$.

Theorem 1 [12]. Let $\{e^{tL}: t \geq 0\}$ be a norm continuous semigroup of completely positive maps on a C*-algebra $A \subseteq B(H)$. Then there exists $K \in A''$ such that

$$e^{tL} \geq e^{tK}(\cdot)e^{tK^*}, \quad t \geq 0.$$

Theorem 1 was first shown independantly for finite dimensional matrix algebras [13] and normal dynamical semigroups on a hyperfinite W*-algebra [1], and some other special cases of this have appeared in [14]. Relatively less is known for semigroups of positive maps, but see [15]. Now if Ψ is a normal completely positive map from a von Neumann algebra A ($\subseteq B(H)$) into $B(H)$, then there exist $V_\alpha \in B(H)$ such that $\Psi(x) = \Sigma V_\alpha^* x V_\alpha$, $x \in A$, [2]. Now if $\{e^{tL}: t \geq 0\}$ is a norm continuous semigroup of completely positive normal unital maps on a von Neumann algebra $A \subseteq B(H)$, let $K \in A$ be given by Theorem 1 so that $\Psi \equiv L-K(\cdot) - (\cdot)K^* \geq 0$, and so $L = K(\cdot) + (\cdot)K^* + \Sigma V_\alpha^*(.)V_\alpha$, for some $V_\alpha \in B(H)$. Since e^{tL} are unital maps, $L(1) = 0$ or $K+K^*+\Psi(1) = 0$. Thus $K = ih - \Psi(1)/2$ for some s.a. h in A. Thus for $x \in A$:

$$L(x) = i \text{ ad } (h)\ (x) + \Sigma V_\alpha^* x V_\alpha - \tfrac{1}{2}\left[\Sigma V_\alpha^* V_\alpha, x\right]_+ \qquad (3)$$

In a symmetric situation (c.f. detailed balance [16, 17]) where

$$\Psi(x) = \tfrac{1}{2}\left(\Sigma V_\alpha^* x V_\alpha + V_\alpha x V_\alpha^*\right), \text{ we can write: } L = \delta - \tfrac{1}{2}\Sigma(\overset{*}{\delta}_\alpha \delta_\alpha + \delta_\alpha \overset{*}{\delta}_\alpha),$$

where δ, δ_α are the derivations $\delta = i\ ad(h)$, $\delta_\alpha = \frac{i}{\sqrt{2}} ad(V_\alpha)$ and $\delta_\alpha^*(x) = \delta_\alpha(x^*)^*$. This corresponds to the diffusion equation in classical probability theory. Such semigroups of type (3) were written down in [18, 19] in a study of quantum stochastic processes. They also arose naturally in [20] (see also [5] for other models) as the reduced dynamics of an N-level system weakly coupled to a thermal reservoir. This particular dynamics is amenable to the work of several authors [18, 21] on relaxation to an equilibrium state on B(H), which is proven under the assumption $\text{lin}(V_\alpha)$ s.a., and $\{V_\alpha\}' = \mathbb{C}$, and the existence of a normal invariant state. Further ergodic and spectral results are discussed in [22], in particular an extension of the Perron-Frobenius theory to finite von Neumann algebras. We should mention that irreversible dynamics and generators of the form (3) have been used to obtain information in purely reversible systems, e.g. characterisation of KMS states using detailed balance conditions [17] or correlation inequalities [23], (c.f. the talk of M. Fannes).

As a consequence of Theorem 1, it can be shown:

Theorem 2 [19, 24]. If $\{e^{tL}: t \geq 0\}$ is a norm continuous dynamical semigroup on a C*-algebra $A \subseteq B(H)$, there exists a Hilbert space H_0, and a strongly continuous contraction semigroup G_t on $H \otimes H_0$ such that $e^{tL}(x)\otimes 1 = G_t(x\otimes 1)G_t^*$; $x \in A$, $t \geq 0$.

Proceeding from this, dilations of some normal dynamical semigroups on von Neumann algebras to groups of automorphisms were obtained in [4, 19, 24]. Dilations of arbitrary families of completely positive maps on C*-algebras, indexed by elements of a group were obtained in [25]. This is in some way analogous to constructing a Markov process from a contraction semigroup in classical probability theory.

There has also been considerable interest in obtaining spatial descriptions of unbounded generators of dynamical semigroups. The following theorem was obtained in the Schrodinger picture for B(H). We then proceed to look at generators of quasi-free type [6, 26-30], concentrating here on the Fermion algebra.

Theorem 3. [31].
(a) Let $\{e^{Kt}: t \geq 0\}$ be a strongly continuous contraction semigroup on a Hilbert space H, and $V_\alpha \in L$ (Dom (K^*), H) such that

$$\langle K^*\varphi, \varphi\rangle + \langle \varphi, K^*\varphi\rangle + \Sigma\langle V_\alpha\varphi, V_\alpha\varphi\rangle = 0, \ \forall\ \varphi \in \text{Dom}\ (K^*).$$

Then there exists a weakly continuous normal dynamical semigroup $\{T_t: t \geq 0\}$ on B(H) such that the infinitesimal generator L_* of the predual action satisfies

$$L_*(\rho) = K^*\rho + \rho K + \Sigma V_\alpha \rho V_\alpha^* \quad \text{on } (1-K^*)^{-1}\ T(H)\ (1-K)^{-1}.$$

(However in general $T_t(1) \neq 1$).

(b) Let $\{T_t: t \geq 0\}$ be a weakly continuous normal dynamical semigroup on B(H), leaving the compacts globally invariant, such that $\langle \cdot\ \Omega, \Omega\rangle$ is a pure invariant state ($\Omega \in H$). Then if $\{e^{Kt}: t \geq 0\}$ denotes the contraction semigroup on H given by $x\Omega \to T_t(x)\Omega$, $x \in B(H)$, there exist $V_\alpha \in L(\text{Dom}(K^*), H)$ such that the generator L_* of the predual action is given by $L_*(\rho) = K^*\rho + \rho K + \Sigma V_\alpha \rho V_\alpha^*$ on $(1-K^*)^{-1}\ T(H)\ (1-K)^{-1}$.

Uniqueness of the standard form $L = K^*(\cdot) + (\cdot)K + \Psi$ of a dynamical semigroup on B(H) has been studied in [13, 32], and, essentially, there is uniqueness if re(K) can be made small in a suitable sense (e.g. when -tr(re(K)) is minimal when dim(H)<∞, or when H is infinite dimensional, there is at most one K with re(K) compact).

If H is a Hilbert space, let A(H) denote the Fermion algebra built on H, generated by operators $\{a(h): h \in H\}$, which are the range of a conjugate linear map a, satisfying the canonical anticommutation relations. If $\{e^{Bt}: t \geq 0\}$ is a strongly continuous contraction semigroup on H, commuting with a positive contraction R on H, then there exists a unique strongly continuous unital dynamical semigroup $\{A_R(e^{tB}) : t \geq 0\}$ on A(H) such that

$$A_R(e^{tB})\,[\,:a^{\#}(f_1)\ \dots\ a^{\#}(f_n):_R\,]$$
$$= \,:a^{\#}(e^{tB}f_1)\ \dots\ a^{\#}(e^{tB}f_n):_R$$

where $:\ :_R$ indicates Wick ordering with respect to the quasi-free state w_R, and $w_R \circ A_R(e^{Bt}) = w_R$. These examples allow us to study dynamical semigroups, which are not norm continuous, and not simply infinite tensor products. If $(\Pi_R, F_R(H), \Omega_R)$ is the GNS decomposition of w_R, then, as in Theorem 3, we can define a contraction semigroup $F_R(e^{Bt})$ on $F_R(H)$ given by $\Pi_R(x)\Omega_R \to \Pi_R(A_R(e^{Bt})(x))\Omega_R$, $x \in A(H)$, such that $\Pi_R\, A_R(e^{Bt})\,(.) \geq F_R(e^{Bt})\Pi_R(.)\, F_R(e^{Bt})^*$, $t \geq 0$, (Ψ), with equality only when $\{e^{Bt} : t \geq 0\}$ are -isometries [28]. If R=0, then $F_0(H)$ is Fock space, and $F_0(e^{Bt}) = 1 \oplus e^{Bt} \oplus \dots$ as usual. If B=ih+k, where h is self adjoint, and k bounded and negative, then it follows from the proof of Theorem 2 (see[4]) that there exists a Hilbert space H_1, a strongly continuous semigroup $\{G_t : t \geq 0\}$ of co-isometries on H_1, an isometry $W : H \to H_1$, (so that $A(H) \subseteq A(H_1)$) and a positive contraction S on H_1 such that WR=SW, $[S, G_t]=0$, and $W e^{Bt} = G_t W$. Then $\Pi_S A_R(e^{Bt})(x) = F_S(G_t)\ \Pi_S(x)\ F_S(G_t)^*$, $x \in A(H)$, $t \geq 0$ is an isometric representation as in Theorem 2. Dilations to groups of automorphisms can be obtained via Sz. Nagy's theorem. One can next ask: given a strongly continuous contraction demigroup $\{e^{Bt} : t \geq 0\}$ on H, commuting with a positive contraction R, when does there exist a strongly continuous contraction semigroup in the representation $\Pi_{R'}$ of another quasi-free state $w_{R'}$ s.t. (2) holds? First consider the following cases:

(a) By Theorem 1, this can be done with $K \in \Pi_{R'}(A(H))''$ if $\{A_R(e^{Bt}) : t \geq 0\}$ is norm continuous. However $\{A_R(e^{Bt}) : t \geq 0\}$ is norm continuous if and only if B is traceclass [27, 28]. In which case $B = -i\ \Sigma\ \mu_n\ h_n \otimes \overline{h}_n - \Sigma\ \lambda_n\ g_n \otimes \overline{g}_n$, where $\{h_n\}, \{g_n\}$ are two complete orthonormal sets in **H**, $\mu_n \in R$, $\lambda_n > 0$, and $\Sigma|\mu_n|$, $\Sigma\ \lambda_n < \infty$, and there exists $0 \leq r_n \leq 1$ with $Rg_n = r_n g_n$. Then if θ is the quasi-free automorphism such that $\theta[a(h)] = -a(h)$, the generator L of $\{A_R(e^{tB}) : t \geq 0\}$ is given by

$$\begin{aligned} L\,(x) &= i\ \Sigma\ \mu_n\ [a^*(h_n)a(h_n), x] - \\ &+ \Sigma\ \lambda_n \{2[(1-r_n)a^*(g_n)+r_n\ a(g_n)]\theta(x)[(1-r_n a(g_n)+r_n\ a^*(g_n)] \\ &- [((1-r_n)a^*(g_n)+ra(g_n))\ ((1-r_n)a(g_n)+ra^*(g_n)),\ x]_+\} \\ &+ 2\ \Sigma\ \lambda_n\ (r_n - r_n^2)\ \{(a^*(g_n) - a(g_n))\theta(x)(a(g_n)-a^*(g_n)) - x\} \end{aligned}$$

so that in fact we can take $K \in \Pi_{R'}(A(H))$ [27, 29].

(b) If $w_{R'}$ is quasi-equivalent to w_R, with R, R' projections, then (4) can be transformed to the representation $\Pi_{R'}$. (i.e. if $R-R'$ is trace class).
For semigroups of Fock type (R=0), we have:

Theorem 4. [30]. If $B \leq 0$ is bounded on H, and p a projection on H such that pBp is trace class then:

(i) The semigroup $\{A(e^{tB}): t \geq 0\}$ extends to a weakly continuous semigroup of completely positive normal maps on $\Pi_p(A(H))''(\simeq B(F(H)))$.

(ii) There exists a strongly continuous self adjoint contraction semigroup G_t on

$F_p(H)$ $(\simeq F(H))$ such that $\Pi_p(A(e^{tB})(\cdot)) \geq G_t \Pi_p(\cdot) G_t$, $t \geq 0$.

References

1. G. Lindblad: Commun. Math. Phys. 48, 119-130 (1976).
2. K. Kraus: Ann. Physics 64, 311-335 (1971).
3. J. Tomiyama: Proc. Japan. Acad. 33, 608-612 (1957).
4. D.E. Evans, J.T. Lewis: Dilations of irreversible evolutions in algebraic quantum theory. Commun. Dubl. Inst. Adv. Studies. 24, 1977.
5. E.B. Davies: Rend. Sem. Mat. Fis. Milano. 47, 165-173 (1977).
6. E.B. Davies: Quantum theory of open systems. Academic Press, 1976. V. Gorini, A. Frigerio, M. Verri, A. Kossakowski, E.C.G. Sudarshan. Rep. Math. Phys. 12, 359 (1977). L. Accardi, V. Gorini, (ed): Mathematical problems in the theory of quantum irreversible processes. 1978 Naples conference (to appear). E.B. Davies: These proceedings.
7. W.F. Stinespring: Proc. Amer. Math. Soc. 6, 211-216 (1955). E. Størmer: Acta. Math. 110, 233-278 (1963).
8. M.D. Choi: Can. J. Math. 3, 520-529 (1972).
9. M. Verri, V. Gorini: J. Math. Phys. 19, 1803-1806. (1978).
10. R.V. Kadison: Ann. Math. 83, 280-293 (1966). S. Sakai: Ann. Math. 273-279 (1966).
11. D.W. Robinson: These proceedings.
12. E. Christensen, D.E. Evans: J. Lon. Math. Soc. (to appear).
13. V. Gorini, A. Kossakowski, E.C.G. Sudarshan: J. Math. Phys. 17, 821-825 (1976).
14. G. Lindblad: Lett. Math. Phys. 1, 219-224 (1976). D.E. Evans: Quart. J. Math. Oxford (2), 28, 271-284 (1977). C.W. Thompson: Commun. Math. Phys. 71-80 (1978) E. Christensen: Commun. Math. Phys. 62, 167-171 (1978).
15. A. Kossakowski: Bull. Acad. Pol. Sci. Sér. Math. Astr. Phys. 20, 1021-1025 (1976) D.E. Evans, H. Hanche-Olsen: J. Funct. Anal. 32, 207-212 (1979).
16. R. Alicki: Rep. Math. Phys. 10, 249-258 (1976).
17. A. Kossakowski, A. Frigerio, V. Gorini, M. Verri: Commun. Math. Phys. 57, 97-110 (1977).
18. E.B. Davies: Commun. Math. Phys. 15, 277-304 (1969); 19, 83-105 (1970); 22, 51-70 (1971).
19. E.B. Davies: Z. Warschein, 23, 261-273 (1972).
20. E.B. Davies: Commun. Math. Phys. 39, 91-110 (1974); Math. Ann. 219, 147-158 (1976).
21. H. Spohn: Rep. Math. Phys. 10, 283-296. A. Frigerio: Lett. Math. Phys. 2, 79-87 (1977); Commun. Math. Phys. 63, 269-276 (1978)
22. D.E. Evans: Commun. Math. Phys. 54, 293-297 (1977). D.E. Evans, R. Høegh-Krohn: J. Lon. Math. Soc. (2). 17, 345-355 (1978). S. Albererio, R. Høegh-Krohn: Commun. Math. Phys. 64, 83-94 (1978)
23. G.L. Sewell: Commun. Math. Phys. 55, 53-61 (1977). M. Fannes, A. Verbeure: J. Math. Phys. 19, 558-560 (1978).
24. D.E. Evans, J.T. Lewis: Commun. Math. Phys. 50, 219-227 (1976)
25. D.E. Evans: D. Phil. Thesis, Oxford, 1975. D.E. Evans: Commun. Math. Phys. 48, 15-22 (1976). E.B. Davies: J. Lon. Math. Soc. 17, 333-338 (1978).
26. L.C. Thomas: D. Phil. Thesis, Oxford, 1971. J.T. Lewis, L.C. Thomas: Z. Wahrschein. 30, 45-55 (1974). E.B. Davies: Commun. Math. Phys. 27, 309-325 (1972). B. Demoen, P. Vanheuverzwijn, A. Verbeure: Rep. Math. Phys. (to appear).

G.G. Emch: Commun. Math. Phys. 49, 191-215 (1976). G.G. Emch: Acta. Phys. Austriaca. Suppl XV, 79 (1976). G.G. Emch, S. Albeverio, J.-P. Eckmann: Rep. Math. Phys. 13, 73-79 (1978). D.E. Evans, J.T. Lewis: J. Funct. Anal. 26, 369-377 (1977). G. Lindblad: Rep. Math. Phys. 10, 393-406 (1976). P. Vanheuverzwijn Ann Inst. H. Poincare (to appear).

27. E.B. Davies: Commun. Math. Phys. 55, 231-258 (1978).
28. D.E. Evans: Commun. Math. Phys. (to appear).
29. D.E. Evans: Completely positive quasi-free maps on the Fermion algebra, in Mathematical Problems in the Theory of Quantum Irreversible Processes. ed. L. Accardi, V. Gorini: Proc. Naples Conference 1978 (to appear).
30. D.E. Evans: Dissipators for symmetric quasi-free dynamical semigroups on the CAR algebra. Preprint 1978.
31. E.B. Davies: Rep. Math. Phys. 11, 169-188 (1977); J. Funct. Anal. (in press).
32. E.B. Davies: Uniqueness of the standard form of the generator of a quantum dynamical semigroup. Preprint 1978.

Pseudospaces, pseudogroups and Pontriagin duality

S.L. Woronowicz
Department of Mathematical Methods in Physics
University of Warsaw, Hoza 74, 00-682 Warszawa
Polska - Poland

0. Introduction

The main result of the Gelfand Naimark theory of commutative C^*-algebras can be stated in the following way. The category of commutative C^*-algebras with unity and unital C^*-homomorphisms is dual to the category of compact topological spaces and continuous maps. It is not difficult to extend this theory in order to include locally compact topological spaces. To this end one has to consider C^*-algebras without unity. If Λ is a locally compact topological space, then $C_\infty(\Lambda)$ denotes the C^*-algebra of all complex continuous functions on Λ tending to 0 at infinity. In this case continuous maps from Λ_1 into Λ correspond to C^*-homomorphisms from $C_\infty(\Lambda)$ into $C(\Lambda_1)$ satisfying certain condition (condition (C) of Section 1) and the theory is comparable with that for compact spaces.

No theory like those above is known for noncommutative C^*-algebras. However we can formally solve the problem considering the category dual to the suitable category of C^*-algebras. This category will be called the category of pseudospaces; its objects are called pseudospaces and its morphisms - continuous mappings of pseudospaces. Clearly the category of pseudospaces includes the category of locally compact topological spaces.

One should point out that any theorem concerning pseudospaces is in fact a theorem of the C^*-algebra theory and nothing more. It means that using the category of pseudospaces we do not create a new theory; we only introduce a new language to the C^*-algebra theory. But this new language provides us new interesting questions and conjectures some solutions. For example in the last section we show that pseudospaces endowed with an associative algebraic rule form the adequate domain for considering generalized Pontriagin duality problems. The theory, we obtain this way should be considered as a C^*-algebraic version of the Kac algebra theory.

1. The category of pseudospaces

As it was mentioned in the introduction, in the theory of locally compact topological spaces both algebras $C_\infty(\Lambda)$ and $C(\Lambda)$ play an important role. Therefore we have to find a noncommutative analog of the correspondence $C_\infty(\Lambda) \longrightarrow C(\Lambda)$. This problem is solved by the well known multiplier functor $A \longrightarrow M(A)$.

To introduce the multiplier algebra of a C^*-algebra A assume for the moment that A is embeded into $B(H)$: $A \subset B(H)$. Then we put

$$M(A) = \{ b \in B(H): ab,\ ba \in A \text{ for all } a \in A\} \tag{1}$$

Clearly $M(A)$ is a C^*-algebra and A is an ideal of $M(A)$.

Let K be a C^*-subalgebra of $M(A)$. We say that K contains an approximate unity for A if there exists an increasing net (e_λ) of elements of K such that $0 \le e_\lambda \le I$ and for any $a \in A$, $e_\lambda a$ and ae_λ converge in norm to a.

Using the definition (1) we can prove the following

Proposition

Let A, B be C^*-algebras and $\alpha : A \longrightarrow M(B)$ be C^*-homomorphism satisfying the following condition:

The image $\alpha(A)$ contains an approximate unity for B (C)

Then there exists unique C^*-homomorphism $\tilde{\alpha}$: $M(A) \longrightarrow M(B)$ extending α.

In particular we see that any C^*-isomorphism $A_1 \longrightarrow A_2$ can be extended to the unique C^*-isomorphism of the corresponding multiplier algebras. This fact shows that the algebra $M(A)$ introduced by (1) is independent of the way we embed A into $B(H)$. If $\alpha : A \longrightarrow M(B)$ and $\beta : B \longrightarrow M(C)$ are C^*-homomorphisms satisfying condition (C) then we can construct C^*-homomorphism $\gamma = \tilde{\beta}\alpha$: $A \longrightarrow M(C)$. It turns out that satisfies condition (C).

We shall consider a category $\mathfrak{C}$ introduced in the following way. Objects of $\mathfrak{C}$ are C^*-algebras; $\mathfrak{C}$-morphisms (say from A into B) are C^*-homomorphisms from A into M(B) satisfying condition (C). The composition of two morphisms $\alpha \in \mathrm{Mor}(A,B)$ and $\beta \in \mathrm{Mor}(B,C)$ is the morphism $\gamma \in \mathrm{Mor}(A,C)$ introduced by the formula $\gamma = \tilde{\beta}\alpha$, where $\tilde{\beta}$: $M(B) \longrightarrow M(C)$ is the unique extention of β.

The category $\mathfrak{P}$ dual to the category $\mathfrak{C}$ described above will be called the category of pseudospaces. Objects of $\mathfrak{P}$ will be called pseudospaces (or more precisely locally compact pseudospaces), and morphisms of $\mathfrak{P}$ will be called mappings (or more precisely continuous mappings)

We shall use the following notation. For any pseudospace Λ the corresponding C^*-algebra will be denoted by $C_\infty(\Lambda)$ and the multiplier algebra $M(C_\infty(\Lambda))$ by $C(\Lambda)$. If $r : \Lambda \longrightarrow \Lambda_1$, then the corresponding $\mathfrak{C}$-morphism will be denoted by $r^* : C(\Lambda_1) \longrightarrow C(\Lambda)$.

A pseudospace Λ is said to be finite (finitedimensional, compact) if the corresponding algebra is finitedimensional (finitely generated, contains unity resp.)

For every pseudospace Λ the identity mapping $\Lambda \longrightarrow \Lambda$ will be denoted by id (or more precisely by id_Λ). We have $\mathrm{id}^*(a) = a$ for any $a \in C_\infty(\Lambda)$.

2. The cartesian product of pseudospaces

Let us remind the definition of the tensor product of C^*-algebras. If C^*-algebras A_1 and A_2 are embeded in $B(H_1)$ and $B(H_2)$ resp. then $A_1 \otimes A_2$ is the C^*-algebra generated by operators of the form $a_1 \otimes a_2 \in B(H_1 \otimes H_2)$, where $a_k \in A_k$, $k = 1,2$. It is known that this algebra is independent of the choice of the embedings $A_k \subset B(H_k)$.

Moreover if α are C^*-homomorphisms from A_k into B_k then there exists unique C^*-homomorphism $\alpha_1 \otimes \alpha_2$ from $A_1 \otimes A_2$ into $B_1 \otimes B_2$ such that $\alpha_1 \otimes \alpha_2 (a_1 \otimes a_2) = \alpha_1(a_1) \otimes \alpha_2(a_2)$ for all $a_k \in A_k$.

Using (1) one can easily show that

$$M(B_1) \otimes M(B_2) \subset M(B_1 \otimes B_2) \tag{2}$$

for any C^*-algebras B_1 and B_2.

Now, let Λ_1 and Λ_2 be pseudospaces. We define their cartesian product $\Lambda_1 \times \Lambda_2$ as the pseudospace corresponding to the tensor produkt of the C^*-algebras associated with Λ_1 and Λ_2 :

$$C_\infty(\Lambda_1 \times \Lambda_2) = C_\infty(\Lambda_1) \otimes C_\infty(\Lambda_2)$$

The canonical projections $p_k : \Lambda_1 \times \Lambda_2 \longrightarrow \Lambda_k$ $(k = 1,2)$ can be introduced as mappings corresponding to the natural embedings $p_k^* : C_\infty(\Lambda_k) \longrightarrow C(\Lambda_1 \times \Lambda_2)$, where

$$p_1^*(a_1) = a_1 \otimes I \qquad a_1 \in C_\infty(\Lambda_1)$$

$$p_2(a_2) = I \otimes a_2 \qquad\qquad a_2 \in C_\infty(\Lambda_2)$$

If r_k are mappings from Λ_k into Λ'_k $(k = 1,2)$, then r_k^* are C^*-homomorphisms from $C_\infty(\Lambda'_k)$ into $C(\Lambda_k)$ and $r_1^* \otimes r_2^*$ is a C^*-homomorphism from $C_\infty(\Lambda'_1 \times \Lambda'_2)$ into $C(\Lambda_1) \otimes C(\Lambda_2)$. According to (2) the latter algebra is contained in $C(\Lambda_1 \times \Lambda_2)$ and $r_1^* \otimes r_2^*$ considered as C^*-homomorphism from $C_\infty(\Lambda'_1 \times \Lambda'_2)$ into $C(\Lambda_1 \times \Lambda_2)$ satisfies the condition (C). The mapping from $\Lambda_1 \times \Lambda_2$ into $\Lambda'_1 \times \Lambda'_2$ corresponding to $r_1^* \otimes r_2^*$ will be denoted by $r_1 \times r_2 : \Lambda_1 \times \Lambda_2 \longrightarrow \Lambda'_1 \times \Lambda'_2$

$$(r_1 \times r_2)^* = r_1^* \otimes r_2^*$$

One can prove that $r_1 \times r_2$ is the only mapping from $\Lambda_1 \times \Lambda_2$ into $\Lambda'_1 \times \Lambda'_2$ such that

$$p'_k(r_1 \times r_2) = r_k p_k \qquad\qquad k = 1,2,$$

where $p_k : \Lambda_1 \times \Lambda_2 \longrightarrow \Lambda_k$ and $p'_k : \Lambda'_1 \times \Lambda'_2 \longrightarrow \Lambda'_k$ are canonical projections

The following statement seems to be interesting:

Proposition

Let Λ be a pseudospace. Then a mapping $d : \Lambda \longrightarrow \Lambda \times \Lambda$ such that $p_k d = id$ $(k = 1,2)$ exists if and only if Λ is the usual space i.e. $C_\infty(\Lambda)$ is commutative.

It means that for pseudospaces (not being spaces) the diagonal map does not exist.

To understand better the ideas used in the next section we shall discuss the following result:

Proposition

Let Λ and Λ_1 be pseudospaces. We assume that Λ is finite and Λ_1 is finitedimensional and compact. Then there exist a finite dimensional compact pseudospace Λ_2 and a mapping $s : \Lambda_2 \times \Lambda \longrightarrow \Lambda_1$ such that for any pseudospace Λ_3 and any mapping $r : \Lambda_3 \times \Lambda \longrightarrow \Lambda_1$ there exists unique mapping $t : \Lambda_3 \longrightarrow \Lambda_2$ such that $r = s(t \times id)$.

To explain, why this result is interesting assume for the moment that we deal with the usual spaces. Then the mapping s from $\Lambda_2 \times \Lambda$ into Λ_1 can be considered as a family of mappings from Λ into Λ_1 indexed by elements of Λ_2 . In the general case we shall speak about pseudofamily of mappings from Λ into Λ_1 . The existence of t means that this pseudofamily is universal i.e. contains all mappings and the uniqueness of t means that every mapping appears in the pseudofamily only once. Therefore Λ_2 can be considered as a pseudospace of all mappings from Λ into Λ_1 .

Example : Let $\Lambda = \Lambda_1$ = two point set. Then Λ_2 is a quite complicated object: $C_\infty(\Lambda_2)$ is the C^*-algebra of all continuous mappings $a : [0,1] \longrightarrow M_2$ such that $a(0)$ and $a(1)$ are diagonal. Here M_2 denotes the algebra of all complex 2×2 matrices.

3. Pseudogroups and duality

Let $S^1 = \{ z \in C : |z| = 1\}$ and Λ be a pseudospace. Assume that $\tilde{u}$ is a unitary element of $C(\Lambda)$. Then, using the functional calculus of normal operators we have C^*-homomorphism

$$C(S^1) \ni f \longrightarrow f(\tilde{u}) \in C(\Lambda)$$

which in turn corresponds to some mapping from Λ into S^1. This way we obtain $1-1$ correspondence between the set of all mappings from Λ into S^1 and the set of all unitaries in $C(\Lambda)$. The mapping corresponding to $\tilde{u}$ will be denoted by u . We shall use this correspondence to define pointwise product of mappings into S^1. Let $u, u_1, u_2 : \Lambda \longrightarrow S^1$. We say that u is the pointwise product of mappings u_1 and u_2 and write $u = u_1u_2$ if $\tilde{u} = \tilde{u}_1\tilde{u}_2$.

This pointwise multiplication is clearly induced by the usual multiplication in S^1. Note however, that pointwise multiplication in the set of mappings into S^1 is not abelian in general.

Definition

Let G be a pseudospace and $\varphi : G\times G \longrightarrow G$. We say that (G,φ) is a pseudo-semigroup if φ obeys the following associativity law: $(\mathrm{id}\times \varphi)\varphi = (\varphi\times \mathrm{id})\varphi$.

Assume that (G,φ) is a pseudo-semigroup, Λ is a pseudospace and $u : \Lambda\times G \longrightarrow S^1$. Denoting by p_1 and p_2 the canonical projections $G\times G \longrightarrow G$ we consider the following three mappings: $u_0 = u(\mathrm{id}\times \varphi)$, $u_1 = u(\mathrm{id}\times p_1)$ and $u_2 = u(\mathrm{id}\times p_2)$ from $\Lambda\times G\times G$ into S^1. We say that (Λ ,u) is a pseudofamily of characters of (G,φ) if u_0 is the pointwise product of u_1 and u_2 :

$$u(\mathrm{id}\times \varphi) = u(\mathrm{id}\times p_1)\, u(\mathrm{id}\times p_2) \qquad (3)$$

Definition

Let (G,φ) be a pseudo-semigroup and $(\hat{G},v)$ be a pseudofamily of characters of (G,φ). We say that $(\hat{G},v)$ is the pseudospace of all characters of (G,φ) if for any pseudofamily (Λ ,u) of characters of (G,φ) there exists unique mapping $r : \Lambda \longrightarrow \hat{G}$ such that $u = v(r\times \mathrm{id})$.

Assume now that (G,φ) is a pseudosemigroup and that $(\hat{G},v)$ is the pseudospace of all characters of (G,φ). We denote by $\hat{p}_1$ and $\hat{p}_2$ the canonical projections $\hat{G}\times\hat{G} \longrightarrow \hat{G}$. Let $v_1 = v(\hat{p}_1\times \mathrm{id})$, $v_2 = v(\hat{p}_2\times \mathrm{id})$ and $v_0 = v_1v_2$ (pointwise product) . Clearly $v_k : \hat{G}\times\hat{G}\times G \longrightarrow S^1$ for $k = 1,2,0$. It turns out that $(\hat{G}\times\hat{G},v_0)$ is a pseudofamily of characters of (G,φ) . According to the previous definition there exists unique mapping $\hat{\varphi} : \hat{G}\times\hat{G} \longrightarrow \hat{G}$ such that

$$v(\hat{\varphi}\times \mathrm{id}) = v(\hat{p}_1\times \mathrm{id})\, v(\hat{p}_2\times \mathrm{id}) \qquad (4)$$

Remembering that the pointwise multiplication is associative one can easily show that $(\hat{G},\hat{\varphi})$ is a pseudo-semigroup. This pseudo-semigroup is said to be dual to (G,φ).

Now, let $s : G\times\hat{G} \longrightarrow \hat{G}\times G$ be the natural mapping permuting G and $\hat{G}$ (i.e. $s^*(a\otimes b) = b\otimes a$). It follows immediately from (4) that

$$vs(\mathrm{id}\times\hat{\varphi}) = vs(\mathrm{id}\times p_1)\, vs(\mathrm{id}\times \check{p}_2)$$

It means that (G,vs) is a pseudofamily of characters of $(\hat{G},\hat{\varphi})$. Therefore, assuming that the double dual $(\hat{\hat{G}},\hat{\hat{\varphi}})$ exists we have the unique mapping $i : G \longrightarrow \hat{\hat{G}}$ such that

$$vs = \hat{v}(i\times \mathrm{id}).$$

The main problem of the theory consists in the following: What additional assumptions one has to impose on (G,φ) in order to prove the existence of the dual and double dual pseudosemigroups and to prove that i is an isomorphism. We belive that these assumptions will distinguish pseudogroups among pseudo-semigroups. For the moment the answer is known for finite **pseudogroups** and partially for compact pseudogroups. We also have the following

THEOREM

Assume that G is a locally compact topological group. Then the dual and double dual of G exist and double dual of G is isomorphic to G (the isomorphism is given by the mapping i introduced above). $\hat{G}$ is a usual space if and only if G is abelian and in this case we have the usual Pontriagin duality.

Selected Issues from

Lecture Notes in Mathematics